Discovering Geometry

An Inductive Approach

Second Edition

Michael Serra

KEY CURRICULUM PRESS
Innovators in Mathematics Education

Author

Michael Serra
George Washington High School
San Francisco, California

Project Team

Project Editor	Dan Bennett
Assistant Editor	Sarah Block
Additional Editorial Development	Masha Albrecht, Casey FitzSimons
Copy Editor	Greer Lleuad
Editorial Assistant	Romy Snyder
Art/Photo Researcher	Ellen Hayes
Production Manager	Luis Shein
Production Coordinator	Susan Parini
Art and Layout	Kirk Mills
Production Assistants	Aran Rasmussen, Ann Rothenbuhler

Interior Design

Mark Ong

Cover Design

Dennis Teutschel

Publisher

Steven Rasmussen

Editorial Director

John Bergez

10 9 8 7 6 5 4   01 00 99 98

Printed in the United States of America   ISBN 1-55953-200-9

Teacher Consultants and Field Testers

Oran Pyle, Tennyson High School, Hayward, California
Wendy Struhl, Oakland Technical High School, Oakland, California
John Dumanske, Philip and Sala Burton Academic High School, San Francisco, California
Tom Swartz, George Washington High School, San Francisco, California
Rodger Gray, Lincoln High School, Stockton, California
Judy Hicks, Standley Lake High School, Broomfield, Colorado
Larry Chiucarello, Nonnewaug High School, Woodbury, Connecticut
Dixie Trollinger, Mainland High School, Daytona Beach, Florida
Genie Dunn, Miami Killian Senior High School, Miami, Florida
Sharon Grand, Baton Rouge High School, Baton Rouge, Louisiana
Carolyn Sessions, Louisiana State University Laboratory School, Baton Rouge, Louisiana
Ralph Bothe, Edward C. Reed High School, Reno, Nevada
Archie Benton, North Buncombe High School, Weaverville, North Carolina
Dave Damcke, Jefferson High School, Portland, Oregon
Jorge Rivera, Saint John's School, San Juan, Puerto Rico
Carol Miller, Tascosa High School, Amarillo, Texas

Advisors and Contributors

Sheldon Berman, Simon Gratz High School, Philadelphia, Pennsylvania
Dudley Brooks, San Francisco, California
Donald Collins, Ph.D., Western Kentucky University, College of Education, Bowling Green, Kentucky
Bob Garvey, Louisville Collegiate School, Louisville, Kentucky
M. Mamikon, Davis, California
Beth Porter, George School, Newtown, Pennsylvania
Doris Schattschneider, Ph.D., Moravian College, Bethlehem, Pennsylvania
Mal Singer, University High School, San Francisco, California

Multicultural and Equity Reviewers

Dean Azule, Confederated Tribes of the Grande Ronde Community, Grand Ronde, Oregon
Joyla Gregory, Chula Vista, California
Regina Heinicke, San Francisco, California
José López, Lawrence Berkeley Laboratory, Berkeley, California
Beatrice Lumpkin, Chicago, Illinois
Charlene Morrow, Mount Holyoke College, South Hadley, Massachusetts
Kimlynne Lee Slagel, Kamehameha Secondary School, Honolulu, Hawaii
Sue Yabuki, Northwest Equals, Portland State University, Portland, Oregon

Teacher Consultants and Field Testers, First Edition

Sam Butscher, Ph.D., Theresa Hernandez-Heinz, Robert Knapp, Jeff Salisbury, Tom Swartz,
Edward Van Pelt, and Li Oi Yu, San Francisco Unified School District, San Francisco, California
Dianne Borchardt, Sandie Gilliam, and Dennis Olson, San Lorenzo Valley High School, Felton, California
Bob Eckland and Katie Makar, Catlin Gabel School, Portland, Oregon
Dave Damcke and John P. Oppedisano, Jefferson High School, Portland, Oregon
Debbie Lindow and Charlene Trachsel, Reynolds High School, Troutdale, Oregon
Connie Callos, Evergreen High School, Vancouver, Washington

Advisors and Contributors, First Edition

Katie Makar, Catlin Gabel School, Portland, Oregon
John Olive, Ph.D., University of Georgia, Athens, Georgia
David Rasmussen, Cabot High School, Nova Scotia, Canada
J. Michael Shaughnessy, Ph.D., Portland State University, Portland, Oregon
Tom Swartz, George Washington High School, San Francisco, California
Richard Wertheimer, School District of Pittsburgh, Pittsburgh, Pennsylvania

Foreword to the First Edition

Michael Serra has written a genuinely exciting geometry book. This book is unique in that the students actually create geometry for themselves as they proceed through the activities and the problems. Concepts are first introduced visually, then analytically, then inductively, and, finally, deductively. The spirit of this text is remarkably consistent with recent research on the development of geometric thinking in adolescents, particularly the levels of thinking in the van Hiele theory.

From the beginning, students participate in the construction of definitions. The author makes excellent use of a "nonexamples and examples" approach to encourage students to build their own definitions. As new geometric figures are introduced, activities are structured so that students can discover properties of these figures. Throughout the book, students are asked to make conjectures about figures and about relationships among figures. These conjectures are brought up again for deductive consideration in the chapters on proof.

In addition to a thorough treatment of all the topics anyone has ever wished to cover in a geometry course, Serra provides a number of extra nuggets and superimposes a creative developmental sequence to his topics. Measurement topics such as area and volume appear earlier than in many geometry texts, prior to formal proof. There is an extensive chapter on transformations and tessellations. The tessellations provide ample opportunity to concretely explore the symmetry patterns that make transformations so powerful. Where some texts pay lip service to these topics, Serra provides us with a full and varied menu. Excellent computer activities are sprinkled throughout the text in strategic places, providing an additional environment for student discoveries. Coordinate topics are woven in wherever appropriate. Most of all, however, this book has super problems, written by a master storyteller. Each chapter has one or more cooperative problem solving activities, which encourage interaction and communication among students. Students will have an opportunity to write and talk about their mathematics and thus will have a better chance of understanding their mathematics.

Formal proof does not appear in this book until the last two chapters. However, by the time students are asked to write proofs, they have already made conjectures for months, have formed and tested their own definitions, have solved logic problems, have developed visualization skills through drawings, constructions, and computer activities, and have studied logic, reasoning, and the nature of proof. *They are ready for proofs!*

In the past we have erred in pushing proofs on students too soon, before they had a handle on shapes and their properties. In the past we have asked many of our students to do two things simultaneously—learn geometric concepts and learn deductive reasoning. I applaud Michael Serra's move to delay proof in his book until students have seen the whole spectrum of geometric concepts. Serra's book gives students a better chance of learning geometry and of learning about proof.

Finally, this is a book for "doers." Students constantly *do* things in this book, both alone and with other students. If you want your students and yourself to become actively involved in the process of learning and creating geometry, then this book is for you.

J. Michael Shaughnessy, Ph.D.
Department of Mathematical Sciences
Portland State University
Portland, Oregon

Foreword to the Second Edition

The first edition of Michael Serra's *Discovering Geometry* provided approaches to teaching and learning geometry that were radically different from those of other geometry textbooks. The second edition is even more exciting. Michael Serra continues to make geometry fun and interesting to both students and teachers.

Serra has added new and richer problem settings and projects that attract and hold the interest and attention of students. Added to the many high-interest fantasy problem settings from the first edition, the second edition includes more real-world applications and examples of geometry from many cultures.

The text can be implemented in a variety of instructional settings, providing opportunities for students to think, discuss, and work together as they actually *do* geometry. Students perform investigations to discover and make conjectures about the wonderful properties of geometry. In addition to traditional construction tools, students can use patty papers to streamline many investigations in this edition. Technology in the form of graphing calculators and computers has been integrated as needed and where appropriate.

This edition of *Discovering Geometry* also comes alive in brilliant color with new cartoons, illustrations, artwork, and pictures from real life and from around the world. Students and teachers will benefit from examples of students' work from Michael Serra's and other teachers' classrooms, which are shared throughout the text.

A big plus of this book, as in the first edition, is that Serra puts proofs in the last chapters. This allows students to first be involved in investigating and conjecturing before they are exposed to formal geometric proofs. This suggests a sequence that has proved to be appropriate and effective.

Michael Serra is to be commended for writing such a unique geometry book for teachers and students.

Bettye Forte
Executive Director, Northeast Instructional Support Team
Mathematics Director, Fort Worth Independent School District
Fort Worth, Texas

What Makes *Discovering Geometry* Different?

Features of *Discovering Geometry*

This book was designed so that you and your teacher can have fun with geometry. In *Discovering Geometry* you "learn by doing." You will learn to use the tools of geometry and to perform geometric **investigations** with them. Your investigations will lead to geometric discoveries. Many of the geometric investigations are carried out in small **cooperative groups** in which you jointly plan and find solutions with other students. I think you will enjoy the **humorous illustrated word problems**. You will help archaeologist Ertha Diggs determine the height of a Mayan jungle temple, help pirate Captain Coldhart bury his treasure, and help Hemlock Bones solve the case of the Belgian stamp theft. I created these problems in the hope of reducing your anxiety about word problems. In the **projects** you will build geometric solids, make kaleidoscopes, design a racetrack, find the height of your school building, and create a mural. There are also several **graphing calculator investigations** and **Geometer's Sketchpad® projects** in the text. In one Sketchpad™ project you will discover the world of a strange class of geometric shapes called fractals! Each chapter closes with a special **cooperative problem solving** activity set at a lunar colony of the future. I think you'll enjoy the extra challenges in the Improving Visual Thinking Skills, Improving Reasoning Skills, and Improving Algebra Skills **puzzles** that I have sprinkled throughout the book.

I have designed the projects, the puzzles, and the calculator and computer activities so that you can do them independently, whether or not your class tackles them as a group. Read through them as you proceed through the book.

Chapter Sequence

I begin with a chapter on geometric art to show you that geometry is found in the art of cultures throughout the world. In Chapter 1, you will learn how to reason inductively. Inductive reasoning is the process that you will use to make geometric discoveries in this book. In Chapter 2, you will use inductive reasoning to create definitions of geometric terms. In Chapter 3, you will use tools of geometry to construct geometric figures. These tools include the compass, the straightedge, and patty papers, which you'll use to make geometric discoveries in the remainder of the book. In Chapter 8, you will learn to create geometric tiling designs similar to some of M. C. Escher's artwork shown on the chapter opening pages. Finally, in the last three chapters, you will learn about another type of reasoning, which is called deductive reasoning or proof.

Please do not feel overwhelmed by the number of chapters in *Discovering Geometry*. It is not possible to cover *all* of the chapters in one school year. Your teacher will guide you through the book to create one of several different types of geometry courses possible using this text.

Suggestions for Success

It is important to be organized when working with *Discovering Geometry*. Keep a notebook with a section for definitions, a section for new geometric discoveries, and a section for daily notes and exercises. Study your notebook regularly. You will need four tools of geometry for the investigations: a compass, a protractor, a straightedge, and a ruler. Keep a calculator handy.

You will find hints and solutions for some key exercises in the back of the book. Exercises that have hints are identified by an asterisk (*). Try to solve the problems on your own without looking at the hints. Refer to the hints to check your method or as a last resort if you can't solve a problem.

Unlike most texts, *Discovering Geometry* will ask you to work cooperatively with your fellow students. This means you should pull your desks together and get to know one another. When you are working together cooperatively, always be willing to listen to one another, to be an active participant, to ask one another questions when you don't understand, and to help one another when asked. When working cooperatively, you can accomplish much more than the sum total of what you can accomplish individually. And best of all, you'll have less anxiety and a whole lot more fun.

Michael Serra
George Washington High School
San Francisco, California

Contents

Chapter 0   Geometric Art

Chapter 1   Inductive Reasoning

Chapter 2   Introducing Geometry

Chapter 3 Using Tools of Geometry

Chapter 4 Line and Angle Properties

Chapter 5 Triangle Properties

Chapter 6 Polygon Properties

Chapter 7 Circles

Chapter 8 Transformations and Tessellations

Chapter 9 Area

Chapter 10   Pythagorean Theorem

Chapter 11   Volume

Modern art includes optical art (op art), which is very geometrical. Victor Vasarely is a famous op artist whose work reflects a strong interest in geometry. In Lesson 0.4, you will learn more about op art, and you will get a chance to create your own op art design.

Dutch artist M. C. Escher is another twentieth-century artist whose work reflects a strong interest in geometry. In Chapter 8, you will make your own artistic creations, following the techniques used by Escher.

Celtic art, *The Celtic Design Book*, Rebecca McKillip, Stemmer House, 1981

Perhaps the most common geometric characteristic found in nature and in art is **symmetry**. The word *symmetry* brings to mind other words like *balance, harmony,* and *equally proportioned.* Flowers, fish, birds, and many other natural objects are symmetric. The human body has what is called **bilateral symmetry**. Both the chambered nautilus and crystals grow symmetrically. Symmetry appears so abundantly in nature—in plants, in animals, and in crystalline structures—that it is not surprising that artists throughout history have taken pleasure in symmetric designs. Visual symmetry is found in the arts and crafts of cultures all around the world.

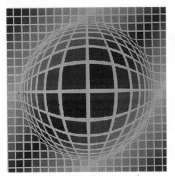 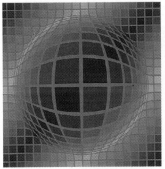 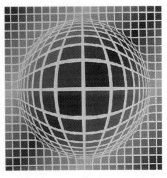

Tsiga I, II, III (1991), Victor Vasarely, courtesy of the artist

You are probably already familiar with two basic types of symmetry, reflectional symmetry and rotational symmetry. A design has **reflectional symmetry** if it can be folded along a line (called the **line of symmetry**) so that all the points on one side of the fold exactly coincide with corresponding points on the other side of the fold. Another test for reflectional symmetry uses a mirror on what appears to be the symmetry line. If the half figure on one side of the mirror line and its reflected image in the mirror re-create the original figure, then it has reflectional symmetry.

Butterfly: one line of reflectional symmetry

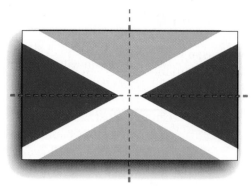

Jamaican flag: two lines of reflectional symmetry

Native American basket: 4-fold rotational symmetry

Japanese boxes. If you ignore colors, the lower left box has 3-fold rotational symmetry. What type of symmetry does the other box have?

Reflectional symmetry is also referred to as *line symmetry* or *mirror symmetry*. An object with just one line of symmetry, like the butterfly pictured above, is said to have **bilateral symmetry**. Many companies use symmetric designs for their corporate logos. Countries throughout the world use symmetry in their national flags. Often the logos and the flags have reflectional symmetry. The Jamaican flag has two lines of symmetry.

A design has **rotational symmetry** if it can be traced and rotated less than a complete cycle about a point so that the tracing can be made to fit exactly onto the original. The number of times the tracing matches the original design in a complete cycle determines what type of rotational symmetry the design has. The Native American basket weaving and the Japanese boxes shown above are examples of designs that have rotational symmetry. A design has point symmetry if it can be made to coincide with itself after a half turn (180 degrees). Logo designs and flags can also have rotational symmetry.

Notice that in addition to its reflectional symmetry, the Jamaican flag also has rotational symmetry. It can be rotated 180° without changing its appearance. The basket and the boxes, however, do not have reflectional symmetry. You will learn more about symmetry in later chapters. (The basket *almost* has reflectional symmetry. Can you see why it doesn't?)

Exercise Set 0.1

In this lesson your goal is to become more aware of the geometry found in nature, the geometric art found in different cultures throughout the world, and the role symmetry plays in both nature and in geometric art.

1. List six natural objects that have geometric shapes. Sketch and name the shapes. Name the type(s) of symmetry for each.

2.* Andy Goldsworthy (b 1956) is a contemporary British artist. He uses materials from nature to create beautiful outdoor sculptures. This artful arrangement of sticks shown at right first appears to have rotational symmetry, but instead it has one line of reflectional symmetry. Can you locate the line of reflection?

3. Bring to class an object from nature that exhibits geometry. Describe the geometry that you find in the object as well as any symmetry the object has.

Outdoor sculpture by Andy Goldsworthy. (For title, see hint to Exercise 2 in Hints section.)

4. Bring an object to school or wear an article of clothing that displays a form of handmade or manufactured geometric art, perhaps a traditional folk art design. Describe any symmetry the object has.

5. The three of diamonds playing card has point symmetry because when rotated 180°, it appears exactly the same. The three of clubs, however, does not have point symmetry. What are the other nonface playing cards that have point symmetry?

Improving Visual Thinking Skills—*Pickup Sticks*

Pickup sticks, a good game for developing motor skills, can be turned into a challenging visual puzzle. In what order are the sticks to be picked up so that you are always removing the top stick?

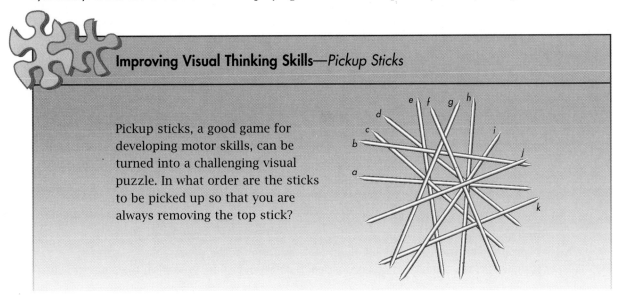

Project

Geometry in Nature and in Art

Read over the four research questions, then agree as a group how to divide the four tasks among you. Go to the library to research your topic. A few days before the assignment is due, meet as a group and have each group member present his or her findings to the group. Ask one another questions and suggest ways to improve one another's reports. Revise your work and hand it in as a group report and/or make a group presentation to class.

1. The four-leaf clover is an example of a plant that exhibits the symmetry of a square. The pentagon (five-sided polygon) is represented in many flowers: spring violets, apple blossoms, wild roses, and forget-me-nots, for example. The hexagon (six-sided polygon) can be found in the petal arrangement of lilies, narcissus, jonquil, and asphodel. Find a book on botany and make a photocopy or a drawing of a flower with four, five, or six petals. Make sure you include the flower's name with the illustration. Write a paragraph describing geometric shapes and symmetries found in that plant.

2. Most nonmicroscopic animals that live on land (including humans) exhibit bilateral symmetry, but many sea creatures have more than bilateral symmetry. The starfish, for example, exhibits the rotational and reflectional symmetry of a regular pentagon. Find a book on zoology and make a photocopy or a drawing of a sea animal that has more than bilateral symmetry. Write a paragraph or two describing its symmetry and naming any geometric shapes the animal resembles. Explain how having symmetry helps the animal survive.

3. Find a photo of a design from some culture within the United States; perhaps an Amish quilt, a Hawaiian basket weaving design, or a Native American blanket. You might start your search in the art or the geography section of the library. Bring a photocopy or a drawing of the design to class. Write a paragraph describing what it is, what culture it's from, and what geometric shapes and symmetries are found in the design.

4. Find a photo of a design from some culture outside the United States; perhaps a knot design from Africa, a mandala from Mexico, a Maori stitched tukutuku panel from New Zealand, or an Islamic wall tile design from Iran. Bring a photocopy or a drawing of the design to class. Write a paragraph describing what it is, what culture it's from, and what geometric shapes and symmetries are found in the design.

5. **Photo or Video Safari** In Lesson 0.1, you saw a few examples of geometry and symmetry in nature and art. If you did the projects above, you found even more examples. Now go out with your group and document examples of geometry in nature and art yourselves. Use a camera or video camera to take pictures of as many examples of geometry in nature and art as you can. Look for many different types of symmetry, and try to photograph art and crafts from many different cultures. Consider visiting museums and art galleries, but make sure it's okay to take pictures when you visit. You might find examples in your home or in the homes of friends and neighbors. If you take photographs, write captions for them that describe the geometry and the types of symmetries you find. If you record video, record your commentary on the soundtrack.

Lesson 0.2

Line Designs

We especially need imagination in science. It is not all mathematics, nor all logic, but it is somewhat beauty and poetry.
— Maria Mitchell

The symmetry and the proportion in geometric designs make them very appealing. Geometric designs are easy to make when you have the tools of geometry. There are four basic tools of geometry: compass, straightedge, ruler, and protractor.

A **compass** is a geometric tool used to construct circles. A **straightedge** is a tool used to construct straight lines. The compass and the straightedge are the classical tools of geometry. All the geometry of the ancient Greeks can be produced with just these two tools. The ancient Greeks laid the foundations of the geometry that you will study in this text.

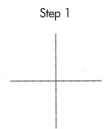

The ruler and the protractor are later inventions. A **ruler** is a tool used to measure the length of a line segment. A ruler has marks used for measurement. A straightedge has no marks. The edge of a ruler can serve as a straightedge. A **protractor** is a tool used to measure the size of an angle in degrees. In the next few lessons on geometric art, you will become familiar with these geometry tools.

Many types of designs can be created with only straight lines. The steps for creating the line designs in Examples A and B are demonstrated below.

Example A

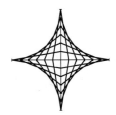

Example B

Step 1	Step 2	Step 3	Step 4
		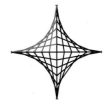	

Step 1	Step 2	Step 3	Step 4

Exercise Set 0.2

1. Each of the line designs below was drawn with straight lines only. Select one design and re-create it on a sheet of paper. Use the steps illustrated for Example A or B to help you.

2. Describe the symmetries of the three designs in Exercise 1. For the third design, describe the different symmetries it has depending on whether or not color is taken into account.

3. Design and complete your own drawing made of straight lines. Be creative. Look ahead to Lessons 0.4, 0.6, and 0.7 for helpful ideas.

4. Geometric shapes can be used to model organic molecules like the benzene molecule. How many different lines of reflectional symmetry does this model appear to have? Does it have rotational symmetry?

Benzene molecule

5. Shah Jahan, Mughal emperor of India from 1628 to 1658, built the beautiful Taj Mahal in memory of his wife, Mumtaz Mahal, who died giving birth to their fourteenth child. The Taj Mahal is described by the poet Rabindranath Tagore as "rising above the banks of the river like a solitary tear suspended on the cheek of time."

   The architect of the Taj Mahal, Ustad Ahmad Lahori, was not only an architect, but also a well-known astronomer and mathematician. He designed the mausoleum with perfect symmetry. Describe two lines of symmetry in the photograph. How does the design of the building's grounds give this view of the Taj Mahal even more symmetry than the building itself possesses?

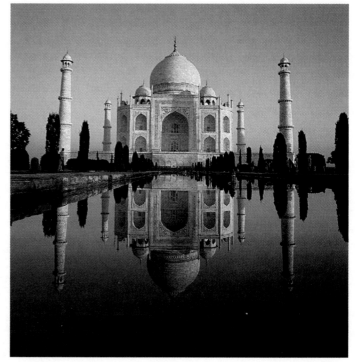

Taj Mahal

Project

Geometry in Home Design

Fallingwater

Frank Lloyd Wright (1885–1959) is often called America's favorite architect. Some of his greatest contributions are in residential architecture. He built homes in thirty-six states: on mountain-tops, nestled in woods, over streams, in the desert, in cities, and in suburbs.

Wright's architecture was based on elements of nature, and he called it organic architecture. Fallingwater, built over a

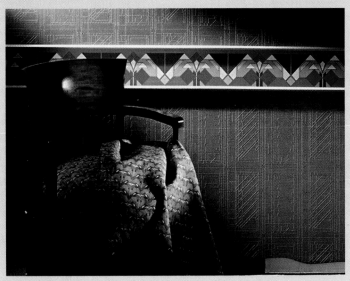

Triangular Tulip Border (1908)

waterfall near Mill Run, Pennsylvania (about fifty miles from Pittsburgh), is a wonderful example of his philosophy of organic architecture. Wright said that Fallingwater was like a giant tree. What would represent the roots of the tree? The trunk of the tree? The branches of the tree? Fallingwater also displays an obvious love of geometry. Describe the geometry that you see in the picture.

Wright often designed not only the buildings but also their skylights and art glass windows, their furniture, carpets, murals, and even their table settings. One of his living rooms was reconstructed in the Metropolitan Museum of Art in New York. Three examples of Frank Lloyd Wright's interior designs are shown here and on the previous page.

The Tree of Life art glass was designed by Wright in 1904 for the Martin House in Buffalo, New York. The Triangular Tulip Border was designed by Wright in 1908 for the Coonley House in Riverside, Illinois. The Robie dining set was designed by Wright in 1908 for the Robie House in Chicago, Illinois.

Tree of Life glass (1904)

Now it is your turn. Design an art glass panel, a wallpaper border pattern, or a piece of furniture in the style of Frank Lloyd Wright. Write a short report about your design, including what geometric shapes and symmetries it has and the geometric tools you used to create it.

Robie dining furniture (1908)

Lesson 0.3

Daisy Designs

The compass is a geometric tool used to construct circles. You can make some very nice designs with only a compass. A daisy is one such simple design. The figures below give you the necessary instructions to make a daisy. Read through the steps for making the design before you begin the exercise set at the end of the lesson.

Step 1

Construct a circle. Then select any point on the circle.

Step 2

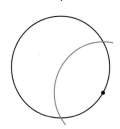

Without changing your compass setting, swing an arc centered at the point selected.

Step 3

Swing an arc from each of the two new points.

Step 4

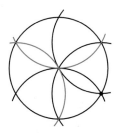

Again, swing an arc from each of the two new points of intersection.

Step 5

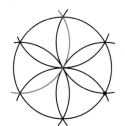

Swing an arc connecting the last two points of intersection.

Step 6

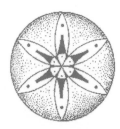

Decorate your daisy design.

The steps for making a daisy are also the steps used to construct a regular hexagon. A **regular hexagon** is a six-sided figure whose sides are all the same length and whose angles are all the same size. There are six petals equally spaced on each daisy. The tips of the daisy touch the circle at points equally spaced, one compass setting apart. If you connect the tips, you will have a regular hexagon. The compass setting (called the radius of the circle) can be marked off exactly six times around the circle.

The six-pointed daisy can be turned into a 12-pointed daisy by making another six-pointed daisy between the petals of the first daisy, as shown in Example A on the next page.

If, instead of stopping at the perimeter of the first circle, you continue by swinging full circles, you end up with a "field of daisies," as shown in Example B on the next page.

Example A

Example B

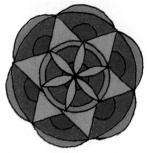

Schuyler Smith, geometry student

Example C

Exercise Set 0.3

1. Create a single six-petal daisy design and color or shade it so that it has rotational symmetry but not reflectional symmetry.

2. Create a single 12-petal daisy design and color or shade it so that it has reflectional symmetry but not rotational symmetry.

3. Using 1 inch for the compass setting (the radius), construct a central regular hexagon and six regular hexagons that each share one side with the original hexagon. It should look similar to, but larger than, the figure at right.

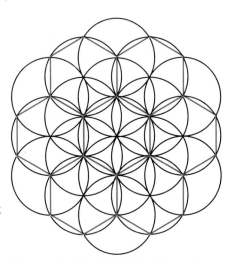

Graphing Calculator Investigation—*Polar Graph Designs*

Certain polar functions have graphs that are similar to daisy designs, though the curves are not arcs of circles. You'll study the mathematics of polar functions in a later course, but in the meantime, this investigation can give you an idea of the role mathematics plays in design.

Set your calculator to polar graphing mode. Graph the function $r = \sin 2\theta$. Sketch this graph in your notebook. Now graph and sketch $r = \sin 3\theta$, $r = \sin 4\theta$, and $r = \sin 5\theta$. In the general form of this function, $r = \sin n\theta$, what does the n tell you?

Graph $r = 2 \sin 2\theta$ and $r = 3 \sin 2\theta$. In the general form of the function $r = a \sin n\theta$, what does the a tell you?

Experiment with polar graphs, using cos instead of sin. How are these graphs similar? How are they different?

CHAPTER 0

Lesson 0.4

Op Art

The most beautiful thing we can experience is the mysterious. It is the true source of all art and science.
— Albert Einstein

Op art (optical art) is a form of abstract art that uses straight lines or geometric patterns to create a special visual effect. The contrasting dark and light regions can, at times, appear to be in motion or to represent a change in surface, direction, and dimension. Victor Vasarely (b 1908) is an artist who can create misleading perceptions with his geometric optical art. Many of Vasarely's works involve transforming grids so that spheres appear to bulge from them, as in the series *Tsiga I, II, III* that appears in Lesson 0.1. *Harlequin,* shown at right, is a rare example of a Vasarely work that includes a human form. Still, Vasarely's trademark sphere is present in the clown's bulging belly.

Op art designs are fun and easy to create. The steps below show how to create one kind of op art design. First, make a design in outline. Run vertical or horizontal parallel lines on your drawing space, varying the spacing to create visual hills and valleys. Finally, shade in alternate spaces.

Harlequin, Victor Vasarely, courtesy of the artist

Step 1

Step 2

Step 3

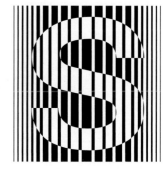

Here are some other examples of optical art.

Example A

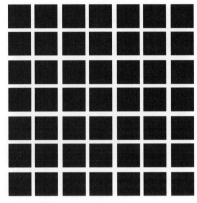

Example B

Example C

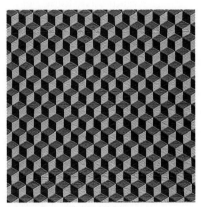

Amish quilt, tumbling block design
Example D

Japanese Op Art, Hajime Juchi, Dover Publications
Example E

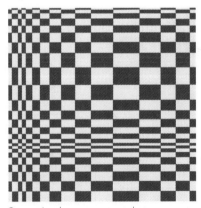

Carmen Apodaca, geometry student
Example F

The steps for creating a design like that shown in Example C are illustrated for you below. First, locate a point on each of the four sides of a square. Each point should be the same distance from a corner, as shown. Your compass is a good tool for measuring equal lengths. Connect these four points to create another square within the first. Repeat until the squares appear to converge on the center. Be careful that you don't fall in!

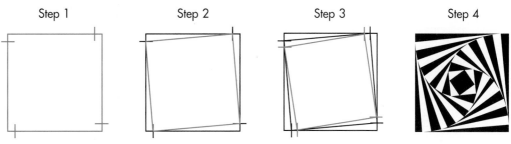

Step 1   Step 2   Step 3   Step 4

Exercise Set 0.4

1. What are the optical effects in Examples A through F?

2. Nature also contains examples of optical art. At first the stripes of a zebra appear to work against it. Certainly on the African savanna the black and white stripes of a zebra would seem to stand out against the golden-brown grasses of the plain. However, they do provide the zebras very effective protection from predators. When and how?

3. Create an optical art design that has reflectional symmetry but not rotational symmetry.

4. Select one of the op art design types from this lesson and create your own version of that type of op art.

5. The Bishop's Palace in Astorga, Spain, was designed by Antoni Gaudí (1852–1926). The classical style of nineteenth-century European art and architecture was changing during the turn of the century, making the climate right for Gaudí's unique and often outlandish style. The palace has Neo-Gothic elements as well as elements of the emerging art nouveau style, but it is still very Gaudí. List as many geometric shapes as you can recognize in the facade of the palace (flat two-dimensional shapes such as rectangles as well as solid three-dimensional shapes such as prisms and cylinders). What type of symmetry do you see in the palace facade?

Project

Drawing the Impossible

You've seen many kinds of optical illusions. Some optical illusions at first appear to be drawings of real objects, but actually they are impossible to make, except on paper. Your first task in this project is to draw on full sheets of paper the four impossible objects shown below.

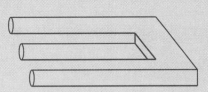

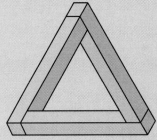

Three prongs from two?

To create this drawing, start with the six parallel lines. Then complete both ends. Try not to look at one end while working on the other.

Penrose triangle

To create the Penrose triangle (named after mathematician and avid puzzle enthusiast Sir Roger Penrose), begin with three equal-sided triangles nested one within the other.

Three towers from four?

Note that the nine vertical edges of the towers in this drawing are not parallel. If you extend the edges upward, you will find that each set of three meets at a point.

Strange shish kebab?

This strange shish kebab is left for you to figure out on your own.

Now that you've practiced by reproducing the objects above, can you create an impossible object of your own? Try it.

designs are illustrated below. Look over the steps before you begin the exercise set.

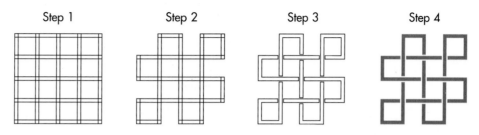

Step 1 Step 2 Step 3 Step 4

A similar approach can be used to create knot designs with rings.

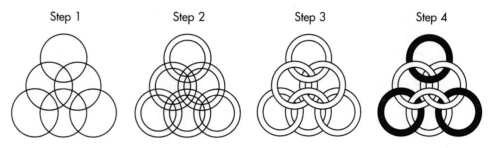

Step 1 Step 2 Step 3 Step 4

Exercise Set 0.6

1. Create a knot design of your own, using only straight lines on graph paper.

2. Create a knot design of your own, using a compass or a circle template.

3. Sketch five rings linked together in such a way that all five can be separated by cutting open one particular ring.

4. One very interesting knot design is known as the Borromean rings. This design appears on the coat of arms of the Borromeo family, who lived during the Italian Renaissance. The design consists of three rings. No two are connected, yet the three together cannot be separated. They are linked together in such a way that if any one ring is removed from the set of three, the remaining two rings are no longer connected. Got that? Good. Sketch the Borromean rings.

5. The Tchokwe storytellers of northeastern Angola are called *Akwa kuta sona* ("those who know how to draw"). When they sit down to draw and to tell their stories they begin by clearing the ground and setting up a grid of points in the sand with their fingertips, as shown below left. The *Akwa kuta sona* then begin to tell a story and, at the same time with one smooth continuous motion, to weave a finger through the sand to create a *lusona* design. Try your hand at creating *sona* (plural of *lusona*). Begin by drawing the correct number of dots. Then, in one motion, re-create one of the *sona* below.

Initial dot grid   Mbemba bird   Rat   Scorpion

6. According to Greek mythology, the Gordian knot was a very complicated knot that no one could undo. Oracles claimed that whoever could undo the knot would become the ruler of Gordium. When Alexander the Great came upon the knot, he simply cut it with his sword and claimed he had fulfilled the prophecy, thus the throne was his. The expression "cutting the Gordian knot" is still used today. What do you think it means?

Knot design examples

Scott Shanks,
geometry student

Tiger Tail,
Diane Cassell, parent of a geometry student

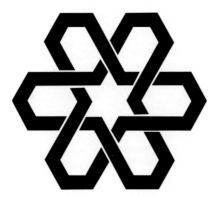

Japanese knot design,
The Patterns of Japan, K•D•C• Co., 1987

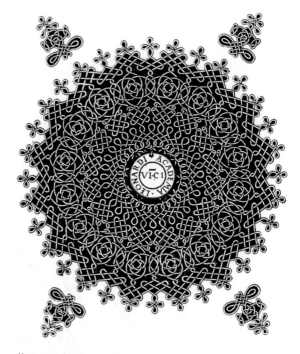

Knot engraving,
Leonardo da Vinci

Russian knot

Islamic Art

Islamic art is rich in geometric forms. Islamic artists were familiar with geometry through the works of Euclid, Pythagoras, and other mathematicians of antiquity, and they used geometric patterns extensively in their art and architecture.

Ceramic tiles, Iran

Western civilization has long looked on Islamic art as merely decorative. However, we have recently come to learn that Islamic art is not only beautiful and geometric, but it is also filled with religious meaning. Islam's prophet Muhammad (A.D. 570–632) preached against idolatry. Many of his followers interpreted his words to mean that the representation of humans or of animals in art was forbidden. Therefore, instead of using human or animal forms for decorations, Islamic artists used intricate geometric patterns. The artists created beautiful artwork while expressing their religious beliefs in a universe ordered by mathematics and reason.

One of the most striking examples of Islamic architecture is the Alhambra, a palace in Granada, Spain. Standing for more than 600 years as a tribute to Islamic artisans, the Alhambra is filled from floor to ceiling with marvelous geometric patterns. The designs in

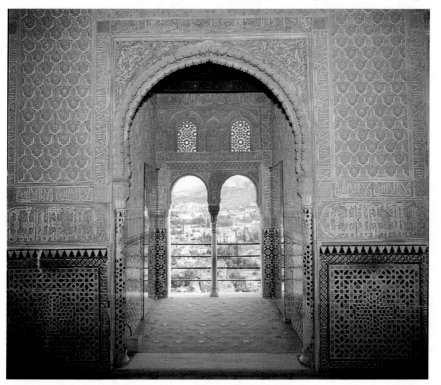

Alcove in the Hall of Ambassadors, Alhambra, a Moorish palace in Granada, Spain

the detail on this page are but a few of the hundreds of intricate geometric patterns found in the tile work and the inlaid wood ceilings of Islamic buildings like the Alhambra. Similar designs can also be found in Islamic carpets and hand-tooled bronze plates.

The geometric patterns often elaborate on basic grids of regular hexagons, equilateral triangles, or squares. Repeating patterns like this are called **tessellations.** You'll learn more about tessellations in Chapter 8. Complex Islamic patterns were constructed with no more than a compass and a straightedge.

You can apply what you learned by creating mandalas and knot designs to create an Islamic-style tessellation. The examples illustrated below show a square-based and a hexagon-based Islamic design.

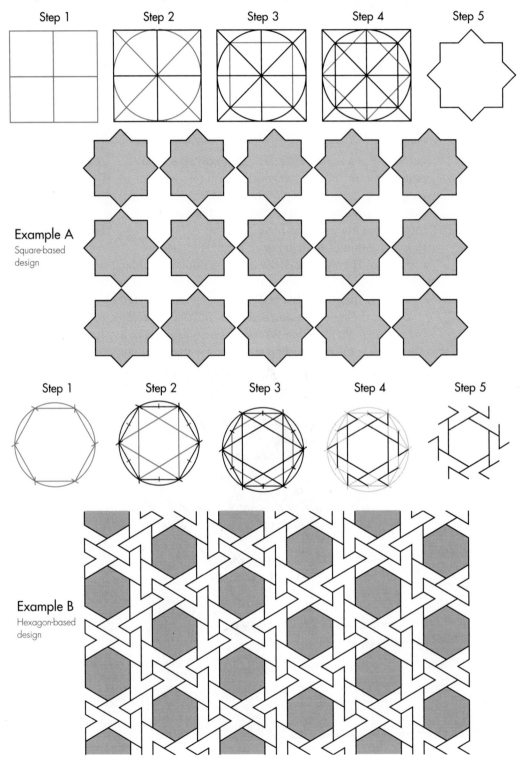

Step 1   Step 2   Step 3   Step 4   Step 5

Example A
Square-based
design

Step 1   Step 2   Step 3   Step 4   Step 5

Example B
Hexagon-based
design

Exercise Set 0.7

In these exercises, you'll create your own Islamic-style tessellation. Use Examples A and B on the previous page as a guide but be creative in inventing your design.

1. Construct a square and use your compass and straightedge to modify and to decorate it, creating a single design like that in Step 5 of Example A.

2. Construct a hexagon and modify and decorate it.

3. Create a tessellation with one of the designs you made in Exercises 1 and 2 by copying and assembling several copies of it. There are several ways you can do this.

   Trace several copies of the design onto tracing paper or patty paper.

   Place a piece of carbon paper between your design and a piece of regular paper, then repeatedly draw over your design to create several copies on the bottom paper.

   If you have access to a photocopy machine, make several copies and cut and paste them together.

   If you have access to a computer with a drawing program, construct your design on the computer and copy and paste to make a tessellation.

4. Add finishing touches to your tessellation by adding, erasing, or whiting out lines as desired. If you want, see if you can interweave a knot design within your tessellation. Color it.

Improving Reasoning Skills—*Bagels I*

In the original computer game of bagels, a player determines a three-digit number (no digit repeated) by making "educated guesses." After each guess, the computer gives a clue about the guess. Here are the clues.

bagels: no digit is correct
pico: one digit is correct but in the wrong position
fermi: one digit is correct and in the correct position

In each of the problems below, a number of guesses have been made, with the clue for each guess shown to its right. From the given set of guesses and clues, determine the three-digit number. If there is more than one solution, find them all.

1.	1 2 3	*bagels*	2.	9 0 8	*bagels*
	4 5 6	*pico*		1 3 4	*pico*
	7 8 9	*pico*		3 8 7	*pico fermi*
	0 7 5	*pico fermi*		2 5 6	*fermi*
	0 8 7	*pico*		2 3 7	*pico pico*
	? ? ?			? ? ?	

Project

Geometry in Sculpture

The sculpture made of silicon bronze shown at right is the *Umbilic Torus* by Helaman Ferguson, created for the Jaime Escalante Award. These sculptures are awarded to outstanding math teachers in Los Angeles schools. (Jaime Escalante was a very dynamic and energetic calculus teacher. The movie *Stand and Deliver* is the story of Mr. Escalante and his heroic efforts at Garfield High School in East Los Angeles.) The sculpture has only one edge. If you were to run your finger along the edge, you would touch the edge at every point and end up back where you started. How many different surfaces do you think the sculpture has?

Umbilic Torus (1990), Helaman Ferguson

Mr. Ferguson has produced many large sculpted pieces in marble and in bronze that are both artistically and mathematically beautiful. If you get a chance to meet him, be sure to ask for your own free personally signed and numbered piece of Ferguson sculpture. He makes these "sculptures" from paper he presses between two granite blocks. Each block has a design carved in it that is the complement of the design carved in the other block. That is, ridges in one block fit into grooves in the other.

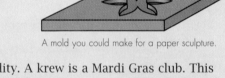

A mold you could make for a paper sculpture.

Geometric objects give Ida Kohlmeyer's sculpture entitled *Krew of Poydras* a dynamic and whimsical quality. A krew is a Mardi Gras club. This sculpture signifies the celebration of the krew from Poydras Street in New Orleans. What shapes can you identify in her sculpture?

Create a geometric sculpture of your own. You can use simple materials like clay, cardboard, styrofoam, and wire to make cubes, cylinders, or other shapes. Remember, your sculpture need not be 40 feet tall like the *Krew of Poydras*! If you want to create something like *Umbilic Torus,* you might practice by making a Möbius strip. Look up Möbius strip in a dictionary or encyclopedia to see how to make one. Or you might choose to carve a simple geometric design and its complement into two surfaces you can use to make pressed paper sculptures.

Write a short report about your sculpture, describing what makes your sculpture geometric, including any symmetry it has and any geometric tools you used to construct it.

Krew of Poydras (1983), Ida Kohlmeyer

Lesson 0.8

Perspective

Art is a lie that makes us realize the truth.
— *Pablo Picasso*

The School of Athens, Vatican fresco by Raphael of Urbino (1483–1520)

Many of the paintings created by European artists during the Middle Ages were commissioned by the Roman Catholic Church. The art was symbolic; that is, people and objects in the paintings were symbols representing religious ideas. Artists were more interested in creating a symbolic scene than in accurately representing people and objects. The background in these religious paintings was usually a solid color, and people were sized according to importance rather than to their distance from the viewer. There was no attempt to draw in perspective.

The symbolic nature of art began to change in the Renaissance. Renaissance artists delighted in nature and in the human form. They began to search for ways to represent the three-dimensional world more naturally on flat, two-dimensional surfaces.

Many Renaissance artists were engineers or architects who were well-versed in mathematics. It is no surprise that these artists turned to geometry to solve the problems of perspective. **Perspective** is the technique of portraying solid objects and spatial relationships on a flat surface so that they appear true-to-life.

Raphael's *School of Athens* on the previous page is a perspective painting that pays homage to science and to philosophy. The two central figures in the painting are Plato and Aristotle. The kneeling figure in the lower left corner is Pythagoras. The figure drawing on a slate in the lower right corner is Euclid, the founder of our geometric tradition. The size of each figure is determined by the distance from the viewer to the figure. The receding arches enhance the realism of the scene.

In their writings on the subject of perspective painting, Renaissance artists were very systematic and mathematical. Leonardo da Vinci even began one of his essays on painting by writing, "Let no one who is not a mathematician read my works."

Perhaps the most influential student of perspective was German artist Albrecht Dürer. Dürer traveled to Renaissance Italy to study the works of earlier painters. Dürer left many woodcuts detailing his method of perspective drawing.

We know from experience that objects closer to us appear larger than similar-

Woodcut from *The Artist's Treatise on Geometry* by Albrecht Dürer (1471–1528)

Perspective study by Jan Vredeman de Vries (1527–1604)

sized objects farther away. You have probably noticed that when looking down railroad tracks toward the horizon, the parallel tracks appear to meet at some point on the horizon line. In a perspective drawing, parallel lines running directly away from the viewer are drawn so that they come together at a point called the **vanishing point**. The vanishing point is located on the horizon line. Can you see an example of a vanishing point in the perspective study by Jan Vredeman de Vries above?

A rectangular solid, or box, is one of the simplest objects to draw in perspective. Follow the steps below to draw a rectangular solid with one face viewed straight on.

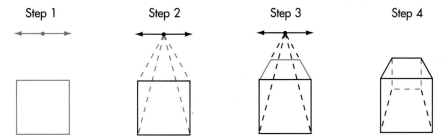

| Step 1 | Step 2 | Step 3 | Step 4 |

Step 1 Begin by drawing a rectangle for the front face. Draw a **horizon line** parallel to the horizontal edges of your rectangle and select a vanishing point on it.

Step 2 To create the edges of the box that recede from view, draw lines, called **vanishing lines**, from the corners of the rectangle to the vanishing point.

Step 3 To create the visible back horizontal edge, draw a line parallel to the horizon line. Use this line to determine the position of the back vertical edges.

Step 4 To complete the figure, draw hidden back vertical and horizontal edges. Erase the horizon line and the unnecessary portions of the vanishing lines.

Notice that only the edges of the box appearing to move away from the viewer are drawn to meet at a vanishing point. The horizontal lines that are parallel to the picture plane are drawn parallel in the picture and are not drawn to the vanishing point. The same is true for the vertical lines that are parallel to the picture plane.

The location of the horizon line in a perspective drawing tells you something about the position of the viewer. If the horizon line is high in the picture, then the viewer is looking from above, perhaps from a hill. If the horizon line is low, then the viewer is on the ground.

One doesn't always view an object straight on. However, as long as the front surface of the object is parallel to the picture plane, only the lines that appear to move directly away from the viewer are drawn to a vanishing point. If an object is viewed from the right, then the vanishing point is placed at the right. If an object is viewed from the left, then the vanishing point is placed at the left.

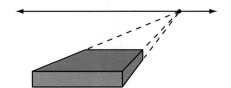

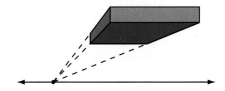

Here the viewer is looking at the rectangular box from above and from the right. Notice that the horizon line is located high in the picture and the vanishing point is placed at the right.

Here the viewer is looking at the rectangular box from below and from the left. Notice that the horizon line is located low in the picture and the vanishing point is placed at the left.

In the first part of this lesson you learned to draw a perspective view of a rectangular object with the front surface parallel to the picture plane. Drawings of this type use **one-point perspective**. If the front surface of a rectangular solid is not parallel to the picture plane, then you need two vanishing points. This is called **two-point perspective**. To illustrate how to draw a figure with two-point perspective, let's look at a rectangular solid with one edge viewed straight on.

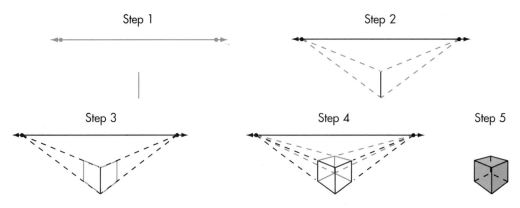

Step 1	Begin by drawing a vertical line segment for the front edge. Draw a horizon line and select two vanishing points on it.
Step 2	To create the edges of the box that recede from view, draw four vanishing lines from the endpoints of the vertical line segment back to the two vanishing points.
Step 3	To create the length and the width of the box, draw vertical line segments intersecting the vanishing lines. The endpoints of these line segments determine the position of the back edges that recede from view.
Step 4	Draw the four remaining vanishing lines and the back hidden vertical edge.
Step 5	Erase the unnecessary portions of the vanishing lines.

In the previous drawings, the lines that recede to the left meet at a vanishing point on the left of the horizon line. The lines that recede to the right meet at a second vanishing point on the right. All the vertical edges in the rectangular solids are parallel to the picture plane and therefore do not meet at a vanishing point—they are drawn parallel to the picture plane.

Sometimes an object is drawn below the horizon line (the viewer is above the object), and occasionally an object is drawn above the horizon line (the viewer is on the ground and the object is in the air). Many times, however, the object is so large or the viewer so close that the object is drawn both above and below the horizon line. Observe the location of the horizon line in the figure below.

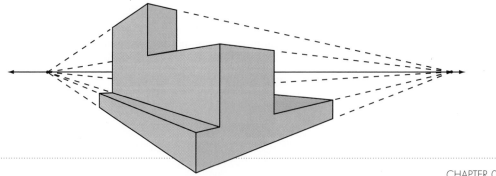

Exercise Set 0.8

1. Use your straightedge and neatly copy the figure at right onto an unlined piece of paper. The easiest, most accurate way to copy a simple figure made with straight lines is to place your paper over the figure and put a dot on your paper at each corner of the figure. Then take the paper off the figure and use your straightedge to connect the dots. After you have copied the figure onto your own paper, locate the vanishing point and draw the horizon line.

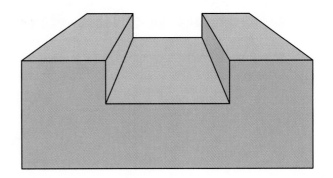

2. Draw a perspective view of a cube. Show all 12 edges. Use dashed lines for the hidden edges.

3. Draw a perspective view of a rectangular box that is viewed from above and from the right.

4.* Draw a perspective view of a rectangular box with a rectangular box removed from the center. (Your drawing will look like a rectangular-shaped donut.)

5. Use your straightedge and neatly copy the figure below onto an unlined piece of paper. Then locate the horizon line by finding the two vanishing points.

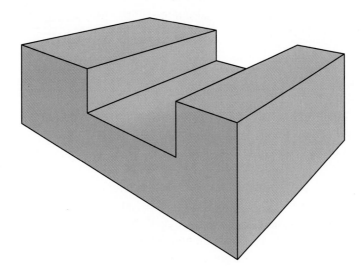

6. Draw a two-point perspective view of a rectangular box viewed from below and from the right.

7. Draw a two-point perspective view of a rectangular box set on top of another rectangular box.

8. Draw a two-point perspective view of a rectangular box with a rectangular window cut out of one of the faces. Show that the box has thickness.

Drawing a High-Rise Complex

Skyscrapers are challenging to draw in perspective because they can be made of many different rectangular solids and thus have many different vanishing lines. Drawing a block of skyscrapers in two-point perspective is illustrated for you below.

Step 1

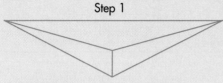

Begin with a horizon line and two vanishing points. Draw the front vertical edge of your first building with all the vanishing lines.

Step 2

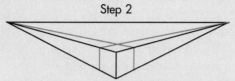

Complete the two-point perspective view of the first building.

Step 3

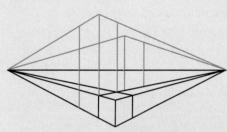

Next, draw in a couple of the taller buildings. Start with the front vertical edge of each building and draw the vanishing lines. Complete the perspective view.

Step 4

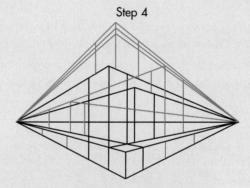

Create additional buildings and use vanishing lines to add architectural details.

Step 5

Erase all unnecessary lines and add other details. See the example of student art on the next page.

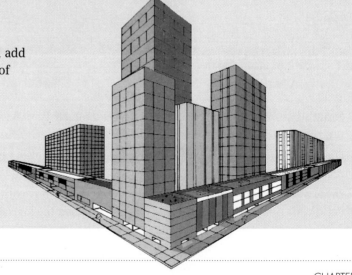

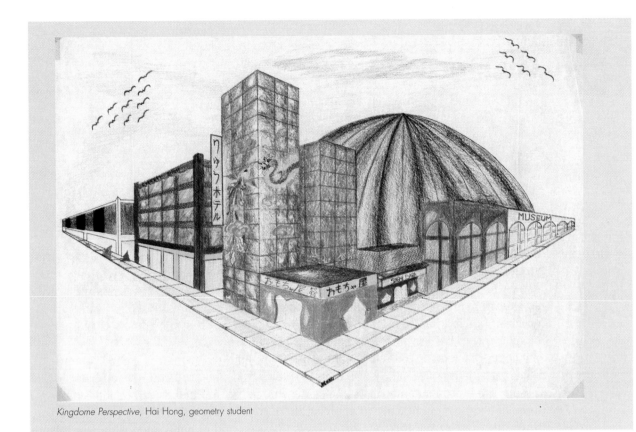

Kingdome Perspective, Hai Hong, geometry student

Improving Visual Thinking Skills—*Match This!*

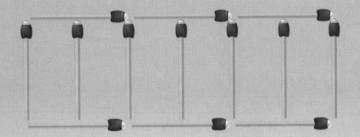

Perhaps you have seen matchstick puzzles before. Here is a classic from British puzzle enthusiast H. Dudeney (1842-1930). The matches represent a farmer's hurdles (portable fence pieces) placed so that they form six pens of equal size. After one hurdle is stolen, the farmer must rearrange the remaining twelve to form another six pens of equal size. How can this be done?

Lesson 0.9

Chapter Review

Let us make a thing of beauty
That long may live when we are gone;
Let us make a thing of beauty
That hungry souls may feast upon;
Whether it be wood or marble,
Music, art or poetry,
Let us make a thing of beauty
To help set man's bound spirit free.
— Edward Matchett

In this chapter, you learned that geometry gives us a vocabulary for describing the world around us. You were asked to describe the geometric shapes and symmetries you see in nature, in everyday objects, in art, and in architecture. You learned that geometry appears in many types of art—ancient and modern, from every culture—and you learned specific ways in which some cultures employ geometry in their art. You also used tools of geometry to create your own works of geometric art.

Looking back on it, you really learned a lot! The end of a chapter is a good time to review and to organize your work. Each chapter in this book will end with a review lesson. This lesson summarizes some of the content of the chapter and includes an exercise set for more practice in applying these concepts. But don't stop there! When you finish the exercise set, there are many ways to assess and to build upon what you gained from the chapter. The section following the exercise set, called Assessing What You've Learned, offers suggestions for how you might review, organize, and communicate what you've learned to others. Whether you follow these suggestions, follow other directions from your teacher, or use study strategies of your own, be sure to take the opportunity to reflect on all you've learned.

Exercise Set 0.9

1. What is the optical effect of the op art design at right?
2. List three of the many cultures that use geometry in their art.
3. Name and describe the uses of the four basic tools of geometry.
4. With a compass, draw a 12-petal daisy.
5. List three things in nature that have geometrical shapes.
6. Each of the cultures listed below made use of a mandala design. What was the mandala design? What purpose did each serve?

   a.* Aztecs   b. Navajos   c. Medieval Europeans

7. Draw an original knot design.
8. Draw a rectangular box with two-point perspective.

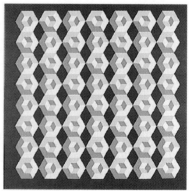

Hot Blocks (1966–67), Edna Andrade

9. Which of the wheels below have reflectional symmetry? How many lines of symmetry does each have?

10. Which of the wheels above have only rotational symmetry? How many rotational symmetries does each have?

11. Did you know that the most read section of the *World Book Encyclopedia* is the flag section? Go to a library and find the flag section of the *World Book Encyclopedia,* or of another reference, and answer the following.

   a. Is the flag of Puerto Rico symmetric?
   b. Does the flag of Kenya have rotational symmetry?
   c. Name a country whose flag has both rotational and reflectional symmetry.

12. Contemporary artist Kunito Nagaoka (b 1940) uses geometry in his work. Nagaoka was born in Nagano, Japan, and was raised near the active volcano Asama. In Japan, he experienced the natural threats of earthquakes and typhoons as well as the manufactured tragedies of Hiroshima and Nagasaki. In 1966, he moved to Berlin, Germany, a city rebuilt in concrete from the

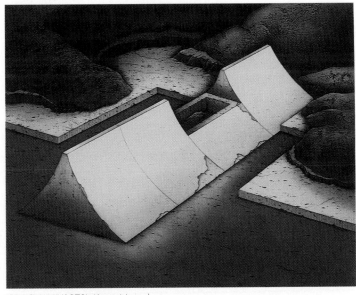

ISEKI/PY XVIII (1978), Kunito Nagaoka

ruins of World War II. These experiences clearly have influenced his work.

   The etching shown above can be viewed as the ruins of a long-forgotten civilization on some strange and faraway planet. Write a paragraph describing what you think might have happened in the scene or what you think the etching might represent. What types of geometric figures are found in the etching? Create your own sketch or painting that uses geometric shapes to evoke a story. Write a one- or two-page story related to the art you made.

Assessing What You've Learned—*Keeping a Portfolio*

An essential part of learning is being able to show yourself and others that you understand the concepts you've encountered and that you can do what you were expected to do. This assessment doesn't just come at the end of a chapter and it isn't limited to tests and quizzes. Assessment isn't even limited to what your teacher sees or what makes up your grade. Every piece of art you make, every construction or investigation you perform, and every project or exercise you complete gives you a chance to demonstrate to somebody—yourself, at least—that you're learning.

Because this chapter is primarily about art, you might organize this chapter's work the way a professional artist does: in a **portfolio**. A portfolio is different from a notebook, both for an artist and for a geometry student. An artist's notebook might contain everything from scratch work to practice sketches to random ideas jotted down. A portfolio is reserved for an artist's most significant or *best* work. It's his or her portfolio that an artist presents to the world to demonstrate what he or she is capable of. The portfolio can also show how an artist's work has changed over time.

Review all the work you've done so far and choose one or more examples of your best works of art and your best projects for inclusion in your portfolio. Your teacher may have specific suggestions for how to select and document your work. One way to document your work is to write a paragraph or two about each piece, addressing the following questions.

- What is the piece?
- What makes this piece representative of your best work?
- What mathematics did you learn or apply in this piece?
- How would you improve the piece if you were to redo or revise it?

Improving Algebra Skills—*Pyramid Puzzle I*

Place four different numbers in the bubbles at the vertices of the pyramid so that the two numbers at the ends of each edge add to the number on that edge.

Cooperative Problem Solving

Designing a Lunar Colony Park

It is the year 2065 on the lunar space port Galileo. The space port has been in continuous operation for the past 25 years. The space port includes the research facility, the recreational and living quarters, the agricultural and life-support divisions, and the lunar landing and surface transportation department.

After years of very successful operation, many residents at retirement have opted to remain at the space port. The executive council of the space port has recently agreed to turn over to the local affiliate of the Senior Citizens Association an undeveloped agricultural pod for development as a park for retired astronauts.

The agricultural pod is a hemispherical reinforced glass dome 70 meters in diameter, which has been set up to reproduce many of the fruits and vegetables native to planet earth. The temperature and climate is computer controlled to create a continuous year-round growing season—this includes a light rain each Tuesday and Thursday evening.

Your job is to submit before the Galileo Planning Commission a proposal for the park design. The budget limit is $NL10,000 (new lunar dollars). Based on the criteria given below, the Commission will make its selection of a park proposal from the oral presentations at the next commission meeting.

1. Safe design in accordance with accepted Galileo traditions

2. Aesthetics, innovation, and optimum use of the area

3. Cost-effectiveness

Here is what your presentation before the Commission must include.

1. A park design (scaled map where 1 cm = 1 m) showing all improvements

2. An itemized list of costs

3. Explanations of how your design satisfies each of the three criteria

A list of materials and their costs have been itemized in the table below. The materials have all been reproduced to appear authentic or have been imported from the home planet at great expense. The costs have been listed in new lunar dollars.

Materials	Cost	
Park bench (1.5 m long)	$NL 125.00	each
Trash receptacle	40.00	each
Picnic table (2 m long)	150.00	each
Bike rack	100.00	each
Stainless steel drinking fountain	300.00	each
Plumbing for drinking fountain	25.00	per m
Iron barbecue pit	380.00	each
Park overhead lights	500.00	each
Redwood gazebo (5 m diameter)	1500.00	each
Bathroom (men/women 3 by 4 m)	2800.00	each
Clubhouse (redwood 5 m square)		
Prefab with electricity	9000.00	each
Prefab without electricity	6500.00	each
Putting green	20.00	per m^2
Community garden	2.00	per m^2
Asphalt (4 cm thickness)	45.00	per m^2
Trees		
Fruit	150.00	each
Shade	100.00	each
Shrubs	25.00	per m^2
Building materials		
Lumber (5 by 10 cm)	4.50	per m
Lumber (10 by 10 cm)	12.00	per m
Siding (1 by 2 m sheet)	16.00	per sheet
Asphalt shingles	6.00	per m^2
Paint	12.00	per liter (covers 15 m^2)
Railroad ties (4 m long)	15.00	each
Cement bricks	1.50	each

For Exercises 27–30, study the pattern and then complete the problem on the basis of your observations. (Please, no calculators here. You would be missing the point. You are looking for patterns, not mere numerical answers.)

27.

$$1 \cdot 1 = 1$$
$$11 \cdot 11 = 121$$
$$111 \cdot 111 = 12321$$
$$1111 \cdot 1111 = 1234321$$
$$11111 \cdot 11111 = -?-$$

28.

$$6 \cdot 7 = 42$$
$$66 \cdot 67 = 4422$$
$$666 \cdot 667 = 444222$$
$$6666 \cdot 6667 = 44442222$$
$$66666 \cdot 66667 = -?-$$

29.

$$12345679 \cdot 9 = 111111111$$
$$12345679 \cdot 18 = 222222222$$
$$12345679 \cdot 27 = 333333333$$
$$12345679 \cdot 36 = -?-$$
$$12345679 \cdot -?- = 555555555$$

30.

$$9 \cdot 0 + 1 = 1$$
$$9 \cdot 1 + 2 = 11$$
$$9 \cdot 2 + 3 = 21$$
$$9 \cdot 3 + 4 = -?-$$
$$-?- = 41$$

31. Which numbers in the last group are Quatros?

4	52
108	36
60	144

Quatros

2	15
29 18 22	
106	6

Not Quatros

| 86 | 42 |
| 737 | 72 |

Which are Quatros?

32. Which numbers in the last group are Semirps?

2	53
13	97
11	71
23	47

Semirps

15	21
209	
25	190
39	

Not Semirps

| 123 | 51 |
| 67 | 27 |

Which are Semirps?

Improving Reasoning Skills—*Puzzling Patterns*

The next few problems are very different. Watch out!

1. 18, 46, 94, 63, 52, 61, –?–
2. O, T, T, F, F, S, S, E, N, –?–
3. 4, 8, 61, 221, 244, 884, –?–
4. 6, 8, 5, 10, 3, 14, 1, –?–
5. B, 0, C, 2, D, 0, E, 3, F, 3, G, –?–

6. A E F H I K L M N T V W
    B C D G J O P Q R S U
    Where do the X, Y, and Z go?

7. 2, 3, 6, 1, 8, 6, 8, 4, 8, 4, 8, 3, 2, 3, 2, 3, –?–

Lesson 1.3

Picture Patterns

We talk too much; we should talk less and draw more.
— *Johann Wolfgang von Goethe*

In this lesson you will combine your inductive reasoning skills with your artistic skills. First you will use inductive reasoning to imagine what the next shape will be in each given pattern. Then you will draw what you visualize. Like an artist, you will be translating your thoughts into pictures. Visual thinking is used by more than just artists: Scientists and mathematicians translate information into graphs; architects turn their clients' ideas into drawings of new buildings; composers turn the patterns of musical sounds into picture patterns.

Graph   Native American ceramic design   Musical notation

Strip patterns or frieze patterns are a type of decorative repeating pattern found in many cultures. You can find frieze patterns on Native American baskets and pottery, on Chinese and Persian carpets, and on the walls, the floors, and the ceilings of ancient Greek, Roman, and Islamic buildings.

Exercise Set 1.3

Draw the next shape in each picture pattern.

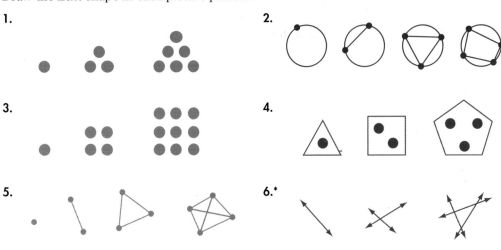

1.

2.

3.

4.

5.

6.*

7.

8.*

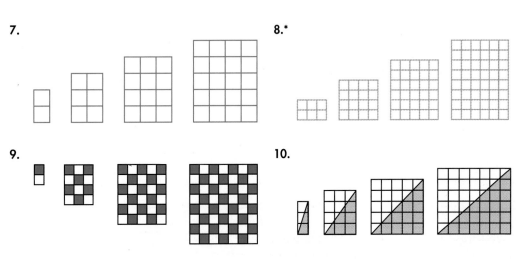

9.

10.

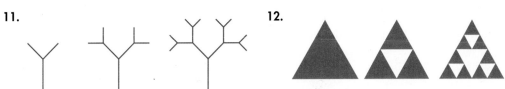

11.

12.

For Exercises 13 and 14, copy all four measures of the musical pattern shown. Draw the location of all the notes in the next two measures, assuming the pattern continues.

13.

14.

For Exercises 15–17, you will need graph paper. Each picture pattern below was created on a grid of squares. Follow the given steps for each pattern.

Step 1   Copy the pattern (2 squares high by 16 squares long) onto graph paper.

Step 2   Continue the pattern in a two by six region at the end of the pattern.

Example

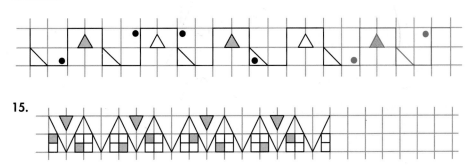

15.

16.

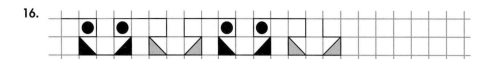

17. Now it's your turn. On graph paper, create a repeating pattern of your own.

In Exercises 18 and 19, trace the frieze pattern onto tracing paper or onto patty paper. Determine what part of the pattern is repeated and how it moves to its next position. Then trace that part onto the end of the frieze pattern.

18. African woven basket design

19. Painted rafter, Maori (New Zealand)

20. The photo below shows blockprinting on handmade paper at Tulsi Meher Ashram in Nepal. Make a rough sketch of the part of the pattern that is carved onto the block.

Copy the graphs in Exercises 21–23 onto graph paper. For each exercise, find two more points on the graph that continue the pattern.

21.

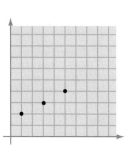

22.

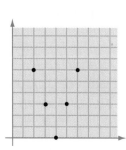

23.

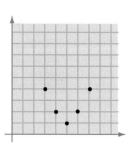

For Exercises 24–27, find the next two terms that continue the pattern.

24. 1, 3, 9, 27, 81, 243, –?–, –?–

25. 1, 5, 17, 53, 161, 485, –?–, –?–

26. 1, 5, 14, 30, 55, 91, –?–, –?–

27. $1, \frac{3}{2}, \frac{9}{4}, \frac{27}{8}$, –?–, –?–

Geometer's Sketchpad Project

Patterns in Fractals

In Lesson 1.2, you discovered patterns that you used to continue number sequences. In most cases, you found each term by applying a simple rule to the term before it. Such rules are called **recursive rules.** Some picture patterns in Lesson 1.3 are also generated by recursive rules. You find the next picture in the sequence by looking at the picture before it and comparing that to the picture before it, and so on. Sketchpad scripts can have recursive steps that tell the script to replay itself on designated parts of the figure. You can use scripts with recursive steps to create initial stages of fascinating geometric figures called **fractals.** Some fractals have strange properties, like an infinite perimeter surrounding a finite area. Fractals exhibit **self-similarity,** meaning that if you zoom in on a part of the figure, it looks like the whole. A true fractal is a theoretical figure—the recursive rule is applied infinitely many times. Making an actual fractal would require infinite time and a computer with infinite memory. In this project, you'll use a Sketchpad script to create the first few stages of a fractal called the **Sierpinski gasket.**

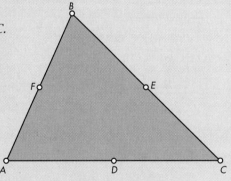

Step 1 Open a Sketchpad sketch and a new script. Click the Record button on the script.

Step 2 Use the segment tool to draw a triangle *ABC*.

Step 3 Select the three vertices of the triangle and construct the polygon interior.

Step 4 Select the three sides of the triangle and construct the midpoints.

Step 5 Select points *A*, *F*, and *D* and click Loop in your script.

Step 6 Repeat Step 5 for triangles *FBE* and *DEC*.

Step 7 Hide midpoints *F*, *E*, and *D*.

Step 8 Stop your script.

This script constructs a triangle, its interior, and midpoints on its sides. The recursive steps, which you recorded by clicking Loop, repeat that process on the three triangles formed by the midpoints with the corners of the original triangle. Create three new points and play your script. Choose a recursion depth of 1. The script constructs a triangle, then it constructs three identical triangles in its corners, covered by the original triangle.

The original triangle is a stage 0 Sierpinski gasket. If you hide it, you'll reveal the stage 1 Sierpinski gasket.

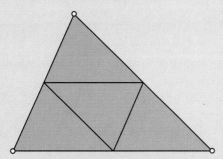 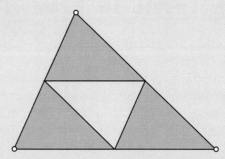

The original triangle is a stage 0 Sierpinski gasket. To see the stage 1 gasket, click in the center of the figure to select the large triangle, then hide it.

The triangle with the center removed is a stage 1 gasket. This gasket consists of three triangles, one in each corner.

1. Play your Sierpinski gasket script with a recursion depth of 4. Your script constructs a stage 4 gasket, but it's covered by stage 3, 2, 1, and 0 gaskets. By hiding layers of triangles, you can reveal successive stages of the gasket. Click in the center of the triangle and hide it, revealing the stage 1 gasket (three triangles). Hold down the Shift key and click in the centers of the three remaining triangles and hide them, revealing the stage 2 gasket. How many triangles make up the stage 2 gasket? Record this number in the table below. (Note that hiding each layer of triangles is like removing area from the gasket.) Make sure your gasket is big enough so that you can easily select triangles and hide the stage 2 layer of triangles. Finally, hide the stage 3 layer of triangles to reveal the stage 4 gasket. (The triangles in the stage 3 layer might be pretty small, so be careful to click right in their centers to select them.) Complete the table to show how many triangles comprise each stage of the gasket.

Stage	0	1	2	3	...	n	...	50
Number of Δs	1	3	-?-	-?-	...	-?-	...	-?-

How many triangles comprise a stage n gasket? A stage 50 gasket? What stage gasket is the gasket shown at the beginning of the project? Write a sentence or two explaining what might be meant when it's said that the Sierpinski gasket exhibits self-similarity.

2. What's happening to the area of the Sierpinski gasket in successive stages? Suppose you start with a stage 0 gasket (just a plain old triangle) with area 1 unit. What would be the area of a stage 1 gasket? A stage 2 gasket? It would be impractical to measure areas in Sketchpad to answer these questions. Instead, look for a pattern. Copy the chart below and fill in the missing values.

Δ Stage	0	1	2	3	...	n	...	50
Area	1	$\frac{3}{4}$	-?-	-?-	...	-?-	...	-?-

Once you've found a rule for the area of a stage n gasket, use a calculator to find the area of a stage 50 gasket. What would be the area of a stage infinity gasket?

3. What about the perimeters of Sierpinski gaskets? Suppose your stage 0 gasket had a perimeter of 6 units. If you consider lengths inside the main triangle, the perimeter of a stage 1 gasket would be 9 units (the sum of the perimeters of the three triangles). What would be the perimeter of a stage 2 gasket? Copy the chart below and fill in the missing values.

Δ Stage	0	1	2	3	...	n	...	50
Perimeter	6	9	-?-	-?-	...	-?-	...	-?-

Once you've found a rule for the perimeter of a stage n gasket, use a calculator to find the perimeter of a stage 50 gasket. What would be the perimeter of a stage infinity gasket?

4. Manipulate your Sierpinski gasket so that it fills a page. If you print three copies of your Sierpinski gasket, you can put the copies together to create a larger gasket of a stage greater than your original. How many copies would you need to print in order to create a gasket two stages greater than your original? Ask your teacher how many copies of your gasket you may print, then print exactly the number of copies you need to combine them into a greater stage gasket. Combine them into a poster or a bulletin board.

5. Experiment with scripts in the Fractals folder in the Sample Scripts that came with The Geometer's Sketchpad.

6. See if you can make other recursive scripts that construct fractals.

7. The word *fractal* was coined by Benoit Mandelbrot (b 1924), a pioneering researcher in this new field of mathematics. He was the first to use high-speed computers to create the figure below, called the **Mandelbrot set**. Describe the self-similarity you see in the figure. Guess what that little dot you can see in the set's tail is.

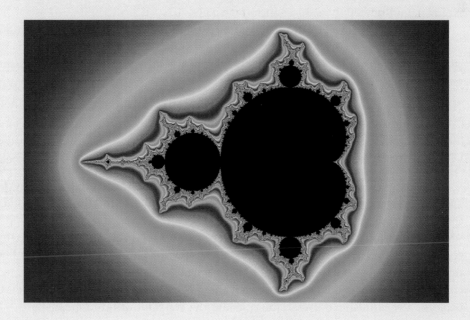

Lesson 1.4

Finding the *n*th Term

If you do something once, people call it an accident. If you do it twice, they call it a coincidence. But do it a third time and you've just proven a natural law.
— *Grace Murray Hopper*

What if you needed to know the value of the 20th, 200th, or 2000th term of a number sequence? You may see the pattern that would give you the 20th term if you knew the 19th. And you could get the 19th if you just knew the 18th . . . . Can you see the problem? You certainly don't want to generate all the values between 1 and 20 just to get one answer. If you knew a rule for calculating *any* term in a sequence, without having to know the previous term, you could apply it to directly calculate the 20th term.

You can create some number sequences by applying a rule, sometimes called a **function**, to consecutive numbers. For example, applying the rule $4n - 1$ to consecutive numbers 1, 2, 3, 4, . . . gives the sequence 3, 7, 11, 15, ... The table below shows this sequence.

Term	1	2	3	4	5	6	7	8	...	20	...	*n*
Value	3	7	11	15	19	23	27	31	...	79	...	$4n - 1$

Each number in the top row represents a term number or a position in the sequence. The number beneath each term number is the value of that term of the sequence. The first term of the sequence has a value of 3, the second term has a value of 7, and so on. The *n*th term has a value of $4n - 1$. The rule for this pattern, $n \rightarrow 4n - 1$, states that to find the value of any term in this sequence, multiply the term number by 4 and subtract 1. Therefore, if you wish to know the value of the 20th term of this sequence, substitute 20 for *n,* multiply 20 by 4, and subtract 1 to get 79.

Generating sequences from rules is fairly straightforward. Finding a rule for a given sequence takes more investigation. What would you do to get the next term for the sequence 32, 39, 46, 53, 60, 67, 74, . . . ? Look at the differences between consecutive terms to see if there is a pattern.

n	1	2	3	4	5	6	7	8	...	20
-?-	32	39	46	53	60	67	74	-?-	...	-?-

Differences   7   7   7   7   7   7

In this case there is a pattern—they always differ by 7. Therefore the next term is 74 + 7, or 81. But what about the 20th term? Or the 200th term? How do we find the rule that gives the *n*th term?

Let's perform an investigation to discover how to find a rule for a sequence that has a constant difference between terms. Look at the next few examples to discover how rules relate to the sequences they generate. In Investigation 1.4.1, use the given rule to complete each table. Then check to see if there is a constant difference between values. Look for a relationship between the rule and the constant difference.

Investigation 1.4.1

Copy and complete each table. Find the differences between consecutive values.

1.

n	1	2	3	4	5	6	7	8
$n - 5$	-4	-3						

2.

n	1	2	3	4	5	6	7	8
$2n + 1$	3	5						

3.

n	1	2	3	4	5	6	7	8
$3n - 2$	1	4						

4.

n	1	2	3	4	5	6	7	8
$5n + 7$	12	17						

Did you spot the pattern? If a sequence has a constant difference of 2, then the number in front of the n (the coefficient of n) is also 2. If the coefficient of n in a rule is 3, then the sequence has a common difference of 3. In general, if the difference between the value of each term of a sequence is a constant, then the coefficient of n in the formula is that constant.

Let's return to the problem from the previous page.

n	1	2	3	4	5	6	7	8	...	20
-?-	32	39	46	53	60	67	74	-?-	...	-?-

Differences   7   7   7   7   7   7

Because the constant difference is 7, you know part of the rule is $7n$. How do you find the rest of the rule? The first term ($n = 1$) of the sequence is 32, but if we apply part of the rule we have so far to $n = 1$, we get $7n = 7(1) = 7$, not 32. So how should we alter the rule? How can we get from 7 to 32? By adding 25, because $7 + 25 = 32$. So the rule should be $7n + 25$. Check it by trying the rule with other terms in the sequence.

n	1	2	3	4	5
$7n + 25$	32	39	46	53	60
	$7(1) + 25$	$7(2) + 25$	$7(3) + 25$	$7(4) + 25$	$7(5) + 25$

Investigation 1.4.2

Rules that generate sequences with a constant difference between terms are called **linear functions**. To see why they're called linear functions, try graphing the values in the table from the beginning of the lesson, shown here.

Term	1	2	3	4	5	6	7	8	...	20	...	n
Value	3	7	11	15	19	23	27	31	...	79	...	$4n - 1$

Use n for the horizontal coordinate and $4n - 1$ for the vertical coordinate. In other words, graph the points (1, 3), (2, 7), (3, 11), (4, 15), and so on. To help you review from algebra, the first two points are graphed. Use a graphing calculator if one is available.

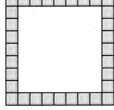

What do you notice about the points you graphed? Try graphing values from one or more of the tables in this lesson. What can you conjecture about the graphs of sequences that have a constant difference between terms? Complete the statement: The graph of a sequence with a constant difference is a set of points that —?—.

Exercise Set 1.4

In Exercises 1–4, use what you have discovered about sequences to find the rule (nth term), then find the value of the 20th term for the sequence.

1.*

1	2	3	4	5	6	...	n	...	20
3	9	15	21	27	33	...	-?-	...	-?-

2.

1	2	3	4	5	6	...	n	...	20
11	23	35	47	59	71	...	-?-	...	-?-

3.

1	2	3	4	5	6	...	n	...	20
-4	4	12	20	28	36	...	-?-	...	-?-

4.

1	2	3	4	5	6	...	n	...	20
0	7	14	21	28	35	...	-?-	...	-?-

5. If the pattern of square donuts continues, what is the rule for the number of squares in the nth donut? What is the number of squares in the 200th donut?

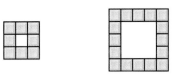

Squares in a square donut

Donut	1	2	3	4	5	6	...	n	...	200
Number of squares	8	16	-?-	-?-	-?-	-?-	...	-?-	...	-?-

6. If the pattern of crosses continues, what is the rule for the number of squares in the nth cross? What is the number of squares in the 200th cross?

Squares in a cross array

Cross	1	2	3	4	5	6	...	n	...	200
Number of squares	1	5	9	-?-	-?-	-?-	...	-?-	...	-?-

7. Graph the values in your tables from Exercises 5 and 6. Which set of points lies on a steeper line? Why?

8. How many diagonals can you draw from one vertex in a polygon with 35 sides? How many triangles are formed when you draw those diagonals?

  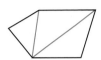

Number of diagonals from one vertex in a polygon

Sides	1	2	3	4	5	6	...	n	...	35
Diagonals from one vertex			–?–	–?–	–?–	–?–	...	–?–	...	–?–
Triangles formed			–?–	–?–	–?–	–?–	...	–?–	...	–?–

9.* If you place 200 points on a line, into how many parts does it divide the line? Wait! Don't start placing 200 points on a line. You need to discover a rule that relates the number of points placed on a line to the number of parts created by those points. Then you can use your rule to answer the question.

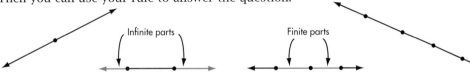

Points dividing a line

Points dividing line	1	2	3	4	5	6	...	n	...	200
Infinite parts	–?–	–?–	–?–	–?–	–?–	–?–	...	–?–	...	–?–
Finite parts (no overlap)	–?–	–?–	–?–	–?–	–?–	–?–	...	–?–	...	–?–
Total parts	–?–	–?–	–?–	–?–	–?–	–?–	...	–?–	...	–?–

For Exercises 10 and 11, study the given information until you discover a relationship that holds true for all the data. Then state your findings by completing the conjecture.

10. $25 \cdot 3 = 75$ $19 \cdot 7 = 133$
     $-17 \cdot 31 = -527$ $-21 \cdot -7 = 147$
     $-1 \cdot 41 = -41$ $103 \cdot -7 = -721$
Conjecture: The product of —?—.

11. $3 \cdot 4 = 12$ $8 \cdot 9 = 72$
     $23 \cdot 24 = 552$ $-3 \cdot -4 = 12$
     $-1 \cdot 0 = 0$ $-15 \cdot -14 = 210$
Conjecture: The product of two consecutive integers is always —?—.

For Exercises 12 and 13, study the pattern, then complete the problem on the basis of your observations. (Again, no calculators please.)

12. $1 + (9 \cdot 0) = 1$
     $2 + (9 \cdot 1) = 11$
     $3 + (9 \cdot 12) = 111$
     $4 + (9 \cdot 123) = -?-$
         $-?- = 11111$

13. $8 + (9 \cdot 0) = 8$
     $7 + (9 \cdot 9) = 88$
     $6 + (9 \cdot 98) = -?-$
     $5 + (9 \cdot 987) = -?-$
         $-?- = 88888$

For Exercises 14 and 15, find the next two terms in the pattern.

14. 1, 2, 6, 24, 120, –?–, –?–

15. 4, 5, 7, 11, 19, 35, 67, –?–, –?–

16. Which numbers in the last group are Palis?

121	33
737	404
40,204	11

Palis

1,211	49
	17
7,373	366
8,229	113

Not Palis

4,014	5,125
42,124	7,131

Which are Palis?

Improving Algebra Skills—*Algebraic Magic Squares*

A magic square is an arrangement of numbers in a square array such that the numbers in every row and every column have the same total. In some magic squares the two diagonals have the same totals as the rows and the columns.

For example, in the magic square at right, the sum of each row is 18, the sum in each column is 18, and the sum in each diagonal is also 18.

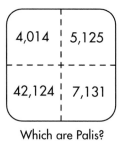

5	10	3
4	6	8
9	2	7

1. Complete the five-by-five magic square below. In this case don't try just any number; use only the numbers in the following list: 6, 7, 9, 13, 17, 21, 23, 24, 27, and 28.

20			8	14
	19	25	26	
	12	18		30
29	10	11		
22			15	16

2. Find *x* in the magic square below. Hint: Because each row and column must add to the same total, set the algebraic sum of one row or column equal to another and solve for *x*.

x	$x - 1$	8
$x + 5$	5	$x - 3$
$x - 2$	$x + 3$	$x + 2$

Lesson 1.5

Figurate Numbers

Einstein was once asked, "What is your phone number?" He answered, "I don't know, but I know where to find it if I need it."

Ancient mathematicians, most notably the Alexandrian mathematician Diophantus, were particularly interested in numbers that corresponded to geometric figures. Such numbers are called **figurate numbers**. The most recognizable of these are the square numbers 1, 4, 9, 16, and so on. You probably recognize these numbers as 1^2, 2^2, 3^2, and 4^2, and if the sequence is continued, the nth term can be found by the rule n^2. We say "n squared" because sets of dots corresponding to these numbers can be arranged in squares.

Square numbers

To find the 20th square number, we apply the rule n^2 to get $20^2 = 400$. Now consider a pattern of rectangles.

Rectangular numbers

This pattern represents the sequence 2, 6, 12, 20, ... Looking at the differences between terms, you can guess how to continue the pattern.

Term	1	2	3	4	5	6	7	8	9	...	n
Number of dots	2	6	12	20	30	42	56	-?-	-?-	...	-?-

Differences   4   6   8   10   12   14

2nd differences   2   2   2   2   2

How do you find the 20th term? The differences between terms aren't constant, so it's not a linear pattern, but the second set of differences *is* a constant. That is, the differences themselves form a linear sequence—4, 6, 8, 10, and so on—that has a constant difference of 2.

Sequences of this type have two linear factors, as the table below shows.

Term	1	2	3	4	5	6	7	8	...	n
Number of dots	2	6	12	20	30	42	56	-?-	...	-?-
Factors	1·2	2·3	3·4	4·5	5·6	6·7	7·8	-?-	...	$n \cdot (n+1)$

Sequences with two linear factors are called **quadratic sequences**. The rule for terms in a sequence of rectangular numbers has two factors because the total number of dots is found by multiplying the number of rows by the number of columns. For example, the third rectangular number in this sequence can be represented by three rows of four dots each, giving a total of $3 \cdot 4 = 12$ dots.

Investigation 1.5.1

Here's a pattern for you to practice on. This time there are no visual clues like a rectangular pattern of dots. Look at the differences between consecutive terms, and look at the difference between these differences. If the second set of differences is constant, try factoring the terms. Try to find a rule for the nth term of this sequence before you read the explanation that follows the table.

Term	1	2	3	4	5	6	7	...	n
Value	1	6	15	28	45	66	–?–	...	–?–

You should have found that the differences between consecutive terms are constantly increasing by 4, that is, the second set of differences is constant. So you find the seventh term by adding 25 to the sixth term.

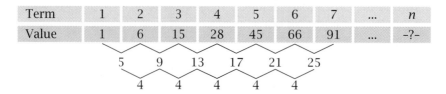

Term	1	2	3	4	5	6	7	...	n
Value	1	6	15	28	45	66	91	...	–?–

How can we use this discovery to find the nth term? Try factoring the values.

Term	1	2	3	4	5	6	7	...	n
Value	1	6	15	28	45	66	91	...	–?–
Factors	$1 \cdot 1$	$2 \cdot 3$	$3 \cdot 5$	$4 \cdot 7$	$5 \cdot 9$	$6 \cdot 11$	$7 \cdot 13$	...	–?–

Notice that there may be many ways to factor the values. In this case, 28 could have been factored as $2 \cdot 14$, and 45 could have been factored as $3 \cdot 15$. But once a pattern emerges, you can usually see which factorization is going to work. Look at the sequences formed by the factors: 1, 2, 3, 4, 5, 6, . . . and 1, 3, 5, 7, 9, ... The first factor is just n. The second factors form a linear sequence with a constant difference of 2. Its rule is $2n - 1$. So the rule for the nth term of the sequence: 1, 6, 15, 28, . . . is $n(2n - 1)$.

Just to be sure, test the rule. If the nth term is $n(2n - 1)$, we should be able to find the fifth term by substituting 5 for n. Does $5(2 \cdot 5 - 1)$ give us the fifth term? Test the rule on another term.

Finally, since the terms of the sequence could immediately be factored to give two linear sequences, you should be able to arrange dots into rectangles to represent the terms of the sequence. Try it. Draw rectangular arrangements representing 1, 6, 15, and 28.

Investigation 1.5.2

Fifteen pool balls can be arranged in a triangle, so 15 is a triangular number.

In this investigation, you'll come up with a rule for describing triangular numbers. Copy and complete the table for the number of black dots in each term of the pattern below.

Term	1	2	3	4	5	6	...	20	...	n
Total dots	2	6	12	-?-	-?-	-?-	...	-?-	...	-?-
Black dots	1	3	6	-?-	-?-	-?-	...	-?-	...	-?-

Were you able to find a rule for the nth term that you could use to find the 20th term? Perhaps you noticed that the black dots formed half of a rectangle. Because the rule for this rectangular pattern of dots was $n(n + 1)$, the rule for the number of black dots is $n(n + 1)/2$. This is the rule describing triangular numbers, and, as you'll see in future lessons, variations of this rule come up often in many types of problems.

Investigation 1.5.3

Copy the chart below. See if you can complete it without any visual clues.

Term	1	2	3	4	5	6	...	20	...	n
Value	1	5	12	22	35	-?-	...	-?-	...	-?-

Did you look at differences? If so, you discovered that the first set of differences isn't constant, but the second set is. That may have helped you find the 6th term, but unless you have plenty of time on your hands and you want to find every term in between, you probably want to find a rule to help you find the 20th term.

Term	1	2	3	4	5	6	...	20	...	n
Value	1	5	12	22	35	51	...	-?-	...	-?-

Differences   4   7   10   13   16

2nd differences   3   3   3   3

A constant in the second set of differences means the rule is quadratic; that is, it has two factors. Look for a pattern in the possible factors.

Term	1	2	3	4	5	6	...	20	...	n
Value	1	5	12	22	35	51	...	-?-	...	-?-
Factors	1·1	1·5	2·6	2·11	5·7	3·17	...	-?-	...	-?-

Not much help. Even if you try other factors, like $3 \cdot 4 = 12$, the pattern just isn't visible. Perhaps the pattern is disguised by a division by 2, like the triangular numbers. To test this, double the pattern and see if you can find a rule for the doubled pattern.

Term	1	2	3	4	5	6	...	20	...	200
Value	1	5	12	22	35	51	...	-?-	...	-?-
Doubled	2	10	24	44	70	102	...	-?-	...	-?-
Factored	1·2	2·5	3·8	4·11	5·14	6·17	...	-?-	...	-?-

$$\begin{matrix} 1 & & 1 & & 1 & & 1 & & 1 \\ & 3 & & 3 & & 3 & & 3 & & 3 \end{matrix}$$

Two patterns emerge: $1, 2, 3, 4, 5, 6, \ldots$ and $2, 5, 8, 11, 14, 17, \ldots$. The nth term for $1, 2, 3, 4, 5, 6, \ldots$ is n. The nth term for $2, 5, 8, 11, 14, 17, \ldots$ is $3n - 1$. Therefore the rule for the doubled pattern is $n(3n - 1)$. Divide this rule by 2 to find the rule for the original pattern: $n(3n - 1)/2$.

This strategy of looking for differences and then factoring, or of doubling the pattern and then factoring, can be used to find rules for a number of quadratic sequences, including the quadratic sequences in Exercise Set 1.5. Does the sequence you just investigated relate to a figurate number? As it turns out, you've found the rule for pentagonal numbers!

Exercise Set 1.5

Find the nth term and the 20th term in the sequence.

1.

1	2	3	4	5	6	...	n	...	20
4	8	12	16	20	24	...	-?-	...	-?-

2.

1	2	3	4	5	6	...	n	...	20
5	7	9	11	13	15	...	-?-	...	-?-

3. If the pattern of rectangles continues, what is the rule for the number of squares in the nth rectangle? What is the number of squares in the 200th rectangle?

Squares in a rectangular array

Rectangle	1	2	3	4	5	6	...	n	...	200
Number of squares	6	12	-?-	-?-	-?-	-?-	...	-?-	...	-?-

Introducing Geometry

Three Worlds, M. C. Escher, 1955

This chapter introduces you to the terms and the symbols of geometry. In this chapter you will write your own definitions of many geometric figures and terms. You will use inductive reasoning to arrive at your definitions. Keep a notebook with a list of all the terms and their definitions. As each new term in this chapter is defined, add its definition to your list. Draw examples next to your definitions.

Building Blocks of Geometry

Nature's Great Book is written in mathematical symbols.
— *Galileo Galilei*

The three building blocks of geometry are points, lines, and planes. Ancient mathematicians tried to define these terms. The ancient Greeks said, "A point is that which has no part. A line is breadthless length." The Mohist philosophers in ancient China said, "The line is divided into parts, and that part which has no remaining part is a point." Those definitions don't help much, do they?

A **definition** is a statement that clarifies or explains the meaning of a word or a phrase. In order to define *point, line,* and *plane,* however, you must use words or phrases that themselves need definition or further explanation.

Have you ever encountered a word that you didn't know? Perhaps you looked it up in a dictionary, only to find another unknown word in the definition. So you looked that word up and found the word you started with. Frustrating, isn't it?

© News America Syndicate, 1985

By permission of Johnny Hart and Creators Syndicate, Inc.

Early mathematicians were similarly frustrated trying to define *point, line,* and *plane.* These terms remain undefined. To help you gain an intuitive understanding of their meaning, however, we offer general descriptions of them. While not precise definitions, these descriptions will give you a sense of what is meant by *point, line,* and *plane.*

A **point** is the basic unit of geometry. It has no size. It is infinitely small. It has only location. A tiny seed is a physical model of a point. A point, however, is smaller than the smallest seed. You use a dot to represent a point. You name a point with a capital letter. The point shown at right is called *P*.

Picture of a point

Physical model of a point: a poppy seed

A **line** is a straight arrangement of points. There are infinitely many points in a line. It has infinite length but no thickness. It extends forever in two directions. A taut telephone wire is a physical model of a line. A line, however, is thinner, longer, and straighter than any wire ever made. You name a line by naming two points on the line. The line shown at right is called line *AB* or line *BA*. You can also use a symbol for line. Place the symbol for a line above the letters and write $\overleftrightarrow{AB}$ or $\overleftrightarrow{BA}$.

Picture of a line

Physical models of lines: phone wires

Picture of a plane landing on a picture of a physical model of a plane

A **plane** has length and width but no thickness. It is a flat surface that extends forever. A wall, a floor, or a ceiling is a physical model of a plane. A flat piece of rolled-out dough is a model of a plane, but a plane is thinner and broader than any piece of dough you could roll. You represent a plane with a four-sided figure, like a piece of paper drawn in perspective. Of course, this represents only part of a plane, as a plane extends forever in length and in width. (No, airplanes have nothing to do with geometric planes.)

Picture of a plane

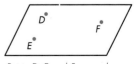

Physical model of a plane: dough rolled flat

Using the undefined terms *point*, *line*, and *plane*, you can define many new geometric figures and terms. Many will be defined for you in this chapter and in those that follow. Others you will define. Keep a definition list (glossary) in your notebook, and each time you encounter a new geometric definition, add it to your list. Illustrate your definition with a simple sketch. Read the definition list daily.

Here are your first definitions. Begin your list.

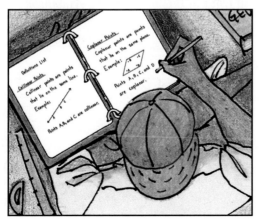

Collinear points *are points that lie on the same line.*

Points *A*, *B*, and *C* are collinear.

Coplanar points *are points that lie on the same plane.*

Points *D*, *E*, and *F* are coplanar.

Name three balls that are collinear. Name three balls that are coplanar but not collinear. Name four balls that are not coplanar.

Here are two more definitions. Add them to your list. (Drawing a picture of space could prove to be difficult.)

In mathematics, "space" means something different from "outer space."

Space *is the set of all points.*

*A **line segment** consists of two points and all the points between them that lie on the line containing the two points.*

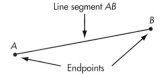

The two points are called the **endpoints** of the line segment. The portion of a telephone wire between two poles is a physical model of a line segment whose endpoints are where the wire attaches to the poles. The poles themselves are also physical models of line segments.

Line segment AB may be written using a symbol as $\overline{AB}$ or $\overline{BA}$.

The length, or measure, of a line segment is the distance between its endpoints. There are two ways to indicate the length of a segment. One way is to indicate the two endpoints without using the line segment symbol. For example, $MN = 2''$ means that the length of segment MN measures two inches. You can also indicate the measure of a segment by putting a lowercase m in front of the line segment symbol, as in $m\overline{PQ} = 4.5$ cm. If no units are used when indicating the length of a segment in a figure, it is understood that the units are arbitrary and that the choice of units is not important in understanding the diagram.

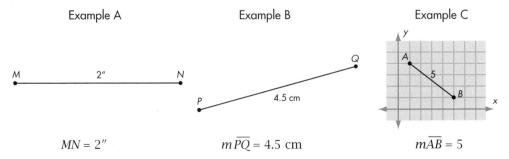

Example A	Example B	Example C
$MN = 2''$	$m\overline{PQ} = 4.5$ cm	$m\overline{AB} = 5$

Here is another very important definition. Add it to your definition list.

Ray AB is the part of line AB that contains point A and all the points on $\overleftrightarrow{AB}$ that are on the same side of point A as point B.

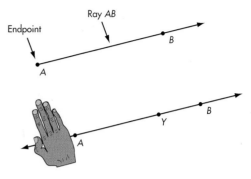

Ray AB

Endpoint

B

A

Physical model of a ray: a laser beam

A ray begins at a point and goes on forever in one direction. Point *A* is the *endpoint* of ray *AB*. A laser beam, which shines a thin beam of light in one direction, is a physical model of a ray. You need two letters to name a ray. The first letter is the endpoint of the ray, and the second letter is any other point that the ray passes through. Ray *AB* may be written symbolically as $\overrightarrow{AB}$. If *Y* is another point that the ray passes through, the ray may also be called $\overrightarrow{AY}$. You can think of a ray as part of a line. If you cut everything to the left of point *A* off line *AB*, you are left with ray *AB*. Note that $\overrightarrow{AB}$ is not the same as $\overrightarrow{BA}$.

Here is a definition that is based upon the definition of a ray. Add it to your definition list. Draw and label a picture of an angle next to your definition.

*An **angle** is two rays that share a common endpoint, provided the two rays do not lie on the same line.*

Physical model of an angle: television antennas

The common endpoint of the two rays that form an angle is called the **vertex** of the angle. The two rays are called the **sides** of the angle. Segments can form the sides of an angle, too, because you can always imagine segments extended into rays. A television antenna is a physical model of an angle. Note that changing the length of the antenna doesn't change the angle. You use three letters to name angles. The middle letter must be the vertex of the angle. The other two letters are points on each of the sides of the angle.

Shown at right is the angle formed by $\overrightarrow{AT}$ and $\overrightarrow{AP}$. Point *A*, the common endpoint of the two rays, is the vertex of the angle. Rays *AT* and *AP* are the sides of the angle. This angle is named angle *TAP* or angle *PAT* or symbolically as ∠*TAP* or ∠*PAT*.

Why do you think the definition for angle includes the phrase "provided that the two rays do not lie on the same line"? What would an angle look like if the two rays that formed it lay on the same line?

Exercise Set 2.1

1. List other examples of physical models of a point, a line, a plane, and a ray, different from the examples in this lesson.

Name each line in two different ways.

2.

3.*

4.

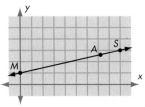

Use a ruler to draw each line. Place two dots on each line to represent points, then label them. Don't forget to put arrowheads on the ends of each line to show that it goes on forever.

5. $\overleftrightarrow{AB}$ **6.** $\overleftrightarrow{KL}$ **7.** $\overleftrightarrow{PU}$

Name each line segment.

8.

9.

10.

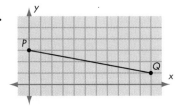

Draw and label each line segment.

11. $\overline{AB}$ **12.** $\overline{PQ}$ **13.** $\overline{TS}$

Use your ruler to determine the measure of each line segment to the nearest tenth of a centimeter. Write your answer in the form $m\overline{AB}$.

14. A ————————————————————————————— B

15. C ———————————————— D

16. E ————————————————————— F

Use your ruler to draw each segment as accurately as you can. Label each segment.

17. $AB = 4.5$ cm   **18.** $CD = 3$ in.   **19.** $EF = 24.8$ cm

Name each ray in two different ways.

20.*

21.

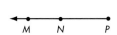

22.

Draw and label each ray.

23. $\overrightarrow{AB}$ **24.** $\overrightarrow{YX}$ **25.** $\overrightarrow{MN}$

Name each angle in two different ways.

26.*

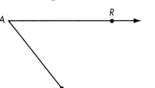

27.

28.

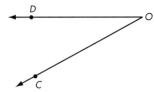

Draw and label each angle.

29. ∠*TAN*   **30.** ∠*BIG*   **31.** ∠*SML*

On occasion, when no confusion results, an angle can be named with just the vertex letter instead of the usual three letters.

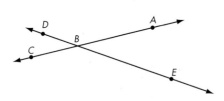

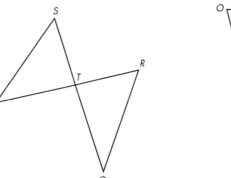

There are four angles with vertex *B*. You must use three letters to identify any of the four angles around vertex *B*.

There is only one angle with vertex *P*, so you may write ∠*P* instead of ∠*RPQ*.

For each diagram, list the angles that can be named using only one vertex letter.

32.*   **33.**   **34.**

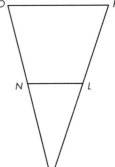

35. Draw a figure that contains at least three angles and requires three letters to name each angle.

36. Draw a plane and four coplanar points, exactly three of which are collinear.

For Exercises 37–39, copy the coordinate grid onto graph paper.

37. Draw segment *AB,* where point *B* has coordinates (2, –6).

38. Draw $\overrightarrow{OM}$ with endpoint (0, 0) that goes through point *M*(2, 2).

39. Draw $\overleftrightarrow{CD}$ through points *C*(-2, 1) and *D*(-2, -3).

40. If the coordinates of collinear points *P*(-6, -2), *Q*(-5, 2), and *R*(-4, 6) were to have their signs reversed, would the three new points still be collinear?

41. Into how many regions do 500 parallel lines in a plane divide that plane?

42. Into how many regions do 500 lines in a plane all passing through the same point divide that plane?

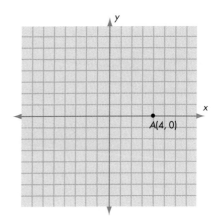

Lesson 2.2

Poolroom Math

Is the angle formed by the two hands of a tiny wristwatch at 9:37 smaller than the angle formed by the two hands of a large clock? No, the lengths of rays as they appear on paper or the lengths of the hands of a timepiece have nothing to do with how "big" an angle is. The actual rays of an angle go on forever, whether they're drawn short or drawn long.

So how *do* we talk about how big or how small angles are? How are angles measured? Picture two rays overlapping. Then move one of the rays by rotating it about the vertex to form an angle. The **measure of an angle** is the smallest

Big Ben of the Houses of Parliament in London, England

Little Benji: a watch

amount of rotation necessary from the overlapping position to the final angle. You measure this rotation in units called **degrees**. (There are other angle measure units, but you won't use them in this book.) One full circle of rotation is 360 degrees (usually written 360°). One-half circle of rotation is 180°, one-fourth is 90°, and so on. According to the definition of an angle, the measure of an angle can be anything between 0° and 180°. When you wish to indicate the size or **degree measure** of an angle, write a lowercase *m* in front of the angle symbol. For example, $m\angle ZAP = 34°$ means the measure of angle *ZAP* is 34 degrees.

The geometric tool used to measure the number of degrees in an angle is a **protractor**. To measure angles with a protractor, follow the three steps below.

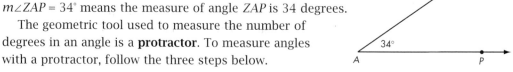

Step 1	Step 2	Step 3
	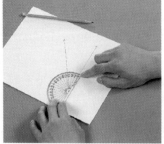	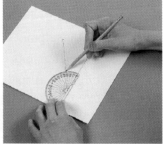
Place the center mark of the protractor on the vertex of the angle.	Rotate the zero-edge of the protractor to line up with one side of the angle.	Read the measure of the angle where the other side of the angle crosses the protractor's scale.

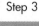

Your protractor has two scales around its edge, so you must be careful to read the correct scale. First, note whether the angle you are measuring looks like its measure is greater than or less than a ninety-degree angle. If the angle measure looks less than ninety degrees, use the smaller of the two numbers on the protractor. If the angle measure looks greater than ninety degrees, use the larger of the two numbers. In the photos, the angle measures less than ninety degrees, so its measure is read from the smaller scale. Keep in mind that whenever you use a protractor to measure an angle or a ruler to measure a segment, the measurement you get is only an approximation of the actual measure of the angle or the segment. It is impossible to measure an angle or a segment exactly.

When you label geometric figures, mark angles and segments that have the same measure with like markings. For instance, in Example A, ∠DOG has the same measure as ∠CAT. Both angles are marked with an arc and one slash. In Example B, $\overline{AB}$ has the same length as $\overline{AC}$. Both segments are marked with two slashes. To indicate the measure of an angle, place the measure in the interior of the angle. To indicate the length of a segment, place its measure near the segment.

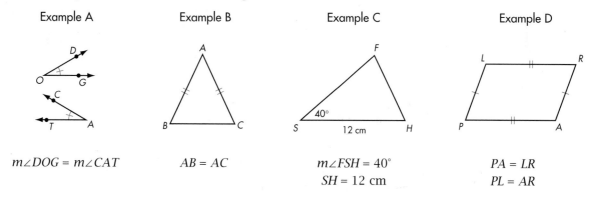

Example A	Example B	Example C	Example D
$m\angle DOG = m\angle CAT$	$AB = AC$	$m\angle FSH = 40°$ $SH = 12$ cm	$PA = LR$ $PL = AR$

Two figures that have the exact same size and shape are **congruent**. For two segments or two angles to have the same size and shape, they would have to have the same measure.

*Two segments are **congruent segments** if and only if they have the same measure.*

*Two angles are **congruent angles** if and only if they have the same measure.*

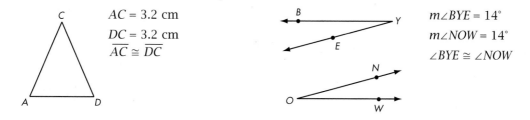

$AC = 3.2$ cm
$DC = 3.2$ cm
$\overline{AC} \cong \overline{DC}$

$m\angle BYE = 14°$
$m\angle NOW = 14°$
$\angle BYE \cong \angle NOW$

The relationship between congruent figures is like the relationship between equal numbers. The symbol for congruence is ≅ and is read "is congruent to." You use the = symbol between equal numbers and the ≅ symbol between congruent figures.

Exercise Set 2.2

For Exercises 1–10, use the protractor pictured to figure out the measure of the angle named. Measure as accurately as you can. The symbol ≈ means "is about equal to."

1. $m\angle AOB \approx$ -?-   **2.** $m\angle AOC \approx$ -?-

3. $m\angle XOA \approx$ -?-   **4.** $m\angle AOY \approx$ -?-

5. $m\angle ROA \approx$ -?-   **6.** $m\angle TOA \approx$ -?-

Use subtraction to find the size of each angle.

7.* $m\angle COB \approx$ -?-   **8.** $m\angle YOX \approx$ -?-

9. $m\angle XOT \approx$ -?-   **10.** $m\angle COY \approx$ -?-

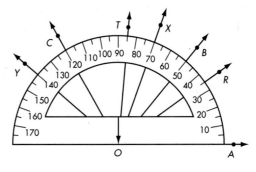

For Exercises 11–18, use your protractor to find the measure of the angle to the nearest degree.

11. $m\angle FAN \approx$ -?-

12. $m\angle IBM \approx$ -?-

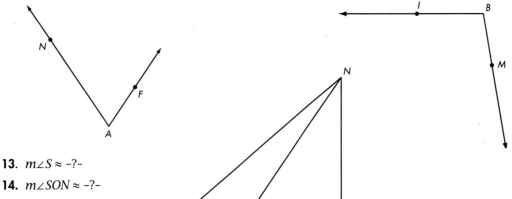

13. $m\angle S \approx$ -?-

14. $m\angle SON \approx$ -?-

15. $m\angle NOR \approx$ -?-

16. $m\angle SNO \approx$ -?-

17. $m\angle ONQ \approx$ -?-

18. $m\angle SNQ \approx$ -?-

19. Which angle has greater measure, $\angle SML$ or $\angle BIG$? Why?

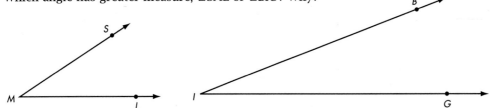

For Exercises 20–22, use your protractor to draw an angle of the size indicated as accurately as you can.

20. $m\angle A = 40°$ **21.** $m\angle B = 90°$ **22.** $m\angle C = 135°$

For Exercises 23–25, draw a clock face with hands to show the time indicated. Be as accurate as you can.

23.* 3:30   **24.** 3:40   **25.** 3:15

For Exercises 15–18, sketch or trace a copy of the African kente cloth shown at right. Add labels to your sketch to name the figures asked for.

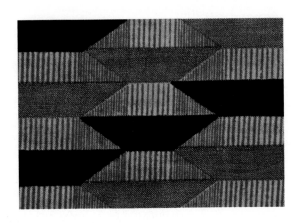

15. A linear pair

16. A pair of complementary angles

17. A pair of parallel segments

18. A pair of perpendicular segments

For Exercises 19 and 20, refer to the graph.

19. Find possible coordinates of a point P so that points P, T, and S are collinear.

20. Find possible coordinates of a point Q so that $\overleftrightarrow{QR} \parallel \overleftrightarrow{TS}$.

21. Write the converse of the statement "If $m\angle D = 40°$ and $m\angle C = 140°$, then angles C and D are a linear pair." Is the converse true? Is the original statement true? Draw a counterexample for each false statement.

22. Draw a counterexample to show that the following statement is false: "If point A is not the midpoint of $\overline{CT}$, then $CA \neq AT$."

In Exercises 23 and 24, set up a table of values, find a rule, and use the rule to answer the questions.

23. How many parallel lines in a plane will divide that plane into 500 regions?

24. How many lines in a plane all passing through the same point will divide that plane into 500 regions?

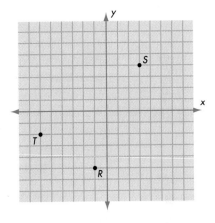

---

1. Find two light coins (dimes) and two dark coins (pennies) and arrange them on a grid of five squares, as shown. Your task is to switch the position of the two light and two dark coins in exactly eight moves. A coin can slide into an empty square next to it or can jump over one coin into an empty space. Record your solution by drawing eight diagrams that show the moves.

2. Find three light coins (dimes) and three dark coins (pennies) and arrange them on a grid of seven squares as shown.
Your task is to switch the position of the three light and dark coins in exactly 15 moves. A coin can slide into an empty square next to it or it can jump over one coin into an empty space. Record your solution by listing in order which color coin is moved. For example, your list might begin DLDLDLDL . . . .

Lesson 2.5

Defining Polygons

It takes a whole village to raise a child.
— *Nigerian saying*

Polygons   Not polygons

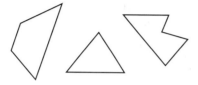   

A ***polygon*** *is a closed geometric figure in a plane, formed by connecting line segments endpoint to endpoint with each segment intersecting exactly two others.*

Each line segment is called a **side** of the polygon. Each endpoint where the sides meet is called a **vertex** of the polygon.

A **convex polygon** is a polygon in which no segment connecting two vertices is outside the polygon. A **concave polygon** is a polygon in which at least one segment connecting two vertices is outside the polygon. From here on, when we speak of a *polygon* in this book, we will always mean a *convex polygon*.

Convex polygons   Concave polygons

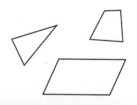

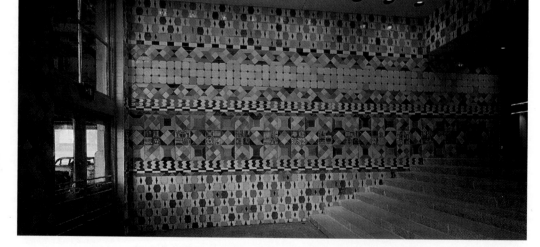

Joyce Kozloff used many different types of polygonal tiles in her design for the Amtrak train station in Wilmington, Delaware. Which are concave?

You classify polygons by the number of sides they have.

Sides	Name	Sides	Name	Sides	Name
3	Triangle	7	Heptagon	11	Undecagon
4	Quadrilateral	8	Octagon	12	Dodecagon
5	Pentagon	9	Nonagon		
6	Hexagon	10	Decagon	n	n-gon

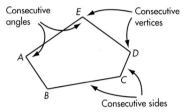

When referring to a specific polygon, list in succession the capital letters representing consecutive vertices. For example, the pentagon at right can be referred to as pentagon *ABCDE*. You can also call it pentagon *DCBAE*. When the polygon is a triangle, you can use the Δ symbol. For example, Δ*ABC* means triangle *ABC*.

If two vertices of a polygon are connected by a side, then they are **consecutive vertices**. If two sides share a common vertex, then they are **consecutive sides**. If two angles share a common side, then they are **consecutive angles**.

Recall from Lesson 2.2 that two segments or two angles are congruent if and only if they have the same measures. Polygons that are exactly the same size and shape are **congruent polygons**.

These flowers are approximately the same size and shape. Can you think of other examples of congruence in nature?

For convex polygons this means two things: If the angles and the sides of one polygon are congruent to the corresponding angles and sides of another polygon, then the two polygons are congruent. For example, if the four angles of quadrilateral *CAMP* are congruent to the four corresponding angles of quadrilateral *SITE*, and if the four sides of quadrilateral *CAMP* are congruent to the four corresponding sides of quadrilateral *SITE*, then quadrilateral *SITE* is congruent to quadrilateral *CAMP*. When you write the symbolic statement of congruence of the two figures, the letters of the corresponding congruent angles should be written in an order that indicates the correspondences.

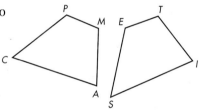

$CA \cong SI$ $\angle C \cong \angle S$
$AM \cong IT$ $\angle A \cong \angle I$
$MP \cong TE$ $\angle M \cong \angle T$
$PC \cong ES$ $\angle P \cong \angle E$
$CAMP \cong SITE$

The definition also means that if two polygons are congruent, then their corresponding angles and sides are congruent. For example, if it is given that quadrilateral *CAMP* is congruent to quadrilateral *SITE*, it follows that their four pairs of corresponding angles and four pairs of corresponding sides are also congruent.

You may recall the definition of the perimeter of a polygon.

*The **perimeter** of a polygon is the sum of the lengths of its sides.*

The polygon shown at right has a perimeter of 37 cm.

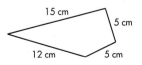

15 cm
5 cm
12 cm
5 cm

Investigation 2.5

Write a good definition of each geometric term or figure below. Once you are satisfied with the definitions you have written, discuss them with others near you. Try to arrive at one common definition your class can agree on for each term. Add these definitions to the definition list in your notebook. Draw and label a picture to illustrate each definition.

1. Define *diagonal of a polygon.*

Diagonals of polygons

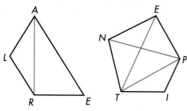

Segments *AR, PN, TE,* and *PT* are diagonals.

Not diagonals of polygons

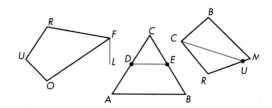

Segments *FL, FO, CU,* and *DE* are not diagonals.

2. Define *equilateral polygon.*

Equilateral polygons

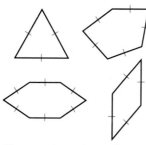

Not equilateral polygons

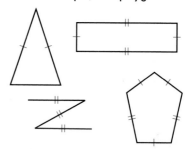

3. Define *equiangular polygon.*

Equiangular polygons

Not equiangular polygons

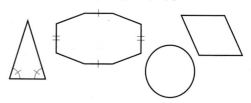

4.* Define *regular polygon.*

Regular polygons

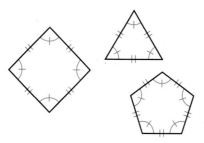

Not regular polygons

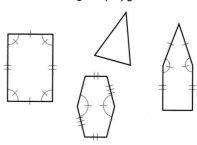

7. Define *altitude of a triangle.*

Altitudes of triangles

Not altitudes of triangles

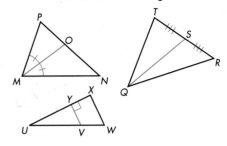

Segments *MN, EG, CD,* and *IK* are altitudes.

Segments *MO, QS,* and *VY* are not altitudes.

In a triangle, the length of the altitude is called the **height**. A triangle has three different altitudes, and, therefore, it has three different heights.

Exercise Set 2.6

In Exercises 1–15, match the symbol or the term on the left with one of the figures on the right.

1. $\overline{AB}$

2. $\overrightarrow{AB}$

3. Isosceles $\triangle ABC$

4. Median of a triangle

5. Altitude of a triangle

6. Angle bisector in a triangle

7. Right $\triangle ABC$

8. Equilateral $\triangle ABC$

9. $\overleftrightarrow{AB} \perp \overleftrightarrow{CD}$

10. $\overleftrightarrow{AB} \parallel \overleftrightarrow{CD}$

11. Vertical angles

12. Complementary angles

13. Supplementary angles

14. Hexagon

15. Octagon

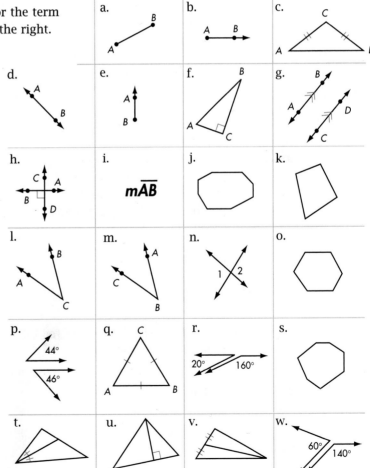

In Exercises 16–19, refer to the photograph of the building under construction. Measure or just estimate angle measures in the photo to identify each type of triangle.

16. An acute scalene triangle

17. An isosceles triangle that is not equilateral

18. Two approximately equilateral triangles

19. Three approximately right triangles

In Exercises 20–29, sketch and carefully label the figure.

20. Acute isosceles $\triangle ACT$ with $AC = CT$

21. Right isosceles $\triangle RGT$ with $RT = GT$, $m\angle RTG = 90°$

22. Scalene $\triangle SCL$ with median $\overline{CM}$

23. Equilateral $\triangle EQL$ with altitude $\overline{QT}$

24. Equilateral octagon $OCTAHDRN$ with $m\angle CTA = 90°$

25. Obtuse scalene triangle FAT with $m\angle FAT = 100°$ and altitude $\overline{FY}$

26. Two different isosceles triangles with a perimeter of $4a + b$

27. An equilateral triangle with a perimeter of $12a + 6b$

28. Two different triangles that are not congruent, each with a side of 6 cm and an angle measuring 40°

29. Two different triangles that are not congruent, each with angles measuring 50° and 70°

For Exercises 30–32, copy the segment onto graph paper. Locate a third point as instructed. Can you find more than one point that satisfies the condition?

30.

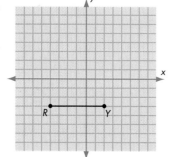

Locate a point L so that $\triangle LRY$ is isosceles.

31.

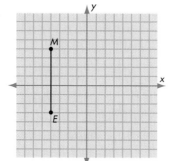

Locate a point O so that $\triangle MOE$ is a right isosceles triangle.

32.

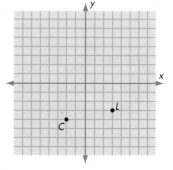

Locate a point R so that $\triangle CRL$ is a right isosceles triangle.

33. Draw a counterexample to show that the converse of the following statement is false: If a triangle is equilateral, then it is isosceles.

34. What type or types of triangle has one or more lines of symmetry?

35. Write a new definition for an isosceles triangle, based on the triangle's reflectional symmetry. Does your definition apply to equilateral triangles? Explain.

36. Write a new definition for an equilateral triangle, based on the triangle's rotational symmetry.

Lesson 2.7

Special Quadrilaterals

The difference between the right word and the almost right word is the difference between lightning and the lightning bug.

— Mark Twain

Lightning   Not lightning

Investigation 2.7

Write a good definition for each geometric term or figure. Discuss your definitions with others in your class. Agree on a common set of definitions for your class and add them to your definition list. Draw and label a picture to illustrate each definition.

1.* Define *trapezoid.*

Trapezoids   Not trapezoids

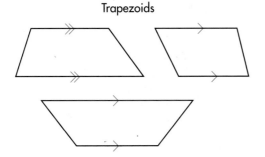 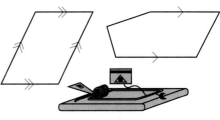

Note: Matching arrows in a diagram indicate parallel segments.

2. Define *kite.*

Kites   Not kites

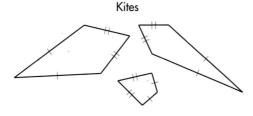 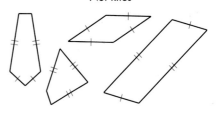

3. Define *parallelogram.*

Parallelograms   Not parallelograms

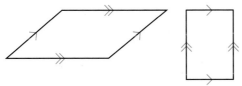 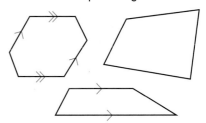

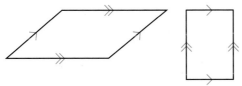 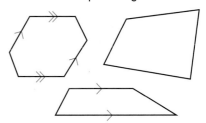

4. Define *rhombus*.

Rhombuses

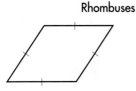

Not rhombuses

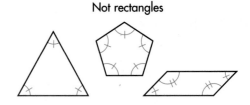

Both polygons are parallelograms.

5.* Define *rectangle*.

Rectangles

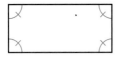

Not rectangles

Both polygons are parallelograms.

6.* Define *square*.

Squares

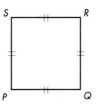
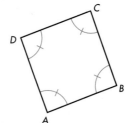

Not squares

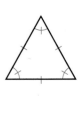
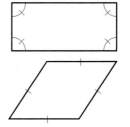

Quadrilateral *ABCD* is a rhombus.
Quadrilateral *PQRS* is a rectangle.

Exercise Set 2.7

In Exercises 1–10, write the words or the symbols that make the statement true.

1. The three undefined elements of geometry are —?—, —?—, and —?—.

2. "Line *AB*" is written in symbolic form as —?—.

3. "Line segment *AB*" is written in symbolic form as —?—.

4. "Ray *AB*" is written in symbolic form as —?—.

5. "Angle *ABC*" is written in symbolic form as —?—.

6. The point where the two rays of an angle meet is called the —?—.

7. The geometric tool used to measure an angle is called a —?—.

8. "Line *AB* is parallel to segment *CD*" is written in symbolic form as —?—.

9. The sentence "Segment *AB* is perpendicular to line *CD*" is written in symbolic form as —?—.

10. The angle formed by a light ray coming into a mirror is —?— the angle formed by a light ray leaving the mirror.

In Exercises 11–22, identify the statement as true or false. For each false statement, sketch a counterexample that proves the statement false.

11. An angle is measured in degrees.

12. An acute angle is an angle whose measure is less than 90°.

13. An obtuse triangle has exactly one angle whose measure is greater than 90°.

14. A diagonal is a line segment that connects any two vertices of a polygon.

15. If the measures of two angles add up to 90°, then they are supplementary.

16. If two lines intersect to form a right angle, then the lines are perpendicular.

17. An angle bisector is any ray or segment that divides an angle into two angles.

18. A median of a triangle is a line segment from a vertex to the midpoint of the opposite side.

19. A trapezoid is a device used to capture zoids.

20. A rhombus is a parallelogram whose sides all have the same measure.

21. A kite is a parallelogram.

22. Every square is a rectangle.

In Exercises 23–28, draw a triangle that fits the name. If impossible, write "not possible." Use your geometric tools to make your drawings as accurate as possible.

23. Scalene triangle

24. Scalene acute triangle

25. Obtuse isosceles triangle

26. Isosceles right triangle

27. Equilateral right triangle

28. Scalene isosceles triangle

29.* Copy the pool table and the balls shown. If the cue ball is hit as indicated, show the path of the five ball as it hits three cushions before coming to rest.

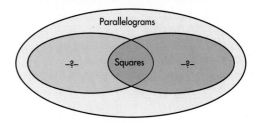

30. At right is a concept map showing the relationships between some members of the parallelogram family. This type of concept map is known as a **Venn diagram**. Fill in the missing names.

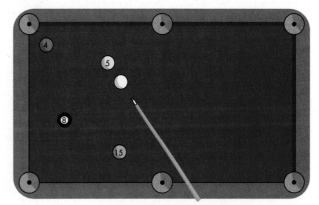

Venn diagram for parallelograms

31. Many of the geometric figures you have defined are closely related to one another. A diagram can help you see the relationships between them. At right is a concept map showing the relationships between members of the triangle family. This type of concept map is known as a **tree diagram** because the relationships are shown as branches of a tree. Copy and fill in the missing branches of the tree diagram for triangles.

32. What types of quadrilaterals have one or more lines of symmetry? Describe the reflectional symmetry of these special quadrilaterals and write new symmetry-based definitions for them.

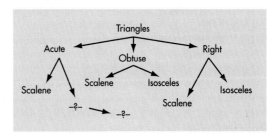

Tree diagram for triangles

33. What type (or types) of quadrilateral has only rotational symmetry?

34. The kite *KYTE* is placed on the number line, as shown, and then is "rolled" side over side until side *KE* is again resting on the number line. At what point on the number line will *E* end up?

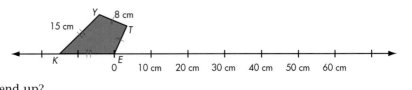

35. Sketch a triangle with an angle measuring 50°, with an angle measuring 70°, and with one of the sides not between them measuring 15 cm. Can you use the same information to make a second triangle that is different from the first triangle?

36. How can you measure exactly 12 inches with just a regular $8\frac{1}{2}'' \times 11''$ piece of paper? Do it. Check to see how accurate you are by measuring against a ruler.

37. If 500 rays are drawn from the vertex of an angle into its interior, into how many regions is that angle's interior divided?

38. If the *x*- and *y*-coordinates of collinear points *P*(3, 5), *Q*(7, 8), and *R*(-1, 2) are reversed, are those three points still collinear?

39. This German house has a number of special quadrilaterals in its design. Sketch an outline of the house or carefully trace it onto a patty paper, then sketch and label special quadrilaterals approximately where they appear. From your diagram, name a trapezoid, an isosceles trapezoid, a rectangle, a parallelogram, a rhombus, and a square.

German house

Improving Visual Thinking Skills—*Scrambled Vocabulary*

Unscramble the letters to form familiar geometry terms.

Example: tubose—obtuse

1. eucat
2. legan
3. extrev
4. indema
5. mysterym
6. dusiar
7. tompindi
8. mohsurb
9. soilsecse
10. raae
11. nestgem
12. sulanun
13. nenagtt
14. megyoter
15. learlinco

Lesson 2.8

Space Geometry

Space: The final frontier
— Star Trek

In the beginning lessons of this chapter you were introduced to *point, line,* and *plane,* and you used those terms to define a wide range of other geometric figures, from rays to polygons. All of your work, however, was done on a single flat surface, a single plane. In this lesson you will learn a little about space geometry: the geometry of objects that are not restricted to a flat surface (not necessarily coplanar). The geometric solid you are probably most familiar with is the basic box, or rectangular solid. You learned how to draw boxes using perspective in Chapter 0. Perspective makes solids look more realistic, so it's often used in art. But for the purpose of doing geometry, it's not usually necessary to draw solids in perspective. Below are tips for drawing rectangular solids without perspective.

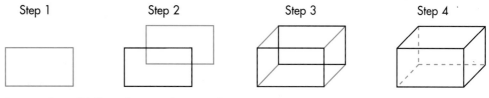

Step 1   Step 2   Step 3   Step 4

Rectangular solid (face closest to the viewer)

Use dashed lines to represent lines that would not be visible if this object were actually solid.

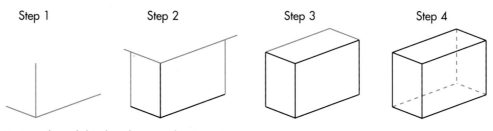

Step 1   Step 2   Step 3   Step 4

Rectangular solid (edge closest to the viewer)

Some other three-dimensional objects or space figures you will study include the six basic geometric solids shown below.

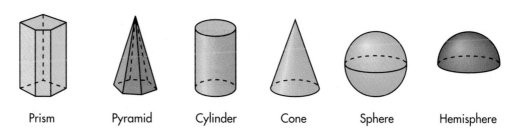

Prism   Pyramid   Cylinder   Cone   Sphere   Hemisphere

Formal definitions of these solids will come in a later chapter. The shapes of the solids are probably already familiar to you even if you are not familiar with their formal names. The ability to draw these geometric solids is an important visual-thinking skill. Here are some drawing tips. Don't forget the hidden lines.

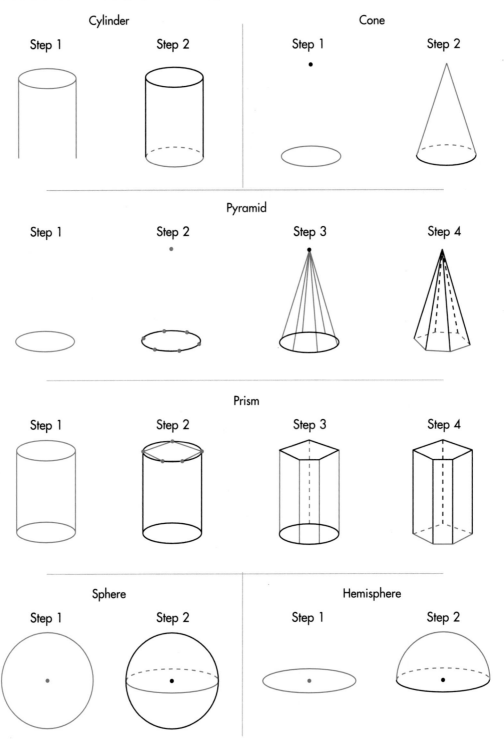

Exercise Set 2.8

In Exercises 1-6, draw the figure. Study the drawing tips provided on the previous page before you start.

1. Cylinder

2. Cone

3. Prism with a hexagonal base

4. Pyramid with a heptagonal base

5. Sphere

6. Hemisphere

7. This small neighborhood police station (called a *koban*) in Tokyo is a prism-shaped building with a pyramid roof and a cylindrical porch. Draw a cylindrical building with a cone roof and a prism-shaped porch.

In Exercises 8-11, draw and label the solid figure described. In Exercises 8-10, make a scaled drawing of the figure. For example, in Exercise 9 you are asked to draw a rectangular solid 2 m by 3 m by 4 m. There is no way that you're going to draw a solid that is actually that big. Draw the solid at a smaller scale so that the dimensions are about 2 units by 3 units by 4 units, then label the figure with meters.

8. A rectangular solid with a base 6 cm by 12 cm, a height of 3 cm, and a face closest to the viewer

9. A rectangular solid 2 m by 3 m by 4 m, sitting on its biggest face, with an edge closest to the viewer

10.* A rectangular solid 3 inches by 4 inches by 5 inches resting on its smallest face. Draw lines on the three visible surfaces showing how the solid can be divided into 60 cubic-inch boxes.

11. A cube (a prism, each face being a square). Draw lines on the three visible faces to show how it can be divided into 27 identical smaller cubes.

In Exercises 12-14, find the lengths x and y. (Each angle on each polygonal side of the block is a right angle.)

12.

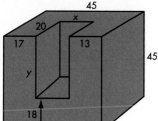

13.

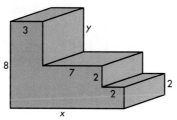

14.

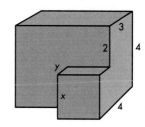

Each figure in Exercises 15–17 represents a card with a wire attached. The three-dimensional figure formed by spinning the two-dimensional figure on the wire between your fingers is called a **solid of revolution**. Sketch the solid of revolution formed when the two-dimensional figure is revolved by spinning the wire.

15.*

16.

17.

When a solid is cut by a plane, the resulting plane figure is called a **section**. In Exercises 18–21, sketch the section formed when the solid is sliced by the plane, as shown.

18. 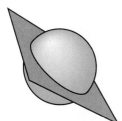   **19.** 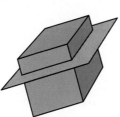   **20.** 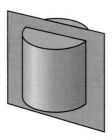   **21.**

This set of exercises will help you visualize relationships between geometric figures in the plane and in space. All the statements in Exercises 22–34 are true except two. Make a sketch or use physical objects to demonstrate each true statement. For the two false statements, produce a counterexample demonstrating that each is false. Pencil tips and thumbtacks can represent points. Rulers, pencils, spaghetti, or stiff wires can represent lines. Sheets of heavy paper, cardboard, or your desk top can represent planes.

22. One and only one distinct line can be drawn through two different points.

23. One and only one distinct plane can be made to pass through three noncollinear points.

24. Exactly one distinct plane passes through one line and a point not on the line.

25. If a line intersects a plane not containing it, then the intersection is exactly one point.

26.* If two lines are perpendicular to the same line, then they are parallel.

27. If two different planes intersect, then their intersection is a line.

28. If a line and a plane have no points in common, then they are parallel.

29. If two coplanar lines are both perpendicular to a third line in the same plane, then the two lines are parallel.

30. If two different planes do not intersect, then they are parallel.

31. If a plane intersects two parallel planes, then the lines of intersection are parallel.

32. If three random planes intersect (no two parallel and no three through the same line), then they divide space into six parts.

33. If a line is perpendicular to two lines in a plane, but the line is not contained in the plane, then the line is perpendicular to the plane.

34.* If two lines are perpendicular to the same plane, then they are parallel to each other.

35. Miriam the Magician placed four cards faceup on her magic table (the four cards shown below left). Blindfolded, she instructed someone from her audience to come up to the stage and turn exactly one card around 180°.

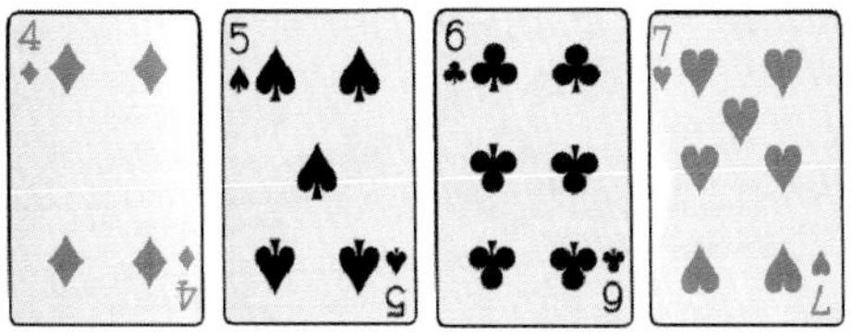

Before turn

After turn

Miriam removed her blindfold and claimed she was able to determine which card was turned 180° by "magically reading the psychic energy emanating from the cards." Of course, it is possible to figure out which card was turned around without magic. Can you figure out which card was turned?

Improving Visual Thinking Skills—*Picture Patterns I*

Here is a sequence you "read" left to right.

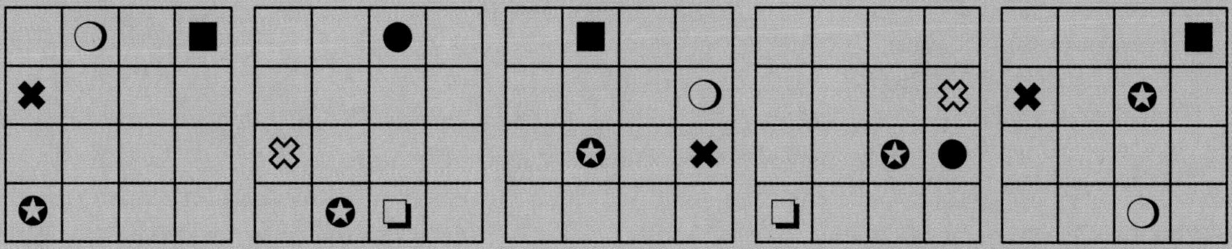

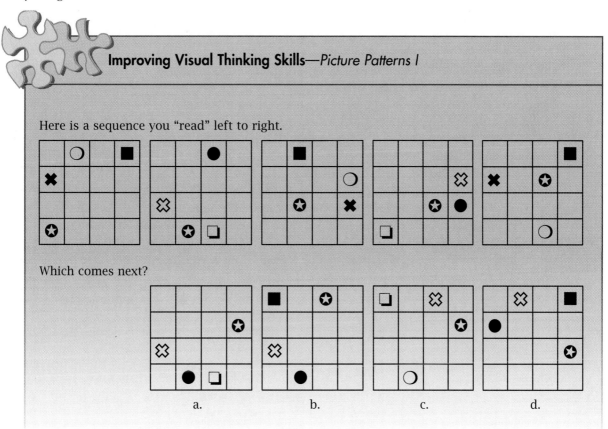

Which comes next?

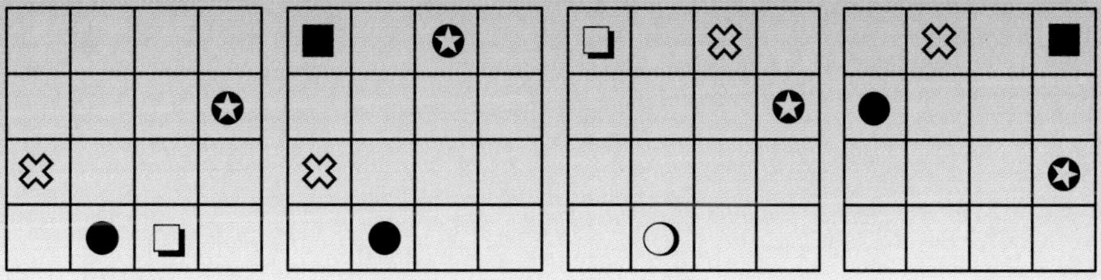

a. b. c. d.

Project

Traveling Networks

The River Pregel runs through the university town of Königsberg (now Kaliningrad in Russia). In the middle of the river are two islands connected to each other and to the rest of the city by seven bridges.

Many years ago, a tradition developed among the college students of Königsberg. They challenged one another to make a round trip over all seven bridges,

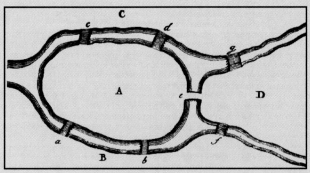

The Seven Bridges of Königsberg

walking over each bridge once and only once before returning to the starting point. For a long time no one was able to do it, and yet no one was able to show that it couldn't be done. In 1735, students finally wrote to Swiss mathematician Leonhard Euler (1707–1783), asking for his help on the problem. Euler (pronounced "oyler") was able to show that the feat cannot be done. (The illustration above is from Euler's manuscript.) The walking is good exercise, but it is impossible to cross all the bridges without crossing at least one of them more than once.

Euler reduced the problem to a network of paths connecting the two sides of the river C and B, and the two islands A and D, as shown in the network at right. Then Euler demonstrated that the task was impossible. Euler's solution began the branch of geometry called **topology**. In this special project you will work with a variety of networks to see if you can come up with a rule to determine whether a network can or cannot be "traveled." A collection of points connected by paths is called a **network**. When we say a network can be traveled, we mean that the network can be drawn with a pencil without lifting the pencil off the paper and without retracing any paths. (Points can be passed over more than once.) Try the networks on the next page to see which ones can be traveled and which are impossible to travel. It will make things easier if you know that the number of paths is not important in determining whether or not a network can be traveled. This leaves only the points to be investigated. There are two types of points in networks: odd points and even points.

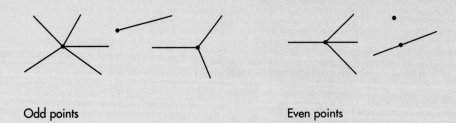

Odd points   Even points

Copy and complete the table below. When you have completed the table, you should find that seven networks cannot be traveled.

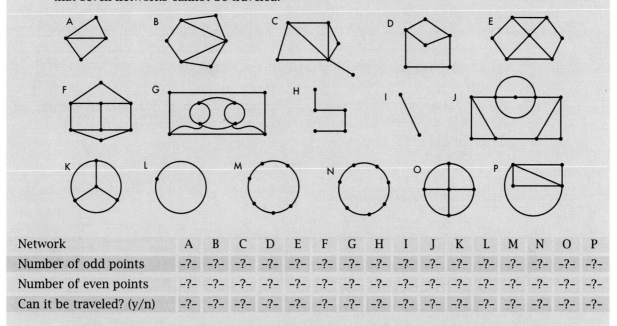

Network	A	B	C	D	E	F	G	H	I	J	K	L	M	N	O	P
Number of odd points	-?-	-?-	-?-	-?-	-?-	-?-	-?-	-?-	-?-	-?-	-?-	-?-	-?-	-?-	-?-	-?-
Number of even points	-?-	-?-	-?-	-?-	-?-	-?-	-?-	-?-	-?-	-?-	-?-	-?-	-?-	-?-	-?-	-?-
Can it be traveled? (y/n)	-?-	-?-	-?-	-?-	-?-	-?-	-?-	-?-	-?-	-?-	-?-	-?-	-?-	-?-	-?-	-?-

Look carefully at networks L, M, N, and P. These four networks have only even points. Notice that the network can be traveled whether you have one even point or seven even points, as long as there are no odd points. It appears that you can have as many even points as you desire and the network can still be traveled. Therefore, it is the odd points that can cause trouble for network travelers! Study the table to see how the number of odd points determines whether or not a network can be traveled. When you see the relationship, you will be able to complete the conjecture below.

Conjecture: A network can be traveled whenever —?—.

By successfully completing the conjecture, you have solved the network problem, using inductive reasoning. After you have solved a problem this way, it's a good idea to test your solution on some more examples, and then, if possible, to verify it by using logical reasoning. Let's take a closer look at the conjecture above by using a bit of logical reasoning.

Let's agree that any network that has all even points can be traveled. Let's look at some networks with odd points that we were able to travel. If a network can be traveled and it has odd points, where do you have to start and where do you end? Look carefully at networks A, B, D, and H. How would you travel them?

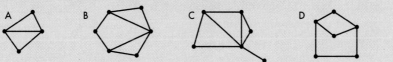

Did you notice that if you start at one odd point, you must end at the other? If you start at one of the even points, you cannot cover all the paths. If there are more than two odd points, as in network C, select one as your starting point (for example, one with three segments leaving it). If you start at this odd point, you must leave it, return to it, and leave it again to cover all lines leaving this point. Meanwhile, because you didn't start at the other odd points, you must also go to each of the other odd points, leave them, and eventually return to each of them to cover all the paths at each of the odd points. But how can you possibly end at more than one point? This leaves only one logical conclusion.

Use your conjecture to solve each problem.

1. Draw the River Pregel and the two islands shown on the first page of this project. Draw an eighth bridge so that the students can travel over all the bridges exactly once if they start at point *C* and end at point *B*.

2. Draw the River Pregel and the two islands. Can you draw an eighth bridge so that students can travel over all the bridges exactly once, starting and finishing at the same point? How many solutions can you find?

3.* Arthur A. Autovit is just about to drive to pick up his kids at day care when he remembers he's left his wallet in the house. He leaps out of the car and slams the door, locking his keys in his running car. No matter, there's a spare set in the house. And, luckily, Arthur forgot to lock his back door. But where are the spare keys? He has forgotten where he put them. They could be in any room of the house. If he enters the house through the door to room A and leaves through the door to room G, can he find a path through all the rooms, passing through each and every door exactly once? If possible, show the route. If impossible, explain why it cannot be done. Draw a network in your explanation. Can it be done if he omits one door from his path?

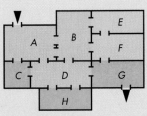

Autovit home

4.* There are eight major roads in Euler County. A county road map is shown at right. Dusty Rhodes, the county highway engineer, has a truck designed to insert reflectors into the middle of the road. Dusty's truck is not very fuel efficient, however, and she wants to find the shortest possible route covering the eight county roads. The truck is kept at point *B* on the map and must return to the same point. Dusty suspects she may have to go over at least one road more than once. From what we have learned, she is correct. What is the shortest distance she must travel to cover all eight county roads?

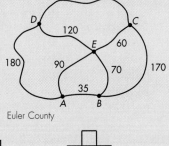

Euler County

5. Did any of the networks have only one odd vertex? Can you create a network that has only one odd vertex? Try it. How about three odd vertices? Or five odd vertices? Can you create a network that has an odd number of odd vertices? Explain why or why not.

6. Can this *lusona* design really be created by following one continuous path? If so, show how. If not, explain why not.

Lusona design

Project

Euler's Formula for Networks

If you did the project preceeding this one, titled Traveling Networks, you learned that networks consist of points and paths (curved lines). Connecting points with paths forms regions between the paths. There is a formula for the relationship among the numbers of points, paths, and regions, and this formula holds for any network that can be drawn on a piece of paper. The formula is called Euler's Formula for Networks. Copy and complete the table below, using the networks from Traveling Networks. If you didn't do that project, you might want to at least read the first two pages before you continue here. When counting regions, include the region that remains on the outside of the network. (The reason for counting the outer region will become clear when you find Euler's Formula for Solids in Chapter 12.)

Leonhard Euler

Network	A	B	C	D	E	F	G	H	I	J	K	L	M	N	O	P
Number of points (P)	–?–	–?–	–?–	–?–	–?–	–?–	–?–	–?–	–?–	–?–	–?–	–?–	–?–	–?–	–?–	–?–
Number of regions (R)	–?–	–?–	–?–	–?–	–?–	–?–	–?–	–?–	–?–	–?–	–?–	–?–	–?–	–?–	–?–	–?–
Number of paths or lines (L)	–?–	–?–	–?–	–?–	–?–	–?–	–?–	–?–	–?–	–?–	–?–	–?–	–?–	–?–	–?–	–?–

Once you have completed your table, search for a pattern that will lead to a conjecture. You are looking for a relationship among the three parts of a network. In other words, given the number of points (*P*) and the number of regions (*R*), you want a formula to calculate the number of paths or lines (*L*). Perhaps by adding, subtracting, multiplying, or some combination of the three, you will find a formula that works for all networks. When you've found your formula, complete the conjecture below.

Conjecture: If *P* stands for the number of points in a network, if *L* stands for the number of lines in the network, and if *R* stands for the number of regions in the network, then an equation that relates all three for any network is —?— (***Euler's Formula for Networks***).

Use your conjecture to solve each problem.

1. If a network has 26 points and 41 paths, how many regions does the network create?

2. If a network has 36 points and 19 regions, how many paths does the network have?

3. Draw a network that has eight points and ten regions (nine enclosed regions). How many paths will you have to draw among points?

Lesson 2.9

A Picture Is Worth a Thousand Words

You can observe a lot just by watching.
— *Yogi Berra*

A picture is worth a thousand words! That expression certainly applies to geometry. A drawing of an object often conveys information more quickly than a long written description. Visual thinking is a very important skill used by many people. Architects create blueprints. Composers create musical scores. Choreographers must visualize and map out sequences of dance steps. Basketball coaches design plays. Interior designers—well, you get the idea.

Football coach Bill Walsh designing a play

Visual thinking is commonly represented in graphs. Bar graphs, circle graphs, and line or coordinate graphs appear in newspapers and magazines. Business people, scientists, and mathematicians are constantly using information from graphs.

The bar graph gives us a lot of information at a glance. This double bar graph not only tells us which city had the largest population in 1990 and which city is projected to have the largest population by the year 2010, but it also tells us by how much each city is expected to grow. You can even determine which of the world's 20 largest cities is expected to have the greatest percentage growth.

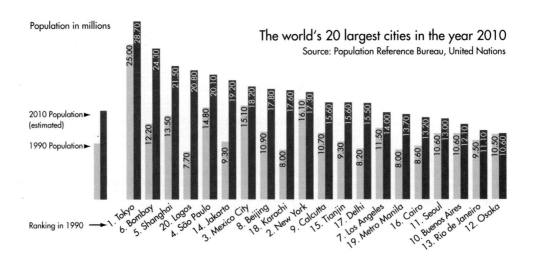

Circle graphs are used to display how a quantity is divided into its parts, usually by percentage. Of all the water on the planet, what percentage is freshwater? Of the freshwater, what percentage is trapped by ice caps and glaciers?

Line graphs are especially useful when comparing two quantities. If you wish to see how the number of species that have become resistant to pesticides has increased over time, you can see it by graphing the *number of resistant species* versus *time*. Graphs give you a visual way to think about data.

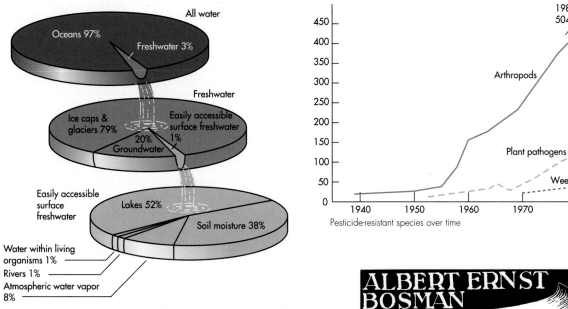

Pesticide-resistant species over time

Drawing diagrams lets you apply visual thinking to problem solving. Let's look at some examples that show how to use visual thinking to solve problems.

Example A

Volumes 1 and 2 of a two-volume set of math books sit next to each other on a shelf. They sit in their proper order: Volume 1 on the left and Volume 2 on the right. Each front and back cover is one-eighth inch thick, and the pages portion of each book is one inch thick. If a bookworm starts at the first page of Volume 1 and burrows all the way through to the last page of Volume 2, how far will she travel?

Take a moment and try to solve the problem in your head.

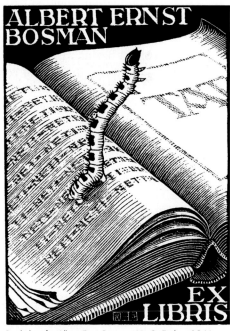

Bookplate for Albert Ernst Bosman, M. C. Escher, 1946

Did you get two and one fourth inches? It seems reasonable, doesn't it?

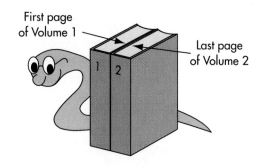

First page of Volume 1

Last page of Volume 2

Guess what? That's not the answer. Let's get organized. Reread the problem to identify what information you are given and what you are trying to find.

You are given the thickness of each cover, the thickness of the page portion, and the position of the books on the shelf. You are trying to find how far it is from the first page of Volume 1 to the last page of Volume 2. Draw a picture and locate the position of the pages referred to in the problem.

Now "look" how easy it is to solve the problem. She just traveled one fourth of an inch through the two covers!

Let's look at another example of how visual thinking helps to solve a word problem.

Example B

In Reasonville, streets that begin with a vowel run north-south unless they end with the letter *d*, in which case they run east-west. All other streets run in either direction. Euclid Street runs perpendicular to Germain Street. Fermat Street runs parallel to Germain Street. In which direction does Fermat Street run?

Make a diagram based on the given information. You can start your diagram with the information from the first sentence. Then you can add to the diagram as new information is revealed. In rereading, you learn that streets beginning with vowels run north-south unless they end with the letter *d*, in which case they run east-west. Euclid Street begins with a vowel but ends with the letter *d*, so it runs east-west. You are trying to find the direction of Fermat Street.

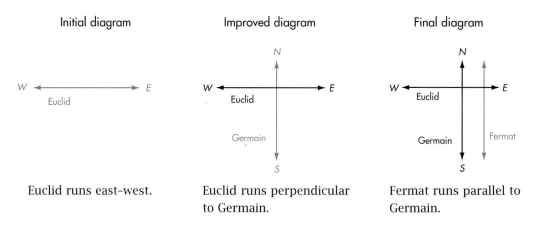

Initial diagram

Euclid runs east-west.

Improved diagram

Euclid runs perpendicular to Germain.

Final diagram

Fermat runs parallel to Germain.

The final diagram reveals the answer. Fermat Street runs north-south.

When we use visual thinking in geometry to describe points that satisfy a set of conditions or the path of something moving according to a given set of instructions, the set of points or the path is called a **locus** of points. Let's look at an example showing how to solve a locus problem.

Example C

Huey, Duey, and Louie are standing on a flat, grassy field reading their treasure map. Huey is standing at one of the features marked on the map, a gnarled tree stump, and Duey is standing atop the other item marked on their map, a large black boulder. The map indicates that the treasure is buried 60 meters from the tree stump and 40 meters from the large black boulder. If Huey and Duey are standing 80 meters apart, what is the locus of all the points where the treasure might be buried?

Start by drawing a diagram based on the information given in the first two sentences, then add to the diagram as new information is revealed. In rereading, you learn the treasure must be 60 meters from Huey and 40 meters from Duey, so the only two possible spots would be where the two circles intersect.

Initial diagram	Improved diagram	Final diagram
	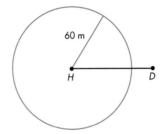	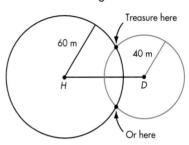
Huey and Duey are standing 80 m apart.	A circle represents all the possible points 60 m from Huey.	Another circle shows all the points 40 m from Duey. The treasure lies at one of the circles' two points of intersection.

Exercise Set 2.9

Match each term on the left with its figure on the right.

1. Angle bisector in a triangle
2. Median in a triangle
3. Altitude in a triangle
4. Diagonal in a polygon
5. Pair of complementary angles
6. Pair of supplementary angles
7. Pair of vertical angles
8. Obtuse angle
9. Acute angle

Use the graphs from this lesson to complete Exercises 10–12.

10. Which city had the largest population in 1990? Which city is projected to have the largest population by the year 2010? By how much is Tokyo expected to grow from 1990 to 2010? By how much is Mexico City expected to grow from 1990 to 2010? Which of these two cities, Mexico City or Calcutta, is expected to have the greater percentage growth?

11. What percentage of all the planet's water is easily accessible surface freshwater? If just one fourth of the fresh water trapped in ice caps and glaciers were made accessible, by how many times would the amount of accessible freshwater be increased?

12. In the 20 years between 1960 and 1980, the number of arthropods (invertebrates such as insects, spiders, and crustaceans) resistant to pesticides increased approximately how many times?

13. **Group Activity** Research a recent almanac or a United States Census Bureau report and create a circle graph or a bar graph that represents some data that interests you.

Now try your hand at some word problems. Read and reread each problem carefully, determining what information you are given and what it is that you are trying to find. Then draw and label a diagram. Finally, solve the problem.

14. Mary Ann has been contracted to design and to build a fence around the outer edge of a rectangular garden plot that measures 25 feet by 45 feet. The posts are to be set five feet apart. How many posts will she need?

15.* F. Freddie Frog is at the bottom of a 30-foot well. Each day he jumps up three feet, but then, during the night, he slides back down two feet. How many days will it take Fast Freddie to get to the top and out?

16. Midway through a 2000-meter race, a photo was taken showing the positions of all five runners. The picture shows Meg 20 meters behind Edith. Edith is 50 meters ahead of Wanda, who is 20 meters behind Olivia. Olivia is 40 meters behind Nadine. At this point in the race, who is ahead? Who is second? Who is third? Who is fourth? Who is holding up the rear? In your diagram, use M for Meg, E for Edith, and so on.

17.* Points A and B lie in a plane. Sketch the locus of points in *the plane* that are equally distant from points A and B. Sketch the locus of points in *space* that are equally distant from points A and B.

18.* Sketch the locus of points in the plane of angle A that are equally distant from the sides of angle A.

19. Line AB lies in a plane. Sketch the locus of points in that plane that are 3 cm from $\overleftrightarrow{AB}$. Sketch the locus of points in space that are 3 cm from $\overleftrightarrow{AB}$.

20. A pair of parallel interstate gas and power lines run 10 meters apart and are equally distant from relay station A. The power company needs to locate a gas-monitoring point on one of the lines exactly 12 meters from relay station A. Draw a diagram showing the locus of possible locations.

21.* Motion-efficiency expert Martha G. Rigsby needs to locate a supply point equally distant from two major work-inspection stations in an electronics assembly plant. The work stations are 30 meters apart and are positioned halfway between a pair of parallel heat-sensitive walls. The walls are 24 meters apart. The supply point must be at least 4, and at most 20, meters from either wall. Draw a diagram of the locus of possible locations.

22. **Computer Activity** Construct $\overline{AB}$ and $\overline{CD}$, where point C is on $\overline{AB}$. Construct the perpendicular bisector of $\overline{CD}$. Construct or trace the locus of this perpendicular bisector as point C moves along $\overline{AB}$. Describe the shape formed by this locus of lines.

23. Beth Mack is lost in the woods 15 km east of the north-south Birnam Woods Road. She begins walking in a zigzag pattern: 1 km south, 1 km west, 1 km south, 2 km west, 1 km south, 3 km west, and so on. Beth walks at the rate of 4 km/h (kilometers per hour). If it is 3:00 p.m. now and the sun sets at 7:30 p.m., will there still be sunlight when she reaches Birnam Woods Road?

Lost in Birnam Woods

Improving Algebra Skills—*Algebraic Magic Squares II*

A magic square is an arrangement of numbers in a square array such that the numbers in every row and every column have the same total.

In some magic squares the two diagonals have the same totals as the rows and the columns. For example, in the magic square at top right, the sum of each row is 21, of each column is 21, and of each diagonal is 21.

In an algebraic magic square the algebraic sums in all rows and columns have the same totals.

In the algebraic magic square at bottom right, the sums in the diagonals are equal to the sums in the rows and columns. Find the value of *x*.

6	11	4
5	7	9
10	3	8

$3x$	$2x-1$	$2x+4$
$2x+1$	15	$3x-1$
14	$3x+1$	$2x$

Lesson 2.10

Chapter Review

A mind that is stretched by a new idea can never go back to its original dimensions.
— *Oliver Wendell Holmes*

In this chapter, you learned about the building blocks of geometry and how to measure lengths and angles. You learned that segments or angles are congruent if they have the same measure. Then you began the important process of defining geometric terms. Even if many of these terms were already familiar to you, in just this one chapter you should have increased your mathematics vocabulary immensely! Here is a list of terms from this chapter that you should know. Most of the terms should appear in your definitions list in your notebook.

point	right angle	concave	right triangle
line	acute angle	triangle	acute triangle
plane	obtuse angle	quadrilateral	obtuse triangle
collinear	midpoint	pentagon	scalene
coplanar	angle bisector	hexagon (etc.)	isosceles
space	parallel	consecutive vertices	median
line segment	perpendicular	consecutive angles	altitude
ray	complementary	consecutive sides	trapezoid
angle	supplementary	perimeter	kite
degree measure	vertical angles	diagonal	parallelogram
congruent	linear pair	equilateral	rhombus
biconditional	polygon	equiangular	rectangle
counterexample	convex	regular polygon	square

In the process of defining polygons, you also learned the definition of congruent polygons. Congruence is an important idea you'll return to several times in this text. Finally, you applied your visual-thinking skills to draw and analyze three-dimensional figures in space and to draw and analyze graphs and diagrams to solve problems.

It may seem that there's a lot to memorize in this chapter. But having defined terms yourself, you're more likely to remember and understand them. The key is to practice using these new terms and to be organized. Do the following exercises, then read Assessing What You've Learned for tips on staying organized.

Exercise Set 2.10

In Exercises 1-20, identify the statement as true or false. For each false statement, explain why it is false or sketch a counterexample.

1. The three basic building blocks of geometry are point, line, and plane.

2. "The ray from point P through point Q" is written in symbolic form as $\overrightarrow{PQ}$.

3. "The line segment from point P to point Q" is written in symbolic form as $\overline{PQ}$.

4. "The length of line segment PQ" is written in symbolic form as PQ.

5. The vertex of angle PDQ is point P.

6. The symbol for the word *perpendicular* is $\perp$.

7. A scalene triangle is a triangle with no two sides the same length.

8. An acute angle is an angle whose measure is less than 90°.

9. If $\overleftrightarrow{AB}$ intersects $\overleftrightarrow{CD}$ at point P, then $\angle APD$ and $\angle BPC$ are a pair of vertical angles.

10. A diagonal is a line segment in a polygon, connecting any two nonconsecutive vertices.

11. If two lines lie in the same plane and are perpendicular to the same line, then they are parallel.

12. If the sum of the measures of two angles is 180°, then the two angles are complementary.

13. A trapezoid is a quadrilateral having exactly one pair of parallel sides.

14. A polygon with ten sides is called a decagon.

15. A rhombus is a parallelogram with all the angles equal in measure.

16. A square is a rectangle with all the sides equal in length.

17. To show that an angle is a right angle, you mark it with a little box.

18. The vertex angle of an isosceles triangle is between the two sides of equal length.

19. An altitude in an acute triangle is a perpendicular line segment connecting a vertex with the opposite side.

20. Exactly four statements in Exercises 1-20 are false.

In Exercises 21-31, sketch and label the figure.

21. Trapezoid $TRAP$ with $\overline{TR} \parallel \overline{AP}$ and $\overline{TR} \perp \overline{AR}$

22. Isosceles right $\triangle ABC$ with $\overline{AB} \cong \overline{BC}$

23. Scalene $\triangle PTS$ with $PS = 3$, $ST = 5$, $PT = 7$, and median $\overline{SO}$

24. Hexagon $REGINA$ with diagonal $\overline{AG}$ parallel to sides $\overline{RE}$ and $\overline{NI}$

25. Trapezoid $TRAP$ with $\overline{AR}$ and $\overline{PT}$ the nonparallel sides. Let E be the midpoint of $\overline{PT}$ and let Y be the midpoint of $\overline{AR}$. Draw $\overline{EY}$.

26. Square $ABCD$ with each side trisected by two points. The order of the points is: A, E, F, B, G, H, C, I, J, D, K, L. Draw $\overline{FG}$, $\overline{HI}$, $\overline{JK}$, $\overline{LE}$.

27. A dodecagon with every other vertex connected by diagonals to form a hexagon. Shade in the triangular regions between the hexagon and the dodecagon.

28. Obtuse isosceles triangle OLY with $\overline{OL} \cong \overline{YL}$ and median $\overline{LM}$

29. A rectangular solid 2 inches by 3 inches by 5 inches resting on its largest face. Draw lines on the three visible faces, showing how it can be divided into 30 identical smaller cubes.

30. A cube with a plane passing through it. The section is rectangle $RECT$.

31.* A cube with a plane passing through it. The section is hexagon $SPIDER$.

In Exercises 32–39, match the term with its figure.

32. Acute isosceles triangle

33. Isosceles right triangle

34. Rhombus

35. Trapezoid

36. Octagon

37. Prism

38. Pyramid

39. Cylinder

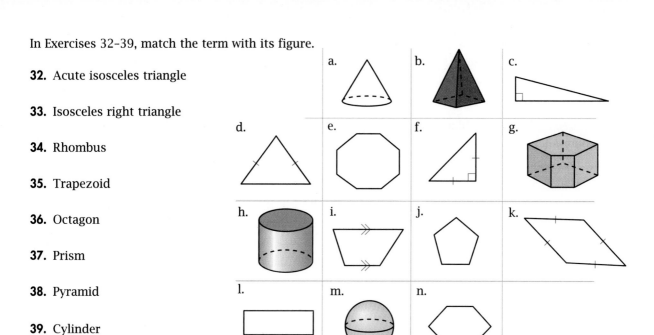

40. Which creatures in the last group are Brewsters?

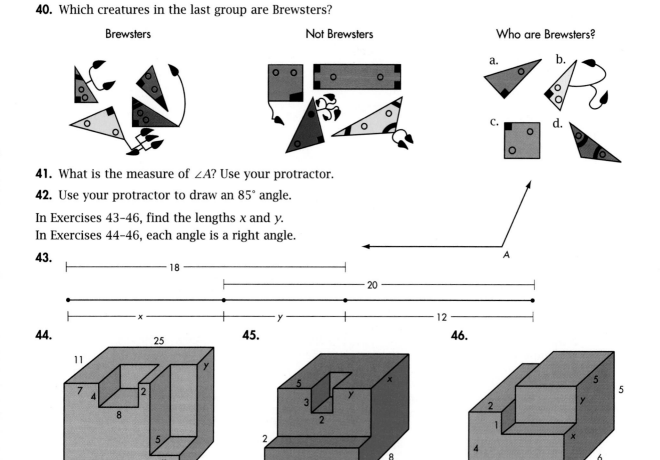

41. What is the measure of ∠A? Use your protractor.

42. Use your protractor to draw an 85° angle.

In Exercises 43–46, find the lengths *x* and *y*.
In Exercises 44–46, each angle is a right angle.

43.

44.

45.

46.

For Exercises 47–49, copy the polygon onto graph paper. Relocate the vertices of the polygon as instructed. Connect the new points to form a new polygon. Is the new polygon congruent to the original?

47.*

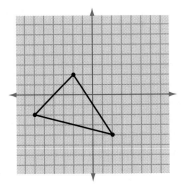

Add 3 to each *x*-coordinate.

48.

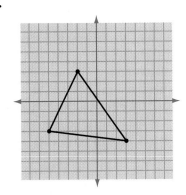

Reverse the sign of each *x*-coordinate.

49.

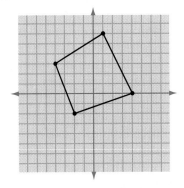

Reverse the sign of each *y*-coordinate.

50. Make a concept map (a tree diagram or a Venn diagram) for quadrilaterals.

51. The box on the right is wrapped with two strips of ribbon, as shown. What is the minimum length of ribbon needed to decorate the box?

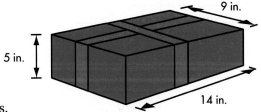

52. Make a set of sketches to show how two quadrilaterals can intersect in exactly one point, two points, three points, four points, five points, six points, seven points, and eight points.

53. At one point in a race, Ringo was 15 feet behind Paul and 18 feet ahead of John. John was trailing George by 30 feet. Paul was ahead of George by how many feet?

54. Jiminey Cricket is caught in a windstorm. At 5:00 p.m. he is 500 cm away from his home. Each time he jumps toward home he leaps a distance of 50 cm, but before he regains strength to jump again he is blown back 40 cm. If it takes a full minute from jump to jump, how long will it take Jiminey to get home?

55. A large aluminum ladder with wheels was resting vertically against the research shed at midnight when it began to slide down the vertical edge of the shed. A burglar was clinging to the ladder's midpoint, holding a pencil flashlight that was visible in the dark. Witness Jane Seymour claimed to see the ladder slide. What did she see? That is, what was the path taken by the bulb of the flashlight? Draw a diagram showing the path. (Devise a physical test to check your visual thinking. You might try sliding a meter stick against a wall, or you might plot points on graph paper.)

56. When all the diagonals possible are drawn from one vertex of a polygon, they divide the polygon interior into 500 triangular regions. How many sides does that polygon have?

57. When all the diagonals possible are drawn from one vertex of a polygon with 500 sides, they divide the polygon interior into how many triangular regions? Write an expression for the number of regions formed by the diagonals from one vertex of an *n*-sided polygon.

Assessing What You've Learned—*Organize Your Notebook*

Is this textbook filling up with folded-up papers stuffed between pages? If so, that's a bad sign! But it's not too late to get organized. Keeping a well-organized **notebook** is one of the best habits you can develop to improve and assess your learning.

In this chapter's first lesson, you started a definition list. Many books include a definition list (sometimes called a glossary) in the back. This book makes you responsible for your own glossary, so it's essential that, in addition to taking good notes, you keep a complete definition list that you can refer to. Whether you've been keeping a good list or not, go back now through each lesson in the chapter and double-check that you've completed each definition and that you understand it. Compare your list with those of your group members. Get help from classmates or your teacher on any definition you don't understand. A good way to review your notebook is to read through it and write a one-page summary of your notes for the chapter. If you can write a good summary that means your notes are complete and well organized. In addition, the summary will be a helpful study aid.

As you progress through the course, your notebook will become more and more important. And if you create a one-page summary for each chapter, the summaries will be very helpful to you when it comes time for midterms and final exams. You'll add sections for investigations and conjectures. Your notebook will become both a document for assessing what you've done in the course and your best reference source. You'll find no better learning and study aid than a good notebook.

Though you may keep them together, your portfolio, journal, and notebook are different types of documents for assessing your learning. If you're also keeping a portfolio and/or a journal, add appropriate entries to these documents.

Write in Your Journal

- You've stepped into the "Twilight Zone" and have turned on the television. The only program on features you doing your math homework. What do you see? Describe your surroundings and how you look while you're working. Write a narration for the program.

- Describe progress you're making toward goals you've set for yourself in this class. Is it necessary to make any changes in how you work to keep you on track for these goals?

Update Your Portfolio

- If you did one of the projects in this chapter, document your work and add it to your portfolio.

- Choose one homework assignment that demonstrates your best work in terms of completeness, correctness, and neatness. Add it (or a copy of it) to your portfolio.

Cooperative Problem Solving

Geometrivia I

The geometry students at the lunar space port Galileo have just completed a geometry puzzle called Geometrivia for their earth friends at the NASA Academy in Atlanta. The students at the space port have been trained to learn in cooperative small groups. Geometrivia has been designed with that in mind. It is a multistep problem-solving puzzle. In this puzzle you are to answer each lettered question, convert the letters of the answer to numbers according to the table found at the end of the Questions section, add the numbers, and, finally, substitute the sum for that letter into the formula found after the table. When all letter variables have been replaced by numbers, calculate the value of x (x is a whole number). Some questions require information you must find in the chapter. Some questions require library research. Still other questions require that you locate a particular word in a sentence from the chapter. Geometrivia demands team effort.

Example

Here is how the first question reads.

> a In a triangle, a line segment connecting a vertex to the midpoint of the opposite side is called a —?—.

Step 1   Find the answer. The answer is median.

Step 2   Substitute numbers for letters according to the table:
For *median*, the substitutions are 12 for m, 4 for e, 3 for d, 8 for i, 0 for a, 12 for n.

Step 3   Add the substituted numbers:
$12 + 4 + 3 + 8 + 0 + 12 = 39$.

Step 4   Your sum is the value for the letter a, so $a = 39$.

Questions

a In a triangle, a line segment connecting a vertex to the midpoint of the opposite side is called a —?—.

b The famous set of math books called the *Elements* was written by —?—.

c The symbol $\overrightarrow{AB}$ represents the —?— AB.

d "Based on this definition, you should have selected —?— C as the only Ork in the last group." (This sentence can be found in one of the first five lessons of this chapter.)

e A —?— is a ten-sided polygon.

f Albert Einstein was born in the month of —?—.

g An angle whose measure is less than 90° is —?—.

h A, B, D, G, K, P, -?-

i A triangle with two sides the same length is —?—.

j Archimedes was a native of the Greek city of Syracuse on the island of —?—.

k A —?— is a quadrilateral with exactly two sides parallel.

l The symbol for —?— is ∥.

m —?— was in third place midway through the race between Meg, Edith, Wanda, Olivia, and Nadine.

n The vertex of angle *SUN* is point —?—.

p Sir Isaac Newton was born on the holiday called —?—.

Table

a	b	c	d	e	f	g	h	i	j	k	l	m	n	o	p	q	r	s	t	u	v	w	x	y	z
0	1	2	3	4	5	6	7	8	9	10	11	12	12	11	10	9	8	7	6	5	4	3	2	1	0

Formula

$$X = \frac{\dfrac{n(j-g)}{k}\left[\dfrac{(a-d)(i-p)}{h}\right]^{\frac{mn}{l}}}{\sqrt{\dfrac{c(l-m)+b(e-f)}{bc}+\dfrac{h(c-n)}{g-c}}}$$

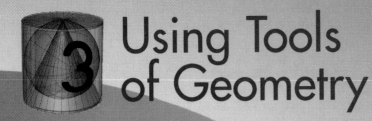

Using Tools of Geometry

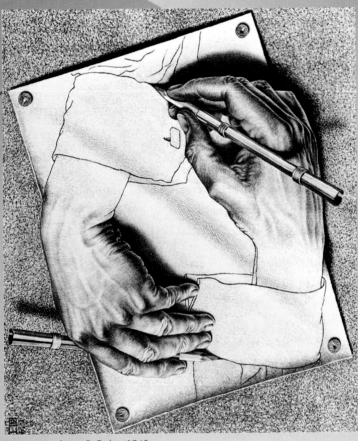

Drawing Hands, M. C. Escher, 1948

This book uses a discovery approach: In it you'll perform investigations to discover for yourself the properties of geometry. In Chapter 1, you learned about inductive reasoning, the reasoning used to perform investigations and to make conjectures. In Chapter 2, you learned the symbols and the vocabulary you will use in your geometric investigations. In this chapter you will use basic tools of geometry—a **compass** and a **straightedge** and **patty papers**—to engage in geometric investigations. You might also use a computer tool for dynamically investigating geometry. Whatever geometric tools you use, your work will go more smoothly and be more enjoyable if you work cooperatively in groups, sharing tasks and comparing results. In your notebook, keep a record of your investigations and a separate list of your conjectures.

Lesson 3.1

Duplicating Segments and Angles

It is only the first step that is difficult.
— Marie de Vichy-Chamrod

The compass, like the ruler, has been a useful geometric tool for thousands of years. The ancient Egyptians used the compass to mark off distances. During the Golden Age of Greece, Greek mathematicians made a game of geometric constructions, using a compass and a straightedge alone. Over the years, people have enjoyed trying to discover just what they could and could not construct with only a compass and a straightedge.

A modern tool of geometry

Patty paper constructions are a variation on the game of geometric constructions. All the geometry that can be constructed with a compass and a straightedge can be created using patty papers. Patty papers are the waxed squares of paper that go between the uncooked hamburger patties at fast-food restaurants. If your teacher does not have a supply of patty papers, you can substitute tracing paper. If you have access to a computer with a dynamic geometry program, you can perform constructions electronically.

In this chapter you'll mostly construct geometric figures. In the previous chapters, you drew and sketched many figures. As you go on in this text, you will make many more geometric figures. You will see the words *sketch, draw,* and *construct* often. Each has a specific meaning.

When you **sketch** an equilateral triangle, you may make a freehand sketch of a triangle that looks equilateral. You need not use any geometric tools.

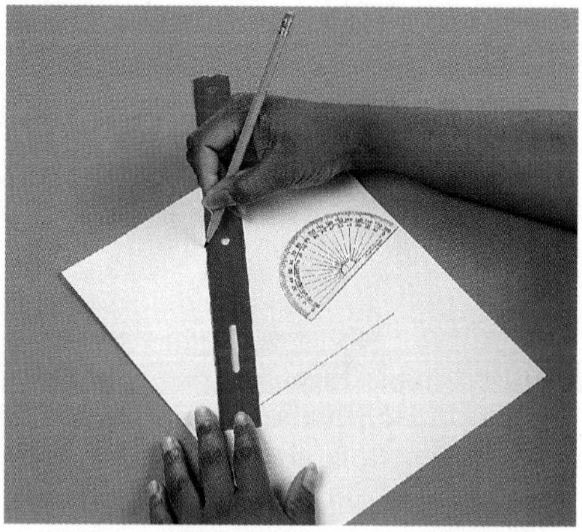

When you **draw** an equilateral triangle, you should draw it carefully and accurately. Use your geometric tools. You may use a protractor to measure angles and a ruler to measure the sides to make them appear equal in measure.

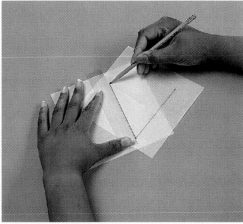

When you **compass-and-straightedge construct** an equilateral triangle, you don't rely on measurements. You must use only a compass and a straightedge. Your construction guarantees that your triangle is equilateral.

The investigation that follows is the first of many geometric investigations you will make using your geometry tools. Create an investigation section in your notebook to keep a record of your geometric investigations. Clearly label each investigation for future reference (this first one is Investigation 3.1.1). Write a statement summarizing the results of your investigation underneath your work. In most investigations this summary statement will be a conjecture.

How would you duplicate a line segment using only your compass and straightedge? Let's find out.

When you **patty-paper construct** an equilateral triangle, you use patty papers or tracing paper to fold and/or trace equal segments. You may use a straightedge to draw a segment, but you may not use a compass or other measuring tools.

Investigation 3.1.1

Draw a line segment. Experiment and discover a method for constructing a congruent segment. Your new segment must be the same length as the original. Remember, you may not use the measurements on your ruler.

Discuss your method with others near you. Write a statement summarizing what you did to construct a segment congruent to a given segment.

You've just discovered how to duplicate segments. Using only a compass and a straightedge, how would you duplicate an angle? In other words, given an angle, how would you construct an angle that is congruent to a given angle? Remember, you may not use your protractor. A protractor is a measuring tool, not a construction tool.

Investigation 3.1.2

Construct an angle. Experiment and devise a method for constructing a second angle, congruent to the first, using only a compass and a straightedge.

Here are a couple of suggestions that might help you in this task. Begin by drawing an angle on the top half of a sheet of paper, then draw a ray at the bottom of the paper. The ray will be one side of the angle. The endpoint of the ray will be the vertex of the duplicate angle. You're half finished! Now construct an arc on the original angle, using the vertex as center. This arc will help you locate the other side of the angle. You're on your own from here!

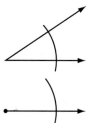

Share your ideas and methods with others near you. Write a few statements summarizing what you did to duplicate an angle.

Exercise Set 3.1

By tradition, neither a ruler nor a protractor is ever used to perform geometric constructions. Rulers and protractors are measuring tools, not construction tools. (You may use a ruler as a straightedge in constructions, provided you do not use its marks for measuring.)

1. Using just a compass and a straightedge, duplicate the three line segments shown below on your own paper. Label them as they're labeled in the book. (Please don't write in this book.)

The ancient Greeks developed their algebra by using a compass and a straightedge. Numbers represented the lengths of line segments. The addition of two numbers was the sum of the lengths of two segments. The subtraction was the difference of the lengths. Use your copies of the segments shown above to construct the segments in Exercises 2 and 3.

2.* Construct a line segment with length $AB + CD$.

3. Construct a line segment with length $AB + 2EF - CD$.

Use a compass and a straightedge to duplicate each angle shown below on your own paper. We've placed an arc in each angle so that you may duplicate it without writing in this book. If you work carefully, you'll leave no clues (holes, arcs) behind to help (or confuse) the student who uses the book next year.

4.*
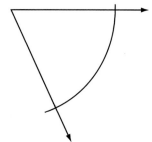

5.
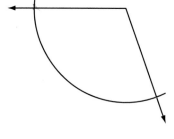

6. Draw an acute angle. Label it *TNY*. Use your compass and straightedge to construct a duplicate by the method you've devised.

7. Draw an obtuse angle. Label it *LGE,* then duplicate it.

8. Draw two acute angles on your paper. Construct a third angle with a measure equal to the sum of the measures of the first two angles. Remember, you cannot use a protractor—use a compass and a straightedge only.

Now, if you can duplicate line segments and angles, you can duplicate polygons.

9. Draw as large an acute triangle as you can make on the top half of your paper. Duplicate it on the bottom half, using your compass and straightedge. Make sure to leave all of your construction marks behind to show your method.

10. Construct an equilateral triangle. Let each side be the length of the segment below.

11. Repeat Exercises 9 and 10 with patty-paper constructions.

12.* Draw a quadrilateral. Label it quadrilateral *QUAD*. Duplicate it by using your compass and straightedge. Label the construction *COPY*.

13. I am standing on the curb facing 24th Street. A half block to my left is J Avenue, and K Avenue is a half block to my right. Numbered streets run parallel to one another and are all perpendicular to lettered avenues. If P Avenue is the northernmost street running east to west, which direction (N, S, E, or W) am I facing?

Improving Visual Thinking Skills—Design I

Use your geometry tools to create one or both of the diagrams below. Then decorate your designs.

Lesson 3.2

Constructing Perpendicular Bisectors

To be successful, the first thing to do is to fall in love with your work.
— Sister Mary Lauretta

A **segment bisector** *is a line that passes through the midpoint of a segment.*

Segment *AB* has a
midpoint *O*.

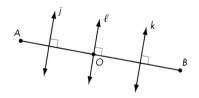

Lines *j*, *k*, and ℓ are
perpendicular to $\overline{AB}$.

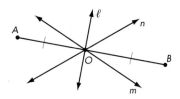

Lines ℓ, *m*, and *n*
bisect $\overline{AB}$.

A **perpendicular bisector** *of a line segment is a
line that divides the line segment into two congruent
parts (bisects it) and that is also perpendicular to the
line segment.*

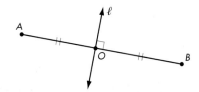

Although by definition segment bisectors and
perpendicular bisectors are lines, you can also
refer to segments and to rays as segment
bisectors or perpendicular bisectors.

There is exactly one midpoint on a segment.
In a plane, there is exactly one line
perpendicular to a segment through a given
point on the segment. Combining these two
ideas, it seems reasonable to conclude that in
a plane, a segment has exactly one
perpendicular bisector.

Investigating perpendicular bisectors with patty papers

In Investigation 3.2.1, you will discover how to construct the perpendicular bisector of
a segment. Let's investigate.

Investigation 3.2.1

Step 1   Draw a segment on a patty paper. Label it $\overline{PQ}$.

Step 2   Fold your patty paper over so that endpoints
         P and *Q* coincide (land exactly on top of each
         other). Crease your paper along the fold.

Step 1

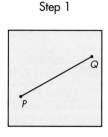

Step 2

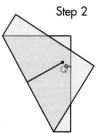

Step 3	Unfold your paper. Draw a line in the crease. Does the line appear to be the perpendicular bisector of $\overline{PQ}$? Check with your ruler and protractor to verify that the line in the crease is indeed the perpendicular bisector of $\overline{PQ}$.

Step 3

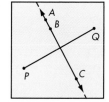

How would you describe the relationship of the points on the perpendicular bisector with the endpoints of the bisected segment? Let's perform one more step in our investigation.

Step 4	Place three points on your perpendicular bisector. Label them *A*, *B*, and *C*. With your compass, compare the distances *PA* and *QA*. Compare the distances *PB* and *QB*. Compare the distances *PC* and *QC*.

Step 4

What do you notice about the two distances from each point on the perpendicular bisector to the endpoints of the segment? Compare your results with the results of others near you. You should now be ready to state your first conjecture.

C-1 If a point is on the perpendicular bisector of a segment, then it is —?— from the endpoints (***Perpendicular Bisector Conjecture***).

Attach (glue, tape, or staple) your folded patty paper to a page in the investigation section of your notebook. Complete the conjecture and write it underneath your investigative work as part of the statement summarizing your investigation.

The Perpendicular Bisector Conjecture will be very useful to you later in this text. You will make many more conjectures as you proceed through this book. The conjectures that are important to keep track of have been numbered and, in most cases, named. Start a list of these numbered conjectures in your notebook. The Perpendicular Bisector Conjecture (C-1) should be the first entry on your list. Sketch an illustration of the conjecture, placing it near the written conjecture. You should now have four parts to your notebook: a section for homework and notes, an investigations section, a definitions list, and now a conjecture list. Because you're responsible for writing your own definitions and conjectures, there are no lists of these in your text. Thus you'll need to refer to the lists in your notebook often.

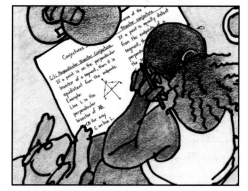

Let's take another look. In Investigation 3.2.1 you discovered one way to find the perpendicular bisector of a segment—by paper folding. How can you find the perpendicular bisector of a segment by using only a compass and a straightedge? You also discovered that if a point is on the perpendicular bisector, then it is equally distant from the two endpoints. What about the converse of this statement? If a point is equally distant from two endpoints, will it be on the perpendicular bisector? Can locating two points that are each equally distant from the endpoints of a segment help you construct the perpendicular bisector? Maybe we're on to something. Let's investigate.

Investigation 3.2.2

Draw a line segment. Start with the steps below.

Step 1

Step 2

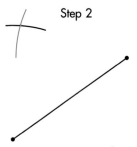

Using one endpoint as center, swing an arc on one side of the segment.

Using the same compass setting but using the other endpoint as center, swing a second arc intersecting the first.

Now you're on your own. You found one point equally distant from the endpoints of the segment. Use your compass to find another. Use these points to construct a line. Is the line the perpendicular bisector of the segment? Use the paper-folding technique of Investigation 3.2.1 to check whether the line you constructed with your compass and straightedge is, in fact, the perpendicular bisector. Try it. Bring the endpoints together and fold. Is the line you constructed the perpendicular bisector of the segment? Beneath your construction work, complete the conjecture below and write a summary of what you did in this investigation.

C-2 If a point is equally distant from the endpoints of a segment, then it is on the —?— of the segment (*Converse of the Perpendicular Bisector Conjecture*).

Exercise Set 3.2

In Exercises 1–6, match the term with its figure at right.

1. Acute scalene triangle

2. Obtuse isosceles triangle

3. Right isosceles triangle

4. Acute isosceles triangle

5. Obtuse scalene triangle

6. Right scalene triangle

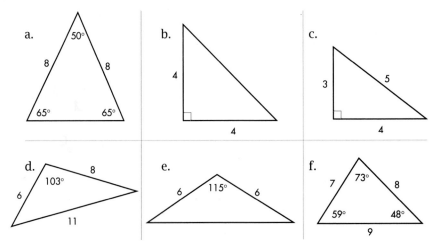

In Exercises 7–11, perform the construction using only a compass and a straightedge.

7. Construct and label $\overline{AB}$. Construct the perpendicular bisector of $\overline{AB}$.

8.* Construct and label $\overline{QD}$. Construct perpendicular bisectors to divide $\overline{QD}$ into four congruent parts.

9.* Construct a line segment so close to the edge of a piece of paper that you can only swing arcs on one side of the segment. Then construct the perpendicular bisector of the segment.

10.* Construct $\overline{MN}$ with length equal to the average length of $\overline{AB}$ and $\overline{CD}$ below.

11.* Using $\overline{AB}$ and $\overline{CD}$ above, construct the distance $2AB - \frac{1}{2}CD$.

12. Perform patty-paper constructions on Exercises 7, 8, 10, and 11.

Study the given information until you discover a relationship that holds for all the data. State your findings by completing the conjecture.

13.
$25 \cdot 3 = 75$ $19 \cdot 7 = 133$
$-17 \cdot 31 = -527$ $-21 \cdot -7 = 147$
$-1 \cdot 41 = -41$ $103 \cdot -7 = -721$
Conjecture: The product of two odd integers is always —?—.

14.
$3 \cdot 4 = 12$ $8 \cdot 9 = 72$
$23 \cdot 24 = 552$ $-3 \cdot -4 = 12$
$-1 \cdot 0 = 0$ $-15 \cdot -14 = 210$
Conjecture: The product of two consecutive integers is always —?—.

15.* Sketch and label a polygon that has exactly three sides of equal length and exactly two angles of equal measure.

16. Sketch two different triangles such that each has sides that measure 5 cm and 9 cm but that are not congruent.

17. Huey, Duey, and Louie are standing on a football field. Huey is standing on one of the 40-yard lines, and Louie is standing on the other 40-yard line. What is the locus of all the points between the two 40-yard lines where Duey can stand and be equally distant to both Huey and Louie?

18.* Draw the new position of △MON if it is reflected about the dotted line. Label the coordinates of the vertices.

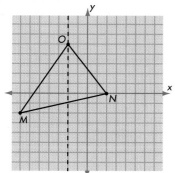

19.* Draw △DAY after it is rotated 90° clockwise about the origin. Label the coordinates of the vertices.

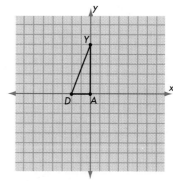

20.* List the letters from the alphabet below that have a horizontal line of symmetry.

A B C D E F G H I J K L M N O P Q R S T U V W X Y Z

Constructing Perpendiculars

The purpose of education . . . is to create in a person the ability to look at the world for himself, to make his own decisions.

— *James Baldwin*

The construction of a perpendicular segment from a point to a line (with the point not on the line) is another very important construction. An altitude of a triangle is a perpendicular from a point to a line.

You already know how to construct perpendicular bisectors. You can use that know-how to construct a perpendicular from a point to a line.

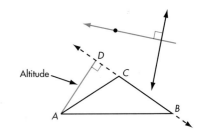

Investigation 3.3.1

Draw a line and a point near but not on the line. Starting with the steps below, devise a method to construct a perpendicular from the point to the line.

Step 1	Step 2	Step 3

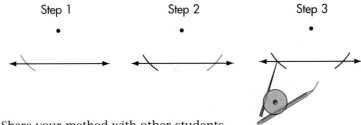

 Experiment! Share your method with other students.

Agree on a best method. You have just discovered a method of constructing a perpendicular from a given point to a given line. Beneath your construction work, write a summary of what you did in this investigation. Let's see what you can discover about this perpendicular segment.

Investigation 3.3.2

On a clean sheet of paper, perform the following steps.

Step 1 Draw and label $\overleftrightarrow{AB}$ and a point P not on $\overleftrightarrow{AB}$.

Step 2 Construct a perpendicular from point P to $\overleftrightarrow{AB}$. Label the point of intersection of the perpendicular and the given line point M.

Step 3 Select three points on $\overleftrightarrow{AB}$ and label them Q, R, and S. Draw segments PQ, PR, and PS.

Step 4 Compare the lengths PM, PQ, PR, and PS.

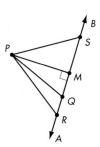

How do the lengths compare? Can you find a segment from point P to $\overleftrightarrow{AB}$ that is shorter than $\overline{PM}$? Compare your results with the results of others near you. You should be ready to state your observations as a conjecture.

C-3 The shortest distance from a point to a line is measured along the —?— from the point to the line (***Shortest Distance Conjecture***).

Write the conjecture as your summary statement beneath your investigative work and add the conjecture to your conjecture list.

The construction of a perpendicular from a point to a line lets you find the distance from a point to a line. The definition below is based on this construction. Add it to your definition list.

The **distance from a point to a line** *is the length of the perpendicular segment from the point to the line.*

Let's take another look. How would you use patty papers to create a perpendicular from a point to a line?

Investigation 3.3.3

On a piece of patty paper, perform the steps below.

Step 1   Draw and label line AB and a point P not on $\overleftrightarrow{AB}$.

Step 2   Fold the line on top of itself and slide until . . . well, you figure out the rest.

Step 3   Once you have a crease that appears to be a perpendicular to the line through the point, test it. Use a corner of another patty paper (a right angle!) to test to see if it is really perpendicular.

Share your results with others near you. Attach your folded patty paper to a page in the investigation section of your notebook. Write a statement summarizing what you did with patty papers to create a perpendicular from a point to a line.

Step 1

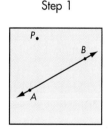

Step 2

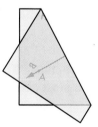

Step 3

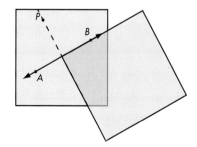

Exercise Set 3.3

In Exercises 1–6, perform compass-and-straightedge constructions.

1. Draw an obtuse angle. Place a point inside the angle. Now construct perpendiculars from the point to both sides of the angle.

2. Draw an acute triangle. Label it *ABC*. Construct altitude $\overline{CD}$. (We didn't forget point *D*. Your job is to locate it!)

3.* Draw obtuse △*OBT* with ∠*O* obtuse. Construct altitude $\overline{BU}$. In an obtuse triangle, an altitude can fall outside of the triangle. To construct an altitude from point *B* in your triangle, you must extend the side $\overline{OT}$.

4. Rondo, the distant cousin of Frodo, is at point *R* near the juncture of the rivers Loudwater and Mitheithel. He has just sensed the approach of a band of angry Orcs. Help him determine the shortest path to a river and save him from becoming the main course at an Orc banquet. Toward which river should he run? Trace the map, then use a construction to demonstrate that you have found the shortest path.

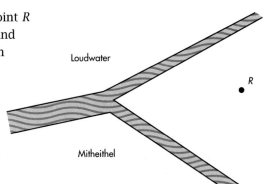

Loudwater

R

Mitheithel

5.* How can you construct a perpendicular to a line through a point that is on the line? Construct a line. Mark a point on your line. Now experiment. Devise a method to construct a perpendicular to your line at the point. This is another important geometric construction.

6. Draw a line. Mark two points on the line and label them *Q* and *R*. Now construct a square *SQRE* with $\overline{QR}$ as a side.

In Exercises 7–10, perform patty-paper constructions.

7. Draw an acute triangle on your patty paper. Place a point inside the triangle. Now construct perpendiculars from the point to all three sides of the triangle.

8. Construct a right isosceles triangle. Label it *ABC* with point *C* the right angle. Fold to construct the altitude $\overline{CD}$.

9.* Draw obtuse △*OBT* with ∠*O* obtuse. Fold to construct the altitude $\overline{BU}$. (Don't forget, you must extend the side $\overline{OT}$.)

10. How would you fold to construct a perpendicular through a point on a line?

Sketch, draw, or construct each figure in Exercises 11–18. Label the vertices with the appropriate letters. When you sketch or draw, use the special marks that indicate right angles, parallel segments, and segments and angles equal in measure.

11. Sketch obtuse triangle *FIT* with $m\angle I > 90°$ and median $\overline{IY}$.

12. Sketch $\overline{AB} \perp \overline{CD}$ and $\overline{EF} \perp \overline{CD}$.

13. Use your protractor to carefully draw a regular pentagon. Draw in all the diagonals. What is the name of the figure formed by the diagonals?

14.* Construct an octagon. Print within the octagon the first word that comes to mind.

15. Sketch an A-frame house with the vertex angle (the peak angle) measuring less than 45° but more than 30°.

16. Sketch a sailboat with one sail in the shape of an obtuse isosceles triangle.

17. Sketch five lines in a plane that intersect in exactly five points. Now do this a second way.

18. Sketch a quadrilateral. Label it *FOUR*. Sketch a second quadrilateral with all four sides twice the length of the four sides of quadrilateral *FOUR*. Label the vertices of your second quadrilateral F_1, O_1, U_1, and R_1 so that vertex F matches vertex F_1, vertex O matches vertex O_1, and so on.

Note: Points F_1, O_1, U_1, and R_1 (read "F sub one, O sub one, U sub one, and R sub one") are examples of points labeled with subscripts. The subscripts are not funny exponents! We occasionally use subscripts to label points or sets of points that are somehow related to another set of points.

19.*A rectangle and a circle can intersect in three points and in six points, as shown at right. Make sketches showing all the other possible ways that a rectangle and a circle can intersect.

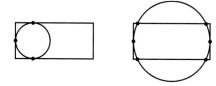

20.*Draw the new position of △*TEA* if it is reflected over the dotted line. Label the vertices' coordinates.

21.*Draw the new position of △*CUP* if it is rotated 180° counterclockwise about the origin. Label the vertices' coordinates.

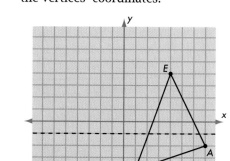

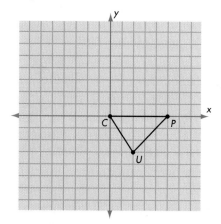

Improving Visual Thinking Skills—*Constructing an Islamic Design*

This Islamic design is based on two intersecting squares that form an eight-pointed star. Most Islamic designs of this kind can be constructed using only a compass and a straightedge. Try it. Use your compass and straightedge to re-create this design or to create a design of your own based on an eight-pointed star.

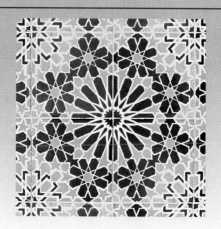

Lesson 3.4

Constructing Angle Bisectors

*An **angle bisector** is a ray that has an endpoint on the vertex of the angle and that divides the angle into two angles of equal measure.*

While the definition states that the bisector of an angle is a ray, you may also refer to a segment on the ray with an endpoint at the vertex as an angle bisector.

In Investigation 3.4.1, that follows, you will take a closer look at the bisector of an angle.

Investigation 3.4.1

Step 1   Step 2   Step 3

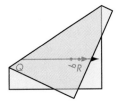

Step 1   On a patty paper, draw a large acute angle. Label it *PQR.*

Step 2   Fold your patty paper over so that $\overrightarrow{QP}$ and $\overrightarrow{QR}$ coincide. Crease your patty paper along the fold.

Step 3   Unfold your patty paper. Draw a ray with endpoint Q along the crease. Is the ray the angle bisector of $\angle PQR$? (Why? Because the fold formed two angles that coincided.)

Clearly, an angle has a bisector. Does an angle have only one bisector? Can you find another ray that also bisects the angle? What else is true about an angle bisector? How would you describe the relationship between the points on the angle bisector and the sides of the angle being bisected? Let's perform two more steps in our investigation.

Step 4   Place a point on your angle bisector. Label it *A.* Compare the perpendicular distances to the two sides. To do this, place one edge of a second patty paper on one of the sides of the angle. Slide the edge of the patty paper along the side of the angle until an adjacent perpendicular side of the patty paper passes through the point. Mark this distance on the patty paper.

Step 4

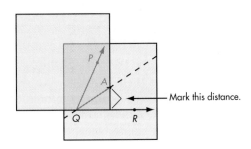

Mark this distance.

Step 5   Compare this distance with the distance to the other side by
         repeating the process on the other ray.

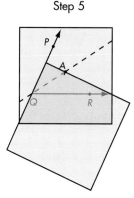

Step 5

         What do you notice about the two distances from a point on
         the angle bisector to the sides of the angle? Try it with other
         points on the angle bisector. Compare your results with the
         results of others. You should now be ready to state a
         conjecture.

C-4   If a point is on the bisector of an angle, then
          it is —?— from the sides of the angle
          (***Angle Bisector Conjecture***).

         Attach your folded patty paper to a page in the investigation section of your note-
book. Complete the conjecture and write it beneath your investigative work as part of the
statement summarizing your investigation. Add the conjecture to your conjecture list.
         You've found the bisector of an angle by patty-paper folding. Now let's take another
look. See if you can construct the angle bisector with a compass and a straightedge.

Investigation 3.4.2

         Draw an acute angle. Devise a method to construct the bisector of the angle.

Step 1   Step 2

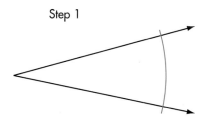   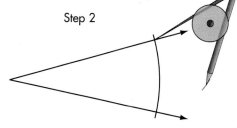

         Experiment! Once you think you have constructed the angle bisector, you can fold
your paper to see if the ray you constructed is actually the bisector. Share your method
with other students. Agree on a best method. Beneath your investigative work, write a
summary of what you did in this investigation.

         One more investigation: In the first lesson of this chapter you constructed an
equilateral triangle. What can you discover about equilateral triangles?

Investigation 3.4.3

Step 1   On a sheet of paper, construct a large equilateral triangle.
Step 2   With your protractor, measure the three angles of the triangle. What is the measure of
          each angle of the equilateral triangle? Try this again. Construct a second equilateral
          triangle. What is the measure of each angle? Compare your results with the results of
          others. You should be ready to state a conjecture.

C-5 The measure of each angle of an equilateral triangle is —?—.

Complete the conjecture and write it beneath your investigative work as part of the statement summarizing your investigation. Add it to your conjecture list.

Exercise Set 3.4

Match each geometric construction with its diagram.

1. Construction of an angle bisector
2. Construction of a median
3. Construction of a perpendicular bisector
4. Construction of an altitude
5.* Construction of an angle measuring 60°

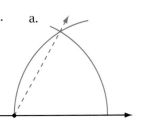

a.

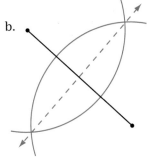

b.

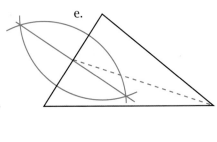

c. d. e.

You now know how to duplicate angles and bisect angles. You can also construct angles that measure 60° and 90°. With this as a start, you can construct angles of many other sizes. In Exercises 6–13, describe how you would construct an angle with the measures given. (Don't actually perform the construction.)

6.* 30° **7.** 45° **8.** 15° **9.** 120°

10. $22\frac{1}{2}°$ **11.** 75° **12.*** $52\frac{1}{2}°$ **13.** 135°

Now for the hard part! In Exercises 14–16, construct an angle with the given measure. Clearly identify which angle in your construction has the desired measure. Remember, you may use only your compass and straightedge. No protractor!

14.* 135° **15.** 105° **16.** $7\frac{1}{2}°$

17. With patty papers, fold to construct an angle with a measure of 15°.

18. Draw an acute triangle. Bisect one angle with a compass and a straightedge. Construct an altitude from a second vertex and a median from the third vertex.

19. Repeat Exercise 18 with patty-paper constructions. Which set of construction tools do you prefer, patty papers or a compass and a straightedge? Why?

20. Write the converse of the Angle Bisector Conjecture. Do you think it's true? Why or why not?

Lesson 3.5

Constructing Parallel Lines

Parallel lines are lines that lie in the same plane and do not intersect.

Though from this photo's perspective these railroad tracks appear to meet, the two tracks are actually the same distance apart everywhere. Imagine what would happen if they weren't!

The first pair of lines shown above intersect. They are clearly not parallel. The second pair of lines do not meet as drawn. However, if they were extended, they would intersect. Therefore they are not parallel. But what about the third pair? Hard to tell, right? You could extend them, but you might need to draw them on a roll of toilet paper to extend them far enough in both directions so that you could be sure they wouldn't meet. If two lines in a plane are parallel, then they must always be the same distance apart. If the distances between them are not the same everywhere, then the lines are not parallel.

You can also say that if two lines are always the same distance apart, then they are parallel. This gives you a way, called the equidistant method, of constructing parallel lines. Investigation 3.5.1 shows how you can construct a line parallel to a given line through a given point. The given line is m and the given point is P.

Investigation 3.5.1

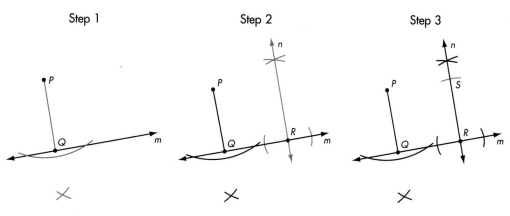

Step 1   Construct a perpendicular from the given point P to the given line m. Label the point of intersection Q (PQ is the distance from point P to line m).

Step 2   Select a point R on line m (as far away from point P as possible) and construct a line through point R perpendicular to line m. Label the perpendicular line n.

Step 3   With your compass, locate a point S on line n (points P and S are on the same side of line m) so that $PQ = RS$.

Step 4　　Construct $\overleftrightarrow{PS}$. Because $PQ = RS$, $\overleftrightarrow{PS}$ and line m are the same distance apart and thus $\overleftrightarrow{PS} \parallel m$.

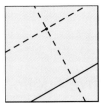

Step 4

Write a statement beneath your construction, summarizing what you did to create a line parallel to a given line through a given point.

Not only is $\overleftrightarrow{PS} \parallel m$, but $\overline{PQ}$ and $\overline{RS}$ appear to be parallel. This should make sense because in Chapter 2, you observed that if two lines in the same plane are perpendicular to the same line, they are parallel to each other. Here is a case that illustrates this. Notice that because $\overline{PQ} \perp m$ and $\overline{RS} \perp m$, then $\overline{PQ} \parallel \overleftrightarrow{RS}$.

This gives us another way to create parallel lines: construct two lines perpendicular to the same line. In Investigation 3.5.2, you will use this idea to construct with patty paper a line through a given point parallel to a given line.

Investigation 3.5.2

Here are the steps in pictures.

Step 1	Step 2	Step 3	Step 4
	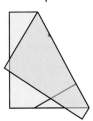		

Line and point.　　Fold perpendicular.　　Fold perpendicular to the perpendicular.　　TA DA!

Share your methods with others near you. Attach your folded patty paper to a page in the investigation section of your notebook. Write a statement describing what you did with patty papers to create a line parallel to a given line through a given point.

Exercise Set 3.5

1.* Sketch trapezoid *ZOID* with $\overline{ZO} \parallel \overline{ID}$, point *T* the midpoint of $\overline{OI}$, and *R* the midpoint of $\overline{ZD}$. Sketch segment *TR*.

2. Draw rhombus *ROMB* with $m\angle R = 60°$ and diagonal $\overline{OB}$.

3. Draw rectangle *RECK* with diagonals $RC = EK = 8$ cm intersecting at point *W*.

4. Sketch $\overleftrightarrow{AB}$, $\overleftrightarrow{CD}$, $\overleftrightarrow{EF}$, and $\overleftrightarrow{GH}$ with $\overleftrightarrow{AB} \parallel \overleftrightarrow{CD}$, $\overleftrightarrow{CD} \perp \overleftrightarrow{EF}$, and $\overleftrightarrow{EF} \parallel \overleftrightarrow{GH}$.

5. Draw octagon *ALTOSIGN* with $\overline{AS} \perp \overline{GT}$.

6. Draw a line and a point not on the line. Use the equidistant method with a compass and a straightedge to construct a second line parallel to the first line that passes through the point.

7. Draw a line on a patty paper and a point on the patty paper not on the line. Fold to construct a second line parallel to the first line that passes through the point.

8. A rectangle can also be defined as a parallelogram with one right angle. You choose the tools—compass and straightedge or patty papers—and then construct rectangle *RECT* with $\overline{RE}$ and $\overline{EC}$ as a pair of consecutive sides.

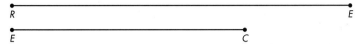

9. Construct a square with *z* the perimeter.

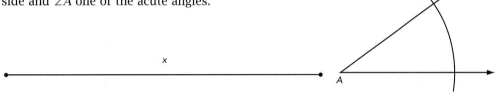

10. Construct a rhombus with *x* the length of each side and ∠*A* one of the acute angles.

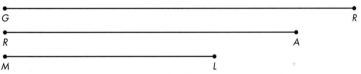

11. Using a compass and a straightedge, construct trapezoid *TRAP* with $\overline{TR}$ and $\overline{AP}$ the two parallel sides and with *AP* the distance between them. (There are many solutions!)

12. Using patty papers or a compass and a straightedge, construct parallelogram *GRAM* with $\overline{RG}$ and $\overline{RA}$ the two consecutive sides and *ML* the distance between $\overline{RG}$ and $\overline{AM}$. (How many solutions can you find?)

13. Using a compass and a straightedge, construct a large acute scalene triangle and label it *ABC*. Locate the midpoint of side $\overline{AB}$ and label it *M*. Locate the midpoint of side $\overline{BC}$ and label it *N*. Construct lines through points *M* and *N* parallel to the sides $\overline{AC}$ and $\overline{AB}$, respectively.

14.* List the letters from the alphabet below that have a vertical line of symmetry.

A B C D E F G H I J K L M N O P Q R S T U V W X Y Z

15. If *D* is the midpoint of $\overline{AC}$ and *C* is the midpoint of $\overline{AB}$, what is the length of $\overline{AB}$ if *BD* = 12 cm?

16.* Sketch the locus of points on the coordinate plane in which the *x*-coordinate is equal to the *y*-coordinate.

17. Sketch the locus of points equidistant from a pair of parallel lines, in the same plane as the lines. Describe the locus of points in *space* equidistant from a pair of parallel lines.

Lesson 3.6

Construction Problems

People who are only good with hammers see every problem as a nail.
— Abraham Maslow

If you can combine a given set of triangle parts (arranged in a given way) to create a triangle, and if all the triangles created with those parts turn out to be congruent, then the given set of parts **determines** a triangle. You can test triangles for congruence by sliding, rotating, or flipping them to see if they fit over each other.

Exercise Set 3.6

In Exercises 1–15, first sketch and label the figure you are going to construct. Second, perform either a compass-and-straightedge construction or a patty-paper construction. Third, write a few sentences describing the steps in your construction. The lowercase letter shown above some segments represents the length of the segment. In Exercises 1–3, the given parts determine a triangle.

1.* Given:

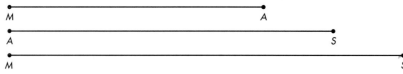

Construct: △*MAS*

2. Given:

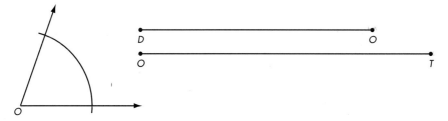

Construct: △*DOT*

3. Given:

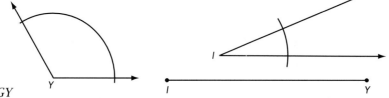

Construct: △*IGY*

4. Given the triangle shown at right, construct another triangle that has angles congruent to the given angles but that has sides *not* congruent to the given sides. (Does this construction demonstrate that three angles do not determine a triangle?)

5.* The two segments and the angle below do not determine a triangle. Construct two different triangles named *ABC* that have the three given parts.

A _____ *B*

B _____ *C*

A

6.* Given:

x

y

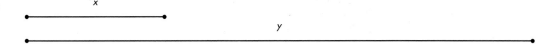

Construct: Isosceles triangle *CAT* with *y* the perimeter and *x* the length of the base

7. Given:

z

Construct: An isosceles right triangle with *z* the length of each of the two congruent sides

8. Given:

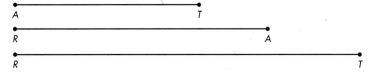

A _____ *T*

R _____ *A*

R _____ *T*

Construct: a. △*RAT*
   b. Median $\overline{TM}$
   c. Angle bisector $\overline{RB}$

9. Given:

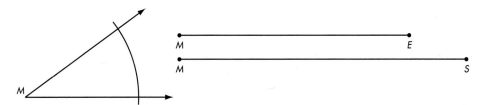

M _____ *E*

M _____ *S*

M

Construct: a. △*MSE*
   b. Perpendicular bisector of $\overline{MS}$
   c. $\overline{OU}$ where *O* is the midpoint of $\overline{MS}$ and *U* is the midpoint of $\overline{SE}$

10. Draw a triangle. Label it *REG*. Then construct another triangle that has a perimeter equal to half the perimeter of △*REG*.

11. Construct a kite.

12. Construct a quadrilateral with two pairs of opposite sides of equal length.

13. Construct a quadrilateral with exactly three sides of equal length.

14. Construct a quadrilateral with all four sides of equal length.

15. Group Activity For this excercise, use a computer drawing program, or patty papers, or a compass and straightedge. Draw a large scalene obtuse $\triangle ABC$ with $\angle B$ the obtuse angle. Construct the median $\overline{BM}$, the angle bisector $\overline{BR}$, and the altitude $\overline{BS}$. What is the order of the points on $\overline{AC}$? Compare your results with the results of others near you. Is the order of points always the same for the foot of the median, the foot of the altitude, and the foot of the angle bisector? Write a conjecture.

16. How many sides does a polygon have if it has 500 diagonals from each vertex?

17.* Sketch the locus of points on the coordinate plane in which the sum of the x-coordinate and the y-coordinate is equal to 9.

Improving Reasoning Skills—*Spelling Card Trick*

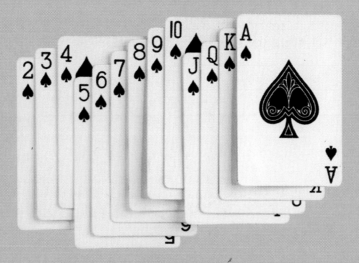

The card trick described below uses one complete suit from a regular deck of playing cards. How must you arrange the cards so that you can successfully complete the trick? Here is what your audience should see and hear as you perform.

1.   As you take the top card and place it at the bottom of the pile, say "A."

2.   Then take the second card, place it at the bottom of the pile, and say "C."

3.   Take the third card, place it at the bottom, and say "E."

4.   You've just spelled *ace*. Now take the fourth card and turn it faceup on the table. The card should be an ace.

5.   Continue in this fashion, saying "T," "W," and "O" for the next three cards. Then turn the next card faceup. It should be a two.

6.   Continue spelling *three, four, . . . , jack, queen, king.* Each time you spell a card, the next card turned faceup should be that card.

Lesson 3.7

Constructing Points of Concurrency

You now can perform a number of constructions in triangles, including angle bisectors, perpendicular bisectors of the sides, medians, and altitudes. In this lesson and the next lesson you will discover special properties of these lines and segments. You choose the tools of discovery, either patty papers or a compass and a straightedge. Or, use a computer with a geometry drawing program.

Investigation 3.7.1   or   or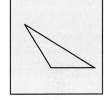

Step 1   Draw a large acute triangle on one patty paper and an obtuse triangle on another. (If you're using a compass and a straightedge instead of patty papers, draw your triangles on the top and the bottom halves of a piece of paper.)

Step 2   Fold (or construct) the three angle bisectors in each triangle.

What do you notice? Compare your results with the results of others near you. State your observations as a conjecture.

   C-6   The three angle bisectors of a triangle —?—.

Do similar investigations with perpendicular bisectors and with lines through the altitudes. Compare your results with others and make conjectures.

   C-7   The three perpendicular bisectors of a triangle —?—.

   C-8   The three altitudes (or the lines through the altitudes) of a triangle —?—.

Were you surprised by the results of your investigations? Complete each of the three conjectures and write them beneath your investigative work as part of the statement summarizing your investigations, then add them to your conjecture list.
The definition below will help you refine your conjectures.

Point of concurrency

Concurrent lines *(segments or rays) are lines that intersect in a single point.*

The point of intersection is the **point of concurrency**. It's no big deal for two lines to be concurrent. It is special, however, for three or more lines to be concurrent.

Each of the points of concurrency that you found in your investigations has a name.

The point of concurrency of the three angle bisectors in a triangle is the **incenter**. The point of concurrency of the three perpendicular bisectors of the sides of a triangle is the **circumcenter**. The point of concurrency of the three altitudes or the lines through the altitudes of a triangle is the **orthocenter**.

In the next investigation you will discover what is special about the incenter and the circumcenter.

Investigation 3.7.2   or  or

Step 1   Compare the distances from the circumcenter to each of the three sides. Are they the same? Compare the distances from the circumcenter to each of the three vertices. Are they the same?

Step 1

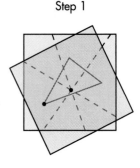

What do you notice? Compare your results with the results of others near you. State your observations as your next conjecture.

   C-9   The circumcenter of a triangle —?—.

If the incenter is on all three angle bisectors, then what must be true? Let's look.

Step 2   Compare the distances from the incenter to each of the three sides. Are they the same? Compare the distances from the incenter to each of the three vertices. Are they the same?

Step 2

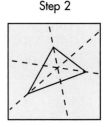

What do you notice? Compare your results with the results of others near you. State your observations as your next conjecture.

   C-10   The incenter of a triangle —?—.

From the conjecture you made earlier about perpendicular bisectors, you know that each point on a perpendicular bisector of a segment is equally distant from the ends. If the circumcenter of a triangle is on all three perpendicular bisectors, then it follows logically that it is equally distant from all three vertices of the triangle. Because it is equally distant from all three vertices of the triangle, you should be able to use your compass to construct a circle that passes through the three vertices of the triangle with the circumcenter as the center of the circle.

From the conjecture you made earlier about angle bisectors, you know that each point on an angle bisector is equally distant from the sides of the angle. Because the incenter of a triangle is on all three angle bisectors, then it follows logically that it is equally distant from all three sides. Because it is equally distant from all three sides, you should be able to use your compass to construct a circle that just touches the three sides with the incenter as the center of a circle.

So, the circumcenter is the center of a circle that passes through the three vertices of a triangle, and the incenter is the center of a circle that touches each side of the triangle. Here are a few more vocabulary terms that help describe these geometric situations.

*A circle is **circumscribed about a polygon** when it passes through each vertex of the polygon. (The polygon is inscribed in the circle.)*

*A circle is **inscribed in a polygon** when it touches each side of the polygon at exactly one point. (The polygon is circumscribed about the circle.)*

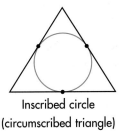

Inscribed circle
(circumscribed triangle)

Circumscribed circle
(inscribed triangle)

Exercise Set 3.7

For Exercises 1–5, make a sketch and answer the question.

1.* Nate Knobbytyre is the dirt-bike officer of Mt. Thermopolis State Park. He wishes to position himself at a point that is the same distance from each of three straight, intersecting (not concurrent) bike paths. Help Nate locate this point so that in an emergency he will be able to get to any one of the paths by the shortest route possible. Which point of concurrency does Nate need to locate?

2. Stained-glass artist Sally Solare wishes to inscribe a circle in a triangular portion of her latest abstract design. Which point of concurrency in the triangular section of her design does Sally need to locate?

3. Rosita is installing a round sink in her new kitchen countertop. She has marked three points on the countertop to indicate points through which a circle must pass so that she can install the sink. Which point of concurrency of the triangle connecting the three points must she locate to construct the circle?

4. Julian Chive is redesigning his kitchen. He wishes to put a butcher block table at a location equally distant from the refrigerator, stove, and sink. The locations of the refrigerator, stove, and sink form what's called a kitchen's work triangle. Which point of concurrency does Julian need to locate?

5. Frankye has built a home for her pet hamsters, Brad and Janet. It is in the shape of a triangular prism. She wants to cut out the largest possible circular entrance from one of the bases. Which point of concurrency in the triangular base does Frankye need to locate in order to construct her entrance for Brad and Janet?

6. **Computer Activity** Is it possible for the midpoints of the three altitudes of a triangle to be collinear? Investigate by using a geometry computer program. Write a paragraph describing your findings.

7. It is the year 2060 on the lunar colony Galileo. The Director of Food Production for Galileo, Ima Gourmet, has been instructed to locate the food production facility that is equally distant from the three residential sections of the lunar colony. Which point of concurrency does Ima need to locate?

8.* Draw a large triangle. Construct a circle inscribed in the triangle.

9.* Draw a triangle. Construct a circle circumscribed about the triangle.

10. Notice that the circumcenter is in the interior of acute triangles and on the exterior of obtuse triangles. Use graph paper to draw a right triangle. Where is the circumcenter of a right triangle?

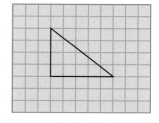

11. Notice that the orthocenter is in the interior of acute triangles and on the exterior of obtuse triangles. Use graph paper to draw a right triangle. Where is the orthocenter of a right triangle?

12. One event at this year's Battle of the Classes will be a pie-eating contest between the sophomore, junior, and senior classes. Five members of each class will be positioned on the football field at the points indicated below. At the whistle, one student from each class will run to the pie table, eat exactly one pie, and run back to his or her group. The next student will then repeat the process. The first class to eat five pies and return to home base will be the winner of the pie-eating contest. Where should the pie table be located so that it will be a fair contest? Describe how the contest planners should find that point.

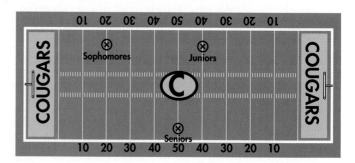

more than one way, using different tools (compass and straightedge versus patty papers, for example)? Look over your conjecture list. Can you perform all of the investigations that led to these conjectures? Demonstrate at least one construction and at least one investigation for a classmate, a family member, or your teacher. Do every step from start to finish and explain what you're doing.

Of course, performance assessment is just one way you can assess what you learned in this chapter.

Organize Your Notebook

- Your notebook should have two new components: an investigation section and a conjecture list. Review the contents of your notebook. Make sure it's complete, correct, and well organized.

- Write a one-page chapter summary from your notes.

Write in Your Journal

- How do doing constructions, looking for patterns, and making conjectures compare to other ways in which you've learned math in the past?

Update Your Portfolio

- If you did a project in this chapter, document your work and add it to your portfolio.

- Choose a construction problem from this chapter that you found particularly interesting and/or challenging. Describe each step, including how you figured out how to move on to the next step. Add this to your portfolio.

Improving Algebra Skills—*Pyramid Puzzle II and III*

Place four different numbers in the bubbles at the vertices of each pyramid so that the two numbers at the ends of each edge add to the number on that edge.

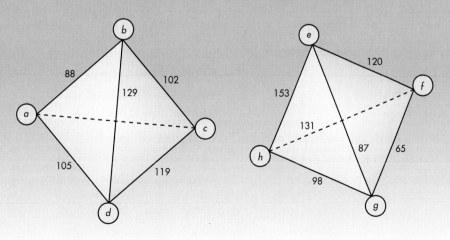

Cooperative Problem Solving

Lunar Survival

You have recently joined the lunar research team aboard the space station *Entropy*. On the way to the uninhabited Lunar Research Outpost, your space station transport vehicle malfunctioned. This forced you and the crew to make an emergency landing on the lunar surface, 150 kilometers from your intended destination. All of your communication devices are inoperable. After the emergency, the crew was able to retrieve from the transport vehicle a number of items (listed below). Because you are at least 1000 kilometers from the lunar space port Galileo, survival depends on the crew reaching the Lunar Research Outpost. Your first survival task is to rank the listed items in terms of their importance to the crew's ability to reach the outpost.

Copy the list of items below onto a clean sheet of paper. Without discussion with fellow crew members, rank the items on your list. Then discuss with your crew the relative importance of each salvaged item. After a thorough group discussion, rank the items by placing the number 1 by the most important item, number 2 by the second most important item, and so on through number 14, the least important. Finally, to the right of each item, give a reason for your ranking. According to NASA scientists, there is a correct ranking.

- Solar-powered FM receiver-transmitter
- Signal flares
- Box of matches
- 20 meters of nylon rope
- First-aid kit containing injection needles
- Two 50-kilogram tanks of oxygen
- Parachute silk
- Food concentrate
- Portable heating unit
- One case dehydrated milk
- 20 liters of water
- Magnetic compass
- Life raft
- Star map of moon's constellations

Line and Angle Properties

Still Life and Street, M. C. Escher, 1937
©1996 M. C. Escher / Cordon Art – Baarn – Holland. All rights reserved.

In this chapter you'll investigate properties of lines and angles. You'll make conjectures about vertical angles, linear pairs of angles, parallel and perpendicular lines, midpoints, and more. In many of the lessons you'll explore these properties in the coordinate plane. You'll see how some of the algebra you've already learned applies to geometry. In the last lesson of the chapter you'll discover connections between geometry and probability.

Lesson 4.1

Discovering Angle Relationships

Discovery consists of looking at the same thing as everyone else and thinking something different.
— Albert Szent-Györgyi

Before you begin this lesson, review the definitions of the following pairs of special angles. The wording of these definitions may not be exactly the same as the wording you used in Chapter 2, but the idea is the same.

*Two angles are **complementary angles** if their measures add to 90°.*

*Two angles are **supplementary angles** if their measures add to 180°.*

*If $\overleftrightarrow{PQ}$ intersects $\overleftrightarrow{RS}$ at point T, then ∠PTS and ∠RTQ are **vertical angles**. Angles PTR and STQ are also a pair of vertical angles.*

*If $\overleftrightarrow{PQ}$ intersects $\overleftrightarrow{RS}$ at point T, then ∠PTS and ∠STQ form a **linear pair of angles**. Three other linear pairs of angles are also formed.*

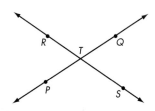

Investigation 4.1.1

On a clean sheet of paper, draw two lines *PQ* and *RS* intersecting at point *T,* as shown. With a protractor, carefully measure all four angles and record their measures.

$$m\angle PTS \approx \text{--?--} \qquad m\angle RTQ \approx \text{--?--}$$
$$m\angle PTR \approx \text{--?--} \qquad m\angle STQ \approx \text{--?--}$$

Do you notice anything special about the measures of the pair of vertical angles ∠PTS and ∠RTQ? What about vertical angles ∠PTR and ∠STQ?

Try this again. On a sheet of paper, draw two lines *JK* and *LM* intersecting at point *N*. Measure all four angles. Compare your results with the results of others near you. State your findings as a conjecture.

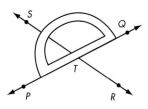

 C-15 If two angles are vertical angles, then —?— (***Vertical Angles Conjecture***).

What about a linear pair of angles? Are they equal in measure? Not necessarily. Are they related at all? What about the sum of their measures?

Investigation 4.1.2

Find the sum of the measures of each linear pair of angles formed by the intersecting lines $\overleftrightarrow{PQ}$ and $\overleftrightarrow{RS}$ that you drew in Investigation 4.1.1.

$m\angle PTS + m\angle STQ \approx$ –?– $m\angle PTR + m\angle PTS \approx$ –?–

$m\angle PTR + m\angle RTQ \approx$ –?– $m\angle RTQ + m\angle STQ \approx$ –?–

Find the sum of the measures of each linear pair of angles formed by the intersecting lines $\overleftrightarrow{JK}$ and $\overleftrightarrow{LM}$ that you drew in Investigation 4.1.1.

$m\angle JNM + m\angle MNK \approx$ –?– $m\angle JNL + m\angle JNM \approx$ –?–

$m\angle JNL + m\angle LNK \approx$ –?– $m\angle LNK + m\angle MNK \approx$ –?–

Compare your results with those of others near you. State a conjecture about angles that form a linear pair. Add it to your conjecture list.

C-16 If two angles are a linear pair of angles, then —?— (***Linear Pair Conjecture***).

By now you've had some practice using inductive reasoning and a variety of tools to perform investigations. In doing so, you've discovered patterns and properties and have made generalizations. Your work as a mathematician isn't done once you've made a conjecture, though. Many times you'll want to test or verify your conjecture by looking again for counterexamples or by performing the investigation in a different way, perhaps using different tools. You might also look back at your work to try to gain more insight into *why* your conjecture is true. Sometimes such insight can lead to further, related discoveries. Finally, when you make a discovery, you'll often want to communicate your findings to others. The section that follows, Take Another Look, gives you a chance to extend, to communicate, and to assess your understanding of the work you did in the investigation.

Take Another Look 4.1

The straightedge-and-protractor method you used in Investigations 4.1.1 and 4.1.2 is one way to confirm the Vertical Angles Conjecture and the Linear Pair Conjecture. Like most conjectures, there are a number of ways to convince yourself and others that the conjectures you made are true. Try one or more of these activities.

1. Use patty papers to demonstrate the Vertical Angles Conjecture in less than thirty seconds!

2. Use a geometry software program to investigate the Vertical Angles Conjecture or the Linear Pair Conjecture.

3.* Use algebra to show how the Vertical Angles Conjecture follows logically from the Linear Pair Conjecture.

Exercise Set 4.1

The small letters in the figures below represent angle measures. Without using a protractor, but with the aid of your two new conjectures, determine each lettered angle measure. You may find it helpful to trace the diagrams onto a sheet of paper so that you can write on them. You do not need to find the angle measures in alphabetical order, but list your answers in alphabetical order to make them easier to check.

1.

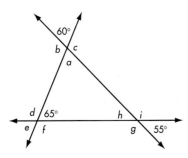

2.

3.

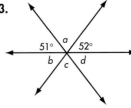

4.*

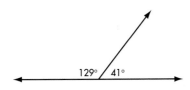

5.

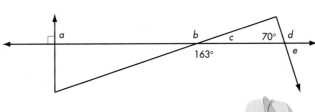

6. The figure at right shows a force applied to a tooth by braces. The force is represented by the blue arrow, called a vector, and it can be broken into horizontal and vertical components. The force shown here would pull the tooth up and in. If the force vector makes a 55° angle with its vertical component, what angle does it make with its horizontal component?

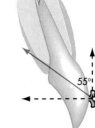

7. What's wrong with this picture?

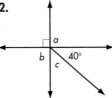

8. A tree on a 30° slope grows straight up. What are the greatest and smallest measures of the angles the tree makes with the hill? Explain.

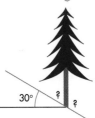

9. Yoshi is building a cold frame for his plants. He wants to cut two pieces of wood strips so that they'll fit together to make a right-angled corner. What can you say about the angles at which he should cut the strips?

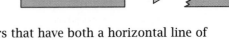

10.* From the alphabet shown below, list the letters that have both a horizontal line of symmetry and a vertical line of symmetry.

A B C D E F G H I J K L M N O P Q R S T U V W X Y Z

11. *From the alphabet shown, list the letters that have 180° rotational symmetry.

12. *If everyone in the town of Skunk's Crossing (population 84) has a telephone, how many different two-way phone conversations can take place in the town?

13. While looking out his front window, Tanakki sees a bus pass by the front of the house, heading towards the setting sun. His house is to the bus driver's left. Which direction is Tanakki facing?

14. Show how to cut a 24-cm piece of ribbon into three pieces so that the longest piece is six times as long as two shorter pieces of equal length.

15. If $\overrightarrow{BD}$ is the angle bisector of $\angle ABC$, and if $\overrightarrow{BE}$ is the angle bisector of $\angle ABD$, what is the $m\angle ABC$ if $m\angle DBE = 36°$?

16. *Sketch the locus of points on the coordinate plane for which the *x*-coordinate minus the *y*-coordinate is equal to 4.

17. If two congruent angles are supplementary, what must be true of the two angles? Make a sketch, then complete the following conjecture: If two angles are both congruent and supplementary, then —?—.

18. *Use algebra to explain why the conjecture from Exercise 17 is true.

19. **Group Activity** In Chapter 3, you discovered how to fold paper to get the perpendicular bisector of a segment. Can you explain why this works? Copy and complete this explanation: The two segments into which the fold divides the original segment are —?— because they coincide. Therefore the crease bisects the segment. Pairs of angles formed by the segment and the crease are congruent because they —?—. The sum of the measures of each linear pair of angles is —?—. Thus each angle measures —?—. If each angle measures —?—, then the crease is —?— to the segment. Therefore the line in the crease is a —?— of the segment.

Improving Algebra Skills—*Number Line Diagrams*

1. Translate the number line diagram into an equation, then solve for the variable.

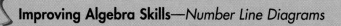

2. Translate the equation into a number line diagram.

    $2(x + 3) + 14 = 3(x - 4) + 11$

Lesson 4.2

Discovering Properties of Parallel Lines

That's the way things come clear. All of a sudden. And then you realize how obvious they've been all along.
— Madeleine L'Engle

*Two or more lines are **parallel** if and only if they are in the same plane and do not intersect.*

A line intersecting two or more other coplanar lines is called a **transversal**. A transversal that intersects two lines creates different types of angle pairs. Three types are listed below.

One pair of **corresponding angles (CA)** is ∠1 and ∠5. Can you find three more pairs of corresponding angles?

One pair of **alternate interior angles (AIA)** is ∠3 and ∠6. Do you see another pair of alternate interior angles?

One pair of **alternate exterior angles (AEA)** is ∠2 and ∠7. Do you see another pair of alternate exterior angles?

When parallel lines are cut by a transversal, there is a special relationship among the angles. Let's investigate.

Investigation 4.2.1

Using the lines on a piece of graph paper or lined paper as a guide, draw a pair of parallel lines. If you don't have ruled paper, use both edges of your ruler or straightedge to create parallel lines. Now draw a transversal that intersects the parallel lines. Label the angles with numbers, as shown in the figure at top right.

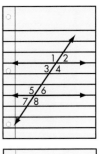

Place a patty paper over the set of angles ∠1, ∠2, ∠3, and ∠4 and copy the two intersecting lines onto the patty paper.

Slide the patty paper down and compare angles 1 through 4 with angles 5 through 8.

Do you notice a relationship among the measures of corresponding angles, alternate interior angles, or alternate exterior angles? Compare your results with the results of others near you. State your next conjecture.

13. If you've watched any submarine movies, you've seen a periscope. It is the optical instrument that permits a sailor to see above the surface of the ocean.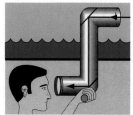

   The periscope shown at right is designed so that the line of sight through the bottom tube is parallel to the light ray at the top tube. The middle tube is vertical (perpendicular to the top and bottom tubes). What are the measures of the incoming and outgoing angles formed by the light rays and the mirrors in this periscope? Are the surfaces of the mirrors parallel? How do you know?

In Exercises 14 and 15, use patty papers to construct each figure.

14. Copy the triangle shown below, then locate the centroid.

15. Copy the triangle shown below, then locate the point that is equally distant from all three sides.

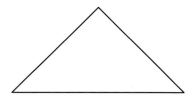

16. Electra Watt is planning to build a battery recharging station somewhere between three intersecting (nonconcurrent) electric car routes. At which point of concurrency should she build so that the station is equally distant from the three routes?

17. Sketch a rectangular solid 2 cm by 3 cm by 4 cm, resting on its smallest face with an edge facing the viewer.

18. Solve for *y*.
   $2x + y = 8$

19. Solve for *y*.
   $8x - 2y = 12$

20. How many squares of all sizes are in an eight-by-eight grid of squares? Count all the one-by-one squares, all the two-by-two squares, etc. Don't forget to count the entire eight-by-eight square.

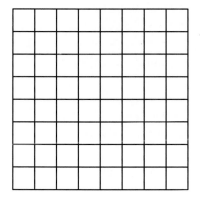

21. The coiled basket shown below is from the Pima, a Native American tribe of New Mexico. How many degrees of rotation are there in each of the basket's rotational symmetries?

Project

"Lines" on a Sphere

We're so used to looking at world maps that it's easy to forget how distorted things can get when our spherical earth is shown on a flat (planar) piece of paper. There are various methods (called projections) for transferring the information on a globe to a map, but they all involve cutting and distortion. One effect of this process is that areas nearer the North and South poles are often greatly enlarged. Greenland, for example, looks far less impressive on a globe than it does on many maps. Some maps, like the one shown below, compensate by squishing polar regions, but that's just another kind of distortion.

Another effect of flattening the globe onto a map is that shortest distances between two locations can be deceiving. Looking at the map below, you might think the shortest distance between Orlando, Florida, and Katmandu, Nepal, is the straight line shown in blue. Likewise, you might think that the route between Bogota, Columbia, and Colombo, Sri Lanka, is the more or less parallel straight line, also shown in blue. But the actual routes that a plane flying the shortest distance would choose are the curves shown in red. Why on earth would a plane flying from Orlando to Katmandu travel over Greenland? To answer that question, you'll need to look at a globe.

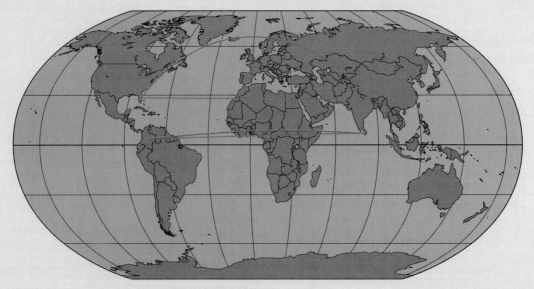

In this project, you'll explore the concepts of "straight lines" and parallels on a sphere. You'll need a globe and a piece of string long enough to go at least halfway around the globe.

1. To find the shortest path between Orlando and Katmandu, hold one end of the string at Orlando, pull tight, and position the string so that it also goes through Katmandu. The tight string shows the shortest path. Now can you see why an airplane flying from Orlando to Katmandu would fly over Greenland? Write a paragraph explaining why a plane would not necessarily fly due east from a location to a destination that is due east of that location.

2. The shortest path (also called a **geodesic line**) between two points on a sphere always lies along what's called a **great circle**. Great circles on a sphere are analogous to lines in the plane. Write a sentence that describes in your own words what a great circle is.

3. Suppose the plane traveled from Orlando to Katmandu, then continued on this path around the world. Suppose another plane traveled around the world on the path between Bogota and Colombo. Would their paths cross? Are there any two around-the-world routes that don't cross?

4. Suppose on a sphere you're given a "line" (great circle) and a point not on the line. How many lines that are parallel to the given line pass through the point? How is a sphere different from a plane in this respect?

Improving Reasoning Skills—*Bagels II*

In the original computer game of bagels, a player determines a three-digit number (no digit repeated) by making "educated guesses." After each guess, the computer gives a clue about the guess. Here are the clues.

bagels: no digit is correct
pico: one digit is correct but in the wrong position
fermi: one digit is correct and in the correct position

In each of the problems below, a number of guesses have been made, with the clue for each guess shown to its right. From the given set of guesses and clues, determine the three-digit number. If there is more than one solution, find them all.

1. 1 2 3   *pico*
    4 5 6   *pico*
    7 8 9   *pico*
    9 4 1   *bagels*
    3 7 5   *pico*
    6 3 8   *pico*
    ? ? ?

2. 1 9 8   *pico fermi*
    7 6 5   *bagels*
    4 3 2   *pico*
    1 2 9   *pico fermi*
    ? ? ?

Lesson 4.3

Midpoint and Slope Conjectures

Some of us are timid. We think we have something to lose so we don't try for the next hill.
— *Maya Angelou*

Surveyors and mapmakers of ancient Egypt, China, Greece, and Rome used various coordinate systems to locate points. Egyptians in particular made extensive use of square grids and used the first known rectangular coordinates at Saqqara around 2650 B.C. By the seventeenth century, the age of European exploration, the need for accurate maps and the development of easy-to-use algebraic symbols gave rise to modern coordinate geometry. Seventeenth-century French philosopher and mathematician René Descartes (pronounced "day cart") is credited with the development of this new system, which is called Cartesian coordinate geometry. (*Cartesian* comes from the Latin form of Descartes's name.)

In algebra you graphed points and lines on the Cartesian coordinate system. So far in this text you have practiced graphing points in the exercise sets. In your algebra class, you were probably given a rule for finding the midpoint of a segment and another rule for finding the slope of a line. In this lesson you'll rediscover these rules. If you already know them, the following investigations should help you better understand the rules. Let's take a look.

How would you find the coordinates of the midpoint of a segment?

Seventeenth-century map

Investigation 4.3.1

Case 1

Find the midpoints of $\overline{AB}$ and $\overline{CD}$.

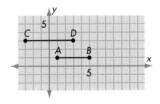

Case 2

Find the midpoints of $\overline{EF}$ and $\overline{GH}$.

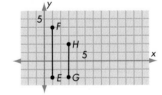

In Cases 1 and 2, the lines are either horizontal or vertical, therefore the midpoint is not difficult to find. What if a segment is slanted with respect to the axes and you know only the coordinates of its endpoints?

Estimate the coordinates of the midpoints of $\overline{IJ}$ and $\overline{KL}$. How would you find the coordinates of

Case 3

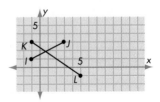

Project

Best-Fit Lines

Why is it important to learn about equations for lines and about the slope-intercept formula? You've already seen a few examples of linear relationships in the real world: When a bicyclist travels at a constant velocity, the graph of her distance traveled plotted against time is a straight line. The relationship between temperature in degrees Celsius and temperature in degrees Fahrenheit is also linear. Linear relationships arise in many fields of research, from economics to biology: cost increases versus time, amount of education versus income, life span of a population of fish versus concentrations of pollutants in the water. Most real-world data doesn't lie on a perfectly straight line, but if the data is approximately linear, a "best-fit" line can be found.

Finding the slope of the line that best fits a set of points allows us to predict values that we cannot easily measure. Researchers measure values for the *x*- and *y*-axes, plot their data, and determine if there is a linear relationship between the two variables. When you hear on the news, for example, that smoking causes cancer or that researchers predict that the cost of college tuition will rise, scientists and researchers are making predictions by using real data and plotting it on a graph. If the points appear close to linear, the researchers fit a line to their points and determine the equation of the line. The results of these studies are used to make decisions about the future or to determine if a particular change in a situation seems to make a difference.

There are statistical methods for finding precise best-fit lines. In this project you'll find approximate best-fit lines by drawing lines that appear to closely fit data. The example below shows a linear relationship between the age of a child and his or her height.

The table below represents the average height in centimeters of children in the United States between the ages of seven and fifteen.

Age	7	9	11	12	13	14	15
Height	119.3	132.0	142.2	147.3	152.4	157.5	162.2

Complete problems 1–7 on your own and compare your results with group members. Then work with your group on the project described in problem 8.

1. On a graph, plot height on the *y*-axis and age on the *x*-axis, using data from the table above.

2. Sketch a line that appears to most closely fit the points.

3. Locate two points on the line and use the Slope-Intercept Conjecture to find an equation for the line.

4. Use your equation to predict the average height of a ten-year-old child. (In other words, use the equation you derived for your line and find y when $x = 10$.)

5. How does your prediction compare to the real measured value of 137.1 cm as the average height of a ten-year-old?

6. What can you conclude about the equation you derived? Can it help you predict the average height for any given age?

7. Can you use your equation to predict the average height of forty-year-olds? How about two-year-olds? Explain.

8. **Research Project** Go to the library and find a source of ordered pairs of data. U.S. Census figures or a world almanac are good sources. Choose data that interests you, such as education versus income, income versus life expectancy, gross national product versus birth rates in different countries, etc. Graph the data to see if it appears linear. If it does, draw an approximate best-fit line and write its equation. Write a paragraph explaining what you can conclude from the graph.

9. **Group Activity** Work with your group to devise a study or an experiment for which you can gather ordered pairs of data. For example, survey friends and family to learn their heights and shoe size, or determine the distance from home to school (blocks) versus the time it takes to walk from one location to the other, or keep track of the hours you spend watching television versus the hours you spend studying. As a group, decide on what data interests you. Each group member should collect data. Plot all the data collected by your group. If the data appears to be linear, draw an approximate best-fit line and find its equation. Discuss your findings with your group members and write an explanation of what the slope of this line represents. If your data does not appear to be linear, discuss theories with your group members about why it might not be linear.

Improving Visual Thinking Skills—*Rotating Gears*

In what direction will gear E rotate if gear A rotates in a counterclockwise direction?

Lesson 4.7

Geometric Probability

We cannot leave the haphazard to chance.
— N. F. Simson

You've just been visiting a friend in jail. Now your only hope of success is buying property on St. James Place. To do that, you'll need a roll of six on your next turn! (That's right, we're on the Monopoly board.) Will you succeed? You ask yourself, "What's the probability of rolling a six on my next roll?"

What are the chances? What are the odds? What's the probability? You've heard these expressions many times and have *probably* used them yourself. Probability theory, an important branch of mathematics, attempts to answer questions like these. Probability is important to people in many occupations, from insurance agents to professional backgammon players. Probability is a measure of the likelihood that an event will have a particular outcome. The **probability** of an outcome is the ratio of the number of ways in which that particular outcome can occur (successful outcomes) to the total number of equally likely possible outcomes.

$$\text{probability} = \frac{\text{number of successful outcomes}}{\text{number of possible outcomes}}$$

A probability is always a number between 0 and 1 inclusive. The ratio tells the likelihood of a successful outcome. A probability of 0 means that there is no chance of a successful outcome. A probability of 1 means that a successful outcome is certain.

Let's look at a few examples of probability problems.

Example A

What is the probability of rolling a 6 with one toss of a single die?

Because there are six possible outcomes (1, 2, 3, 4, 5, and 6) and only one 6, the probability of rolling a 6 is $\frac{1}{6}$. In fact, each of the numbers 1 through 6 has an equal probability of $\frac{1}{6}$.

Example B

What is the probability of the spinner landing on one of the colored sectors?

The angle measures of the colored sectors total 120°, or $\frac{120}{360}$, of the complete circle. Therefore the probability of the spinner landing on either color is $\frac{1}{3}$.

Example C

What is the probability of rolling a sum of
6 with a pair of dice?

Let's organize our counting by setting
up a table.

Roll (total on both dice)	Ways of getting total	Number of ways to get total
2	(1, 1)	1
3	(1, 2) (2, 1)	2
4	(1, 3) (2, 2) (3, 1)	3
5	(1, 4) (2, 3) (3, 2) (4, 1)	4
6	(1, 5) (2, 4) (3, 3) (4, 2) (5, 1)	5
7	(1, 6) (2, 5) (3, 4) (4, 3) (5, 2) (6, 1)	6
8	(2, 6) (3, 5) (4, 4) (5, 3) (6, 2)	5
9	(3, 6) (4, 5) (5, 4) (6, 3)	4
10	(4, 6) (5, 5) (6, 4)	3
11	(5, 6) (6, 5)	2
12	(6, 6)	1

By organized counting we see that there are five ways to roll a total of 6 and there are
36 total rolls possible. Therefore the probability of rolling a 6 is $\frac{5}{36}$.

Example D

What is the probability of randomly selecting one of the longer
diagonals from among all the diagonals of a regular hexagon?

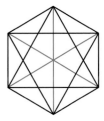

By organized counting you can see that there are three longer
diagonals and six shorter diagonals, for a total of nine diagonals.
Therefore the probability of selecting one of the longer diagonals
is $\frac{3}{9}$ or $\frac{1}{3}$.

In many geometric probability problems, you will not be able to count the events.
Example E demonstrates a problem that involves more than organized counting.

Example E

Given $\overline{AB}$ with midpoint M, what is the probability of randomly selecting a point on the
segment closer to point A or to point B than to point M?

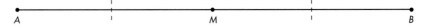

All the points to the left of the midpoint of $\overline{AM}$ are closer to point A than to point M—
these points make up one fourth of the entire segment. All the points to the right of the
midpoint of $\overline{MB}$ are closer to point B than to point M—this is also one fourth of the
entire segment. Therefore the probability of randomly selecting a point closer to either
points A or B than to point M is $\frac{1}{2}$.

Exercise Set 4.7

After you've finished your homework and have eaten dinner, you're offered the following game of chance, which features the spinner shown at right.

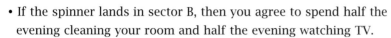

- If the spinner lands in sector A, then you agree to spend the rest of the evening cleaning your room.

- If the spinner lands in sector B, then you agree to spend half the evening cleaning your room and half the evening watching TV.

- If the spinner lands on sector C, you're free to spend the rest of the evening reading.

- If the spinner lands on sector D, then you can listen to music as loud as you want for the entire evening.

- You can't do more than one thing at once.

Assuming that the spinner is fair and that you agree to the rules of the game, determine the probability in Exercises 1–4.

1. What is the probability that you will get to listen to loud music all evening?

2. What is the probability that you will either get to read or listen to music all evening?

3. What is the probability that you'll agree to clean your room for at least half of the evening?

4. What is the probability that you will not have to spend the whole evening cleaning your room?

When possible, use organized counting to solve these probability problems. When it is not possible to use organized counting, try drawing diagrams.

5. The faces of the octahedral die shown at right are numbered 1 through 8. If each face is equally likely to appear on top of this die when thrown, what is the probability of rolling a sum of 10 with a pair of octahedral dice?

6. If two of the points $L, I, N, E, A,$ and R are selected at random, what is the probability that the distance between them will be exactly one unit?

7. If two of the points $R, O, U, N,$ and D are selected at random, what is the probability that the two points will be consecutive (next to each other on the circle)?

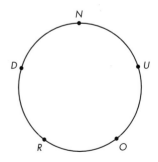

8.* What is the probability of randomly selecting one of the longest diagonals from among all the diagonals in a regular octagon?

9. What is the probability of randomly selecting one of the shortest diagonals from among all the diagonals of a regular octagon?

10. Given $\overline{AB}$, what is the probability of randomly selecting a point on the segment closer to point A than to point B?

11. Given $\overline{AB}$ with midpoint M, what is the probability of randomly selecting a point on the segment closer to point M than to point A?

12.* Given $\overline{AB}$ with midpoint M, what is the probability of randomly selecting a point on the segment closer to point M than to either point A or point B?

13.* If three different points are selected randomly from six equally spaced points on a circle, what is the probability that the triangle formed by connecting these three points will be equilateral?

14.* Adventurer Dakota Davis is carefully walking across an ancient 120-foot long rope bridge that is stretched across a canyon whose walls rise 90 feet above a river. If the rope breaks where he is holding it (a random point), what is the probability that one end of the rope bridge will hit the piranha-infested waters below?

15. In the game of Igba-ita, played by the Ibo people of Africa, players toss handfuls of cowrie shells into a circle to see if they land right side up or upside down. The outcome depends on the combinations of up and down cowries. Suppose a cowrie is equally likely to land up or down. What is the probability that four cowries tossed into the circle will all land the same way (up or down)? What is the probability that three land one way and the fourth lands the other way?

Determine each lettered angle measure shown below.

16.

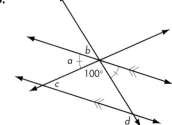

17.

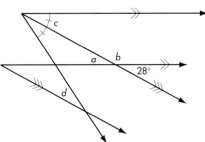

Project
Making a Clinometer

If I have seen farther than others it is because I have stood on the shoulders of giants.
— Isaac Newton

One of the most important applications of angle measurement comes from navigation by way of astronomy. Sailors at sea, where there are no visible landmarks, depend on the location of stars or the sun to indicate where they are. For example, in the Northern Hemisphere, Polaris (the North Star), stays more or less fixed at the same angle above the horizon for a given latitude. By noting whether Polaris appears higher overhead or closer to the horizon, sailors can tell whether their course is taking them north or south.

The earliest known navigation tool was used by the Polynesians. It was an extremely simple tool, and, in fact, it didn't measure angles. Early Polynesians carried several different length hooks made from split bamboo and shells. Navigators simply held a hook at arm's length, positioned the bottom of the hook at the horizon line, and sighted Polaris

through the top of the hook. The length of the hook required to sight the star measured the star's height relative to the horizon and told the navigators their approximate latitude. Early Polynesians were remarkable navigators. In addition to these hooks, they relied on much skill and knowledge to find their way among the islands scattered throughout the South Pacific.

A clinometer is related to other early tools used to measure the location of stars. It is a protractor-like tool used to measure an angle of inclination (also called angle of elevation), the angle at which an object or a slope rises above horizontal. Clinometers are useful in many areas besides navigation. In fact, because a clinometer relies on a plumb line, it doesn't work too well at sea unless it is very calm. United States Forest Service Ranger Al Sousi is shown using a clinometer on his compass to measure the angle of a mountain slope. In many snow conditions, a slope steeper than 35° can pose a high avalanche hazard. You'll learn about many more applications of a clinometer in Chapter 13 when you use your clinometer to indirectly measure heights.

In this project, you'll make a clinometer and use it to measure angles of inclination. Two types of clinometers are described here.

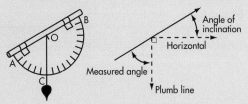

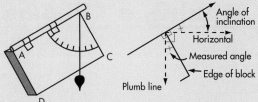

One type of clinometer uses a protractor with a viewing tube. The tube is attached to the zero-edge ($\overline{AB}$) and a plumb line is attached to the vertex point (O). Hold the device in such a way that when the top of the object is sighted through the viewing tube, the plumb line crosses the angle measurements on the protractor (C), forming an acute angle ($\angle AOC$). This angle is the complement of the desired angle.

Another type of clinometer is made from a rectangular block of wood or piece of cardboard. Align the zero-edge of the protractor with the front of the block and align the protractor's 90°-line with the top of the block. Suspend a plumb line from the front top corner. When you sight along the top edge, the angle between the plumb line and the front edge is equal to the desired angle.

Make one of these clinometers or one of your own design. Then use it to do one or more of these activities.

1. Find a number of different inclines to measure—hills, stairways, angles to high objects—but before you measure, try to guess what the angle is. When you're sighting along an incline, be sure to take the height of your eye into account. You may get more accurate measures if you crouch low. Compare your measurements to your guesses. Were your guesses too high or too low?

2. Write a paragraph explaining how your clinometer works. That is, explain how the angle formed by the plumb line is related to the angle of inclination when you use that type of clinometer. If you made the first type, why is this angle measured on the clinometer equal to the complement of the angle of inclination? If you made the second type, why is the measure angle on the clinometer equal to the angle of inclination?

3. **Research Project** Assuming you're in the Northern Hemisphere, sight Polaris with your clinometer and measure its angle of inclination. You'll need to find a star chart. Look for one in astronomy books at the library. Polaris is only moderately bright and is found in the Little Dipper constellation. The handle of the Big Dipper points to Polaris. Choose another easy-to-identify star far from Polaris and measure its angle of inclination. Measure both stars again at least two hours later. (You might not want to do this on a school night!) Has the angle for either star changed? Record your measurements.

4. **Group Activity** Find a globe or a map that shows latitude lines and find the approximate latitude where you live. How does it compare to the angle of inclination you measured for Polaris? Correspond with students who live at different latitudes and describe your experiment. Ask them to measure the angle of inclination of Polaris at their latitude. Where would someone living near the equator see Polaris? How about someone living near the North Pole?

Lesson 4.8

Chapter Review

If it takes one kid 36 seconds to blow up a balloon, how long will it take four kids trying to blow up the same balloon?
— *Anonymous*

If a lot of this chapter seemed familiar to you, that's good! In this chapter you investigated (or rediscovered) line and angle relationships and discussed properties of segments and lines in coordinate geometry. Coordinate geometry is one of many places where algebra and geometry come together. That's why some of this chapter may be review from last year's math class. But you can always gain new insights and learn new applications by returning to a concept you've seen before.

There were ten conjectures in this chapter. How many can you state from memory? Making sketches might jog your memory. Once you've tried stating conjectures from memory, review your conjecture list to make sure it's complete and that you understand it.

Can you name all the types of angle pairs you learned about in this chapter that are congruent? How about angle pairs that are supplementary? Historically, what applications led to the development of coordinate systems? What are some applications of linear equations? When you solve two linear equations simultaneously, what does your solution represent? What are some examples of geometry that apply to probability? Review the chapter to make sure you can answer these questions.

Exercise Set 4.8

For Exercises 1–24, identify each statement as true or false. For each false statement, sketch a counterexample or explain why the statement is false.

1. If two angles are vertical angles, then they are congruent.

2. If two angles are a linear pair of angles, then they are congruent.

3. If two parallel lines are cut by a transversal, then corresponding angles are supplementary, alternate interior angles are supplementary, and alternate exterior angles are supplementary.

4. If two lines are cut by a transversal to form pairs of congruent corresponding angles, congruent alternate interior angles, or congruent alternate exterior angles, then the lines are parallel.

5. The x-coordinate of the midpoint of a segment is the average of the x-coordinates of the segment's endpoints.

6. If (a, b) and (c, d) are the coordinates of two points on a line, then the slope m of the line is given by the equation $m = \frac{d-b}{c-a}$.

7. The probability of randomly selecting a point closer to either endpoint than to the midpoint of a segment is $\frac{1}{2}$.

8. In a coordinate plane, two lines are perpendicular if and only if their slopes are reciprocals of each other.

9. In a coordinate plane, if s is the slope of the line and t is the y-intercept of the line, then the slope-intercept form of the equation of the line is $y = sx + t$.

10. When you make a conjecture from observations, you use inductive reasoning.

11. The total number of diagonals in a polygon of n sides is $n(n - 2)$.

12. *Point, line,* and *plane* are undefined terms in our geometry.

13. A protractor is a tool used to find the degree measure of an angle.

14. The three angle bisectors of a triangle intersect at the centroid.

15. If lines ℓ, m, and n are in the same plane, and $\ell \parallel m$ and $m \parallel n$, then $\ell \perp n$.

16. If lines ℓ, m, and n are not all in the same plane, but $\ell \parallel m$ and $m \parallel n$, then $\ell \parallel n$.

17. If lines ℓ, m, and n are in the same plane, and $\ell \perp m$ and $m \perp n$, then $\ell \parallel n$.

18. If lines ℓ, m, and n are not in the same plane, but $\ell \perp m$ and $m \perp n$, then $\ell \parallel n$.

19. If the two rays of one angle are parallel to the two rays of a second angle, then the two angles must be either congruent or supplementary.

20. If two angles are both congruent and a linear pair, then each angle must be a right angle.

21. If point A is $(0, 0)$, point B is $(3, 2)$, point C is $(6, 9)$, and point D is $(10, 3)$, then $\overleftrightarrow{AB} \perp \overleftrightarrow{CD}$.

22. An equation of the line through point $(0, 4)$ with a slope of $\frac{2}{3}$ is $y = \frac{2}{3}x + 4$.

23. If the measures of one pair of consecutive angles in a quadrilateral are 48° and 132°, then the quadrilateral is either a parallelogram or a trapezoid.

24. There are twice as many true statements as false statements in Exercises 1–24.

25. Find each lettered angle measure in the diagram shown below.

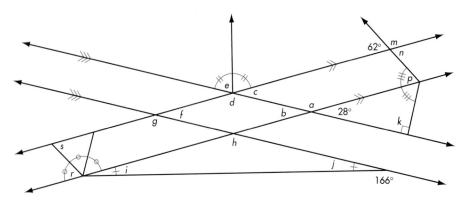

In Exercises 26–29, find the midpoint of the segment with the given points as endpoints.

26. $(6, 4), (12, -2)$ 27. $(5, -8), (-3, -8)$ 28. $(15, 0), (7, 5)$ 29. $(-3, 4), (6, 7)$

In Exercises 30–33, find the equation of the line through each pair of points.

30. $(0, 1), (1, 0)$ 31. $(2, 3), (-2, -3)$ 32. $(4, 5), (3, 5)$ 33. $(-1, -2), (-1, -1)$

In Exercises 34–37, determine whether the lines are parallel, perpendicular, or neither. If they intersect, find their point of intersection.

$A(1, 0)$ $B(3, 2)$ $C(5, 1)$ $D(8, 4)$ $E(3, 4)$

34. $\overleftrightarrow{AB}$ and $\overrightarrow{BC}$ **35.** $\overleftrightarrow{AB}$ and $\overleftrightarrow{CD}$ **36.** $\overleftrightarrow{AE}$ and $\overleftrightarrow{BC}$ **37.** $\overleftrightarrow{AE}$ and $\overleftrightarrow{ED}$

For Exercises 38–41, refer to the graph below right. The graph shows how a 200-ft steel bridge expands in hot weather and contracts in cold weather.

38. At what temperature does the bridge actually measure 200 feet?

39. About how long is the bridge when the temperature is 105°F?

40. Find the slope of the line.

41. If the steel bridge were 2000 ft long instead of 200 ft, about how much expansion should engineers plan for on a site where temperatures range from 0°F to 105°F?

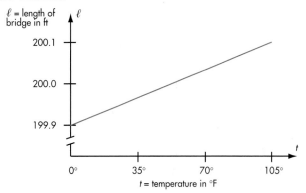

42. The vertices of $\triangle ABC$ have the coordinates $A(6, 0)$, $B(-1, 1)$, and $C(6, 8)$. Find the coordinates of the centroid and the circumcenter. Then find the equation of the Euler line.

43. Tickets for the school play cost $3 for students with student-body cards and cost $5 for all others. The school sold 234 tickets and collected $898. How many students with student-body cards bought tickets?

44. What's wrong with this picture?   **45.** What's wrong with this picture?

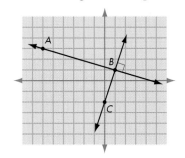   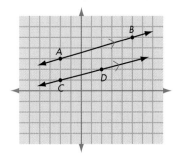

Assessing What You've Learned—*Open-Ended Investigations*

In Chapter 3 you learned how to use various tools to perform constructions and discover geometry. As you become more familiar with this process, you may be ready to devise investigations of your own or to pursue challenging extensions to the investigations. That's what the Take Another Look section is about. You'll find Take Another Look suggestions at the end of many lessons, starting with Lesson 4.1, just before the exercise sets. Your teacher may already have assigned some of them. You can use these open-ended investigations to assess the depth of your understanding of the guided investigations in the lesson.

Take Another Look activities can be quite challenging—you certainly wouldn't want to try all of them. In consultation with your teacher, choose one or more activities to try. Most of these activities call for an investigation or an explanation. Make sure you communicate clearly what you've done and what you've discovered, either in writing or in a presentation or performance assessment. If you write up your Take Another Look work, choose one or more pieces of work that you think are most significant and add them to your portfolio.

Whether or not you choose to add Take Another Look activities to your portfolio, don't overlook other assessment opportunities.

Organize Your Notebook

- However you've decided to organize your notebook, it should contain virtually all of your mathematical output for the year: homework, class notes, investigations, conjectures, definitions, and now, possibly, Take Another Look activities. Review it to be sure it's complete and well organized.

- Write a one-page summary for Chapter 4.

Write in Your Journal

- If you've done a Take Another Look activity, how did that experience differ from doing an investigation in a lesson? Did you work with a group or by yourself? Did you find it easier or harder? How do you think Take Another Look adds to your understanding of the concepts in the lesson?

- When you get stuck on a problem or get anxious about keeping up (and everybody does at some point!), what do you do about it? How do you get help if you need it, and from whom? What other help do you wish were available?

- Make a graph that represents your geometry learning over time, from the start of the year until now. Write an explanation of the behavior of your graph.

Update Your Portfolio

- Your work on an open-ended Take Another Look investigation or question is an excellent candidate for your portfolio. Choose one, document your work, and add it to your portfolio.

- Choose another piece of work to add to your portfolio: an investigation, a project, an exercise set, a test, whatever you think best represents the current quality of your work.

Performance Assessment

- Your teacher may actually want to observe you doing an open-ended investigation from one of the Take Another Look sections, especially if you worked in a group. Show your teacher what you did, explain why you're doing it that way, and describe your findings.

Cooperative Problem Solving

Geometrivia II

The geometry students at the lunar space port Galileo have just received a geometry puzzle called Geometrivia II from their earth friends at the NASA Academy in Atlanta, Georgia. The students at the NASA Academy enjoyed the Geometrivia puzzle they had received from their lunar friends at Galileo a few months ago, so they have reciprocated. The NASA Academy students have also been trained to learn in cooperative small groups, and Geometrivia II has been designed with that in mind.

Geometrivia II is another multistep problem-solving puzzle: In the Network Puzzle, find each lettered angle measure. The answers to the Search and Solve problems can be found in this chapter. Find the answers to the Library Research problems in the math section of your library. Substitute all the values you find into the formula at the end of this section. When all the variables in the formula have been replaced, calculate x. It is a whole number.

Network Puzzle

$a = $ -?-

$b = $ -?-

$c = $ -?-

$d = $ -?-

$e = $ -?-

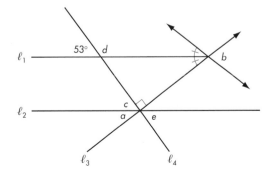

$\ell_1 \parallel \ell_2$

$\ell_3 \perp \ell_4$

Search and Solve

f The smallest answer from Exercise 8, Lesson 4.1

g Their compass bearing when the hiking family leaves camp

h The distance the bicyclist in Example D, Lesson 4.3, can travel in 1 hr 45 min

i The value of *z* in Exercise 6, Lesson 4.2

j The value of *x* if *y* = 31 in Exercise 7, Lesson 4.5

Library Research

k The number of books in the original set of Euclid's *Elements*

l The numerical value for the Egyptian hieroglyphic symbol ⋂ (heel bone)

m Age of a new octogenarian, in years, divided by the number of diagonals in a pentagon

n The year of Johannes Kepler's birth minus the year of Galileo Galilei's birth

p Goldbach's conjecture claims that every even integer greater than or equal to four can be represented as the sum of —?— prime numbers.

Formula

$$X = \dfrac{\left[\sqrt{\dfrac{lm-(k+d)}{\dfrac{f}{g}+\dfrac{c}{e}}}\right]^{\left(\frac{f}{h-g}\right)}}{\sqrt{\dfrac{(m-g)(b-f)}{(np-k)}+\dfrac{(d-i)}{j}+\dfrac{(m-np)}{(a-h)}}}$$

Triangle Properties

Symmetry drawing E103, M. C. Escher, 1959

In this chapter you will discover properties of triangles: the sum of the measures of interior angles and exterior angles of triangles, the inequality relationships between the sides and the angles, and the properties of isosceles triangles. You'll learn why triangles are rigid figures (this rigidity makes triangles so useful in structures). You will also discover the conditions that guarantee two triangles are congruent. Congruence is very important in our technological society. Modern assembly-line production relies on identical or congruent parts. In the assembly of an automobile, for example, parts must be produced so that any particular part can be used in the same location of any car coming down the assembly line.

Lesson 5.1

Triangle Sum Conjecture

Triangles have certain properties that make them useful in all kinds of structures, from bridges to high-rise buildings. One such property of triangles is their rigidity. Nail three boards together end to end and they won't sway, whereas four boards nailed end to end will sway and collapse. The rigid triangles in the bridge shown below are what give it its strength.

Another property of triangles is that identical triangles can be arranged to form parallel lines, as you can also see in the bridge. In this lesson, you'll discover that this property is related to the sum of the angle measures in a triangle.

Investigation 5.1.1

Step 1   On a clean sheet of paper, draw two large acute triangles and two large obtuse triangles.

Step 2   Measure the three angles of each triangle as accurately as possible with your protractor.

Step 3   Find the sum of the measures of the three angles in each triangle.

Compare your results with the results of others near you. Do you always seem to get close to the same result? What appears to be the sum of the three angle measures in every triangle? Before you make a conjecture, check the sum another way.

Step 4	Write the letters *a*, *b*, and *c* in the interiors of the three angles of one of the acute triangles, and carefully cut out the triangle.
Step 5	Tear off the three angles.
Step 6	Arrange the three angles so that their vertices meet at a point.

Step 4   Step 5   Step 6

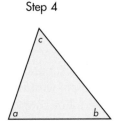   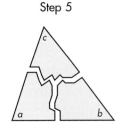   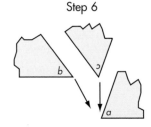

How does this arrangement show the sum of the angle measures? What is the sum of the measures of the three angles of any triangle? Compare your results with the results of others near you. State a conjecture.

 C-25 The sum of the measures —?— (***Triangle Sum Conjecture***).

Add this very important conjecture to your conjecture list and illustrate it. The next conjecture is an immediate result of the Triangle Sum Conjecture.

Investigation 5.1.2

If two angles of one triangle have the same measures as two angles of another triangle, and if the measures of all three angles sum to 180° in each triangle, what can you conclude about the third pair of angles?

Let's check this with patty papers. Draw a triangle on one patty paper. Create a second triangle on another patty paper by tracing two of the angles of your original triangle, but make the side between your new angles different in length. How do the third angles in the two triangles compare? Check with others near you. You should be ready for your next conjecture.

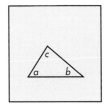

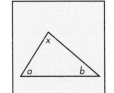

 C-26 If two angles of one triangle are equal in measure to two angles of another triangle, then the third angle in each triangle is —?— (***Third Angle Conjecture***).

Take Another Look 5.1

Try one or more of these follow-up activities to further investigate the Triangle Sum Conjecture and the Third Angle Conjecture.

1. Use a geometry computer program or patty papers to investigate the Triangle Sum Conjecture.

2.* Use algebra and write a paragraph explaining how the Third Angle Conjecture follows logically from the Triangle Sum Conjecture.

3.* Draw a triangle and construct a line through one vertex, parallel to the opposite side. How do the three angles formed at that vertex compare with the angles in the triangle? Explain how the Triangle Sum Conjecture follows logically from the Parallel Lines Conjecture.

4.* Explain why each angle in an equiangular triangle must measure 60°.

5.* Explore the Triangle Sum Conjecture on a sphere or a globe. Can you draw a triangle that has two or more obtuse angles? Three right angles? Write an illustrated report of your findings.

Exercise Set 5.1

Use the Triangle Sum Conjecture to determine each lettered angle measure. You might find it helpful to trace the more complicated diagrams.

1.* $x = $ –?–

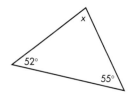

2. $y = $ –?–

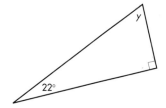

3.* $z = $ –?–

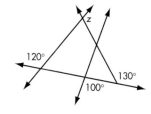

4. $w = $ –?–

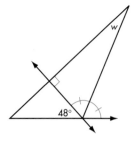

5.* Find the sum of the measures of the marked angles.

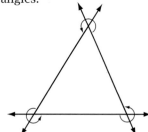

6.* Find the sum of the measures of the marked angles.

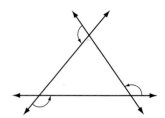

7.* $a = $ –?–
$b = $ –?–
$c = $ –?–
$d = $ –?–
$e = $ –?–

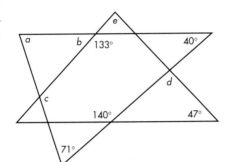

8. $m = $ –?–
$n = $ –?–
$p = $ –?–
$q = $ –?–
$r = $ –?–
$s = $ –?–
$t = $ –?–
$u = $ –?–

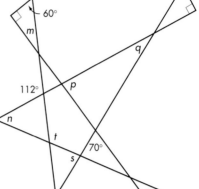

In Exercises 9–11, perform each construction. Remember, in geometric constructions you are not permitted to use a protractor or a ruler, only a compass and a straightedge. Arcs have already been constructed through the angles. You can use the arcs to duplicate the given angles without having to write in this book.

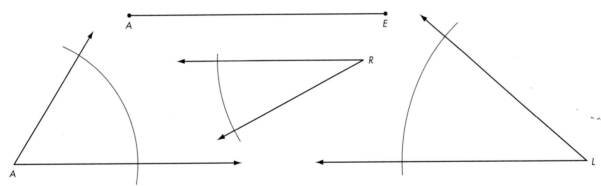

9.* Given ∠A and ∠R of △ARM, construct ∠M.

10.* In △LEG, ∠E and ∠G are equal in measure. Given ∠L, construct ∠G.

11. Given ∠A, ∠R, and side $\overline{AE}$, construct △EAR.

12. Repeat Exercises 9–11 with patty paper constructions.

In Exercises 13–15, find the equation of the line containing each segment indicated. The vertices of △ABC have coordinates $A(0, 0)$, $B(12, 8)$, and $C(6, 12)$.

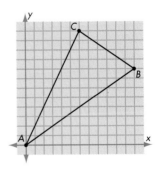

13. Median AM

14. Altitude BK

15. Perpendicular bisector of $\overline{AB}$

One story (two cards)

Two stories (seven cards)

Three stories (fifteen cards)

16. Fast-Fingered Francie is a blackjack dealer in Lost Wages, Nevada. While waiting for customers to arrive at her table, she builds houses of cards. What is the tallest house (number of stories) she can build with two decks of 52 cards? One-story, two-story, and three-story houses are shown above.

17. The graph at right shows changes in the global mean annual surface temperature. Despite the graph's many fluctuations up and down, you should notice an overall trend. Graph a line that approximates this overall trend. Use the slope of this line to estimate what the change in global temperature will be between the years 2000 and 2050 if this trend continues.

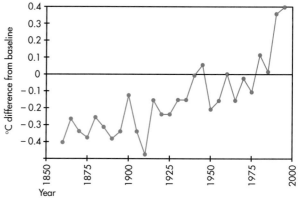

Source: Climate Research Unit, University of East Anglia, UK

18. In 1990, the United Nations Intergovernmental Panel on Climatic Change (IPCC) estimated that the world will, on average, be 3°C warmer by the year 2070. Is this prediction consistent with the trend you graphed in Exercise 17? Scientists believe climatic change will be influenced by a "greenhouse effect" caused by the emission of certain gases into the atmosphere. How might this theory explain the difference between the IPCC's prediction and the trend you graphed in Exercise 17?

Improving Visual Thinking Skills—*Dissecting a Hexagon I*

Trace the hexagon at right twice.

1. Divide the first hexagon into four identical parts. Each part will be a trapezoid.

2. Divide the second hexagon into eight identical parts.

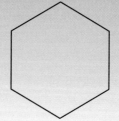

CHAPTER 5

Project

Building a Sextant

In 1699, Sir Isaac Newton conceived of the idea of using double reflection to measure angles. For some reason, he kept his idea a secret. It wasn't for another 30 years that others, most notably John Hadley, a London mathematician and instrument maker, rediscovered the idea and began to produce navigational tools that operated on the principle of double reflection. The most commonly known such tool is called a **sextant**.

A sextant, like the clinometer described in Chapter 4, is used by sailors to determine the angle between the horizon and a reference object—usually a star.

The advantage a double-reflection instrument like a sextant has over other angle-measuring tools is that, with a double-reflection tool, you can view a star by aligning it with the horizon. You don't have to try to look in two places at once. And, unlike a clinometer, a sextant doesn't rely on a plumb line, which swings uncontrollably in anything but the calmest of seas.

In this project you will discover the geometric principles involved in double-reflection instruments, then you'll make one of your own.

How Double Reflection Works

As you saw in Chapter 2, the angle of incidence (the incoming angle of a ray of light) in any flat mirror is equal to the angle of reflection (the outgoing angle). The diagram below demonstrates how the double-reflection principle works when two mirrors are parallel to each other. The solid lines represent light traveling from an object to an observer's eye, bouncing off two parallel mirrors. Use your knowledge of the properties of parallel lines to find the answers to problems 1 and 2.

1. Find the lettered angle measures.

2. Is angle measure *a* equal to angle measure *c*? What does that tell you about the relationship between the line to the object and the horizon line?

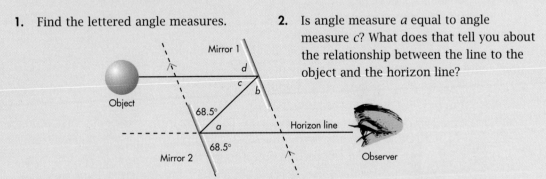

By turning mirror 1 so that it is no longer parallel to mirror 2, an observer can sight objects above the horizon. To find answers to problems 3 and 4, you'll use the Triangle Sum Conjecture and the fact that the angle of incidence is congruent to the angle of reflection.

3. In the figure below, the angle on the sextant measures 21°. Find the lettered angle measures, including y, the desired angle of elevation.

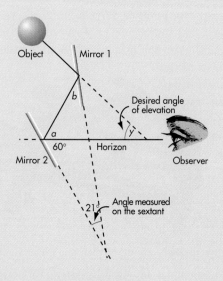

4. Is the angle of elevation always double the angle measured on the sextant? Use algebra to show that in the figure below, $y = 2x$.

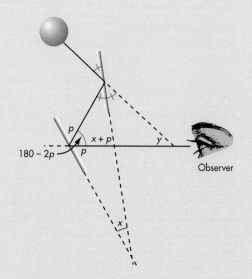

Building Your Own Sextant

On an actual sextant, the arc on which the angle measures are marked forms one sixth of a circle, hence this instrument's name. The "sextant" you'll make is a somewhat simpler double-reflection instrument. You will need two small mirrors, two rulers or sticks, tape, and a protractor. Put together your instrument as shown in the diagram at right.

Once you've built your instrument, practice using the two mirrors to sight objects above the horizon. Face the object whose angle of elevation you're going to measure. Hold the instru-

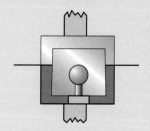

ment so that you can look into mirror 2 and see the horizon. Then adjust the mirrors so that you can see mirror 1 in mirror 2. Continue to adjust the mirrors until you can see the object in mirror 2. You're actually seeing the reflection of a reflection of the object. That is, light from the object is reflected off mirror 1 to mirror 2, then off mirror 2 and into your eye. Align the object with the horizon.

The protractor shows the angle between the two rulers. How does this compare to the object's angle of elevation? If you made a clinometer in Chapter 2, you can use it to compare angles measured with your sextant with angles measured with a clinometer. You can also use your sextant to measure angles other than angles of elevation. Try using it to measure the angle formed by lines of sight to other objects.

Write a short report about the angles you measured and how your sextant works.

Improving Visual Thinking Skills—*Alpha-ominos*

The alphabet can be divided into three physical types: "ups," "downs," and "middle roaders." Ups include *b, d, f, h,* and *k.* Downs include *g, j,* and *p.* Middle roaders include *c, i, m, n, r, s, v,* and *w.*

In the patterns at right, the letters of familiar geometric terms have been replaced by rectangles (for ups and downs) and squares (for middle roaders). Consonants are lightly shaded and vowels are solid black. For the purpose of this puzzle, *y* is always a vowel and all letters are lowercase. Use visual thinking to determine the familiar geometric terms represented by each pattern of rectangles and squares.

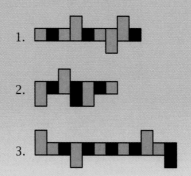

Improving Reasoning Skills—*Container Problem I*

You have an unmarked 9-liter container and an unmarked 4-liter container, and an unlimited supply of water. In table, symbol, or paragraph form, describe the process necessary to end up with exactly 3 liters in one of the containers.

Lesson 5.2

Discovering Properties of Isosceles Triangles

*An **isosceles triangle** is a triangle with at least two sides of the same length.*

In an isosceles triangle, the angle between the two sides of equal length is called the **vertex angle**, and the other two angles are called the **base angles**. The side between the two base angles is called the **base** of the isosceles triangle.

In this lesson you are going to construct isosceles triangles and investigate the measures of the base angles. The investigations work best in small cooperative groups of four or five people. If you are working in groups of four, pair off. One pair will use isosceles triangles that have acute vertex angles, and the other pair will use isosceles triangles with obtuse vertex angles. This way of working is called *pair share*. After your pair has performed your investigation, share your results with the other pair.

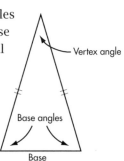

The faces of the Transamerica Pyramid in San Francisco, California, are isosceles triangles.

Investigation 5.2.1

Step 1	Step 2	Step 3	Step 4
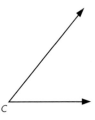		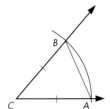	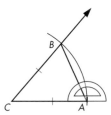
Draw an angle. Label the vertex *C*. This angle will be the vertex angle of your isosceles triangle.	Place the pointer of your compass on point *C* and swing an arc passing through the two sides of ∠*C*.	Label the two points *A* and *B*. Construct $\overline{AB}$. You have constructed isosceles △*ABC* with base $\overline{AB}$.	Use your protractor to measure the base angles of isosceles △*ABC*.

Compare your results with the results of the other pair in your group. What relationship do you notice between the base angles of each isosceles triangle? State your observations as your next conjecture.

Project

Triangles at Work

Throughout this chapter you have studied the properties of triangles. Because the triangle is rigid it is used to add strength to structures. But the triangle is also used in mechanisms that move by changing the length of one side. The cranes at left are one example. Lengthening or shortening the top cable lowers or raises the boom. In this project you are going to investigate the properties of a few mechanisms that take advantage of the rigidity of the triangle while allowing one side to vary in length. Dump trucks, reclining deck chairs, and a car jack are the three devices that you will investigate in this project. To build models of these devices you will need Geostrips® or cardboard strips with paper fasteners, or small wood strips with nuts and bolts.

Modeling a Dump Truck

To unload a dump truck, the driver must tilt the bed of the truck so that the contents will slide out. A dump truck operates with a bed that begins in a horizontal position but then rotates so that the truck's load will slide off the back of the now-tilted bed. On such a truck, the steepness of the bed's angle (∠A) depends on the length x of

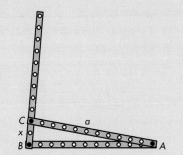

the hydraulic ram that extends to push up the bed. The more friction there is between the surface of the bed and the contents of the dump truck, the greater the measure of ∠A must be before the contents will slide out. The driver can control how much the hydraulic ram gets extended (the value of x) and can stop the ram's movement when the truck's load slips out.

To model the action of the hydraulic ram, set up three of your sticks with paper fasteners as shown in the figure. Adjust x (the number of spaces between your fasteners at points B and C), and carefully measure ∠A with your protractor for each value of x.

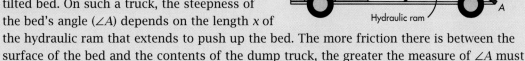

Enter your results into a table like the one below. Use your table to make a graph of $m\angle A$ versus x. Draw a smooth curve connecting the points. You will use your graph to answer the questions below.

x	2	3	4	5	6	7	8	9	10
$m\angle A$	-?-	-?-	-?-	-?-	-?-	-?-	-?-	-?-	-?-

1. What is the greatest value $m\angle A$ can attain for your dump truck model? What length should your hydraulic ram be to achieve that angle?

2. Using your graph, answer the following: About how long must the ram be extended in order to dump (1) a load of bricks (lots of friction with the bed of the truck) that require a 45° angle before they will slide out, (2) a load of sand that will slide out when a 20° angle is achieved, or (3) water.

3. Suppose the ram can only extend to a length of 4 units. Can the dump truck successfully dump a load of wood that requires an angle of 30° to slide out? Explain.

Modeling a Reclining Deck Chair

Deck chairs are designed to support a person's upper body at different angles, so the rigid but variable triangle is again used. To adjust the angle of the back of the chair, the person lifts the lower end (point C) of the support strut ($\overline{BC}$) out of the slot along the base of the recliner ($\overline{AC}$). It would be nice to have set positions along the base of the chair so that the back of the chair can be reclined at different angles in equal increments. Does this mean the slots along

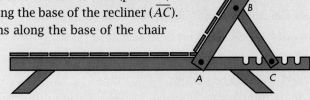

the base should be spaced equally? Let's look and see. Use your sticks to model a chair with an adjustable reclining angle. In your model, let $AB = CB = 4$ units. Move point C and use your protractor to measure reclining angles ($\angle BAC$) of 15°, 30°, 45°, 60°, and 75°. What values of AC correspond to these angles? If you find the spacing is no longer equal, use a ruler and measure the length of x in centimeters for each angle. Make a graph of your results.

AC	-?-	-?-	-?-	-?-	-?-
$m\angle A$	15°	30°	45°	60°	75°

1. Use your graph to predict what value x should be to make an angle of 20°. What about 40°?

2. Compare your results with the dump truck model. Write a brief description that explains the difference between changing the angle at regular intervals and changing the length of one side at regular intervals.

3. Find another tool or piece of equipment that uses a changing triangle to form a support. Make a diagram that explains the way it works, and model it with your sticks.

Modeling a Car Jack

A car jack is used to raise one corner of a car, usually so that a tire can be changed. There are a variety of different types of car jacks. The one shown at right is a rhombus with a variable diagonal that forms two triangles. The diagonal in the jack is a threaded bolt that can be made shorter or longer by turning a crank on one end of it. Making the diagonal longer lowers the jack, and making it shorter raises the jack. Is there a direct or linear relationship between the number of turns of the crank and the height of the jack? Let's investigate.

Use your sticks to model the situation, or even better, bring to class a car jack of the type shown. Close the car jack as far as possible (make the diagonal as long as possible). Record the height in a table as shown. Give the handle two full turns to raise the jack. (If you're using sticks instead of a real jack, shorten the diagonal by one unit.) Record the new height. Give the handle another two full turns (or shorten the diagonal by one more unit), raising the jack further. Record the new height. Repeat until the jack is all the way up. Graph the results you recorded in your table.

Turns	0	2	4	6	...
Height	-?-	-?-	-?-	-?-	...

Improving Algebra Skills—*An Algebra Mind-Reading Trick I*

Mind-reading tricks often employ algebra. Try the following mind-reading trick. Use algebra to explain why the trick works.

Double the number of the month you were born. Subtract 16 from your answer. Multiply your result by 5, then add 100 to your answer. Subtract 20 from your result, then multiply by 10. Finally, add the day of the month you were born to your answer. The number you end up with shows the month and the day you were born! For example, if you were born March 15th, your answer will be 315. If you were born December 7th, your answer will be 1207.

Lesson 5.4

SSS, SAS, and SSA Congruence Shortcuts?

A building contractor has just assembled two massive triangular trusses that will support the roof of a recreation hall. Before the crane hoists them into place, the contractor wants to verify that the two triangular trusses are identical. According to the definition of congruent polygons, if the three sides and three angles of one triangle are congruent to three sides and three angles of another, then the two triangles are congruent. Does that mean the contractor must measure and compare all six parts of both triangles? There must be a shortcut to determine whether triangles are congruent. It would seem that only a few of the measurements are needed, but how many and which ones? Let's see if we can answer this question.

Side

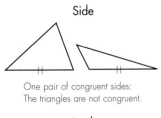

One pair of congruent sides:
The triangles are not congruent.

Angle

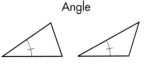

One pair of congruent angles:
The triangles are not congruent.

Wouldn't it be nice if the contractor could measure only one side in each triangle, find that they're congruent, and know that the triangles themselves are congruent? In other words, if she knew that one side of a triangle is congruent to one side of another triangle, could she state that the triangles must be congruent? Clearly they might be. But it only takes one counterexample to demonstrate that they don't have to be congruent. Top left are two triangles with one pair of congruent sides. The triangles are clearly not congruent. Below left are two triangles with a pair of congruent angles. Clearly the triangles are not congruent to each other.

It is clear from these counterexamples that the contractor will have to measure more than one pair of sides or more than one pair of angles to determine that the triangles are congruent.

If two different triangles have two pairs of congruent parts (two pairs of congruent sides, or two pairs of congruent angles, or a pair of congruent sides and a pair of congruent angles), must the two triangles be congruent? A few counterexamples will demonstrate that they need not be.

Side-Side

Two pairs of congruent sides:
The triangles are not congruent.

Angle-Angle

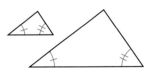

Two pairs of congruent angles:
The triangles are not congruent.

Side-Angle

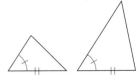

Pair of congruent sides and congruent angles: The triangles are not congruent.

So, if there is a shortcut that the contractor can use to show that two triangles are congruent, it involves measuring and comparing at least three parts of each triangle. What are the combinations of three parts of a triangle that she could measure? The six different combinations of congruent pairs of three parts are diagrammed below.

Side-Side-Side
(SSS)

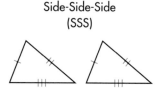

Three pairs of congruent sides

Side-Side-Angle
(SSA)

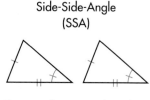

Two pairs of congruent sides and one pair of congruent angles (angles not between the pairs of sides)

Side-Angle-Side
(SAS)

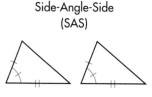

Two pairs of congruent sides and one pair of congruent angles (angles between the pairs of sides)

Angle-Side-Angle
(ASA)

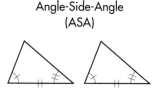

Two pairs of congruent angles and one pair of congruent sides (sides between the pairs of angles)

Side-Angle-Angle
(SAA)

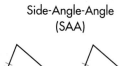

Two pairs of congruent angles and one pair of congruent sides (sides not between the pairs of angles)

Angle-Angle-Angle
(AAA)

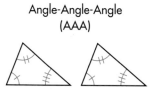

Three pairs of congruent angles

Fortunately, the contractor doesn't have to wait for you to investigate those combinations. She knows some congruence shortcuts. She's measured only those parts necessary to determine that the triangular trusses are congruent. The crane is already lifting them into position.

In this lesson you and your classmates are going to investigate three (SSS, SAS, SSA) of the six possible combinations of three parts. Your class should divide up the three investigations so that each group performs only one of the investigations.

After your group completes its investigation, be ready to share your results with the other groups. If your group is to investigate the SSS case, go to Investigation 5.4.1. If your group is to investigate the SAS case, go to Investigation 5.4.2. If your group is to investigate the SSA case, go to Investigation 5.4.3.

Investigation 5.4.1

THE SSS CASE: If the three sides in one triangle are congruent to the three sides in another triangle (SSS), must the two triangles be congruent? Let's find out. Pick your tools of discovery: either patty papers or a compass and a straightedge.

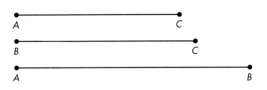

Step 1 Use the three segments to construct a triangle on your own sheet of paper.

Step 2 Compare your triangle with △*ABC*. Is it congruent or is it different? (One way to see if the triangles are congruent is to place them on top of each other and see if they coincide. You may have to flip one of them over to get them to match.)

Step 3 Let's check this with another triangle. First have one person in the group draw a triangle and label it *CBS*. Each group member should copy the three segments $\overline{CB}$, $\overline{BS}$, and $\overline{CS}$ onto his or her own sheet of paper or patty paper. Finally, each group member should construct a triangle with the three segments.

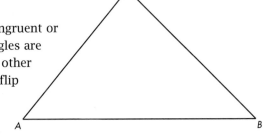

Step 4 Compare your triangles. Were you able to make one that is of a different size or shape?

If three sides of a triangle are congruent to three sides of another, must the triangles be congruent? Make a conjecture.

 C-33 If the three sides of one triangle are congruent to the three sides of another triangle, then —?— (***SSS Congruence Conjecture***).

This is an important conjecture. Write it below your investigative work and add it to your conjecture list. Beneath your conjecture, copy and complete the picture statement below. It's a graphic way to illustrate the SSS Congruence Conjecture.

If you know this:

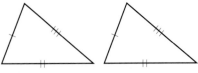

then you also know this:

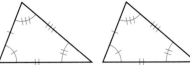

After your group has performed the investigation, be ready to share your results with the groups that did the other investigations. Write down the results of the other groups' investigations in the investigation section of your notebook and add the new conjectures to your conjecture list.

Investigation 5.4.2

THE SAS CASE: If two sides and the angle between them in one triangle are congruent to two sides and the angle between them in another triangle (SAS), must the two triangles be congruent? In other words, given two sides of a triangle and the angle between them, can you construct a different triangle? Pick your tools of discovery, either patty papers or a compass and a straightedge.

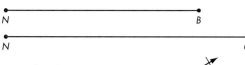

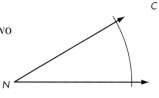

The two segments above are two sides of △NBC. The acute angle is the angle between them.

Step 1 Construct a triangle on a sheet of paper, using these two segments as sides and the angle as the angle between them.

Step 2 Compare your triangle with the given △NBC and the triangles constructed by others in your group. Were you able to make at least one that is of a different size or shape, or are they all congruent?

Step 3 Let's check this with another triangle. First, have one person in the group construct a triangle and label it *TBS*. Each group member should copy ∠B and the two segments $\overline{TB}$ and $\overline{BS}$ onto his or her own sheet of paper or patty paper. Finally, each group member should construct a triangle with the three parts, making sure that the given angle is always constructed between the two given segments.

Step 4 Compare your triangles. Are they all of the same size and shape?

If two sides and the included angle in one triangle are congruent to two sides and the included angle of another, must the triangles be congruent? Make your conjecture.

C-34 If two sides and the angle between them in one triangle are congruent to two sides and the angle between them in another triangle, then —?—
(*SAS Congruence Conjecture*).

This is an important conjecture. Add it to your conjecture list and illustrate it with a picture statement like the one shown at the end of Investigation 5.4.1.

After your group has performed the investigation, be ready to share your results with the groups that did the other investigations. Write down the results of the other groups' investigations in the investigation section of your notebook and add the new conjectures to your conjecture list.

Investigation 5.4.3

THE SSA CASE: If two sides and an angle not between them in one triangle are congruent to the corresponding two sides and an angle not between them in another triangle (SSA), must the two triangles be congruent or, with these parts, can you construct a triangle that is of a different size or shape? Pick your tools of discovery: either patty papers or a compass and a straightedge.

The two segments above right are two sides of △*HBO* and the angle is an angle not between them.

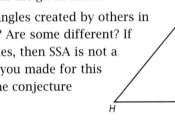

Step 1   Construct a triangle, using these two segments as sides and the angle as an angle that is not between the given sides.

Step 2   Compare your triangle with the triangles created by others in your group. Are they all congruent? Are some different? If you can create two different triangles, then SSA is not a shortcut. Beneath the construction you made for this investigation, complete and copy the conjecture below.

Conjecture: If two sides and an angle that is not between the two sides in one triangle are congruent to the corresponding two sides and an angle that is not between the two sides in another triangle, then the two triangles (are/are not) necessarily congruent.

After your group has performed the investigation, be ready to share your results with the groups that did the other investigations. Write down the results of the other groups' investigations in the investigation section of your notebook and add the new conjectures to your conjecture list.

Take Another Look 5.4

There are other ways to convince yourself that the conjectures you made in this lesson are true. Try one or more of these follow-up activities.

1.   Use a geometry computer program or patty papers to investigate the SSS or SAS Congruence Conjecture.

2.*   In Chapter 3, you discovered how to construct the bisector of an angle. Perform this construction. Use the SSS shortcut for congruence to explain why the construction works.

3.   Use a geometry computer program or other tools to test the following variation of the SSA case: If two sides and the angle opposite the longer of the two sides in one triangle are congruent to two sides and the corresponding angle in another triangle, then the triangles are congruent. Can you find a counterexample?

4.* Is there a conjecture similar to the SSS Congruence Conjecture that you can make about congruence between quadrilaterals? For example, is SSSS a shortcut for quadrilateral congruence? Or, if three sides and a diagonal of one quadrilateral are congruent to the corresponding three sides and diagonal of another quadrilateral, must the two quadrilaterals be congruent (SSSD)? Investigate. Write a paragraph explaining how your conjectures follow from the triangle congruence conjectures you've learned.

Exercise Set 5.4

1.* What conjecture tells you that △LUZ is congruent to △IDA?

2. What conjecture tells you that △CAV is congruent to △CEV?

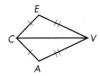

3. What conjecture tells you that △ARC is congruent to △ERN?

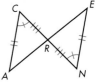

From the information given, complete each statement. If the triangles cannot be shown to be congruent from the information given, write "Cannot be determined" and redraw the figures to show that the triangles are clearly not congruent. Do not assume that segments or angles are congruent just because they appear to be congruent.

4.* △ANT ≅ △ –?–

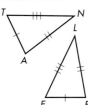

5. △GIT ≅ △ –?–

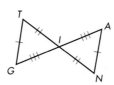

6. △RED ≅ △ –?–
Why?

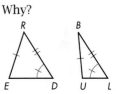

7. U is the midpoint of both $\overline{FE}$ and $\overline{LT}$.
△FLU ≅ △ –?–

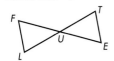

8. $\overline{BW} \cong \overline{ET}$
$\overline{OW} \cong \overline{IT}$
△BOW ≅ △ –?–

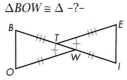

9. $\overline{AM}$ is a median.
△CAM ≅ △ –?–

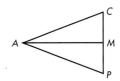

10. △ARE ≅ △ –?–
Why?

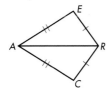

11. △FSH ≅ △ –?–
Why?

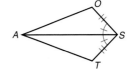

12. △SAT ≅ △ –?–
Why?

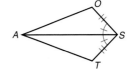

13.* △SUN ≅ △–?–
Why?

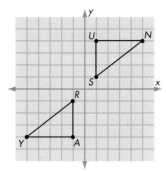

14. △OIL ≅ △–?–
Why?

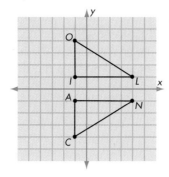

15. $\overline{AL} ≅ \overline{FH}$, △HAF ≅ △–?–
Why?

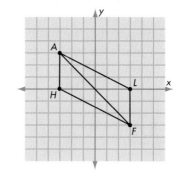

16. Explain why nailing boards diagonally in the corners of this wooden gate makes the gate stronger and prevents it from changing shape.

In Exercises 17–19, use a compass and a straightedge or patty papers.

17. Draw a triangle. Use the SSS shortcut to construct a second triangle congruent to the first.

18. Draw a triangle. Use the SAS shortcut to construct a second triangle congruent to the first.

19.* Construct two triangles that are *not* congruent, even though two sides and an angle not between them in one triangle are congruent to two sides and an angle not between them in the other triangle.

20. Find the point of intersection of the lines $y = \frac{2}{3}x - 1$ and $3x - 4y = 8$.

21. Trace the figure at right. Calculate the measure of each lettered angle.

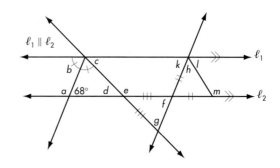

Lesson 5.5

ASA, SAA, and AAA Congruence Shortcuts?

In this lesson you will continue your search for shortcuts that enable you to determine whether two triangles are congruent without having to measure all six parts of each triangle. You and your classmates are going to investigate the remaining three (ASA, SAA, AAA) of the six possible combinations.

Angle-Side-Angle (ASA)	Side-Angle-Angle (SAA)	Angle-Angle-Angle (AAA)
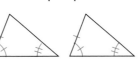		
Two pairs of congruent angles and one pair of congruent sides (sides between the pairs of angles)	Two pairs of congruent angles and one pair of congruent sides (sides not between the pairs of angles)	Three pairs of congruent angles

Divide up the three investigations so that each group performs only one of the investigations. After your group completes its investigation, be ready to share your results with the other groups. If your group is to investigate the ASA case, go to Investigation 5.5.1. If your group is to investigate the SAA case, go to Investigation 5.5.2. If your group is to investigate the AAA case, go to Investigation 5.5.3.

Investigation 5.5.1

THE ASA CASE: If two angles and the side between them in one triangle are congruent to two angles and the side between them in another triangle (ASA), must the two triangles be congruent? Let's investigate. The two angles below are two angles of acute △*MTV*, and the segment is the side between them.

Step 1   Construct a triangle, using these two angles and the segment as the side between them.

Step 2   Compare your triangle with the given △*MTV* and the triangles constructed by others in your group. Are they congruent?

Step 3   Let's check this with another triangle. First have one person in the group construct a triangle and label it *VHI*. Each group member should copy ∠*V* and ∠*H* and side $\overline{VH}$ onto his or her own sheet of paper. Finally, each group member should construct a triangle with the three parts, making sure that the given side is always constructed between the two given angles.

Step 4   Compare your triangles. Are they all of the same size and shape?

If two angles and the included side in one triangle are congruent to two angles and the included side of another, must the triangles be congruent? Make your conjecture.

C-35 If two angles and the side between them in one triangle are congruent to —?— (*ASA Congruence Conjecture*).

Add this new conjecture to your conjecture list. Beneath your conjecture, copy and complete a picture statement as you did in Lesson 5.4.

After your group has performed the investigation, be ready to share your results with the groups that did the other investigations. Write down the results of the other groups' investigations in the investigation section of your notebook and add the new conjectures to your conjecture list.

Investigation 5.5.2

THE SAA CASE: If two angles and a side not between them of one triangle are congruent to the corresponding two angles and side of another triangle (SAA), must the two triangles be congruent? The two angles below are two angles of acute $\triangle AMC$, and the segment is a side not between them.

Step 1   Construct a triangle, using these two angles and the segment as a side not between them. (The measure of the third angle is 180° less the sum of the measures of the two given angles.)

Step 2   Compare your triangle with the given $\triangle AMC$ and with the triangles constructed by others in your group. Are they all congruent?

Step 3   Let's check this with another triangle. First have one person in the group draw a triangle and label it *QVC*. Each group member should copy $\angle V$ and $\angle C$ and side $\overline{CQ}$ onto his or her own sheet of paper or patty paper. Finally, each group member should construct a triangle with the three parts, making sure that the given side is not constructed between the two given angles and that the order of the three parts is always the same.

Step 4   Compare your triangles. Are they all of the same size and shape?

If two angles and a side not between them in one triangle are congruent to the corresponding two angles and side of another triangle, must the triangles be congruent? Make your conjecture.

C-36 If two angles and a side that is not between them in one triangle are congruent to the corresponding —?— (*SAA Congruence Conjecture*).

Add this new conjecture to your conjecture list and illustrate it with a picture statement, as you did at the end of Investigation 5.5.1.

After your group has performed the investigation, be ready to share your results with the groups that did the other investigations. Write down the results of the other groups' investigations in the investigation section of your notebook and add the new conjectures to your conjecture list.

Investigation 5.5.3

THE AAA CASE: In this last of the congruence shortcut investigations, you will determine whether three pairs of congruent angles (AAA) force two triangles to be congruent. (Of course, the measures of the three angles must add up to 180°. Otherwise no triangle would be possible.) The three angles below are the angles of acute △*USA*.

Step 1   Use the three angles shown below to construct a triangle on a sheet of paper.

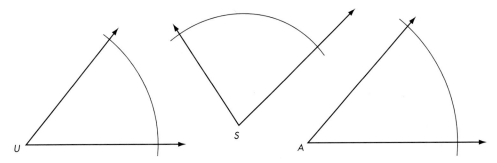

Step 2   Compare your triangle with the triangles constructed by others in your group. Are they all congruent or are some different? Can you construct triangles that use the three angles above and that are *not* congruent?

Based on your observations, complete the conjecture below.

Conjecture: If the three angles of one triangle are congruent to the three angles of another triangle, then the two triangles (are/are not) necessarily congruent.

Write this conjecture beneath your investigative work as your summary statement. After your group has performed the investigation, be ready to share your results with the groups that did the other investigations. Write down the results of the other groups' investigations in the investigation section of your notebook and add the new conjectures to your conjecture list.

Take Another Look 5.5

Try one or more of these follow-up activities.

1. Use a geometry computer program to investigate the ASA or SAA Congruence Conjecture.

2.* In Chapter 3, you discovered how to construct the perpendicular bisector of a segment. Now use what you've learned about shortcuts for congruence to explain why this construction method works.

3.* Explain why the SAA Congruence Conjecture follows logically from the ASA Congruence Conjecture.

4. The SAA Congruence Conjecture includes the word *corresponding.* That word was unnecessary in the ASA Congruence Conjecture because there's only one possible side between two given angles. Likewise, correspondence is guaranteed in the SAS and SSS Congruence Conjectures. Why is it important to state that parts must correspond in the SAA Congruence Conjecture? Draw figures that illustrate your explanation.

5.* Is there a conjecture you can make about congruence between quadrilaterals that is similar to the SAS Congruence Conjecture? For example, is SASA a shortcut for quadrilateral congruence? That is, if two consecutive sides and two consecutive angles (such that one angle is between the two consecutive sides) in one quadrilateral are congruent to the corresponding two sides and two angles of another quadrilateral, must the two quadrilaterals be congruent? If the answer is yes, show why; if not, show a counterexample. How about an SASAS shortcut?

Exercise Set 5.5

In Exercises 1–15, complete each statement from the information given. If the triangles cannot be shown to be congruent from the information given, write "Cannot be determined."

1. Which conjecture tells you that △*BOX* is congruent to △*CAR*?

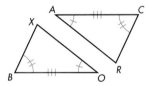

2. Which conjecture tells you that △*KAP* is congruent to △*AKR*?

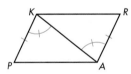

3. Which conjecture tells you that △*FAD* is congruent to △*FED*?

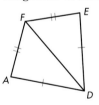

4.* Which conjecture tells you that △*GAS* is congruent to △*IOL*?

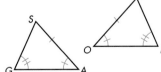

5. △*WOM* ≅ △ –?–
Why?

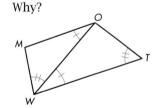

6. △*BLK* ≅ △ –?–
Why?

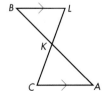

7.* $\triangle MAN \cong \triangle$ -?-
Why?

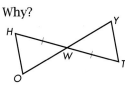

8. $\overline{AW} \perp \overline{WL}$ and $\overline{WL} \perp \overline{KL}$
$\triangle LAW \cong \triangle$ -?-
Why?

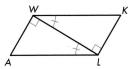

9.* $\overline{PO} \perp \overline{OI}$
$\triangle POL \cong \triangle$ -?-
Why?

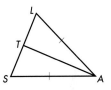

10. $\overline{HO} \parallel \overline{YT}$
$\triangle HOW \cong \triangle$ -?-
Why?

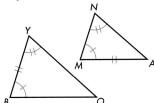

11. $\triangle COT \cong \triangle$ -?-
Why?

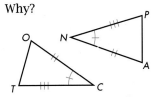

12. $\overline{AT}$ is an angle bisector.
$\triangle LAT \cong \triangle$ -?-

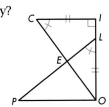

13. $PO = PR$
$\triangle POE \cong \triangle$ -?-
$\triangle SON \cong \triangle$ -?-

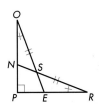

14. $\triangle$ -?- $\cong \triangle$ -?-
Why?

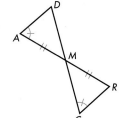

15.* $\triangle SLN$ is equilateral.
Is $\triangle TIE$ equilateral?

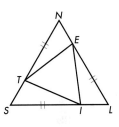

In Exercises 16–18, determine whether the segments or triangles in the coordinate plane are congruent and explain your reasoning.

16. Use slope conjectures to show $\overline{AB} \perp \overline{BC}$, $\overline{CD} \perp \overline{DA}$, and $\overline{BC} \parallel \overline{DA}$. $\triangle ABC \cong \triangle$ -?- Why?

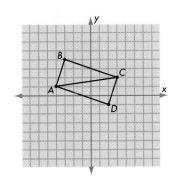

17. Use the Coordinate Midpoint Conjecture to show that O is the midpoint of $\overline{PR}$ and $\overline{SD}$. $\triangle DRO \cong \triangle$ -?- Why?

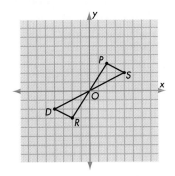

18. Use the Parallel Slope Conjecture to show that $\angle P \cong \angle T$ and $\angle R \cong \angle S$. Is $\triangle PQR$ necessarily congruent to $\triangle TQS$? Explain.

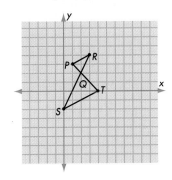

In Exercises 19–22, use a compass and a straightedge or patty papers to perform each construction.

19. Construct a triangle. Use the ASA Congruence Conjecture to construct a second triangle congruent to the first. Write a paragraph to justify your steps.

20. Construct a triangle. Use the SAA Congruence Conjecture to construct a second triangle congruent to the first. Write a paragraph to justify your steps.

21.*Construct two triangles that are *not* congruent, even though the three angles of one triangle are congruent to the three angles of the other triangle.

22.*Construct two triangles that are *not* congruent, even though the three angles of one triangle are congruent to the three angles of the other triangle, and a side in one triangle is congruent to a side in the other triangle.

23. Trace the figure. Calculate the measure of each lettered angle.

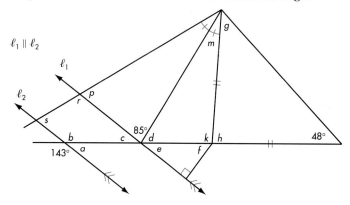

24. Isosceles right triangle *ABC* has vertices with coordinates *A*(-8, 2), *B*(-5, -3), and *C*(0, 0). Find the coordinates of the orthocenter.

Improving Reasoning Skills—*Container Problem II*

You have a small, graduated, cylindrical measuring glass with a maximum capacity of 250 ml. All the graduation marks have worn off except the 150-ml and 50-ml marks. You also have a large unmarked container. It is possible to fill the large container with exactly 350 ml by filling the graduated cylinder twice to the 150-ml mark and once to the 50-ml mark. It is also possible to fill the large container with exactly 350 ml in only two measurings. How?

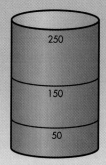

Lesson 5.6

Flow-Chart Thinking

If you can only find it, there is a reason for everything.
— *Traditional saying*

In Lessons 5.4 and 5.5, you discovered four shortcuts (SSS, SAS, ASA, and SAA) for showing that two triangles are congruent. A common technique for showing that two segments or two angles are congruent is to first show that they are corresponding parts of congruent triangles.

The definition of congruent triangles states: If two triangles are congruent, then the *corresponding parts of those congruent triangles are congruent.* We'll use the letters **CPCTC** to refer to the second part of the definition. Let's see how you can use congruent triangles and CPCTC to show that two segments are congruent.

Example A

Is $\overline{AD} \cong \overline{BC}$? If you think so, you are correct. But can you give a logical argument to show that they are congruent?

Here is one possible logical argument.

Because $\angle A \cong \angle B$, $\overline{AM} \cong \overline{BM}$, and $\angle 1 \cong \angle 2$, then $\triangle AMD \cong \triangle BMC$ by the ASA Congruence Conjecture. If $\triangle AMD \cong \triangle BMC$, then $\overline{AD} \cong \overline{BC}$ by CPCTC.

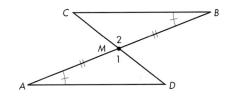

The logical argument presented here is called a paragraph proof. Sometimes a logical argument or a proof is long and complex, and a paragraph might not be the clearest way to present all the important details of your argument.

In Chapter 2, you used concept maps to visualize the relationships between sets of polygons. A concept map can also help you plan and visualize logical thinking. A **flow chart** is a concept map that shows a step-by-step procedure through a complicated system. Boxes represent actions. Arrows connect the boxes to show the flow of the action. Computer programmers use flow charts to plan the logic in programs.

Flow charts can make your logic visible and can help others follow your reasoning. In Example B, a flow chart presents a logical argument that shows two angles are congruent. A logical argument presented in the form of a flow chart is called a **flow-chart proof**.

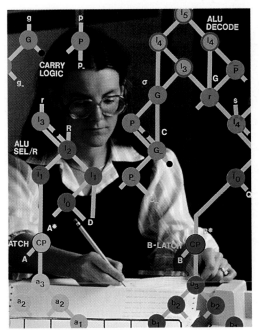

Software engineers use flow charts to visualize logical relationships in complicated computer programs.

Example B

Given: $\overline{AR} \cong \overline{ER}$
$\overline{EC} \cong \overline{AC}$

Show: $\angle E \cong \angle A$

Flow-chart proof:

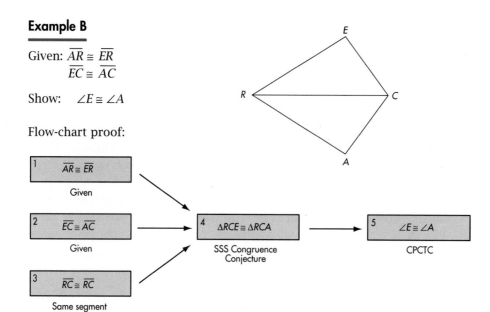

```
1  AR ≅ ER                                      5
   Given                                            ∠E ≅ ∠A
2  EC ≅ AC      4  △RCE ≅ △RCA                       CPCTC
   Given           SSS Congruence
                   Conjecture
3  RC ≅ RC
   Same segment
```

In this example the logical argument flows, according to the arrows, from the information that is given to the conclusion that you are trying to demonstrate. The logical reason supporting each statement is written beneath its box.

Of course, in an argument such as this, your proof is only as good as the conjectures you use to make your argument. The proof demonstrates that if you accept the given information and if you accept the conjectures in the argument, then the conclusion logically follows. You will learn more about writing proofs in later chapters.

Exercise Set 5.6

For Exercises 1–9, copy the figures onto your paper and mark them with the given information. To demonstrate whether or not the segments or the angles indicated are congruent, determine that two triangles are congruent. Then state which conjecture (SSS, SAS, ASA, or SAA) proves them congruent.

1.* $\angle A \cong \angle C$
$\angle ABD \cong \angle CBD$
Is $\overline{AB} \cong \overline{CB}$?
Why?

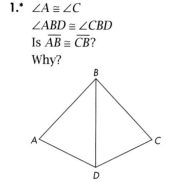

2. $\overline{CN} \cong \overline{WN}$
$\angle C \cong \angle W$
Is $\overline{RN} \cong \overline{ON}$?
Why?

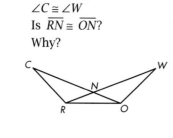

3.* $\overline{SA} \parallel \overline{NE}$
$\overline{SE} \parallel \overline{NA}$
Is $\overline{SA} \cong \overline{NE}$?
Why?

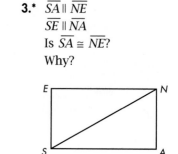

4.* $\angle E \cong \angle T$
M is the midpoint of $\overline{TE}$.
Is $\overline{MI} \cong \overline{MR}$? Why?

5. $\overline{CS} \cong \overline{HR}$
$\angle 1 \cong \angle 2$
Is $\overline{CR} \cong \overline{HS}$? Why?

6.* $\overline{MN} \cong \overline{MA}$
$\overline{ME} \cong \overline{MR}$
Is $\angle E \cong \angle R$? Why?

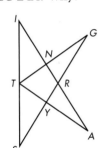

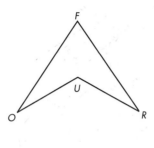

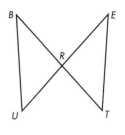

7.* $\angle S \cong \angle I$
$\angle G \cong \angle A$
T is the midpoint of $\overline{SI}$.
Is $\overline{SG} \cong \overline{IA}$? Why?

8.* $\overline{FO} \cong \overline{FR}$
$\overline{UO} \cong \overline{UR}$
Is $\angle O \cong \angle R$? Why?

9.* $\overline{BT} \cong \overline{EU}$
$\overline{BU} \cong \overline{ET}$
Is $\angle B \cong \angle E$? Why?

10. Copy the flow chart. Provide each missing reason or statement in the proof.

Given: $\overline{SE} \cong \overline{SU}$
         $\angle E \cong \angle U$

Show: $\overline{MS} \cong \overline{OS}$

Flow-chart proof:

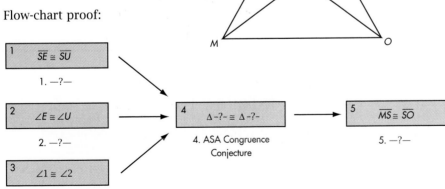

11. Sometimes it's more convenient to write a flow chart from top to bottom. Copy the flow chart. Provide each missing reason or statement in the proof.

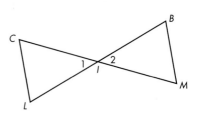

Given: *I* is the midpoint of $\overline{CM}$.
        I is the midpoint of $\overline{BL}$.

Show: $\overline{CL} \cong \overline{MB}$

Flow-chart proof:

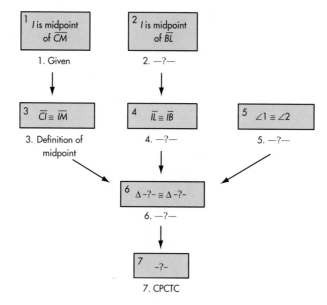

12. Complete the flow-chart proof below to logically demonstrate the argument. There is always more than one flow-chart proof that can be written for a logical argument. One possible proof for this argument has been started for you.

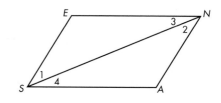

Given: $\overline{SA} \parallel \overline{NE}$
        $\overline{SE} \parallel \overline{NA}$

Show: $\overline{SA} \cong \overline{NE}$

Flow-chart proof:

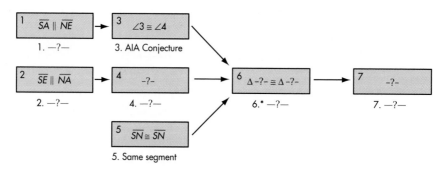

13.* Now it's your turn. Write a flow-chart proof to demonstrate the argument below. You may find it very helpful to work with others on the proof, sharing your ideas.

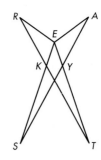

Given: $\angle S \cong \angle T$
$\angle R \cong \angle A$
$\overline{RE} \cong \overline{AE}$

Show: $\overline{RT} \cong \overline{AS}$

For Exercises 14–16, assume right angles and that the lengths of horizontal and vertical segments are given. Provide a short argument that demonstrates whether or not the segments or angles indicated are congruent.

14.* Is $\overline{FR} \cong \overline{GT}$? Why?

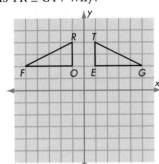

15. Is $\angle OND \cong \angle OCR$? Why?

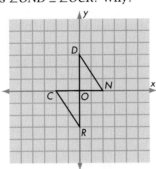

16.* Is $\overline{AT} \cong \overline{TO}$? Why?

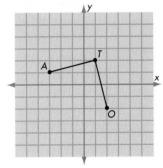

17.* Trace the figure below. Calculate the measure of each lettered angle.

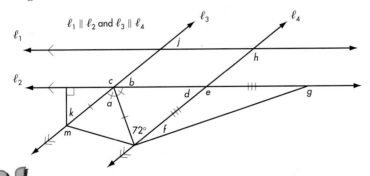

$\ell_1 \parallel \ell_2$ and $\ell_3 \parallel \ell_4$

Improving Reasoning Skills—*Pick a Card*

Nine cards are arranged in a three-by-three array. Every jack borders on a king and on a queen. Every king borders on an ace. Every queen borders on a king and on an ace. (The cards border each other horizontally or vertically but not diagonally.) There are at least two aces, two kings, two queens, and two jacks. Which card is in the center position of the three-by-three array?

Project

Buried Treasure

Only the most foolish of mice would hide in a cat's ear.
But only the wisest of cats would think to look there.
— Andrew Mercer

For problems 1–4, trace onto a clean sheet of paper the outline of the island that appears on the treasure map and the points *A*, *B*, *D*, *H*, *M*, and *S*. On your tracing, find the locations of the buried treasures.

1.* Pirate Alphonse is standing at the edge of Westend Bay (at point *A*), and his cohort, pirate Beaumont, is 60 paces to the north at point *B*. Each pirate can see Captain Coldhart off in an easterly direction, burying a treasure. With his sextant, Alphonse measures the angle between the captain and Beaumont and finds that it measures 85°. With his sextant, Beaumont measures the angle between the captain and Alphonse and finds that it measures 48°. Alphonse and Beaumont mark their positions with large boulders and return to their ship, confident that they have enough information to return later and recover the treasure.

Can they recover the treasure? Which congruence conjecture (SSS, SAS, ASA, or SAA) guarantees they'll be able to find it? If possible, use your geometric tools to locate the position of the treasure on the map. Mark it with an X.

2. Captain Coldhart, convinced someone in his crew stole his last treasure, has decided to be more careful about burying his latest loot. He gives his trusted first mate, Dexter, two ropes, the lengths of which only the captain knows. The captain instructs Dexter to nail one end of the shorter rope to Hangman's Tree (at point *H*) and to secure the longer rope through the eyes of Skull Rock (at point *S*). The captain, with the ends of the two ropes in one hand and the treasure chest tucked under the other arm, walks away from the shore to the point where the two ropes become taut. He buries the treasure at the point where the two ropes come together, collects his ropes, and returns confidently to the ship.

Has Captain Coldhart given himself enough information to recover the treasure? Which congruence conjecture (SSS, SAS, ASA, or SAA) ensures the uniqueness of the location? On your map, locate and mark the position of this second treasure if the two ropes are 100 paces and 150 paces in length.

3. After the theft of two very important ropes from his locker and the subsequent disappearance of his trusted first mate, Dexter, Captain Coldhart is determined that no one will find the location of his latest buried treasure. The captain, with his new first mate, Endersby, walks out to Deadman's Point (at point *D*). The captain instructs his first mate to walk inland along a straight path for a distance of 130 paces. There Endersby is to drive his sword into the ground for a marker (at point *M*), turn and face in the direction of the captain, turn at an arbitrary angle to the left, continue to walk for another 60 paces, stop, and wait for the captain. The captain measures the angle formed by the lines from Endersby to himself and from himself to the sword. The angle measures 26°. The captain places a boulder where he is standing (at point *D*), walks around the bay to Endersby, and instructs him to bury the treasure at this point. Alas, poor Endersby is buried with the treasure. Has the captain given himself enough information to locate the treasure? If he has, determine the unique location of the treasure. If he does not have enough information, explain why not. How many possible locations for the treasure are there? Find them on the map.

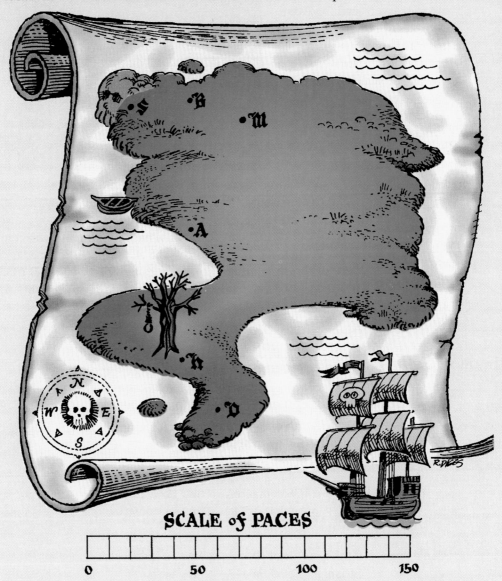

SCALE of PACES

0   50   100   150

4. Now make a map of your own. Identify necessary landmarks and write a story describing how you could locate a place to bury—and later find—a treasure. Use one of the triangle congruence conjectures.

Lesson 5.7

Isosceles Triangles Revisited

The boathouse shown at right is a remarkably symmetrical structure with its isosceles triangle roof and the identical doors on each side. The rhombus-shaped attic window is located on the line of symmetry of this face of the building. What properties does the line of symmetry in an isosceles triangle have?

In this lesson you will make a conjecture about a special segment in isosceles triangles. Then you will use logical reasoning to support your conjecture. In $\triangle ARC$, $\overline{CD}$ is the altitude to the base $\overline{AR}$, $\overline{CE}$ is the angle bisector of the vertex $\angle ACR$, and $\overline{CF}$ is the median to the base $\overline{AR}$. From the diagram it is clear that the angle bisector, the altitude, and the median can all be different line segments. But is this true for all triangles? (Apparently not, because we are asking, right?) Can they all be the same segment? Let's investigate.

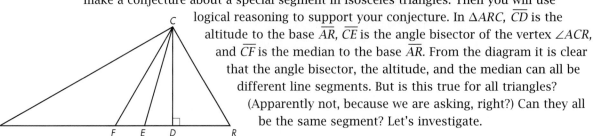

Investigation 5.7

Step 1   Construct a large isosceles triangle on a sheet of unlined paper. Label it *ARK* with *K* the vertex angle.

Step 2   Construct angle bisector $\overline{KD}$ (with point *D* on $\overline{AR}$).

Step 3   With your protractor, measure $\angle ADK$ and $\angle RDK$.

How do $\angle ADK$ and $\angle RDK$ compare? Do they have equal measures? Are they supplementary? Beneath your construction, copy and complete the conjecture below.

Conjecture: The bisector of the vertex angle of an isosceles triangle is also the —?— to the base.

Step 4   With your compass, measure $\overline{AD}$ and $\overline{RD}$.

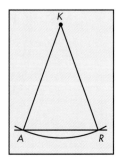

Is D the midpoint of $\overline{AR}$? If D is the midpoint, then $\overline{KD}$ is what type of special segment? Beneath your construction, copy and complete the conjecture below, based on your observations from Step 4.

Conjecture: The bisector of the vertex angle of an isosceles triangle is also the —?— to the base.

Compare your results from both conjectures above with the results of others near you. You should be ready to combine the two conjectures into one.

C-37 In an isosceles triangle, the bisector of the vertex angle is also the —?— to the base and the —?— to the base (***Vertex Angle Bisector Conjecture***).

Take Another Look 5.7

Try one or more of these follow-up activities.

1. Devise a patty-paper investigation to explore the Vertex Angle Bisector Conjecture.

2.* Use a geometry computer program to investigate the converse of the Vertex Angle Bisector Conjecture.

3.* Can you construct a triangle in which exactly two of the three segments—angle bisector, median, and altitude—coincide? Choose your tools of investigation.

4.* What do you think is true of the medians to the congruent sides of an isosceles triangle? Investigate, make a conjecture, and provide a convincing argument to support your conjecture.

Exercise Set 5.7

In Exercises 1–3, $\triangle ABC$ is isosceles with $\overline{AC} \cong \overline{BC}$.

1.* Perimeter $\triangle ABC = 48$
$AC = 18$
$AD = $ -?-

2. $m\angle ABC = 72°$
$m\angle ADC = $ -?-

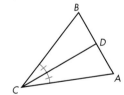

3. $m\angle CAB = 45°$
$m\angle ACD = $ -?-

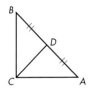

4.* Sixty concurrent lines in a plane divide the plane into how many regions?

5.* While talking on the phone, Janice doodled a number of hexagons and octagons onto a sheet of paper. She counted a total of 140 sides in all. She drew at least five of each type of polygon. How many of each polygon did she sketch? Find all the possible solutions.

6. What is the measure of the angle formed by the hands of a clock at 3:15?

7. Trace the figure below. Calculate the measure of each lettered angle.

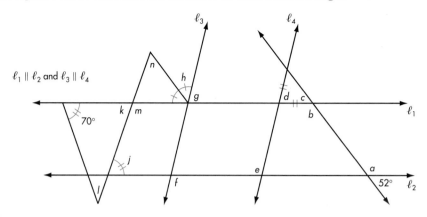

$\ell_1 \parallel \ell_2$ and $\ell_3 \parallel \ell_4$

8.* How many minutes after 3:00 will the hands of a clock be together?

9. If two vertices of a triangle have coordinates $A(1, 3)$ and $B(7, 3)$, find the coordinates of a point so that the triangle formed by the three points is a right triangle. Can you find a different point that also creates a right triangle?

10. Find the equation, in slope-intercept form, of the line through point C parallel to side $\overline{AB}$ in $\triangle ABC$ with vertices $A(1, 3)$, $B(4, -2)$, and $C(6, 6)$.

In Exercises 11 and 12, you'll use a flow chart to logically demonstrate the first part of the Vertex Angle Bisector Conjecture. You'll complete the second part of the conjecture in Exercise 13. Conjecture A in Exercise 11 is used to prove Conjecture B in Exercise 12. You will also need to use Conjecture A in Exercise 13.

11. Copy the flow chart below and supply the missing statement and the missing reason to show by logical reasoning that the conjecture is true.

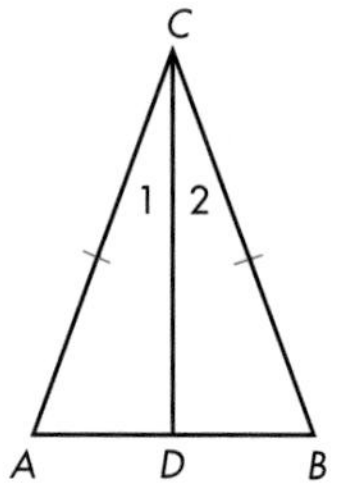

Conjecture A: The bisector of the vertex angle in an isosceles triangle divides the isosceles triangle into two congruent triangles.

Given: $\triangle ABC$ is isosceles with $\overline{AC} \cong \overline{BC}$. $\overline{CD}$ is the bisector of $\angle C$.

Show: $\triangle ADC \cong \triangle BDC$

Flow-chart proof:

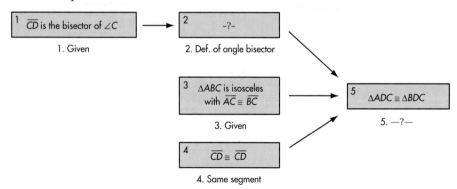

12. Copy the flow chart and supply the missing reasons and the missing statement to show by logical reasoning that the bisector of the vertex angle in an isosceles triangle is also the altitude to the base.

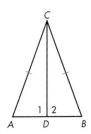

Conjecture B: The bisector of the vertex angle in an isosceles triangle is also the altitude to the base.

Given: $\triangle ABC$ is isosceles with $\overline{AC} \cong \overline{BC}$.
$\overline{CD}$ is the bisector of $\angle C$.

Show: $\overline{CD}$ is an altitude.

Flow-chart proof:

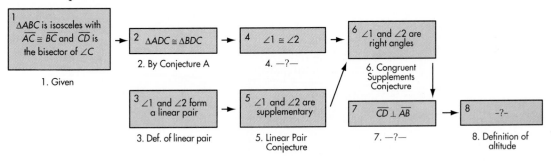

13. You've proven one part of the Vertex Angle Bisector Conjecture. Now prove the other part. Write a flow-chart proof or a paragraph proof for Conjecture C below.

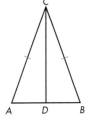

Conjecture C: The bisector of the vertex angle in an isosceles triangle is also the median to the base.

Given: $\triangle ABC$ is isosceles with $\overline{AC} \cong \overline{BC}$.
$\overline{CD}$ is the bisector of $\angle C$.

Show: $\overline{CD}$ is a median.

14. Builders of ancient Egypt used a tool called a *plumb level* in building architectural wonders such as the Egyptian pyramids. With a plumb level, builders can use basic properties of isosceles triangles to determine if a surface is level. In the figure below, $\triangle ABC$ is isosceles. A weight (called the *plumb*) hangs from a string attached at point C. A surface is level if the string is perpendicular to $\overline{AB}$. When the string is perpendicular to $\overline{AB}$, it passes through the midpoint of $\overline{AB}$. Write a flow-chart proof to show that if D is the midpoint of $\overline{AB}$, then $\overline{CD}$ is perpendicular to $\overline{AB}$.

Given: $\triangle ABC$ is isosceles with $\overline{AC} \cong \overline{BC}$.
D is the midpoint of $\overline{AB}$.

Show: $\overline{CD} \perp \overline{AB}$

Napoleon's Theorem

The discovery you'll make in this project is attributed to French emperor Napoleon Bonaparte (1769-1821). This discovery, called Napoleon's theorem, has to do with equilateral triangles constructed on the sides of an arbitrary triangle.

In Steps 1–5, you'll create a script to construct an equilateral triangle and its centroid.

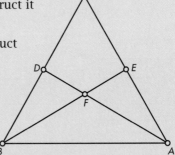

Step 1   Open a new script and click Record.

Step 2   Construct equilateral $\triangle ABC$. (Use a script or construct it from scratch.)

Step 3   Construct the midpoints of $\overline{AC}$ and $\overline{BC}$ and construct two medians.

Step 4   Construct point F, the point of intersection of the two medians. Point F is the centroid of $\triangle ABC$.

Step 5   Hide the medians and the midpoints. Stop your script.

In Steps 6–8, you'll construct equilateral triangles on the sides of an arbitrary triangle.

Step 6   Open a new sketch. Construct a new $\triangle ABC$.

Step 7   Play your equilateral triangle script on the endpoints of each of the three sides. If an equilateral triangle falls inside your triangle, undo and try again, selecting the two given points in reverse order.

Step 8   Connect the centroids of the equilateral triangles. Triangle GLQ is called the outer Napoleon triangle of $\triangle ABC$.

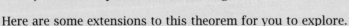

Drag the vertices and the sides of $\triangle ABC$. What can you say about the triangle connecting the centroids of the equilateral triangles? Write what you think Napoleon's theorem might be.

Here are some extensions to this theorem for you to explore.

1.   Construct segments connecting each vertex of your original triangle with the most remote vertex of the equilateral triangle on the opposite side. What do you notice about these three segments? (This discovery was made by Dutch graphic artist M. C. Escher.)

2.   Construct the inner Napoleon triangle by reflecting each centroid across its corresponding edge in the original triangle. Measure the areas of the original triangle and of the outer and inner Napoleon triangles. How do these areas compare?

45. The measure of an angle formed by the bisectors of two angles in a triangle, as shown below, is 100°. What is the measure of the third angle?

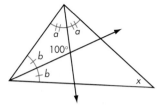

46. The measure of an angle formed by altitudes from two vertices of a triangle, as shown below, is 132°. What is the value of angle measure x?

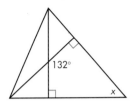

47. Huey hears an explosion three seconds after he sees the flash. Duey hears the explosion five seconds after he sees the flash. Huey and Duey are 1.5 kilometers apart. The sound, traveling at 340 meters/second, came from a northerly direction. Is there enough information for Huey and Duey to locate the site of the explosion? Make a sketch and label all the distances that are known or can be calculated.

48. Find each lettered angle measure in the diagram below.

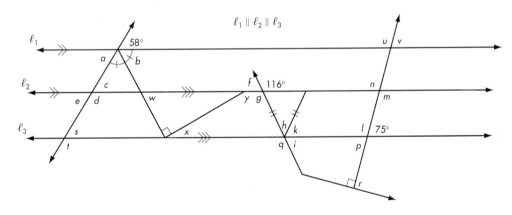

Assessing What You've Learned—*Writing Test Items*

It's one thing to be able to do a math problem. It's another to be able to make one up. If you were writing a test for this chapter, what would it include? Do you understand the content of this chapter well enough to make up problems of your own? Naturally, you should know how to solve any problem you make up. And if it's a fair problem, other students in the class should be able to solve it, too. If you choose to try writing test items, you should work with your group members.

   Start by having a group discussion to identify the key ideas in each lesson of the chapter. Then divide the lessons among group members and have each group member write one problem for each lesson assigned. (For example, this chapter has eight lessons. In a group of four people, each person should write two problems.) Make sure you can do every problem that you write yourself. Try to create a mix of problems in your group, from simple one-step exercises that require you to recall facts to more complex, multistep problems that require more thinking. An example of the first type of problem might be finding a missing angle measure in a triangle. A more complex problem could be a flow chart for a logical argument, or a word problem that requires using geometry to model a real-world situation.

Share your problems with your group members so that everyone has a chance to try out one another's problems. Then discuss the problems in your group: Were they representative of the content of the chapter? Were some too hard or too easy? Would they be fair questions to ask on a test? Writing problems of your own and trying problems written by your classmates should be an excellent way for you to assess and review what you've learned in this chapter. Maybe you can even persuade your teacher to use one or more student-written items on a real test!

Whether or not you decide to try your hand at writing test items, you can also choose to follow up on one or more other assessment options.

Organize Your Notebook

- Review your notebook to be sure it's complete and well organized. It's always a good habit to write a one-page chapter summary based on your notes.

Write in Your Journal

- Write a paragraph or two about what you did in this class that gave you the greatest sense of accomplishment. It doesn't have to be from this chapter. What was it? What did you do? What did you learn from it? What about the work makes you proud of it?

Update Your Portfolio

- Choose a piece of work from this chapter to add to your portfolio: an investigation, a project, a Take Another Look investigation, an exercise set, a test—whatever you think best represents the current quality of your work. Be sure to document the work, explaining what it is and why you chose it for your portfolio.

Performance Assessment

- While a classmate, a friend, a family member, or your teacher observes, perform an investigation that demonstrates one or more of the congruence conjectures. Explain what you're doing at each step, including how you arrived at the conjecture.

Improving Algebra Skills—*Algebraic Sequences I*

Find the next two terms of each algebraic sequence.

$x + 3y$, $2x + y$, $3x + 4y$, $5x + 5y$, $8x + 9y$, $13x + 14y$, –?–, –?–

$x + 7y$, $2x + 2y$, $4x - 3y$, $8x - 8y$, $16x - 13y$, $32x - 18y$, –?–, –?–

Construction Games from Centauri

The geometry students of the lunar colony have received a communication from the Centauri star system. The Centaurian geometry students have sent two of their classic games of geometric construction. Centaurian mathematicians developed a geometry almost identical to that of the geometry studied on Earth; however, they did develop alternative geometric constructions. Like the ancient Greeks of Earth, the early Centaurian geometers played with a compass and a straightedge, as if they were playing a game, to see what geometric figures they could create. The geometers also went on to develop other construction games. In one construction game, they tried to see what they could create by using only a right triangle. In another construction game, they tried to see what they could create by using only a pair of parallel lines.

You and your cooperative problem solving (CPS) team, like the students at the lunar colony, are going to play the two Centaurian games of construction. In each game, you must play by the rules and "construct" different geometric figures. When you play the Centaurian construction games, the term *construct* does not mean "construct with a compass and a straightedge." Instead, it means "create with only the tool or tools allowed by the game." Once you have figured out a method to construct each figure, use your geometric definitions and conjectures to support your method. Explain why your construction works. Good luck!

Game One: Constructions with a Right Triangle

In this first construction game, you are permitted to use only a right triangle. With your right triangle tool you can construct straight lines, construct perpendiculars, duplicate the acute angles of the right triangle, and duplicate the lengths of the legs of the right triangle. (You are not permitted to assume to know the measures of the angles or the relationship between the lengths of the sides.) If you do not have a plastic drafting right triangle, make a right triangle for yourself out of poster board or heavy cardboard.

1. Construct a perpendicular from a given point to a given line.
2. Construct a perpendicular through a given point on a given line.
3. Construct a rectangle.
4. Construct a rhombus.
5. Construct a perpendicular bisector of a segment.
6. Construct an angle bisector.

Can you develop any new right triangle constructions to send back to the geometry students on Centauri as a challenge?

Game Two: Constructions with Parallel Lines

In this second construction game, you are permitted to use only the two parallel edges of a straightedge or a ruler. If you use a ruler, you cannot use the markings on it. With your parallel line tool, you can construct parallel lines that are a fixed distance apart. (With parallel lines come all the properties of parallels and parallelograms!)

1. Construct a rhombus.
2. Construct a rectangle.
3. Construct the angle bisector of a given angle.
4. Construct a perpendicular bisector of a given segment (where the length of the segment is greater than the distance between your parallel lines).
5. Construct a perpendicular through a given point on a given line.
6. Construct a perpendicular bisector of a given segment (where the length of the segment is less than the distance between your parallel lines).
7. Construct a line parallel to a given line through a point not on the given line.
8. Construct a perpendicular to a given line through a point not on the given line.

Can you develop any new parallel line constructions to send back to the geometry students on Centauri as a challenge?

Polygon Properties

In Chapter 2, you defined the term *polygon* and many specific types of polygons. How many different polygons can you name that appear in the portion of M. C. Escher's *Metamorphosis III* shown at left?

In this chapter, you'll discover many properties of polygons, including relationships among their angles, sides, and diagonals. You'll also discover properties special polygons possess that give them many real-world applications.

Metamorphosis III (detail), M. C. Escher, 1967–1968
©1996 M. C. Escher / Cordon Art – Baarn – Holland.
All rights reserved.

Lesson 6.1

Polygon Sum Conjecture

In this lesson you'll investigate the sum of the angle measures in quadrilaterals, pentagons, and other polygons. Then you'll look for a pattern to see if you can make a conjecture about the sum of the angle measures in *any* polygon.

To do this you will need to sum angle measures in a lot of polygons to see if these sums cluster around numbers that form a pattern. This sounds like a task to jigsaw among different groups; that is, each group will work on a different part of the puzzle.

This arrangement of regular polygons is typical of tilings found in many Moorish palaces in Northern Africa. From the tiling, can you determine the measure of each angle of the hexagons?

Investigation 6.1

Each group should investigate a different type of polygon. Turn to the investigation section of your notebook and draw the polygon assigned to your group. Use a protractor to carefully measure all the angles of your polygon, then use your calculator to sum them. Share your results with your group. Throw out the highest and lowest sums of the group and find the average of the remaining sums.

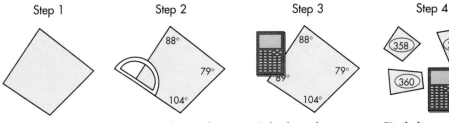

Step 1	Step 2	Step 3	Step 4
Make a polygon.	Measure the angles.	Calculate the sum.	Find the average.

Were your group members' results close? Would you conjecture that the average you found is the actual sum for any polygon of the type you investigated? Copy the table below onto your notebook's investigation page and complete the table as each group reports their results for their polygon. From the completed table, look for a pattern to discover a formula that relates the number of sides of a polygon and the sum of the measures of that polygon's angles.

Number of sides of polygon	3	4	5	6	7	8	...	n
Sum of measures of angles	180°	-?-	-?-	-?-	-?-	-?-	...	-?-

You can now make some conjectures.

C-38 The sum of the measures of the four angles of every quadrilateral is —?— (*Quadrilateral Sum Conjecture*).

C-39 The sum of the measures of the *n* angles of an *n*-gon is —?—
(*Polygon Sum Conjecture*).

Take Another Look 6.1

Try one or more of these follow-up activities.

 1. Use a geometry computer program or patty papers to investigate the sum of the angle measures in different polygons.

2. Draw a polygon onto a sheet of heavy paper. Cut the polygon and tear off its angles. On another sheet of paper, draw a line and a point on the line. Arrange the torn-off angles around the point. Write a paragraph that explains how this activity demonstrates the Polygon Sum Conjecture for this polygon.

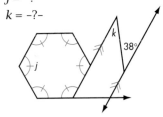

3. Draw several polygons that have four or more sides. In each, draw diagonals from a single vertex. Explain how the Polygon Sum Conjecture follows logically from the Triangle Sum Conjecture. Does the Polygon Sum Conjecture apply to concave polygons?

 4. On a sphere, investigate the Polygon Sum Conjecture. Once you make a conjecture, explain how it is related to the Triangle Sum Conjecture on a sphere. Be sure to test your conjecture on the smallest and largest possible polygons.

Exercise Set 6.1

In Exercises 1–6, use your conjectures to determine each lettered angle measure. You might find it helpful to trace the more complicated diagrams.

1.* $a = $ –?–

2. $b = $ –?–

3.* $c = $ –?–
$d = $ –?–

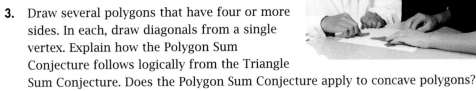

4. $e = $ –?–
$f = $ –?–

5.* $g = $ –?–
$h = $ –?–

6. $j = $ –?–
$k = $ –?–

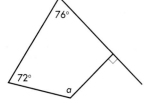

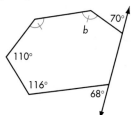

7. Trace the figure below. Calculate each lettered angle measure.

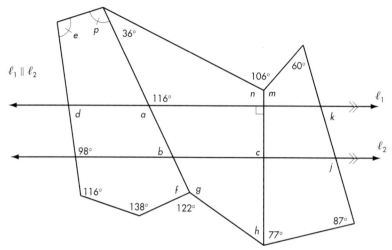

8. What is the sum of the measures of the angles of a decagon?

9. What is the sum of the measures of the angles of a 25-gon?

10. The figure at right is a detail of one vertex of the tiling shown at the beginning of this lesson. Find the missing angle measure x.

11.* How many sides does a polygon have if the sum of its angle measures is 2700°?

12.* Recall that an equiangular polygon is a polygon with all angles equal in measure. What is the measure of each angle of an equiangular decagon?

13. Archaeologist Ertha Diggs has uncovered a piece of a ceramic plate. The original plate appears to have been in the shape of a regular polygon. If the original plate was a regular 16-gon, it was probably a ceremonial dish from the third century. If it was a

regular 18-gon, it was probably a palace dinner plate from the twelfth century. Ertha measures each of the sides of her piece and finds that each side has the same length. She then conjectures that all the sides of the original whole plate had the same length. She measures each of the angles of her piece and finds that they all have the same measure. She then conjectures that all the angles of the original whole plate had equal measures. If each angle measures 160°, from what century did the plate originate?

14. You've been asked to build a window frame for an octagonal window like the one shown at right. To make the frame, you'll cut identical trapezoid pieces. What are the measures of the angles of the trapezoids? Explain how you found these measures.

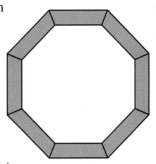

15. Computer Activity Construct a quadrilateral and the midpoints of its four sides. Construct segments connecting the midpoints of opposite sides. Construct the point of intersection of the two segments. Drag a vertex or a side and distort the quadrilateral. Observe these segments and make a conjecture.

16. What is the equation of the perpendicular bisector of the segment with endpoints (–12, 15) and (4, –3)?

17. What is the equation of the median to the side *AB* in △*ABC* with *A*(0,0), *B*(–4, –2), and *C*(8, –8)?

18. What's wrong with this picture?

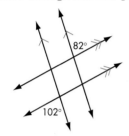

19. What's wrong with this picture?

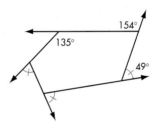

Improving Visual Thinking Skills—*Picture Patterns II*

Draw the next picture in each pattern. Then write the rule for the total number of squares in the *n*th picture of the pattern.

1.

2.

Lesson 6.2

Exterior Angles of a Polygon

You will do foolish things, but do them with enthusiasm.
— *Sidonie Gabriella Colette*

In Lesson 5.3, you extended a side of a triangle to create an exterior angle. If you extend each side of a polygon to form one exterior angle at each vertex, you create a set of exterior angles for the polygon.

In Lesson 6.1, you discovered a formula for the sum of the measures of the *interior* angles of any polygon. In this lesson you will discover a formula for the sum of the measures of a set of *exterior* angles of a polygon.

Let's investigate the sum of the measures of a set of exterior angles for five or six different polygons. Again, this investigation will be easier if you work in groups of four or five. Each group should work on the same polygon they worked on in Investigation 6.1.

Set of exterior angles

Investigation 6.2

Each person in your group should draw a large polygon onto her or his sheet of paper and then perform Steps 1-5.

Step 1   Draw a set of exterior angles.

Step 2   With a protractor, measure all the interior angles of the polygon except one.

Step 3   Use the Polygon Sum Conjecture to calculate the measure of the remaining interior angle.

Step 4   Use the Linear Pair Conjecture to calculate the measure of each exterior angle.

Step 5   Calculate the sum of the measures of the set of exterior angles.

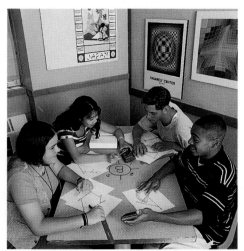

Share your results with your group members. Copy the table below onto your notebook's investigation page and complete the table as each group reports their results for their polygon. From the completed table, find the formula that relates the number of sides of a polygon to the sum of the measures of that polygon's exterior angles.

Number of sides of polygon	3	4	5	6	7	8	...	n
Sum of measures of exterior angles	360°	-?-	-?-	-?-	-?-	-?-	...	-?-

C-40   The sum of the measures of one set of exterior angles —?—
(***Exterior Angle Sum Conjecture***).

With a little logical reasoning you can extend what you discovered in Lesson 6.1 about interior angle measure sums in polygons to make a conjecture about the interior angles in equiangular polygons. The measure of each angle of an equiangular polygon can be found by finding the sum of the measures of all the interior angles and dividing by the number of angles. For an *n*-gon this is —?—.

But now, with the Exterior Angle Sum Conjecture, you can find the measure of each angle in an equiangular polygon by another method. If the sum of the measures of the exterior angles of an *n*-gon is 360°, then what is the measure of each exterior angle of an equiangular polygon? Because each interior angle is the supplement of each exterior angle, the measure of each interior angle of an equiangular *n*-gon can also be expressed as —?—.

State your conjecture.

 C-41 The measure of each angle of an equiangular *n*-gon can be found by using either of the following expressions: —?— or —?— (***Equiangular Polygon Conjecture***).

Take Another Look 6.2

Try one or more of these follow-up activities.

 1. Use patty papers to demonstrate the Exterior Angle Sum Conjecture.

2.* Draw a polygon that has one set of exterior angles. Label the exterior angles. Cut out the set of exterior angles and arrange all the angles about a point. Explain how this activity demonstrates the Exterior Angle Sum Conjecture.

 3. Use the ray tool in a geometry computer program to construct a polygon with one set of exterior angles. Measure the angles and calculate the sum. Dilate the polygon toward a single point. What are you left with when the polygon has shrunk to a single point? How does this activity demonstrate the Exterior Angle Sum Conjecture?

4. Because you know from the Equiangular Polygon Conjecture how to determine the measure of each angle of any regular polygon, you should be able to use a protractor and a straightedge to accurately draw any regular polygon. Try to draw a regular hexagon, octagon, or other polygon. Or, use a geometry computer program to construct regular polygons by rotations.

5. Explain why the Exterior Angle Sum Conjecture follows logically from the Polygon Sum Conjecture and the Linear Pair Conjecture.

6. Use algebra to demonstrate that the two algebraic expressions for the measure of each angle of an equiangular polygon, $\frac{(n-2)180}{n}$ and $180 - \frac{360}{n}$, are equivalent.

7.* Is the Exterior Angle Sum Conjecture also true for concave polygons? Choose your tools and investigate.

 8. Investigate exterior angle sums for polygons on a sphere. Be sure to test cases of the smallest and largest polygons you can draw.

Exercise Set 6.2

In Exercises 1–6, use your conjectures to calculate each lettered angle measure.

1.
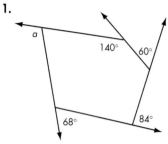

2.
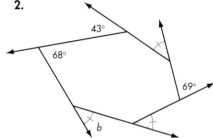

**3.*
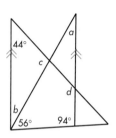

4.
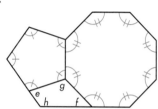

5.
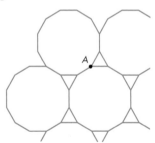

**6.*
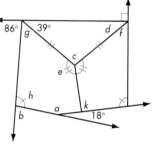

7. Name the regular polygons that appear in the tiling below. Find the measures of the angles that surround point *A* in the tiling.

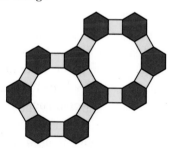

8. Name the regular polygons that appear in the tiling below. Find the measures of the angles that surround any point in the tiling.

9. What is the sum of the measures of a set of exterior angles of a decagon?

10. Four exterior angles of a pentagon measure 63°, 67°, 58°, and 64°. What is the measure of the remaining exterior angle?

11.* What is the measure of each exterior angle of a regular hexagon?

12.* How many sides does a regular polygon have if each exterior angle measures 24°?

13. What is the sum of the measures of the interior angles of a dodecagon?

14. How many sides does a polygon have if the sum of its interior angle measures is 7380°?

15. What is the measure of each interior angle of a regular octagon?

16.* How many sides does a regular polygon have if each of its interior angles measures 165°?

17. The **aperture** of a camera is the opening that limits the amount of light coming through the camera's lens. In many cameras, the aperture is shaped like a regular polygon surrounded by thin sheets that form a set of exterior angles. These sheets move together or apart to close or open the aperture. Explain how the sequence of closing apertures shown below demonstrates the Exterior Angle Sum Conjecture.

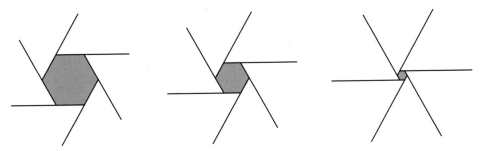

18.* Is there a maximum number of obtuse exterior angles that any polygon can have? If so, what is the maximum? If not, why not? Is there a minimum number of acute interior angles that any polygon can have?

19. Six equilateral triangles can fit about a point ($6 \times 60° = 360°$) without gaps or overlaps. Can you do this with any triangle? Try it. Draw a triangle. Make five copies. Label the interiors of the three angles *a*, *b*, and *c* correspondingly in each triangle. Cut the triangles out. Try arranging them about a point. If you had more triangles, do you think you could continue this process to fill the page?

20. Four rectangles can fit about a point ($4 \times 90° = 360°$) without gaps or overlapping. Can you do this with any quadrilateral? Try it. Draw a quadrilateral. Make three copies. Label the interiors of the four angles *a*, *b*, *c*, and *d* correspondingly in each quadrilateral. Cut the quadrilaterals out. Try arranging them about a point. If you had more quadrilaterals, do you think you could continue this process to fill the page?

21.* Copy all five measures of the musical pattern shown. Assuming the pattern continues, draw in the location of all the notes in the next measure.

22. Draw a counterexample to show that the following statement is false: If a triangle is isosceles, then the base angles cannot be complementary.

23. Find the coordinates of three more points that lie on the line passing through the points (2, –1) and (–3, 4).

For Exercises 24–26, find the algebraic rule that turns the ordered pair on the left into the ordered pair on the right.

24. * (2, –3) → (–2, 3)
(4, 6) → (–4, –6)
(–5, –2) → (5, 2)
(–6, 9) → (6, –9)
(0, –1) → (0, 1)
(x, y) → (?, ?)

25. * (2, 2) → (1, 4)
(4, –1) → (2, –2)
(0, 4) → (0, 8)
(–6, 0) → (–3, 0)
(1, 5) → (0.5, 10)
(x, y) → (?, ?)

26. (4, 1) → (–2, 4)
(–2, –4) → (8, –2)
(3, 2) → (–4, 3)
(–5, 0) → (0, –5)
(–1, –3) → (6, –1)
(x, y) → (?, ?)

27. First grade teacher T. Chumwell is helping his students decorate their classroom for a birthday party. The children want to string crepe paper from each desk to each other desk in the classroom. The fourteen desks are arranged at points evenly spaced around a circle. How many different lengths of crepe paper will be needed, and how many of each length will be needed?

 **Improving Reasoning Skills—***How Did the Farmer Get to the Other Side?*

A farmer was taking her daughter's large pet rabbit, a basket of prize-winning baby carrots, and her small—but hungry—rabbit-chasing dog to town. She came to a river and realized she had a problem. The little boat she found tied to the pier was big enough to carry only herself and one of the three possessions. She couldn't leave her dog on the bank with the little rabbit (the dog would frighten the poor rabbit), and she couldn't leave the rabbit alone with the carrots (the rabbit would eat all the carrots). She had to figure out how to cross the river safely with one possession at a time, ensuring that the dog and the rabbit were not left alone together and that the rabbit and the carrots were not left alone together. How could she safely move back and forth across the river to get the three possessions to the other side?

 **Improving Visual Thinking Skills—***Dissecting a Hexagon II*

Trace the hexagon at right six times. Then divide each hexagon into twelve identical parts. Divide each hexagon differently.

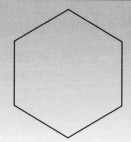

Lesson 6.3

Discovering Kite and Trapezoid Properties

Imagination is the highest kite we fly.
— Lauren Bacall

Recall the definition of a kite from Chapter 2.

*A **kite** is a quadrilateral with exactly two pairs of distinct congruent consecutive sides.*

If you construct two different isosceles triangles on opposite sides of a common base and then remove the base, you have constructed a kite. In an isosceles triangle, the angle between the two congruent sides is called the vertex angle. Therefore let's call the two angles between each pair of congruent sides of a kite the **vertex angles** of the kite and let's call the other pair the **nonvertex angles**.

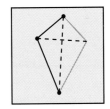

Nonvertex angles

Vertex angles

What properties can you discover about kites? Let's investigate.

Investigation 6.3.1

In this investigation you will look at the diagonals of a kite. Perform the following steps (the first two give you a kite), then compare your results with your group.

Step 1   On a patty paper, draw two segments of different lengths as shown.

Step 2   Fold through the endpoints and trace the two segments on the back of the patty paper.

Step 3   Unfold the patty paper and draw both diagonals.

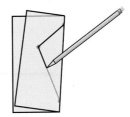

Step 1   Step 2   Step 3

How do the diagonals intersect? Share your observations with others near you. State your observations as your next conjecture.

C-42   The diagonals of a kite are —?— (***Kite Diagonals Conjecture***).

Let's continue. What else seems to be true about the diagonals of kites?

Step 4 How do the diagonals divide each other? Does either one bisect the other? Test by comparing the lengths of the segments on both diagonals.

Share your observations with others near you. State your next conjecture.

C-43 The diagonal connecting the vertex angles of a kite is the —?— of the other diagonal (***Kite Diagonal Bisector Conjecture***).

What else seems to be true about kites?

Step 5 Compare the size of each pair of opposite angles in your kite by folding an angle onto the opposite angle. Do this for both pair of opposite angles.

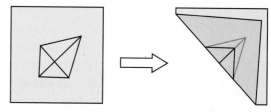

Are both pairs of opposite angles congruent? One pair? Which pair? Are any pairs complementary? Supplementary? Share your observations with others near you. State your observations as your next conjecture.

C-44 The —?— angles of a kite are —?— (***Kite Angles Conjecture***).

Here's one more kite investigation.

Step 6 How do the diagonals divide the opposite angles? Does either diagonal bisect one of the angles? Test by folding on both diagonals.

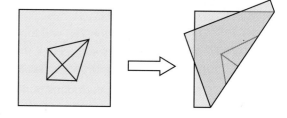

Share your observations with others near you. State your next conjecture.

C-45 The —?— angles of a kite are —?— by a —?— (***Kite Angle Bisector Conjecture***).

Let's move on to trapezoids. Recall the definition of a trapezoid from Chapter 2.

*A **trapezoid** is a quadrilateral with exactly one pair of parallel sides.*

In a trapezoid, the parallel sides are called **bases**. A pair of angles that share a base as a common side are called a pair of **base angles**.

In Investigation 6.3.2, you will discover a property of trapezoids.

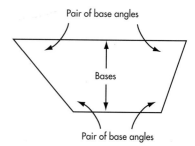

Pair of base angles

Bases

Pair of base angles

Project

Building an Arch

The arches in this Roman aqueduct are typical of arches you can still find throughout what was once the Roman empire. However, arches are not just a feature of ancient constructions. They are a basic architectural element, intended to provide openings and structural support. In this project, you'll design and build your own arch. You may work alone, with a group, or as an entire class.

An arch is a curved structure built out of wedge-shaped blocks whose ends are trapezoids. (These blocks are called voussoirs.) Each block supports, and is supported by, the blocks surrounding it.

The earliest arch was the triangular relieving arch built by the Mycenaeans. Mycenae was an ancient city (2000 to 1100 B.C.) in what is now Greece. The Etruscans of what is now western Italy built the first true arches with radiating voussoirs around 700 B.C. Later, the Romans used the classical arch design in bridges, aqueducts, and buildings. The classical arch is a simple design based on a semicircle. The interior radius is the rise of the arch, and the thickness of the blocks is the depth. The arch is often placed on columns or pillars to increase the height.

The center of the arch is called the keystone. The arch will not hold together until the keystone is in place, so in masonry arches a semicircular wooden support is built to hold the voussoirs in place until the keystone is placed. The abutments on the sides of the arch keep the arch from spreading out and falling down. In the arch above, the trapezoidal blocks are all the same size and shape, except for the keystone.

The first thing you should do to design your arch is to make a list of the decisions you will need to make.

- What size arch would you like to build? Do you want to build an arch that is large enough for you to stand under, or do you want to build an arch that fits on your desk top? Write down the rise of your arch.

- If you decide to build a large arch you may want to do this as a classroom project and have everyone build an identical voussoir.

- How many voussoirs should there be? You will need an even number of voussoirs and one keystone.

- How thick should the pieces be?

Next, draw a scale diagram to determine the size and angle of the voussoirs and the keystone. Once you've decided on side and angle measures, you can make a template for your voussoirs. The example template at right can be folded into a trapezoidal block. Be sure that the template you make has the correct angles for the number of blocks you plan to make. For example, the arch in the diagram makes half a regular 18-gon. You'd need to calculate the measure of each angle in a regular 18-gon and use that to calculate the measures of the angles in your trapezoid. You should have your trapezoid approved by your group or your teacher before construction begins.

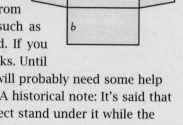

You will need to brace the bottom pieces to keep the arch from spreading apart and falling. Be sure to use light material such as cardboard, and design each piece so that it is easy to build. If you have access to a wood shop, you could build solid wood blocks. Until the keystone is in place, the arch will not stay in place. You will probably need some help holding the pieces in place until the final piece is laid down. A historical note: It's said that when the Romans made an arch, they would make the architect stand under it while the wooden support was removed. That was one way to be sure architects carefully designed arches that wouldn't fall!

Improving Reasoning Skills—*Scrambled Arithmetic*

In the addition statement $65 + 28 = 43$, all the digits are correct but they are in the wrong places! The correct equation should be $23 + 45 = 68$. In each of the three equations below, the operations and the digits are correct, but some of the digits are in the wrong places. Find the correct equations.

1. $11 + 66 = 457$ **2.** $39 \times 11 = 75$ **3.** $\frac{783}{52} = 31$

Lesson 6.4

Discovering Properties of Midsegments

Research is formalized curiosity. It is poking and prying with a purpose.
— *Zora Neale Hurston*

The segment connecting the midpoints of two sides of a triangle is called a **midsegment** of a triangle. In this lesson you will discover special properties of midsegments. In Investigation 6.4.1, you will discover two properties about a midsegment of a triangle.

Investigation 6.4.1

Step 1	Draw a triangle onto a patty paper, pinch to locate midpoints of the sides, and draw in the midsegments.
Step 2	Label the angles as shown. Place a second patty paper over the first and copy one of the four triangles.
Step 3	Compare all four triangles by sliding the copy of one of them over the other three.

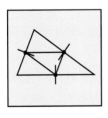

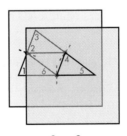

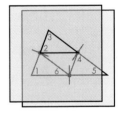

Step 1   Step 2   Step 3

Are the four triangles congruent? Compare your results with the results of others near you. State your observations as your next conjecture.

 C-49 The three midsegments of a triangle divide the triangle into —?—.

A midsegment connects the midpoints of two sides of a triangle. Can you see any relationships between a midsegment and the triangle's third side (the side not containing either endpoint)? Discuss this with others in your group. State your conjecture and explain why you think it's true.

 C-50 A midsegment of a triangle is —?— to the third side and —?— the length of —?— (***Triangle Midsegment Conjecture***).

The line segment connecting the midpoints of the two nonparallel sides of a trapezoid is called the **midsegment** of the trapezoid. In Investigation 6.4.2, you will discover two properties about the midsegment of the trapezoid.

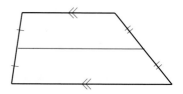

Investigation 6.4.2

Step 1   Draw a small trapezoid on the left side of a patty paper. Pinch to locate the midpoints of the nonparallel sides. Draw in the midsegment.

Step 2   Label the angles as indicated. Place a second patty paper over the first and make a copy of the trapezoid and the midsegment.

Step 3   Compare a pair of base angles with the corresponding angles at the midsegment by sliding the copy up over the original.

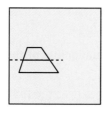 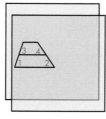 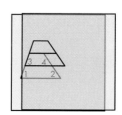

| Step 1 | Step 2 | Step 3 |

Is ∠1 ≅ ∠3? If so, then the midsegment is parallel to the bases. Compare your results with the results of others near you. Then, beneath your construction, copy and complete the conjecture below.

Conjecture: The midsegment of a trapezoid is —?—.

What else can you discover about the midsegment of a trapezoid? In a triangle, the midsegment is half the length of the third side. How does the length of the midsegment compare to the two bases? Let's investigate.

Step 4   On the original trapezoid, extend the longer base to the right by at least the length of the shorter base.

Step 5   Slide the second patty paper under the first. Show the sum of the lengths of the two bases by marking a point on the extension of the longer base.

Step 6   Check how many times the midsegment fits on the segment representing the sum of the lengths of the two bases.

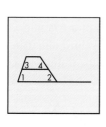 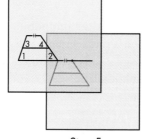 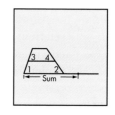

| Step 4 | Step 5 | Step 6 |

What do you notice about the length of the midsegment and the sum of the lengths of the two bases? Compare your results with the results of others. Beneath your construction, copy and complete the conjecture.

Conjecture: The midsegment of a trapezoid is equal in length to —?—.

Now combine the two trapezoid midsegment conjectures into one and add it to your conjecture list.

C-51   The midsegment of a trapezoid is —?— to the bases and is equal in length to —?— (*Trapezoid Midsegment Conjecture*).

Take Another Look 6.4

Try one or more of these follow-up activities.

1. Use a geometry computer program or a compass and a straightedge to confirm your midsegment conjectures.

2.* Use coordinate geometry to confirm that the midsegment of a triangle is parallel to the third side and that the midsegment of a trapezoid is parallel to the two bases.

3.* Are the converses of the Triangle Midsegment Conjecture and the Trapezoid Midsegment Conjecture true? Choose your tools and devise investigations to find out.

Exercise Set 6.4

Use your new conjectures to solve Exercises 1–6.

1.* What is the perimeter of $\triangle TOP$?

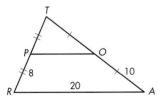

2. $x = $ -?-
       $y = $ -?-

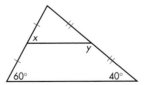

3.* $z = $ -?-

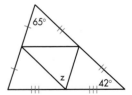

4. What is the perimeter of $\triangle TEN$?

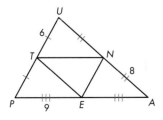

5. $m = $ -?-
       $n = $ -?-

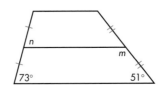

6. $q = $ -?-

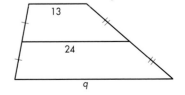

7. Trace the figure below. Calculate each lettered angle measure.

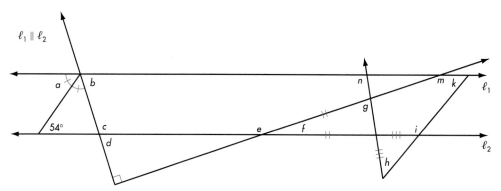

8. *CART* is an isosceles trapezoid. What are the coordinates of point *T*?

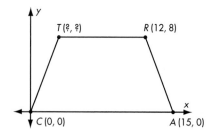

9. *HRSE* is a kite. What are the coordinates of point *R*?

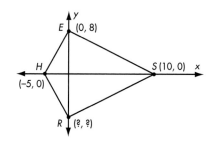

10. How many midsegments are in a triangle? How many are in a trapezoid?

11. Draw a triangle onto a patty paper. Pinch to locate all three midpoints of the three sides. Construct one of the midsegments. Construct a median that crosses the midsegment. How does the midsegment divide the median?

12. Draw another triangle onto a patty paper. Again, patty-paper construct one of the midsegments. Fold to construct an altitude that crosses the midsegment. How does the midsegment divide the altitude?

13. Repeat Exercise 12 with a trapezoid. That is, construct a trapezoid and its midsegment, then fold to construct an altitude between the two bases. How does the midsegment divide the altitude?

14. When you connected the midpoints of the three sides of a triangle, you created four congruent triangles. What type of quadrilateral do you get when you connect the midpoints of the four sides of an isosceles trapezoid?

15.* The term we have given for the line segment connecting the midpoints of the nonparallel sides of a trapezoid, *midsegment,* is not yet the standard term used. In some geometry texts, this segment is called *the median in a trapezoid.* Which do you think is a better name for the segment and why? Explain.

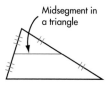

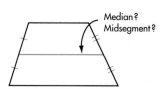

16. Deep in the tropical rain forest of Pana Rica, archaeologist Ertha Diggs and her assistant researchers, Brad and Janet, have uncovered a pyramid similar to the pyramids they'd found on earlier trips. The square-based pyramid is really a truncated pyramid (a square pyramid with a smaller square pyramid removed from the top), but this one has a striking difference. A petrified wooden post is still standing exactly in the middle of the top square base. The post is decorated with symbols and clearly was used in ceremonies. Ertha recalls from her research that descendants of the Maya used such a pole in festivals. In the festivals, priests would string vines of flowers from the top of the pole down to the four corners of the base. They then would continue stringing the vines, in a straight line, down the four edges of the pyramid, thus forming the straight lateral edges of a square-based pyramid. The distances from each of the four corners of the top base to the four corners of the bottom base were all equal to the distances from the four corners to the top of the pole. Knowing this and some geometry, Ertha determines the length of the edges of the top base. She measures the length of one of the bottom bases: 60 meters. With this information she is able to calculate the length of the edge on the top base without having to climb up and measure it. Can you? What is the length of the top edge? How do you know?

Ertha Diggs and the mystery pyramid

17. Monica and Denee are sitting on the front deck of their cabin. They pride themselves on their estimation skills and are taking turns estimating distances between objects in their surroundings. There are two large redwood trees visible from where they are sitting, one on the left side and one on the right side of their property. Denee claims that the two trees are 180 feet apart, and Monica says they are 275 feet apart.

The problem is, they can't measure the distance to see whose estimate is closest because the cabin is located between the trees. All of a sudden, Denee recalls her geometry: "Oh yeah, the Triangle Midsegment Conjecture!" She runs to get a tape measure, a hammer, and a stake. What is she going to do?

Graphing Calculator Investigation—*Drawing Regular Polygons*

In this investigation, you'll learn how to draw regular polygons with your graphing calculator by entering values for the central angle measure of the polygon you wish to draw. A regular polygon's **central angle** is formed by segments from that polygon's center to its consecutive vertices. The measure of each central angle of a hexagon is 60°.

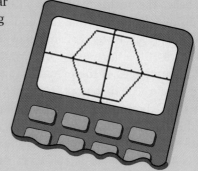

Set your calculator's mode to degrees and parametric. **Parametric equations** give the *x*- and *y*-coordinates of a point in terms of a third variable, or parameter, *t*. (You'll learn more about parametric equations in a later math course.) Set a friendly window with a *y*-range of at least –3.1 to 3.1. Set *t*-min = 0, *t*-max = 360, and *t*-step = 60. Enter the equations $x = 3\cos t$ and $y = 3\sin t$, then graph them. You should get a hexagon. (If your hexagon doesn't appear to be regular, you probably don't have a friendly window.)

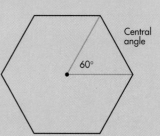

Central angle

60°

The equations you graphed are actually the parametric equations for a circle, but the calculator draws a circle by computing a finite number of points and connecting them. By using a *t*-step of 60 for *t*-values from 0 to 360, you told the calculator to compute only six points on the circle. The *t*-step value determines the measure of each central angle of the polygon.

1. Choose different *t*-steps to draw different regular polygons, such as an equilateral triangle, a square, a regular pentagon, and so on. Complete the statement: The measure of each central angle of an *n*-gon is —?—.

2. Complete the statement: As the measure of each central angle of a regular polygon decreases, the number of sides —?—.

3. As you draw polygons with a greater number of sides, you'll find you soon reach the limits of the calculator's resolution. For example, it's hard to tell a regular 12-gon from a circle. Does this give you an idea for an alternative definition of a circle?

4. Experiment with rotating your polygons by choosing different *t*-min and *t*-max values. For example, try setting *t*-min = –45 and *t*-max = 315, then drawing a square.

5. See if you can draw star polygons on your calculator. Here's a hint: Try *t*-max values of 360*n* and choose *t*-step values that evenly divide your *t*-max but that don't evenly divide 360. Can you explain how this works?

Lesson 6.5

Discovering Properties of Parallelograms

A **parallelogram** is a quadrilateral whose opposite sides are parallel.

In this lesson you will discover some special properties of parallelograms.

Investigation 6.5

Step 1 Using the lines on a piece of graph paper as a guide, draw a pair of parallel lines that are at least 6 cm apart.

Step 2 Using the parallel edges of your straightedge or the points on the graph paper, make a parallelogram. Label your parallelogram *LOVE*.

Step 3 Use a protractor to measure ∠ELO. Then calculate the remaining angles. Compare a pair of opposite angles.

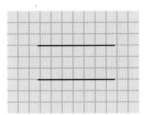 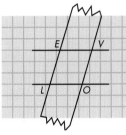 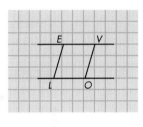

| Step 1 | Step 2 | Step 3 |

Compare your results with the results of others near you. State your observations as your next conjecture.

C-52 The —?— of a parallelogram are —?—.

You have learned that two angles of a polygon that share a common side are called consecutive angles. In parallelogram *LOVE*, ∠LOV and ∠EVO are a pair of consecutive angles. The consecutive angles of a parallelogram are not equal in measure, but there is a relationship between them.

Step 4 Find the sum of the measures of each pair of consecutive angles in your parallelogram.

Compare your observations with the observations of others near you. State a conjecture about the consecutive angles in a parallelogram.

C-53 The —?— of a parallelogram are —?—.

Step 5   With your compass, compare the lengths of the opposite sides of the parallelograms you made.

Compare your results with the results of others near you. State a conjecture about the opposite sides of a parallelogram.

C-54 The —?— of a parallelogram are —?—.

Step 6   Construct in your parallelograms the diagonals $\overline{LV}$ and $\overline{EO}$, as shown below. Label the point of intersection of the two diagonals point M.

Does $LM = VM$? Does $MO = ME$?
If $LM = MV$, then M is the —?— of diagonal LV.
If $MO = ME$, then M is the —?— of diagonal OE.
Does $\overline{EO}$ bisect $\overline{LV}$? Does $\overline{LV}$ bisect $\overline{EO}$?

Compare your observations with the observations of others near you. State a conjecture about the diagonals of a parallelogram.

C-55 The —?— of a parallelogram —?—.

Parallelograms are used in vector diagrams, which have many applications in science. A **vector** is a quantity that has both direction and magnitude. Quantities from physics—such as velocity, acceleration, and force—can be described with vectors. Vectors are often represented as arrows. The length and direction of an arrow represents the magnitude and direction of the vector. Many physics problems involve two or more vector quantities acting on the same object. The resultant vector of two or more vectors is the single vector that has the same effect. One way to draw a **resultant vector**, or **vector sum**, of two vectors is to draw the two vectors tail-to-tail, then complete a parallelogram with those vectors as sides. The resultant vector is the diagonal of the parallelogram from the two vectors' tails.

Air velocity and wind velocity

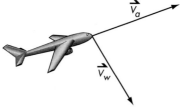

Resultant vector shows ground velocity.

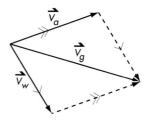

Take Another Look 6.5

Try one or more of these follow-up activities.

1. Use a geometry computer program or patty papers to confirm that your parallelogram conjectures are correct.

2.* Use coordinate geometry to verify one of your parallelogram conjectures.

3.* Are the converses of the parallelogram conjectures true? Choose a tool and devise an investigation to determine whether the converses are true or false.

4.* Choose one of the conjectures from this lesson and explain how it follows logically from earlier conjectures.

Exercise Set 6.5

Use your new conjectures to solve the problems below. In Exercises 1–6, each figure is a parallelogram.

1. $a = $ -?-
   $b = $ -?-

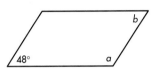

2. $c = $ -?-
   $d = $ -?-

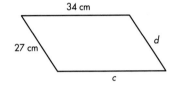

3. $e = $ -?-
   $f = $ -?-

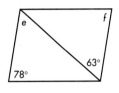

4.* What is the perimeter?

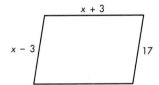

5. $g = $ -?-
   $h = $ -?-

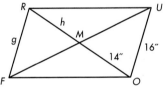

6.* $VF = 36$ m
   $EF = 24$ m
   $EI = 42$ m
   What is the perimeter of
   $\triangle NVI$?

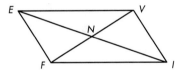

7. Trace the figure below. Calculate each lettered angle measure.

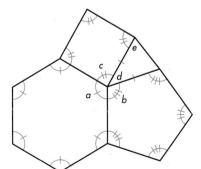

8. Find the lettered angle measures in this tiling of regular polygons.

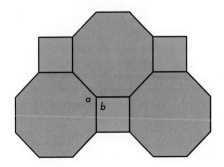

9. Find coordinates of midpoints *E, Z*. Show that the slope of midsegment $\overline{EZ}$ is equal to the slope of $\overline{YT}$.

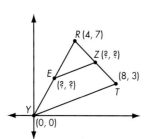

10. Find the coordinates of point *M* in parallelogram *PRAM*.

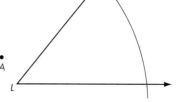

11. Given side $\overline{LA}$, side $\overline{AS}$, and $\angle L$, construct parallelogram *LAST*.

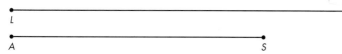

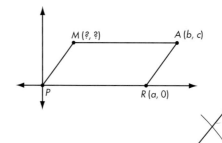

12. * Given side $\overline{DR}$ and diagonals $\overline{DO}$ and $\overline{PR}$, construct parallelogram *DROP*.

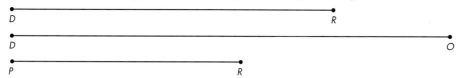

13. Group Activity Construct a quadrilateral. Make a copy of it. Draw a diagonal in the first quadrilateral. Draw the opposite diagonal in the duplicate quadrilateral. Cut each quadrilateral into two triangles along the diagonals drawn. Arrange the four triangles into a parallelogram. Explain how you know this figure is a parallelogram.

In Exercises 14 and 15, copy each pair of vectors and draw the resultant vector.

14.

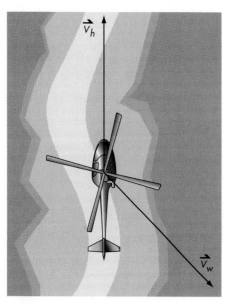

15. *

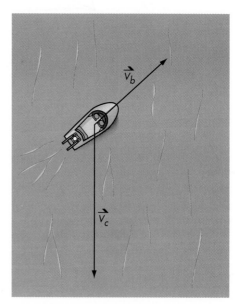

Lesson 6.6

Discovering Properties of Special Parallelograms

The legs of the lifting platforms shown at right form rhombuses. Can you imagine how such a lift would look and work differently if the legs formed parallelograms that weren't rhombuses? In this lesson you will discover some properties of rhombuses, rectangles, and squares. What you discover about the diagonals of a rhombus will help you understand why these lifts work the way they do.

In Investigation 6.6.1, you will discover the special parallelogram formed when you use just the parallel edges of a straightedge.

Investigation 6.6.1

Step 1	Step 2	Step 3
	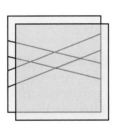	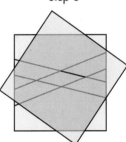
On a patty paper, use a double-edged straightedge to draw two pairs of parallel lines that intersect each other.	Continue by placing a second patty paper over the first and copy one of the sides of the parallelogram.	Compare the length of the side on the second patty paper with the lengths of the other three sides of the parallelogram.

How do the lengths of the four sides compare? Compare your results with the results of others near you. State a conjecture about the type of parallelogram formed when one set of parallel lines are intersected by a second pair of parallel lines that are the same distance apart.

C-56 If two parallel lines are intersected by a second pair of parallel lines the same distance apart as the first pair, then the parallelogram formed is a —?—.

This is a useful conjecture because now you have an easy way to construct a rhombus (an equilateral parallelogram). You don't need a compass or patty papers; you just need a normal ruler with parallel edges. However, if you wish to construct a parallelogram that is not a rhombus, you will need to use two rulers of different widths.

Rhombus

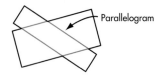

Parallelogram

Now let's discover some properties of rhombuses.

Investigation 6.6.2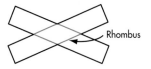

Step 1   Draw in both diagonals of the rhombus you created in Investigation 6.6.1.

Step 2   Place the right-angled corner of a second patty paper into one of the angles formed by the intersection of the two diagonals.

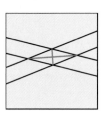

Step 1

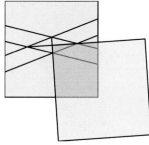

Step 2

What kind of angle is formed? Compare your results with the results of others near you. Conjecture 55 tells us that the diagonals of a parallelogram bisect each other. A rhombus is a parallelogram. Therefore, by Conjecture 55, the diagonals of a rhombus must also bisect each other. But you have also just discovered that the diagonals of a rhombus intersect each other in another special way. Combine these two ideas into your next conjecture.

 C-57 The —?— of a rhombus are —?— of each other.

Use patty papers or a protractor to compare the angles formed by the diagonals and the sides of the rhombus at each vertex. What do you observe? Compare your results with the results of others. State a conjecture about the pair of angles at each vertex.

 C-58 The —?— of a rhombus —?— the angles of the rhombus.

Now let's turn our attention to rectangles. What conjectures can you make?

*A **rectangle** is an equiangular parallelogram.*

CHAPTER 6

Geometer's Sketchpad Project

Star Polygons

If you arrange a set of points roughly around a circle or oval then connect each point to the next with segments, you should get a convex polygon like pentagon *ABCDE*. But what do you get if you connect every second point with segments? You get a star polygon like the 5-pointed star *ABCDE* or the 6-pointed star *FGHIJK* below. (Notice that with a pencil you could draw the five-pointed star without picking up your pencil. The six-pointed star would require you to pick up your pencil.) In this project, you'll investigate properties of star polygons, particularly their angle measure sums.

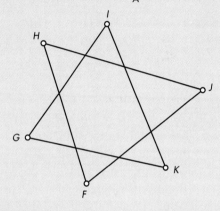

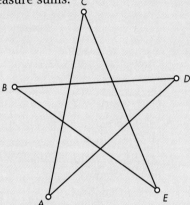

Step 1 Draw five points *A* through *E* in clockwise order so that they roughly lie on a circle. (They don't actually need to be on a circle.)

Step 2 Connect every other point with segments: $\overline{AC}$, $\overline{CE}$, $\overline{EB}$, $\overline{BD}$, and $\overline{DA}$.

Step 3 Measure the five angles *A* through *E* at the star points.

Step 4 Use the calculator to sum the star point angle measures.

Drag each vertex of the star and observe what happens to the angle measures and the calculated sum. Does the sum change? Copy the chart on the next page and enter your findings for the angle measure sum in a 5-pointed star formed by connecting every second point. If you draw little figures in your chart it will help you see the pattern. Connecting every point gives you a polygon. The chart has been started for you with angle measure sums for pentagons and hexagons. You should be able to complete the rest of that column using the Polygon Sum Conjecture.

What if you connect every third point of your five points? If you look back at your five-pointed star, you'll see that connecting every third point ($\overline{AD}$, $\overline{DB}$, etc.) gives you the same star.

Now repeat the investigation for a 6-pointed star. Describe the degenerate star you get by connecting every third point. What would be the sum of the angle measures in this star?

Number of star points	Angle measure sums by how the star points are connected				
	Every point	Every 2nd point	Every 3rd point	Every 4th point	Every 5th point
5	540°	–?–	–?–	–?–	–?–
6	720°	–?–	–?–	–?–	–?–
7	–?–	–?–	–?–	–?–	–?–
8	–?–	–?–	–?–	–?–	–?–
9	–?–	–?–	–?–	–?–	–?–
10	–?–	–?–	–?–	–?–	–?–
11	–?–	–?–	–?–	–?–	–?–
12	–?–	–?–	–?–	–?–	–?–
⋮	⋮	⋮	⋮	⋮	⋮
n	–?–	–?–	–?–	–?–	–?–

To collect data for the rest of your chart, you may want to divide tasks among group members. Make sure that every time you draw a star you drag different vertices to see what happens to the angle measures and calculations. Observe the conditions that cause your calculations to change. Check in with your group members often to share observations and data. Look for patterns emerging in the chart. At some point, instead of using Sketchpad to determine an angle measure sum, you might want to guess what the sum will be first and use Sketchpad to test your guess.

See if you can write the rule for n-point stars. Here's a hint: Use the absolute value symbol. For example, if $n = 4$, $|n - 6| = |-2| = 2$.

Write a short report describing the conjectures you made and the patterns you found in the chart. If you can explain why the angle measure sums are what they are for any particular polygon stars, do so.

Improving Algebra Skills—*Substitute and Solve*

1. If $2x = 3y$, $y = 5w$, and $w = \frac{20}{3}z$
find x in terms of z.

2. If $7x = 13y$, $y = 28w$, and $w = \frac{9}{26}z$
find x in terms of z.

Lesson 6.7

Chapter Review

As best I can recall, I can't recall.
— Oliver North

See if you recall the definitions of these terms related to polygons: *kite, isosceles trapezoid, parallelogram, rhombus, rectangle, square, midsegment of a triangle, midsegment of a trapezoid, vector, vector sum.* If you're unsure, review your definition list. State as many conjectures about these figures as you can, then review your conjecture list.

What do you think is the most important or useful polygon property you learned about in this chapter? How can you find the number of sides of a regular polygon by measuring only one of its interior angles or one of its exterior angles? How do you construct a rhombus by using only a ruler? How can you use Conjecture 60 to determine if the corners of a room are right angles? In what kind of structure might you find isosceles trapezoids? Review this chapter to find answers to these questions.

Exercise Set 6.7

In Exercises 1–20, identify each statement as true or false. For each false statement, sketch a counterexample if possible, or explain why the statement is false.

1. The sum of the measures of one set of exterior angles of a polygon is always less than the sum of the measures of interior angles.

2. A pair of base angles of an isosceles trapezoid are supplementary.

3. If a quadrilateral has two congruent angles, then it has two congruent sides.

4. The diagonals of a parallelogram are congruent.

5. The diagonals of a rhombus bisect the angles of the rhombus.

6. The diagonals of a rectangle are perpendicular bisectors of each other.

7. The sum of the measures of the five angles of a pentagon is 540°.

8. The measure of each angle of a regular dodecagon is 150°.

9. If the vertex angles of a kite measure 48° and 36°, then the nonvertex angles each measure 138°.

10. The sum of the measures of any two angles of a triangle is always greater than the supplement of the third angle.

11. The sum of the measures of any two consecutive angles of a trapezoid is greater than the sum of the measures of any pair of base angles.

12. If a quadrilateral has three congruent angles, then the quadrilateral is a rectangle.

13. Every square is a rhombus.

14. If $\overline{CD}$ is the midsegment of trapezoid *PLYR* with $\overline{PL}$ one of the bases, then $CD = \frac{1}{2}(PL + YR)$.

15. The line through the points (2, -4) and (4, -2) is perpendicular to the line $y = -x + 4$.

16. If the sum of the lengths of two consecutive sides of a kite is 48 cm, then the perimeter of the kite is 96 cm.

17. Every equilateral rectangle is a square.

18. The centroid of a triangle is equally distant from the three sides.

19. If (0, 0), (10, 2), and (9, 7) are the coordinates of three vertices of a rectangle, the fourth vertex has coordinates (-1, 5).

20. Exactly nine statements in Exercises 1–20 are true.

21. Copy and complete the table below by placing a yes or no in each empty space.

	Kite	Isosceles trapezoid	Parallelogram	Rhombus	Rectangle
Each pair of opposite sides ∥					
Opposite sides ≅					
Opposite ∠'s ≅					
Diagonals bisect each other					
Diagonals ⊥					
Diagonals ≅				No	
Exactly one line of symmetry	Yes				
Exactly two lines of symmetry					

22. Copy the table below. For each space sketch a quadrilateral with the properties stated in its row and column. If no quadrilateral is possible for that space, write "None." One space has been completed for you.

		Number of pairs of congruent sides		
		0	1	2
Number of lines of symmetry	0			
	1			
	2			

23. Copy the table below. For each space sketch a quadrilateral with the properties stated in its row and column. If no quadrilateral is possible for that square, write "None." One space has been completed for you.

	Number of pairs of parallel sides		
	0	1	2
0			
1	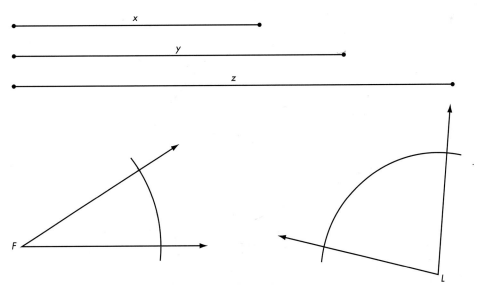		
2			
3			
4			

(Left axis label: Number of right angles)

In Exercises 24–26, use the segments and the angles given below to construct each figure. Use either patty papers or a compass and a straightedge. The lowercase letter above each segment represents the length of the segment in units.

24. Construct kite *FLYR* given ∠*F*, ∠*L*, and *FL* = *x*.

25. Given ∠*F*, *FR* = *z*, and *YD* = *x*, construct two trapezoids *FRYD* that are not congruent to each other.

26. Construct rhombus *SQRE* with *y* and *x* the lengths of the two diagonals.

From the information given, determine which triangles, if any, are congruent. State the congruence conjecture that supports your congruence statement. If the triangles cannot be shown to be congruent from the information given, write "Cannot be determined."

27.* *STARY* is a regular pentagon.

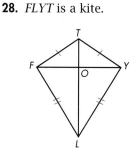

28. *FLYT* is a kite.

29. *PART* is an isosceles trapezoid.

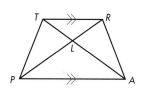

30. Find each lettered angle measure in the figure.

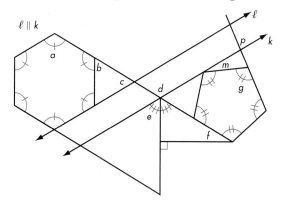

$\ell \parallel k$

31.* Given bases $\overline{OP}$ and $\overline{EN}$, nonparallel side $\overline{ON}$, and $\angle O$, construct trapezoid *OPEN*.

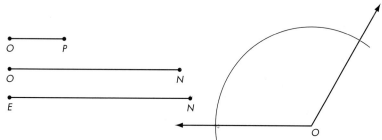

In Exercises 32 and 33, find the *n*th term in each sequence.

32.

1	2	3	4	5	6	...	*n*
1	5	9	13	17	21	...	–?–

33.*

1	2	3	4	5	6	...	*n*
0	1	6	15	28	45	...	–?–

34. Sketch and label two different isosceles trapezoids with a perimeter of $4a + 3b$.

35.* Sketch all the possible different ways for five lines in a plane to intersect.

36. Kite *ABCD* has the vertices $A(-3, -2)$, $B(2, -2)$, $C(3, 1)$, and $D(0, 2)$. Find the coordinates of the point of intersection of the diagonals.

37. Find the equation of the perpendicular bisector of the segment with endpoints $(0, 3)$ and $(6, 1)$.

38. *An airplane is heading north at 900 km/hr. However, a 50 km/hr wind is blowing from the east. Use a ruler and a protractor to make a scale drawing of these vectors. Measure to find the approximate resultant velocity, both speed and direction (measured from north).

39. Nellie Flowers applied for the position of gardener for the estate of the retired, eccentric mathematics professor Ben Dannett. Professor Dannett posed the same problem to each gardening applicant. The first applicant who was clever enough to solve the problem would get the gardening position—with a very handsome salary! Professor Dannett asked each applicant to draw the plans for a garden walk from the following description, showing all dimensions: The outer edges of the walk form a regular polygon, with each side measuring 36 meters. The inner edges of the walk are parallel to the outer edges, and they form the same type of regular polygon with each side measuring 30 meters. Each interior angle of the regular polygons measures 160°. Needless to say, no applicant had solved the professor's conundrum before Nellie. Nellie solved the problem, and she got the job. Could you have solved the problem? How many sides do the two regular polygons have?

Assessing What You've Learned—*Giving a Presentation*

Giving a presentation is one of the most powerful and demanding ways to demonstrate your understanding of a topic. Presentation skills are also among the most useful skills you can develop in preparation for almost any career. The more practice you can get in school, the better.

There are many different types of audiences for presentations. Performance assessment, as described in Chapter 3, usually involves a presentation to an audience of one. By now you may be ready to present some of your mathematical findings to your class. There are a number of things you can do to make your presentation go smoothly.

- Work with a group. Acting as a panel member instead of as a sole presenter can take off some of the pressure. Make sure your group presentation involves all group members so that it's clear everyone contributed equally.

- Choose a topic carefully. Investigations, projects, and Take Another Look activities can all be good subjects for a presentation. Even some exercises, particularly those labeled Group Activity, can serve as good presentation topics.

- Prepare thoroughly. Don't try to memorize everything you're going to say but do prepare a good outline of important points you plan to cover. You may also wish to prepare visual aids—like posters, models, handouts, and overhead transparencies—ahead of time.

- Communicate clearly. Speak up loud and clear. Don't speak in a monotone; vary the pace, pitch, and cadence of your voice to communicate your interest in the subject. Look at everyone in your audience, and don't look down or off in one direction. If you're using an overhead projector, don't talk to it or hide behind it. The clarity of your presentation depends on both well-organized content and the way you deliver it.

Giving a presentation is just one of many possible assessment options. Consider assessing what you've learned in one or more of these other ways.

Organize Your Notebook

• Your conjecture list should be growing fast! Review your notebook to be sure it's complete and well organized. Write a one-page chapter summary.

Write in Your Journal

• Write an imaginary dialogue between your teacher and one or two parents or guardians about your performance and progress in geometry. Try to imagine what your teacher might say about your strengths and weaknesses and about how you might improve. Write what your parent or guardian might say in response and what questions he or she might have for your teacher.

Update Your Portfolio

• Choose a piece of work from this chapter to add to your portfolio: an investigation, a project, a Take Another Look investigation, an exercise set, a test—whatever you think best represents the current quality of your work. Be sure to document the work, explaining what it is and why you chose it for your portfolio.

Performance Assessment

• While a classmate, a friend, a family member, or a teacher observes, carry out one of the investigations from this chapter. Explain what you're doing at each step, including how you arrived at the conjecture.

Improving Visual Thinking Skills—*Coin Swap III and IV*

Arrange four light coins (dimes) and four dark coins (pennies) on a grid of nine squares, as shown. Switch the position of the four light and four dark  coins in exactly 24 moves. A coin can slide into an empty square next to it or can jump over one coin into an empty space. Record your solution by listing, in order, which color coin is moved. For example, your list might begin DLDLDDLLDDD . . . .

Now try it with five light coins (dimes) and five dark coins (pennies). Arrange the coins on a grid of eleven squares, as shown.  Switch the position of the five light and five dark coins in exactly 35 moves. Record your solution as you did above.

How many moves does it take to switch the position of six light and six dark coins?

Cooperative Problem Solving

The Geometry Scavenger Hunt

The lunar colony geometry class from 2065 has just arrived back in our time. They are on a math history scavenger hunt. They were given the scavenger list of mathematics and geometry items shown below. There are four categories in the list: objects from books or magazines, objects from a high school, objects from a home, and objects from a community of our time. The class would like your help.

Scavenger Hunt Rules

Your task is to locate, at most, three items from each category in the time period designated by your teacher. All objects must be brought to class when they are due. The points assigned for each item are shown in parentheses. The team with the most points wins.

1. **Objects from books or magazines**

    • A color photo showing a triangle, a rectangle, a trapezoid, and a circle all in the same photo (20)

    • An advertisement showing a pyramid or a cylinder (10)

    • An article about a scientist or a mathematician and how he or she uses geometry (15)

    • A photo of a sculpture that includes at least four geometric shapes (10)

    • A photo of a building whose design includes at least four different types of quadrilaterals (10)

- An article from a newspaper or a magazine that features mathematics (5) or geometry (10)
- A cartoon about mathematics (5) or geometry (10)

2. **Objects from school**
   - A signed statement from a nonmathematics teacher claiming to have used mathematics (5) or geometry (10) during the past week
   - A nonmathematics textbook that uses geometry to explain a concept (10)
   - A description of a familiar object at school that has 4-fold rotational symmetry (15)
   - An object from the physical education department that has rotational symmetry but not reflectional symmetry (10)
   - A tool or a device from the science department that uses geometry (15)

3. **Objects from your home**
   - A pair of hexahedral dice (5) or a pair of octahedral dice (10)
   - A cooking utensil that has two hemispheres (10)
   - A cylindrical storage container containing a hexagonal prism (5)
   - An object that has rotational symmetry but not reflectional symmetry (5)
   - A carpenter's measuring tool with English units (5) or metric units (10)
   - An abacus or a slide rule (5)
   - An object with reflectional symmetry from a non-European culture (10)

4. **Objects from your community**
   - An object from nature that has rotational but not reflectional symmetry (5)
   - A postcard from near where you live, showing at least two object that have rotational symmetry. On the back of the card, describe the objects and the symmetry that each possesses. (10)
   - A brochure or a signed statement from an employee of an art gallery or of an art or science museum, verifying that you visited the gallery or museum. List at least three things you saw that could be called geometric. (15)
   - A sketch or a photo of a public building whose design includes a cylinder, a hemisphere, and a cone. Include with the sketch a signed note from an adult verifying that you visited the site. (10)
   - An adult willing to visit your class to describe how he or she uses geometry in his or her job (25)

Investigation 7.3.1

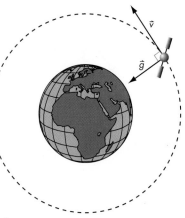

Step 1   Construct a large circle. Label the center O.

Step 2   Using your straightedge, draw a line that appears to touch the circle at only one point. Label the point T. Construct $\overline{OT}$.

Step 3   Use your protractor to measure the angles at T.

Compare your results with the results of others near you. State your observations as a conjecture.

 C-66   A tangent to a circle is —?— to the radius drawn to the point of tangency (***Tangent Conjecture***).

The Tangent Conjecture comes up in many applications related to circular motion. A satellite, for example, maintains its velocity in a direction tangent to its circular orbit. This velocity vector is perpendicular to the force of gravity, which keeps the satellite in orbit. If gravity were suddenly "turned off" somehow, the satellite would travel off into space on a straight line tangent to its orbit.

In Investigation 7.3.2, you will discover something about the lengths of segments tangent to a circle from a point outside the circle.

Investigation 7.3.2

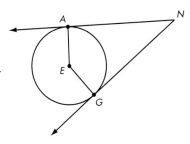

Step 1   Construct a circle. Label the center E.

Step 2   Choose a point outside the circle and label it N.

Step 3   Draw two lines through point N that appear to be tangent to the circle. Mark the points where these lines appear to touch the circle and label them A and G.

Step 4   Use your compass to compare segments NA and NG. (Segments NA and NG are called **tangent segments**.)

Compare your results with the results of others near you. State your observations as your next conjecture.

 C-67   Tangent segments to a circle from a point outside the circle are —?— (***Tangent Segments Conjecture***).

Take Another Look 7.3

1. Use a geometry computer program or patty papers to confirm one of your tangent conjectures.

2.* Choose one of the conjectures in this lesson and explain how it follows logically from earlier conjectures.

3. Investigate and state a conjecture about the quadrilateral formed by two tangent segments to a circle and the two radii to the points of tangency. Explain why you think your conjecture is true.

4.* State a conjecture for segments in space tangent to a sphere. Make a sketch. Test your conjecture with physical objects and explain why you think your conjecture is true.

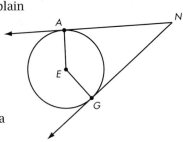

Exercise Set 7.3

1.* Rays *m* and *n* are tangents. *w* = –?–

2.* Rays *r* and *s* are tangents. *x* = –?–

3. Ray *k* is a tangent. *y* = –?–

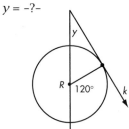

4.* Line *t* is a tangent to both circles. *z* = –?–

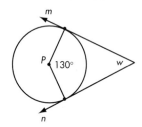

5.* Quadrilateral *POST* is circumscribed about circle *Y*. *OR* = 13 and *ST* = 12. What is the perimeter of *POST*?

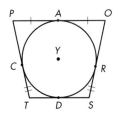

6. Quadrilateral *SHOW* is circumscribed about circle *X*. *WO* = 14, *HM* = 4, *SW* = 11, and *ST* = 5. What is the perimeter of *SHOW*?

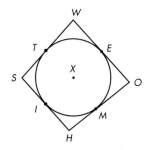

7. Find a real-world example (different from the examples shown in the text) of two internally tangent circles and of two externally tangent circles. Either sketch the examples or make photocopies from a book or a magazine to put into your notebook.

8. Sketch a kite circumscribed about a circle.

9. Sketch a rhombus circumscribed about a circle.

10. Sketch a rectangle circumscribed about a circle. What's special about this rectangle?

For Exercises 11–16, make a sketch of what you are trying to construct and label it. This helps clarify your task. Use the segments below with lengths *r*, *s*, and *t*.

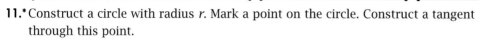

11.*Construct a circle with radius *r*. Mark a point on the circle. Construct a tangent through this point.

12. Construct a circle with radius *t*. Choose three points on the circle and label them *X*, *Y*, and *Z* (see the figure at right). Construct a triangle that is circumscribed about the circle and tangent at points *X*, *Y*, and *Z*.

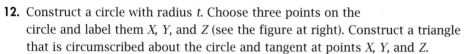

13. Construct two congruent, externally tangent circles with radius *s*.

14. Construct a third circle that is both congruent and externally tangent to the two circles you constructed in Exercise 13.

15. Construct two internally tangent circles with radii *r* and *t*.

16. Construct a third circle with radius *s* that is externally tangent to both the circles you constructed in Exercise 15.

17. In Chinese philosophy, all things are divided into two natural principles, yin and yang. Yin represents the earth, characterized by darkness, cold, or wetness. Yang represents the heavens, characterized by light, heat, or dryness. The two principles combine to produce the harmony of nature. The symbol for yin and yang is shown at right. Construct your own yin and yang symbol. Start with one large circle. Then construct two circles with half the diameter that are internally tangent to the large circle and externally tangent to each other. Finally, construct small circles that are concentric to the two inside circles. Shade or color your yin and yang symbol.

18. What's wrong with this picture? $\overrightarrow{TA}$ and $\overrightarrow{TB}$ are tangent to circle *O*.

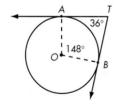

19. Explain how you could use only a T-square, like the one shown, to find the center of a Frisbee™.

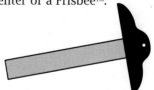

20. Circle *P* is centered at the origin. $\overleftrightarrow{AT}$ is tangent to circle *P* at *A*(8, 15). Find the equation of $\overleftrightarrow{AT}$.

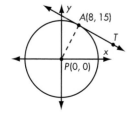

21. Circle *U* passes through points (3, 11), (11, –1), and (–14, 4). Find the coordinates of the center of the circle. Explain your method.

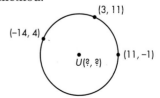

22. Splash Gordon is taking target practice, using water balloons tied to the end of a string. He swings the string and balloon over his head in circles, then he releases the string at the precise moment that will cause the balloon to travel in a straight line to the target. Draw an overhead view that shows the balloon's circular path, the string at the moment Splash releases it, and the balloon's straight path to the target's bull's-eye.

Graphing Calculator Investigation—*Graphing Circles and Tangents*

In this investigation, you'll use algebra to find the equation for a tangent line given the coordinates of a point on a circle. Then you'll use a similar process to graph tangent lines on your graphing calculator.

1. Consider a circle centered at the origin and a point on the circle with coordinates (3, 4). What's the slope of the radius drawn to that point?

2. What's the slope of a tangent line through that point?

3. To graph the tangent line, you need to know the slope, *m*, and the *y*-intercept, *b*. You just found *m*. To find *b*, solve the equation $y = mx + b$ for *b* and substitute the values you already know for *y*, *m*, and *x*. Once you've found *b*, write the equation for the tangent line.

Now that you've practiced finding an equation for a tangent line, here's how to graph circles and tangent lines on the calculator.

- Get a friendly window for your calculator with a *y*-range of at least –3 to 3. Enter the equations $y_1 = \sqrt{(9 - x^2)}$ and $y_2 = -\sqrt{(9 - x^2)}$. When you press GRAPH, you should get a circle. (You'll learn why these equations graph a circle in Chapter 10.)

- Trace and observe the coordinates of a point on the circle. Press the X,T,θ key. What does the value displayed by your calculator represent? Press ALPHA Y. Now what does your calculator display?

- Press (-) X,T,θ ÷ ALPHA Y and store this in ALPHA M. What does the value you just stored in M represent?

- Press ALPHA Y − ALPHA M X,T,θ and store this in ALPHA B. What does the value you just stored in B represent?

- Enter a third equation: $y_3 = Mx + B$ (using the ALPHA key as necessary). Graph the three equations. Describe your result.

- Trace to a different point on the circle and repeat the experiment.

Lesson 7.4

Arcs and Angles

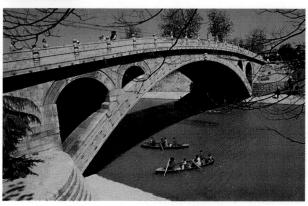

Many arches that you see in structures are semicircular, but the Chinese long ago discovered that arches don't have to be semicircular. The Zhaozhou bridge, shown at left, was completed in A.D. 605. It is the world's first stone arched bridge in the shape of a minor arc, predating similar structures by about 800 years.

In this lesson you'll discover properties of arcs and the angles associated with them.

Recall that the measure of a minor arc is the measure of its central angle. The measure of a major arc is 360° minus the measure of the minor arc making up the remainder of the circle. How does the measure of an inscribed angle compare with the measure of its intercepted arc? Let's investigate.

Investigation 7.4.1

Step 1   Measure ∠COR with your protractor and determine m$\widehat{CR}$.

Step 2   Measure ∠CAR. How does m∠CAR compare with m$\widehat{CR}$?

Step 3   Construct a circle of your own with an inscribed angle and its corresponding central angle.

Step 4   Measure the central angle. What is the measure of the intercepted arc?

Step 5   Measure the inscribed angle. How does the measure of the inscribed angle compare with the measure of its intercepted arc?

Compare your results with the results of others. State a conjecture.

C-68   The measure of an inscribed angle in a circle —?— (***Inscribed Angle Conjecture***).

In the figure at right, ∠AQB and ∠APB both intercept $\widehat{AB}$. Angles AQB and APB are both inscribed in $\widehat{APB}$. Angles AQB and APB appear congruent. Can you find angles inscribed in the same arc that are not congruent? Let's investigate.

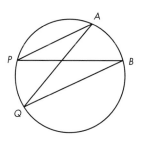

Investigation 7.4.2

Step 1   Construct a large circle.

Step 2   Select two points on the circle. Label them *A* and *B*.

Step 3   Select a point *P* on the major arc and construct inscribed ∠APB.

Step 4   With your protractor, measure ∠APB.

Step 5   Select another point *Q* on major arc *APB* and construct inscribed ∠AQB.

Step 6   Measure ∠AQB. How does the measure of ∠AQB compare with the measure of ∠APB?

       Repeat Steps 1-6 with points *P* and *Q* selected on the minor arc *AB*. Compare the measure of ∠AQB with the measure of ∠APB. Compare your results with the results of others near you. Do you think you can find an angle inscribed in $\overset{\frown}{APB}$ that is not congruent to ∠APB? State your observations as a conjecture.

 C-69   Inscribed angles that intercept the same arc are —?—.

Next you will discover a property of angles inscribed in semicircles.

Investigation 7.4.3

Step 1   Construct a large circle.

Step 2   Construct a diameter.

Step 3   Inscribe three angles in the same semicircle.

Step 4   Measure each angle with your protractor.

       Compare your results with the results of others and make a conjecture.

 C-70   Angles inscribed in a semicircle are —?—.

Now you will discover a property of the angles of a quadrilateral inscribed in a circle.

Investigation 7.4.4

Step 1   Construct a large circle.

Step 2   Construct an inscribed quadrilateral.

Step 3   Measure each of the four inscribed angles. Write the measure in each angle.

It is unlikely that any of the angles are congruent, but there is a special relationship between some pairs of angles. Compare your observations with the observations of those near you. State your findings as your next conjecture.

 C-71 The —?— angles of a quadrilateral inscribed in a circle are —?—.

*A quadrilateral inscribed in a circle is called a **cyclic** quadrilateral.*

Next you will discover a property of arcs formed by parallel lines intersecting a circle.

Investigation 7.4.5

Step 1   On a piece of patty paper, construct a large circle.

Step 2   Lay your straightedge across the circle so that its parallel edges pass through the circle. Draw lines along both edges of the straightedge. Label one arc *AD* and the other *BC*, as shown.

Step 3   Fold your patty paper to compare arcs *AD* and *BC*.

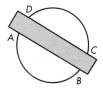

What can you say about arcs *AD* and *BC*? Repeat these steps, using either lined paper or another object with parallel edges. Compare your results with the results of others near you. State your observations as a conjecture.

 C-72 Parallel lines intercept —?— arcs on a circle.

Take Another Look 7.4

 1. Use a geometry computer program to construct a circle and an inscribed angle. Use a point on the circle besides the radius point for the vertex of the angle. Measure the angle and drag the vertex. What conjecture does this illustrate?

 2. Extend Investigation 7.4.4 to see what other properties of cyclic quadrilaterals you can discover. A geometry computer program is a good tool for this activity, or use other tools of your choice.

3.* Choose one of the conjectures in this lesson and explain how it follows logically from earlier conjectures.

4.* State Conjecture 71 in "if-then" form. Then state the converse of the conjecture in "if-then" form. Is the converse also true?

 5. On the surface of a sphere, investigate one or more of Conjectures 68 to 71.

6. Which of the following are always cyclic quadrilaterals: kites, isosceles trapezoids, rhombuses, rectangles, or squares? Explain why each is or is not cyclic.

Exercise Set 7.4

Use your new conjectures to solve Exercises 1–16.

1. $a = -?-$

2. $b = -?-$

3.* $c = -?-$

4. $d = -?-$
$e = -?-$

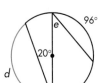

5. $f = -?-$
$g = -?-$

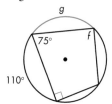

6.* $h = -?-$

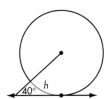

7. *JUST* is a rhombus.
$w = -?-$

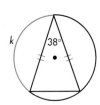

8. *CALM* is a rectangle.
$x = -?-$

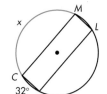

9. *DOWN* is a kite.
$y = -?-$

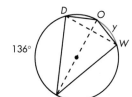

10. $k = -?-$

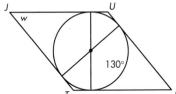

11. $m = -?-$
$n = -?-$

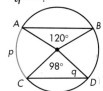

12.* $\overline{AB} \parallel \overline{CD}$
$p = -?-$
$q = -?-$

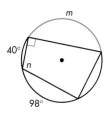

13. $r = -?-$
$s = -?-$

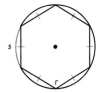

14.* What is the sum of
$a + b + c + d + e$?

15.* $y = -?-$

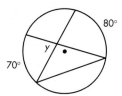

16. What's wrong with this picture?

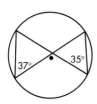

17. Find the lettered angle measures and arc measures.

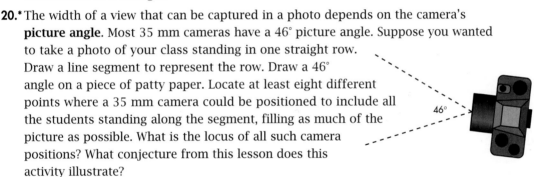

$\ell_1 \| \ell_2$, and ℓ_3 and ℓ_4 are tangents.

18. How can you find the center of a circle by using just the corner of a piece of paper?

19.* Chris Chisholm, while a high school student in Whitmore, California, used Conjecture 70 (Angles inscribed in a semicircle are . . . .) to discover a new and simpler way to find the orthocenter in a triangle. He noticed that if you construct a circle, using one of the sides of the triangle as the diameter of the circle, then you can immediately find an altitude to each of the other two sides! Try it. Use Chris's method to find the orthocenter of a triangle.

20.* The width of a view that can be captured in a photo depends on the camera's **picture angle**. Most 35 mm cameras have a 46° picture angle. Suppose you wanted to take a photo of your class standing in one straight row. Draw a line segment to represent the row. Draw a 46° angle on a piece of patty paper. Locate at least eight different points where a 35 mm camera could be positioned to include all the students standing along the segment, filling as much of the picture as possible. What is the locus of all such camera positions? What conjecture from this lesson does this activity illustrate?

21. Computer Activity Construct a circle and a diameter. Construct a point on one of the semicircles, and construct two chords to create a right triangle inscribed in the semicircle. Locate the midpoint of each of the two chords. What is the locus of the two midpoints as the vertex of the right angle is moved around the circle? Try visualizing this first. Sketch the locus by hand, then use your computer to animate the point on the circle and trace the locus of the two midpoints. What do you get?

Improving Reasoning Skills—*Think Dinosaur*

If the letter in the word *dinosaur* that is three letters after the word's second vowel is also found before the sixteenth letter of the alphabet, then print the word *dinosaur* horizontally. Otherwise print the word *dinosaur* vertically and cross out the second letter after the first vowel.

Lesson 7.5

The Circumference/Diameter Ratio

Here's a nice puzzle. One of two quarters remains motionless while the other rotates around it, never slipping and always tangent to it. When the rotating quarter has completed a turn around the stationary quarter, how many turns has it made around its own center point?

Did you guess one? Two? Three? The solution to the puzzle is not as obvious as it first appears. The best way of seeing the solution is to actually roll one coin around the other. Mark both coins with a felt tip pen or a pencil and try it.

The distance around a polygon is called the perimeter. The distance around a circle is called the **circumference**. In the puzzle above, one quarter is rolling along the circumference of the other quarter. In first thinking about the puzzle, you probably thought that because the circumferences of the two coins are the same, one coin would rotate once in its trip around the other. This was not the case, was it?

Here is another nice visual puzzle. Which is greater, the height of a tennis ball can or the circumference of the can? The height is approximately three tennis ball diameters tall. The diameter of the can is approximately one tennis ball in diameter. If you have a tennis ball can handy, try it. Wrap a string around the can to measure its circumference, then compare this measurement with the height of the can. Surprised?

In this second puzzle you should have discovered that the circumference of the can is greater than three diameters of the can. In this lesson you are going to discover (or perhaps rediscover) the relationship between the diameter and the circumference of every circle. Once you know this relationship, you can measure a circle's diameter and calculate its circumference.

If you measure the circumference and diameter of a circle and divide the circumference by the diameter, you get a number slightly larger than three. The more accurate your measurements, the closer your ratio will come to a special number called π (pi), pronounced like one of your favorite desserts.

In Investigation 7.5, you will experimentally determine an approximate value of π by measuring circular objects and calculating their circumference/diameter ratio. Work in groups of four or five, sharing the tasks of measuring, recording, and calculating. Let's see how close you come to the actual value of π.

R. DAGS

Investigation 7.5

For this investigation you will need the following special materials.

- Round objects you collected in Lesson 7.1 (the larger the objects the better)
- Meter stick or metric sewing tape
- Sewing thread or thin string to measure the circumference of each round object

Step 1 With the thread and the meter stick (or the sewing tape), measure the circumference and diameter of each round object to the nearest millimeter (tenth of a centimeter).

Step 2 Make a table similar to the one below and record the circumference (C) and diameter (D) measurements for each round object.

Name of object	-?-	-?-	-?-	-?-
Circumference (C)	-?-	-?-	-?-	-?-
Diameter (D)	-?-	-?-	-?-	-?-
C/D	-?-	-?-	-?-	-?-

Step 3 Calculate $\frac{C}{D}$ and record the answers in your table.

Step 4 Calculate the average of your $\frac{C}{D}$ results.

Compare the average of your $\frac{C}{D}$ results with the $\frac{C}{D}$ averages of other groups. Are the $\frac{C}{D}$ answers close? You should now be convinced that $\frac{C}{D}$ is very close to 3 for every circle. We define the ratio: $\frac{C}{D} = \pi$. If you solve this formula for C, you get a formula for finding the circumference of a circle in terms of the diameter. Because the diameter is twice the radius ($D = 2r$), you also can get a formula for finding the circumference in terms of the radius. State your conjecture.

C-73 If C is the circumference and D is the diameter of a circle, then there is a number π such that $C =$ —?—. Because $D = 2r$ where r is the radius, then $C =$ —?— (***Circumference Conjecture***).

The number π is an irrational number. Its decimal form never ends and no pattern in it ever repeats. The symbol π is a letter from the ancient Greek alphabet. Mathematicians began using it in the eighteenth century to represent the exact value of the circumference/diameter ratio.

Perhaps no other number has more fascinated and intrigued mathematicians throughout history. Mathematicians in ancient Egypt used $(4/3)^4$ as their approximation of π. Early Chinese and Hindu mathematicians used $\sqrt{10}$. By A.D. 408, Chinese mathematicians were using $355/113$. Today, computers have calculated π to millions of decimal places.

Such accurate approximations for π have been more of intellectual interest than for practical purposes. Still, what do you think a carpenter would say if you asked her to cut a board so that it was 3π feet long? Most calculators have a π button that gives π to eight or ten decimal places. You can use it for most calculations, then round your answer to a specified decimal place. If your calculator doesn't have a π button, or if you ever find yourself without access to a calculator, use the value 3.14 for π. If you're asked for an exact answer instead of an approximation, state your answer in terms of π.

How do you use the Circumference Conjecture? Let's look at two examples.

Example A

If a circle has a circumference of 12π meters, what is the radius?

$$C = 2\pi r$$
$$12\pi = 2\pi r$$
$$r = 6$$

The radius is 6 meters.

Example B

If a circle has a diameter of 3.0 meters, what is the circumference? Use a calculator and state your answer to the nearest 0.1 meter.

$$C = \pi D$$
$$C = \pi(3.0)$$

The circumference is about 9.4 meters.

Take Another Look 7.5

1. Use a geometry computer program to construct a circle. Construct a diameter. Measure the circumference and the diameter. Calculate the ratio of the circumference to the diameter. Change the size of your circle (thus changing the size of both the diameter and the circumference) and observe the ratio of the circumference to the diameter. Does it change? Explain how this confirms the Circumference Conjecture.

2. Locate high-quality spherical balloons, bow calipers, string, and meter sticks. Blow one large breath into a balloon. With the string, measure the balloon's circumference. With the bow calipers, measure the balloon's diameter. Repeat these two steps after blowing into the balloon a second full breath, a third full breath, until the balloon is near breaking. Find the ratio—the circumference to the diameter—for each pair of measurements. Does this confirm the Circumference Conjecture? Explain.

3. Use graph paper or a graphing calculator to graph the data collected from Investigation 7.5 or from Activities 1 and 2 in this Take Another Look. Graph the diameter on the x-axis and the circumference on the y-axis. What is the slope of the best-fit line through the data points? Does this confirm the Circumference Conjecture? Explain.

Exercise Set 7.5

Use the Circumference Conjecture to solve Exercises 1–13. In Exercises 1–8, do not use an approximation for π.

1. If $r = 5$ cm, find C.

2. If $C = 5\pi$ cm, find D.

3. If $C = 24$ m, find r.

4.* If $D = 5\pi$ m, find C.

out that the runner in lane 2 must have a head start of 2π meters over the runner in lane 1; the runner in lane 3 must have a 2π-meters head start over the runner in lane 2; and the runner in lane 4 must also have a 2π-meters head start over the runner in lane 3. With these head starts, each runner will travel 100π meters.

Is the head start always 2π meters? Investigate other tracks to find out. In the table at bottom left, the lane distance is the same, but the inner radius of the track has changed from the first example. In the table at bottom right, the lane width has changed, but the inner radius is the same as the radius of the track on the left. Copy and complete the tables below or make up some of your own. All distances are in meters.

Circular track with inner radius of 65 m and lane width of 1 m

Lane 1	Lane 2	Lane 3	Lane 4
$r = 65$	$r = 66$	$r = -?-$	$r = -?-$
$c = 130\pi$	$c = -?-$	$c = -?-$	$c = -?-$

Circular track with inner radius of 65 m and lane width of 1.5 m

Lane 1	Lane 2	Lane 3	Lane 4
$r = 65$	$r = 66.5$	$r = -?-$	$r = -?-$
$c = 130\pi$	$c = -?-$	$c = -?-$	$c = -?-$

From these examples you may be able to answer the first two of our questions. Most tracks go around a playing field and have straightaways. What about the length of the straightaways (S)? Copy and complete the tables below to calculate the total distance (T) for each lane in the type of track shown at right. Clearly, $T = 2S + 2\pi r$.

Lane 1	Lane 2	Lane 3	Lane 4
$r = 30$	$r = 31$	$r = 32$	$r = 33$
$S = 100$	$S = 100$	$S = 100$	$S = 100$
$T = 200 + 60\pi$	$T = -?-$	$T = -?-$	$T = -?-$

Lane 1	Lane 2	Lane 3	Lane 4
$r = 30$	$r = 31$	$r = 32$	$r = 33$
$S = 200$	$S = 200$	$S = 200$	$S = 200$
$T = 400 + 60\pi$	$T = -?-$	$T = -?-$	$T = -?-$

From these tables you may be able to answer our third question. If you can answer all three questions, you are ready to start designing your racetrack.

Again, your task in this project is to design a four-lane oval track with straightaways and semicircular ends. The semicircular ends must have inner diameters of 50 meters so that the distance of one lap in the inner lane is 800 meters. You determine a width for the lanes. Draw starting and stopping segments in each lane so that an 800-meter race can be run in all four lanes.

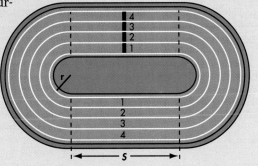

On your mark, get set, GO!

Geometer's Sketchpad Project

Turning Wheels

Imagine a bug gripping your bicycle tire as you ride down the street. What would the bug's path look like? What path does the moon make as it rotates around the earth while the earth rotates around the sun? With Sketchpad's animation capabilities, you can model a rotating wheel. In this project, you'll investigate the path of a point on a wheel as it rolls along the ground or around another wheel. You'll start by constructing a stationary circle with a rotating spoke.

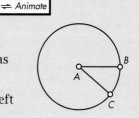

These circles in the sand were created by the wind blowing blades of grass as if they were spokes on a wheel.

Step 1 Construct a small circle with center point *A* and radius-defining point *B*. Construct radius $\overline{AB}$.

Step 2 Construct a second radius, $\overline{AC}$.

Step 3 Select point *C* and the circle and use the Edit menu to create an action button that animates point *C* around the circle quickly.

Double-click the animation button. So far you have a circle with one spoke that rotates in a counterclockwise direction.

How can you make a wheel that will roll? You can't roll your circle with the spinning spoke because if the circle moves, the spoke would have to move with it. But you can make a different circle and, as you move it, use circle *A* to rotate it. Here's how.

Step 4 Construct a long horizontal segment *DE* going from right to left and construct a point *F* on the segment.

Step 5 Select points *A* and *F*, in order, and choose Mark Vector in the Transform menu. Select circle *A*, point *C*, and $\overline{AC}$, and use the Transform menu to translate by the marked vector.

Step 6 Construct a line *FC′* overlapping $\overline{FC'}$.

Step 7 Construct a point *G* anywhere on the line. Hide the line and construct $\overline{FG}$.

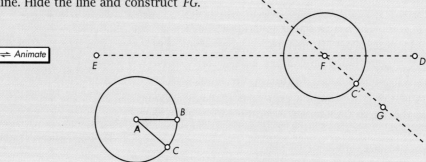

39.* One nautical mile was originally defined to be the length of one minute of arc of a great circle of the earth. (There are 60 minutes of arc in each degree. A great circle on a sphere is the intersection of the sphere and a plane that cuts through the sphere's center.) However, scientists discovered that the earth is not a perfect sphere. It is fatter through the equator than through the poles. Defined as one minute of arc, therefore, one nautical mile could take on a range of values. It would be shortest along a great circle through the poles and longest along the great circle called the equator. To remedy this, an international unit of one nautical mile was established. The international nautical mile is defined as 1.852 kilometers (about 1.15 miles). Given that the polar radius of the earth is 6357 kilometers and that the equatorial radius of the earth is 6378 kilometers, use the original definition to calculate one nautical mile near a pole and one nautical mile near the equator. Show that the international nautical mile is between both values.

Assessing What You've Learned—*Self Assessment*

In each of Chapters 0 through 6, you were introduced to a different way of assessing what you learned. You can't be expected to have tried all seven of these ways. More likely, your teacher has adapted the assessment methods he or she finds most appropriate for your class. From this chapter on, no new methods of assessment will be introduced. Instead you'll read brief suggestions of ways to use the assessment methods you've read about so far.

Whatever assessment methods you've used in your class, you should be getting the idea that assessment means more than a teacher giving you a number or a grade. Communicating what you've learned to others (like your teacher) gives you a chance to assess your own learning. All the methods presented so far could be described as self-assessment techniques. Many are also good study habits. Being aware of your *own* learning and progress is the best way to stay on top of what you're doing and to achieve the best results. Keep this in mind as you try one or more of the following suggestions.

Write in Your Journal

• You may be near the end of your school year's first semester or just getting started in the second semester. Look back over the first semester and write about what you perceive were your strengths and weaknesses. What grade would you have given yourself for the semester? How would you justify that grade?

• Set new goals for the new semester or for the remainder of the year. Write them in your journal and compare them to goals you set at the beginning of the year. How have your goals changed? Why?

Organize Your Notebook

• Review your notebook to be sure it's complete and well organized. Look at the last numbered conjecture in this chapter. Do you have all this chapter's conjectures on your list? Write a one-page chapter summary.

Update Your Portfolio

• Choose a piece of work from this chapter to add to your portfolio. Document the work according to your teacher's instructions.

Performance Assessment

• While a classmate, a friend, a family member, or a teacher observes, carry out one of the investigations or Take Another Look activities from this chapter. Explain what you're doing at each step, including how you arrived at the conjecture.

Write Test Items

• Divide the lessons from this chapter among group members and write at least one test question per lesson. Try out the test questions written by your classmates and discuss them.

Give a Presentation

• Give a presentation on an investigation, a Take Another Look activity, a group activity from an exercise set, a project, or even one of the more challenging puzzles. Give it with your group, or try giving a presentation on your own.

Improving Visual Thinking Skills—*Picture Patterns III*

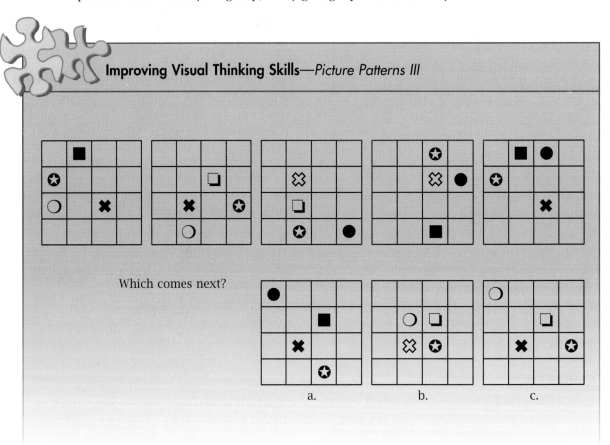

Which comes next?

a.　　　　b.　　　　c.

Cooperative Problem Solving

Designing a Theater for Galileo

You and your architectural engineering team are competing for the contract to design a new circular theater with a revolving center stage. The theater is to be built beneath the great dome R-3 at the lunar space port Galileo. The Arts Director of Galileo has asked each potential engineering team to submit its design and its calculations for the new theater. Although overall design is important, the job will go to the team that produces the design with the greatest seating capacity. The director has given you the following restrictions and guidelines.

1. The theater must contain only one level and seat at least 1000 people.

2. The stage should be at least 10 meters in diameter.

3. The outer diameter of the theater interior should be at most 42 meters.

4. The seating should be divided into sections by equally spaced aisles radiating from center stage. There should be no fewer than four radial aisles and no more than eight radial aisles. Each radial aisle should be at least 1 meter in width.

5. There should be two concentric aisles. The innermost concentric aisle around the stage should be at least 1 meter wide and at most 2 meters wide. The outer concentric aisle should be at least 2 meters wide and at most 4 meters wide and should ring the perimeter of the theater.

6. Safety codes at the lunar colony require that each seat be at least 60 centimeters wide and that each seating position be at least 90 centimeters in depth.

7. Safety codes require that there be no more than 30 seats in any row.

Your job is to draw a scaled plan of the theater, including seating, aisles, and the stage. Your plan should maximize seating capacity. You should be able to support your design with calculations that verify your seating capacity. You will need to calculate the following.

1. The number of rows and the number of seats in each row

2. The width at the stage end of each radial aisle and the width at the back end of each radial aisle; the width of the concentric aisles

3. Total seating capacity for your plan

8 Transformations and Tessellations

Magic Mirror, M. C. Escher, 1946

In this chapter you will discover some basic properties of transformations and symmetry. As you learned in Chapter 0, symmetry is found in art, architecture, and nature. It is also found in dance, music, poetry, and, of course, mathematics. You will use transformations to create special tiling patterns called tessellations. The hexagon pattern in the honeycomb of the bee is a tessellation of regular hexagons. Tessellations are found in Islamic designs. After observing the Islamic designs in the Alhambra in Granada, Spain, Dutch artist M. C. Escher (1898–1972) began creating tessellations using recognizable shapes such as birds, fish, reptiles, and humans. In this chapter you will learn a few of his techniques and create your own designs.

Lesson 8.1

Transformations

Geometry is not only the study of figures; it is also the study of the movement of figures. If you move all the points of a geometric figure according to set rules, you can create a new geometric figure. Such a motion establishes a correspondence between the points of the original figure and the points of the new figure. This new figure is called the **image**. If each point of a plane figure can be paired with exactly one point of its image on the plane, and if each point of the image can be paired with exactly one point of the original figure, then the correspondence is called a **transformation**.

A transformation that preserves size and shape is called a **rigid transformation**, or an **isometry** (from the Greek for "same measure"). The image is always congruent to the original in an isometry.

Three types of isometries are translation, rotation, and reflection. More familiar words for these motions are *slide, turn,* and *flip*. You have been working informally with translations, rotations, and reflections in your patty-paper investigations and in some motion exercises on the coordinate plane.

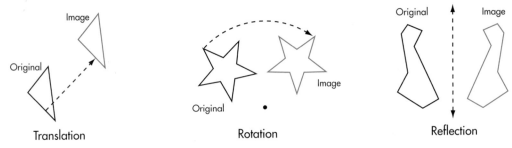

Translation   Rotation   Reflection

A **translation** is the simplest type of isometry. If you copy a figure onto a piece of paper, then slide the paper along a straight path without turning it, your slide motion models a translation isometry. In a slide, points in the original figure move an identical distance along parallel paths to the image. That is, all the points in the original figure are equidistant from their images. A distance and a direction together define a translation. You can use an arrow, representing a **translation vector**, to show distance and direction. The length of the translation vector from starting point to tip represents the distance, and the direction in which the arrow is pointing represents the direction of the translation.

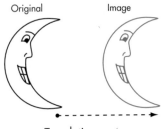

Translation vector

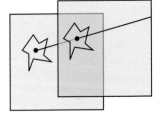

Patty-paper translation

A translation can also be represented on a coordinate grid by using an ordered-pair rule. For example, the ordered pair rule $(x, y) \rightarrow (x + 5, y + 3)$ moves each point of the original figure five units to the right and three units up. We sometimes abbreviate the rule by simply writing the ordered pair (5, 3), which we read as "translation: (5, 3)." The figure below left shows a translation: (5, 3). What translation is shown in the figure below right?

A prime mark (') is often used with the label of an image point. The image of point A under a transformation of any type is often called point A' (read "A prime").

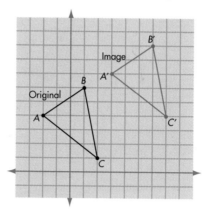
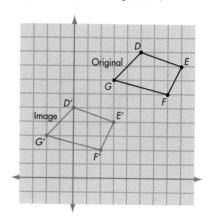

A **rotation** is a second type of isometry. In a turning motion, points in the original figure rotate or turn an identical number of degrees about a fixed center point. A center of rotation together with an amount and a direction of rotation (clockwise or counter-clockwise) define a rotation. If no direction is given, the direction of rotation is assumed to be counterclockwise.

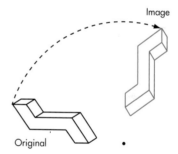

Original

Image

If you place a piece of patty paper over a figure and trace it (making an image of the original), then place a dot on the patty paper, put your pencil point on the dot, and rotate the patty paper about the point, your turn motion models a rotation isometry.

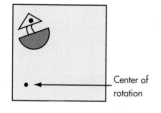
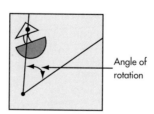
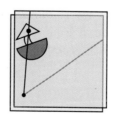
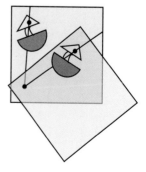

Center of rotation

Angle of rotation

Patty-paper rotation

A **reflection** is a third type of isometry. If you draw a figure onto a piece of paper, place the edge of a mirror onto your paper so that the mirror is perpendicular to the paper, and look at the figure in the mirror, you will see the figure reflected. A reflection produces a figure's "mirror image." A **line of reflection** (also called a **mirror line**) defines a reflection. The line of reflection is the perpendicular bisector of every segment joining a point in the figure with the image of the point. If a point in the figure is on the line of reflection, then the point is its own image. Each point in a figure is reflected or flipped over the reflection line as if the reflection line were a mirror.

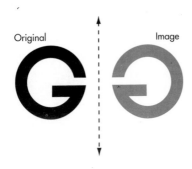

Original   Image

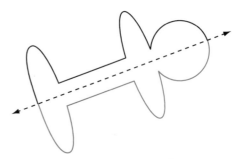

A reflection can be represented on a coordinate grid by using an ordered pair rule. For example, the ordered pair rule $(x, y) \rightarrow (x, -y)$ reflects each point of the original figure to the opposite side of the x-axis.

You can also model a reflection with patty paper. Sketch a figure onto a sheet of patty paper. Draw a line of reflection onto your patty paper. Fold the patty paper along the line of reflection. Trace the figure onto the folded portion of your patty paper. Unfold.

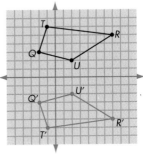

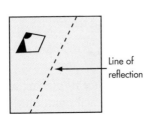

Line of reflection

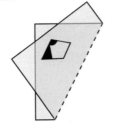

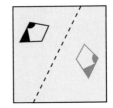

Patty-paper reflection

Combining a translation with a reflection gives a special two-step isometry called a **glide reflection**. A sequence of footsteps is a common example of a glide reflection. In Lesson 8.2, you'll experiment with this and other combined transformations.

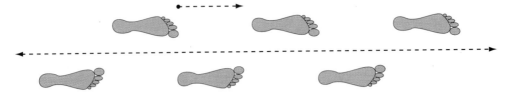

A third name for reflectional symmetry is **mirror symmetry** because half a figure with reflectional symmetry is a mirror image of the other half. If you place a mirror perpendicular to the plane of the figure on that figure's line of symmetry, the half-figure and its image produce a complete figure.

If a figure can be rotated about a point in such a way that its rotated image coincides with the original figure after turning less than 360°, then the figure has **rotational symmetry**. The logo design on the previous page has 6-fold rotational symmetry. You can trace a figure and test it for rotational symmetry. Place the copy exactly over the original, put your pen or pencil point on the center to hold it down, and rotate the copy. Count the number of times the copy and the original coincide until the copy is back in the position it started in. The letter Z has 2-fold rotational symmetry; when it is rotated 180° and 360° about a center of rotation, the image coincides with the original figure. A figure that has 2-fold rotational symmetry has **point symmetry**. A figure is said to be point-symmetric if it has a 180° rotational symmetry.

The basket design shown at right has 4-fold rotational symmetry. Many designs, like that of the basket, have both reflectional and rotational symmetry. Others, like the logo design shown on the previous page, have only rotational symmetry. Of course, every image is identical to the original figure after a rotation of some multiple of 360°. However, we don't call a figure symmetric if this is the only symmetry it has.

Basket design, Botswana
(4-fold rotational symmetry)

You've just seen two types of symmetries: reflectional symmetry and rotational symmetry. Two other symmetries, translational symmetry and glide-reflectional symmetry, really exist only for infinite patterns. If a finite design is said to have translational or glide-reflectional symmetry, it is understood that this would be true only if the particular design were to continue indefinitely.

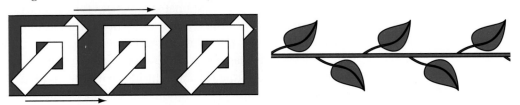

Translational symmetry   Glide-reflectional symmetry

If a pattern can be translated a given distance in a given direction in such a way that the image coincides with the original, then the pattern has **translational symmetry**. The given distance and the given direction are identified by the translation vector. In the pattern shown above left, any vector that has the same length as the vectors shown (or a multiple of the length) and that is parallel to the vectors shown can be a translation vector. If a design can undergo a glide-reflection isometry in such a way that the image coincides with the original, then the design has **glide-reflectional symmetry**.

Some polygons have no symmetry. Others have a single symmetry. Regular polygons, however, have many symmetries. A square, for example, has four reflectional symmetries and four rotational symmetries.

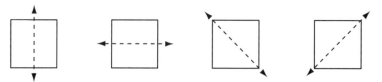

The four reflectional symmetries of a square

A square can be rotated 90°, 180°, 270°, and 360° about its center, and it will coincide with itself. A square has 4-fold rotational symmetry.

The four rotational symmetries of a square

Does the number of sides of a regular polygon determine the number of symmetries of each type? Let's investigate.

Investigation 8.3

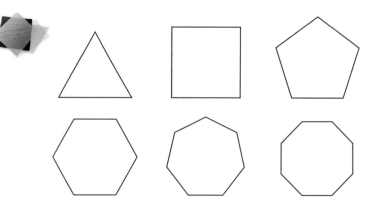

Find the number of reflectional and rotational symmetries for each of the regular polygons listed in the table below. If necessary, use a mirror to locate the lines of symmetry for each of the regular polygons. To find the number of rotational symmetries, you may wish to trace each regular polygon above onto patty paper. Place the tracing over the original and rotate until the tracing coincides with the original. In counting the rotational symmetries, include the 360° rotation (but not the 0° rotation).

Number of sides of regular polygon	3	4	5	6	7	8	...	n
Number of reflectional symmetries	-?-	4	-?-	-?-	-?-	-?-	...	-?-
Number of rotational symmetries (≤360°)	-?-	4	-?-	-?-	-?-	-?-	...	-?-

When you've completed the table, make a conjecture about the number of reflectional and rotational symmetries in a regular n-gon.

C-78 A regular polygon of n sides has —?— reflectional symmetries and —?— rotational symmetries. The measure of the smallest angle of rotation for a rotational symmetry of a regular polygon of n sides is —?—.

Exercise Set 8.3

In Exercises 1–3, identify the type (or types) of symmetry in each design.

1.

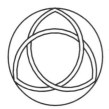

2.

3.

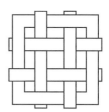

4. All of the woven baskets from Botswana shown below have rotational symmetry and most have reflectional symmetry. Which have (or has) 7-fold symmetry? 9-fold symmetry? Which has rotational symmetry but not reflectional symmetry? What-fold rotational symmetry does it have?

5. The Palestinian silk cloth at right has two pairs of square-based designs. Describe the symmetry of each of the two designs. How does this symmetry differ from the symmetry of the whole cloth? Why?

6. List three objects in your home that have rotational symmetry but not reflectional symmetry. List three objects in your classroom that have reflectional symmetry but not rotational symmetry.

7. Have you noticed any differences in visual effects between figures with reflectional symmetry and figures with only rotational symmetry? It seems that objects with reflectional symmetry seem to be more static, while figures with rotational symmetry convey motion. Why do you think this is?

8. The word DECODE remains unchanged when it is flipped about its horizontal line of symmetry. Find another such word with at least five letters. Hold a contest in your group or your class to see who finds the longest word with a horizontal line of symmetry.

◄--DECODE--►

↑
┆
T
O
M
A
T
O
┆
▼

9. The word TOMATO has a vertical line of symmetry when it is written in column form. Thus when a mirror is placed on the line of symmetry, TOMATO is seen in the mirror. Find another such word with at least five letters. Hold a contest in your group or your class to see who finds the longest word with a vertical line of symmetry.

10.* Have you noticed that some letters have both horizontal and vertical symmetries? Have you also noticed that all the letters that have both horizontal and vertical symmetries also have point symmetry? Is this a coincidence? Not at all. Use what you learned in Lesson 8.2 to explain why.

11. Write what would appear on the T-shirt shown at right if you were not seeing its image in a mirror.

12. How many reflectional symmetries does an isosceles triangle have?

13. How many reflectional symmetries does a rhombus have?

14. Sketch a figure that has point symmetry but not line symmetry.

15.* Name a polygon that has a 36° rotational symmetry.

16. Is it possible for a triangle to have exactly one line of symmetry? Exactly two? Exactly three? Support your answers with sketches.

17. Is it possible for a quadrilateral to have exactly one line of symmetry? Exactly two? Exactly three? Exactly four? Support your answers with sketches.

By Holland. © 1976, Punch Cartoon Library

18.* A curve is point symmetric if it coincides with itself when rotated 180° about a point. Draw two points onto a piece of paper and connect them with a curve that is point symmetric.

19. Design a logo with rotational symmetry for Happy Time Ice Cream Company. (Or design a logo for your group or a made-up company, perhaps using your own name.)

20. Your colonizing spaceship has just landed on Tao, the small earthlike planet of the binary star system Gemini in the Andromeda galaxy. You have been instructed by Captain James Kirk IV to design a flag for the new colony. Design the flag with reflectional symmetry.

21. The design below left comes from *Inversions,* a book by Scott Kim. Not only does the design spell the word *mirror,* it does so with mirror symmetry! Scott is able to turn *m*'s into *r*'s, and you can do almost the same thing. Find a word (with more than five letters) or words that have vertical symmetry—use only letters of the alphabet that have vertical lines of symmetry. Some palindromic words (words that spell the same forward and back) have vertical symmetry. If all the letters in the palindrome have vertical lines of symmetry, then the palindromic word will have mirror symmetry. Words like MOM, WOW, TOOT, and OTTO all look like their images when seen in a mirror.

22. Can you make out the fancy word above right? It's *symmetry,* and it not only spells symmetry, it *is* symmetric. Rotate the design 180° and it is identical to the original. This design also comes from Scott Kim's *Inversions.* Scott is able to write many words so that they possess rotational or reflectional symmetry. MOW can be written so that it is point symmetric. If you can design an *R* so that when rotated 180°, it can pass for a *y*, then you can write the name Roy so that it has point symmetry. Try it! Another word Scott suggests we can all write with point symmetry is *chump.* Try writing *chump* (in script) so that when rotated 180°, it reads the same.

23. Scott Kim suggests that symmetry can also be found in handshakes. There is a nice rotational symmetry to a handshake between two people. What about handshakes among three people? The illustrations below show three different symmetric handshakes among three people. Now it is your turn. Find two partners and create a symmetric three-person handshake different from the three shown. Make a sketch or take a photo of your group handshake—or be ready to demonstrate your three-person handshake to the rest of the class.

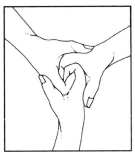

24. **Group Activity** In this lesson you discovered symmetry properties of regular polygons. You can use pattern blocks to explore symmetries of other polygons. Most pattern-block sets come with the six shapes shown at right. Each block has one- or two-inch sides and has angles whose measures are multiples of 30°. Use the blocks to complete these exercises.

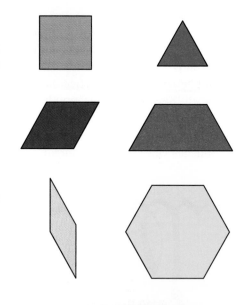

   a. Find the reflectional and rotational symmetries of a rhombus.

   b. Find the reflectional and rotational symmetries of an isosceles trapezoid.

   c. Combine two pattern blocks to form a rectangle. Find its reflectional and rotational symmetries.

   d. Combine two pattern blocks to form a parallelogram. Find its reflectional and rotational symmetries.

   e. Combine three pattern blocks to form a kite. Find its reflectional and rotational symmetries.

25. This Turkish plate at right, held by artist Nurten Sahin, displays different symmetries in the rings surrounding the center circle. What type of symmetry does the center circle have? What type of symmetry does the outermost ring have?

26. Assume the design below continues infinitely. What type or types of symmetry does it have?

Frieze of bowmen from The Palace of Artaxerxes II in Susa, Iran

27. Assume the design below continues infinitely. What type or types of symmetry does it have?

Ornamental Chinese lattice design

28. What is the coordinate rule that transforms $\triangle TRY$ onto $\triangle T'R'Y'$? $(x, y) \rightarrow (?, ?)$. What is the equation of the line of reflection?

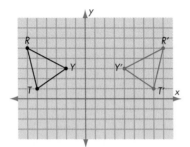

29. What is the coordinate rule that transforms $\triangle VRY$ onto $\triangle V'R'Y'$? $(x, y) \rightarrow (?, ?)$. What is the equation of the line of reflection?

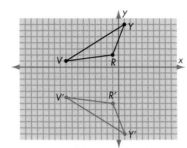

30. What is the coordinate rule that transforms $\triangle HRD$ onto $\triangle H'R'D'$? $(x, y) \rightarrow (?, ?)$. What are the equations of the two lines of reflection?

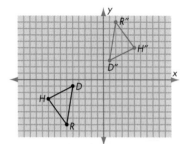

Improving Reasoning Skills—*Checkerboard Puzzle*

1. Four checkers are arranged on the corner of a checkerboard, as shown at right. With exactly three horizontal or vertical jumps, remove all three red checkers, leaving the single black checker. Any checker can jump any other checker. To record your solution, copy and complete the table at far right.

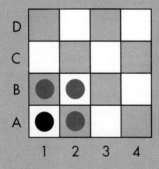

From	To
–?–	–?–
–?–	–?–
–?–	–?–

2. Now with exactly seven horizontal or vertical jumps, remove all seven red checkers, leaving the single black checker. Copy and complete the table to record your solution.

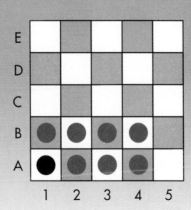

From	To
–?–	–?–
–?–	–?–
–?–	–?–
–?–	–?–
–?–	–?–
–?–	–?–
–?–	–?–

Project

Kaleidoscopes I

You've discovered that the composition of two reflections over intersecting reflection lines is equivalent to a rotation. This idea is used in kaleidoscopes. Let's take a look.

1. Hinge two mirrors together with tape so that the angle between them is capable of being adjusted to different sizes, as shown. Place the hinged mirrors onto a design, such as a postcard or one of your geometric art designs. Adjust the size of the angle of the hinged mirror. This is a simple kaleidoscope.

2. Place a small object, such as a coin or a thumbtack, between the hinged mirrors. Adjust the mirrors until you see only the original object and three images. What is the angle formed by the mirrors? Adjust the mirrors until you see only the original object and five images. What is the angle formed by the mirrors? At what angle should the mirrors be adjusted to see a total of eight objects (one original and seven images)?

3. Place a vertex of one of the pattern blocks at the hinge of the mirrors and adjust the mirrors so that they align with the sides of the block. Sketch the shape made by the block and its images. Repeat for all six pattern blocks.

4. Place a dot and a line onto a piece of paper, as shown. Place the edge of one mirror through the dot and adjust the second mirror so that the reflections of the segment between the mirrors form a regular polygon. First form an equilateral triangle, then a square, and so on. After forming each regular polygon, measure the angle between the mirrors. Complete the table below.

Regular polygon	3	4	5	6	7	8	9	10	11	12
Mirror angle	-?-	-?-	-?-	-?-	-?-	-?-	-?-	-?-	-?-	-?-

5. See if you can find two different mirror angles that make the same regular polygon. Write a paragraph describing two different ways to form the image of a regular polygon from a single segment using a pair of hinged mirrors. For each of the two methods, relate the mirror angle to the number of sides in the polygon. Describe how the mirrors must be positioned.

Project

Kaleidoscopes II
Building a Tube Kaleidoscope

Many people have looked through kaleidoscopes. But how many people know how kaleidoscopes are made or how they work? If you did the Kaleidoscopes I project, you already know something about how they work. The tube-shaped kaleidoscope you may be familiar with was invented by physicist Sir David Brewster in 1816. There are many varieties of tube kaleidoscopes. Some scopes have colored glass or plastic that tumbles about the end chamber, creating an infinite series of symmetric patterns. Some have nothing in the end chamber but a lens, and the designs created depend on the objects at which the kaleidoscope is aimed. Some have a marble at the end. Some modern scopes have colored, nonmixing liquids in the end chamber that create flowing symmetric designs. Some handcrafted kaleidoscopes sell for well over $100.

Here is your group project.

1. Design a kaleidoscope.

2. Make a diagram that shows your kaleidoscope plan.

3. Collect the raw materials and tools necessary to create your kaleidoscope.

4. Build your kaleidoscope.

5. Write a report that includes diagrams describing what you did, a list of the materials you used, problems you encountered and how you solved them, and how you modified your design from your original plans and why. Describe who in your group did which tasks.

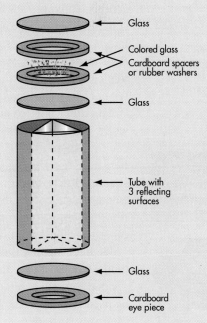

Glass

Colored glass
Cardboard spacers
or rubber washers

Glass

Tube with
3 reflecting
surfaces

Glass

Cardboard
eye piece

The figure at right is an exploded view of a typical kaleidoscope. This entire scope must be rotated in order for the colored pieces to tumble into the different designs. To create a kaleidoscope with an end chamber that turns while the body remains stationary is a little trickier. This process involves using a slightly larger piece of tube for the rotating end chamber. The material for a kaleidoscope can vary, depending on what your group can find and afford. The tube can be a cardboard cylinder found around the home. For a kaleidoscope to last your lifetime, use plastic plumbing pipe (PVC). The round clear pieces can be glass, a lens, clear plastic, or clear acetate. The three reflecting surfaces can be mirror glass, glass painted black on one side, or clear vinyl painted black on one side. The colored pieces that go into the end chamber can be almost anything translucent. Don't put too many colored pieces between the spacers! If you use too many, they don't tumble well and they may block out too much light.

Lesson 8.4

Tessellations with Regular Polygons

I see a certain order in the universe and math is one way of making it visible.
— May Sarton

Honeycombs like the one shown at left are remarkable geometric structures. The hexagonal cells bees make are ideal because they fit together perfectly without any gaps. People often use regular hexagons as floor tiles for the same reason. The regular hexagon is one of many shapes that can completely cover the plane without gaps or overlaps. Mathematicians call such an arrangement of shapes a **tessellation** or a **tiling**. The word *tessellation* comes from the small square ceramic tiles, called *tesserae*, that the Romans used to create mosaic tile designs. When a tessellation uses only one shape, as in a honeycomb, it's called a **monohedral tiling**.

Tessellations are found in many natural forms other than honeycombs. The molecular structures of some crystals show tessellating patterns. The scales of snakes and the close packing of corn kernels are both very close to being tessellations (but not exactly, because of small gaps and overlaps). Where are other tessellations in nature?

Tessellations can be found in designs from cultures throughout the world. The tiling shown at right from the seventeenth-century Topkapi Palace in Turkey combines regular hexagons and equilateral triangles. The Renaissance floor mosaics of St. Mark's Basilica in Venice, Italy, exhibit many different types of tessellations.

Regular hexagons and equilateral triangles combine in this tiling from the seventeenth-century Topkapi Palace in Istanbul, Turkey.

You don't have to go around the world in search of tessellations. You can find them in every home. Floor tiles have tessellating patterns of squares. Brick walls or fireplaces and wooden decks often display creative tessellations of rectangles. Where are other tessellations in your home?

We have already mentioned that squares (floor tiles) and regular hexagons (honeycombs) create monohedral tessellations. Because each regular hexagon can be divided into six equilateral triangles, it follows logically that equilateral triangles can also be used to create a monohedral tessellation. Will other regular polygons tessellate? Let's look at it logically.

For shapes to fill the plane edge to edge without gaps or overlaps, their angles, when arranged around a point, must have measures that add to exactly 360°. If the sum were less, there would be a gap. If the sum were more, the shapes would overlap. Six 60° angles from six equilateral triangles sum to 360°, as do four 90° angles (squares) and three 120° angles (regular hexagons). What about regular pentagons? Each angle in a regular pentagon measures 108°.

Three angles would sum to only 324°, leaving a gap too small for a fourth angle. Three angles from regular heptagons would overlap. In any regular polygon with more than six sides, each angle has a measure greater than 120°. Thus no more than two can fit about a point without overlapping.

Therefore the only regular polygons that create monohedral tessellations are equilateral triangles, squares, and regular hexagons. The tessellations in which tiles are congruent regular polygons whose edges exactly match are called **regular tessellations**.

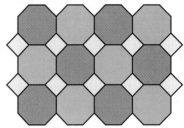

Tessellations can involve more than one type of shape. The octagon-square combination at right is a commonly seen example.

In this tessellation, two regular octagons and a square meet at each vertex point. When the same combination of regular polygons (of one or more kinds) meet in the same order at each vertex of a tessellation, it is called a **semiregular tessellation**. Below are two more examples of semiregular tessellations.

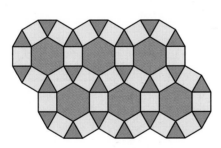

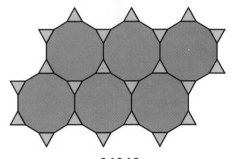

3.4.6.4

The same order of polygons appears at each vertex. The sum of the measures of the angles at a vertex is 360° because 60° + 90° + 120° + 90° = 360°.

3.12.12

The same order of polygons appears at each vertex. The sum of the measures of the angles at a vertex is 360° because 60° + 150° + 150° = 360°.

There are eight different semiregular tessellations. Semiregular tessellations are identified by listing the number of sides of the polygons about each vertex, as shown under the two illustrations above. Start with the polygon with the least number of sides, then list the number of sides of each polygon as you move (clockwise or counterclockwise) about the vertex. How would you identify the octagon-square tessellation?

You've seen three of the eight semiregular tessellations. Let's find the other five. In Investigation 8.4, you'll need to use the set of regular polygons below—all with sides of the same length. Unless you have a plastic template that has outlines of all (or most) of the regular polygons pictured, you can either make a set of poster-board polygons to use as templates or use patty papers to copy those below. One set of templates can be shared by a group of four or five students.

Investigation 8.4

To make each regular polygonal template: Copy each polygon below onto a sheet of patty paper. You now have a template of each of the regular polygons pictured, and you're ready for the investigation. (If you wish to have poster-board templates, place your patty paper on top of your poster board. Using a sharp pencil point, make a mark at the vertices through your patty paper onto the poster board. Use your straightedge to connect the marks. Cut out the polygonal shape.)

Step 1   After you have created your set of regular polygons, investigate which combinations of two regular polygons can be used to create a semi-regular tessellation. Remember, the same combination of regular polygons must meet in the same order at each vertex for the tiling to be semiregular.

Step 2   Next investigate which combinations of three regular polygons can be used to create a semiregular tessellation.

Step 3   Finally, investigate all other combinations of four or more regular polygons that can be used to create a semiregular tessellation.

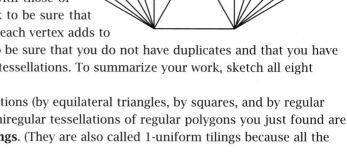

Compare your findings with those of others in your group. Check to be sure that the sum of the measures at each vertex adds to 360°. Check your findings to be sure that you do not have duplicates and that you have found all eight semiregular tessellations. To summarize your work, sketch all eight semiregular tessellations.

The three regular tessellations (by equilateral triangles, by squares, and by regular hexagons) and the eight semiregular tessellations of regular polygons you just found are called the **Archimedean tilings**. (They are also called 1-uniform tilings because all the vertices are identical.)

If the arrangement at each vertex in a tessellation of regular polygons is not the same, then the tessellation is called a **demiregular tessellation**. If there are two different types of vertices, the tiling is called 2-uniform. If there are three different types of vertices, the tiling is called 3-uniform. Two demiregular tessellations are pictured below.

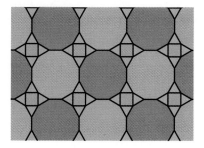

$3.4.3.12/3.12^2$

In the 2-uniform tessellation $3.4.3.12/3.12^2$, two arrangements of polygons appear at vertices.

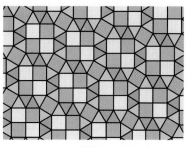

$3^3.4^2/3^2.4.3.4/4^4$

In the 3-uniform tessellation $3^3.4^2/3^2.4.3.4/4^4$, three arrangements of polygons appear at vertices.

Like semiregular tessellations, demiregular tessellations are identified by listing the number of sides of the polygons about a vertex. Unlike semiregular tessellations, however, more than one combination occurs. Each vertex sequence is separated from other sequences with a slash (/), and repetitions of polygons are abbreviated with exponents. There are 20 different 2-uniform tessellations of regular polygons. The number of 4-uniform tessellations of regular polygons is still an unsolved problem. Let's look at two more examples.

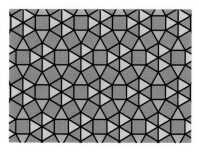

$3^6/3^2.4.3.4$

In the 2-uniform tessellation $3^6/3^2.4.3.4$, two arrangements of polygons occur at the vertices.

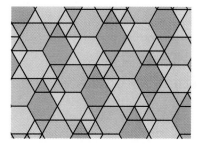

$3^4.6/3^2.6^2/3.6.3.6/6^3$

In the 4-uniform tessellation $3^4.6/3^2.6^2/3.6.3.6/6^3$, four arrangements of polygons occur.

Exercise Set 8.4

1. List two objects or designs in your home that are monohedral tessellations of polygons.

2. List two objects or designs outside of your math classroom that are semiregular tessellations of polygons.

In Exercises 3–5, give the numerical symbol for the semiregular tessellations.

3.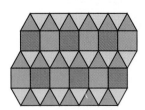

4.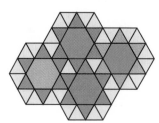

5.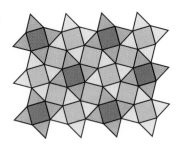

In Exercises 6–8, use templates of regular polygons or use pattern blocks.

6. Sketch the 3.6.3.6 tessellation.

7. Sketch the 4.6.12 tessellation.

8.* Using your templates, show that two regular pentagons and a regular decagon fit about a point but that 5.5.10 does not create a semiregular tessellation.

In Exercises 9–11, give the numerical symbol for the demiregular tessellations.

9.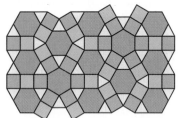

10.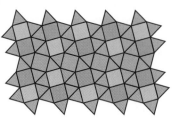

11.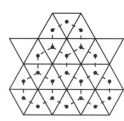

12. Using your templates, create the demiregular tessellation $3.12^2/3.4.3.12$. Draw your design onto a full sheet of paper. Decorate and color your design.

13.* There is more than one distinct form for the demiregular tessellation $3^3.4^2/4^4$. Can you find two? Using your templates or pattern blocks, create one version of the demiregular tessellation $3^3.4^2/4^4$ on the top half of a full sheet of paper and create a second on the bottom half of your paper. Color the top tessellation design so that it retains all its symmetries. Color the bottom tessellation design so that if you take color into account, it retains a reflectional but not rotational symmetry.

14. A tessellation of equilateral triangles is shown at right. When you connect the center of each triangle across the common sides of the tessellating triangles, you get another tessellation. This new tessellation is called the **dual** of the original tessellation. Every tessellation of regular polygons has a dual. The dual of the equilateral triangle tessellation is the regular hexagon tessellation. Draw a square tessellation and a hexagon tessellation. Make the dual of each.

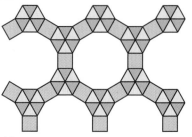

15. You can make dual tessellations of semiregular tessellations, but they may not be tessellations of regular polygons. Try it. Sketch the dual of 4.8^2, shown at right.

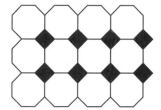

16. Tiling with rectangles is perhaps one of the most common tessellation patterns. All of the tessellations you have seen so far in this lesson were edge-to-edge tilings. However, with non-edge-to-edge tiling, the number of patterns are literally infinite and quite interesting. Wooden decks, for example, can be built in a variety of non-edge-to-edge designs, as shown below. Use graph paper to create your own wooden deck design different from these designs.

  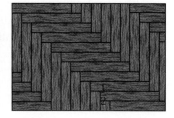

17. Reflect $y = 2x + 3$ over the y-axis and determine the equation of the new image line.

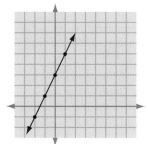

18. Reflect $y = \frac{1}{2}x - 4$ over the x-axis and determine the equation of the new image line.

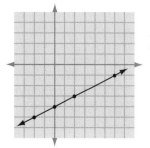

In Exercises 19 and 20, copy the position of the balls and the holes onto patty papers. Fold to locate the reflected images, then locate the point (or points) on the pool-table cushions or miniature-golf walls at which the ball should be aimed.

19.* The miniature-golf player Duffer Dolores needs a hole in one on this last hole to win the tournament. She recognizes that it is going to take more than one wall bounce to sink the ball. How should she hit the ball from the tee (T)? To what point (or points) on the walls should she aim so that the golf ball bounces off the walls and goes into the hole at point H?

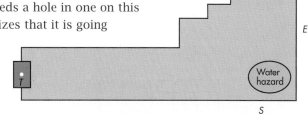

20.* The pool player known by many as "Frisco Fats" needs to sink the eight ball into the NW corner pocket, but he seems to be trapped. Can he hit the cue ball to some point on the N cushion so that the cue ball bounces out, strikes the S cushion, and finally taps the eight ball into the corner pocket? To what point on the N cushion should he aim? To what point on the S cushion?

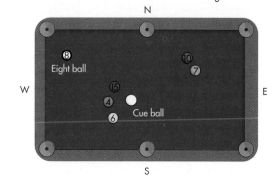

Lesson 8.5

Tessellations with Nonregular Polygons

In Lesson 8.4, you tessellated with regular polygons. You learned that there are only three regular tessellations and eight semiregular tessellations. What about tessellations of nonregular polygons? Will any scalene triangle tessellate? Let's investigate.

Investigation 8.5.1

Stack three pieces of paper together and fold them in half. Draw a scalene triangle on one half of the top half-sheet and cut it out, cutting through all six half-sheets of paper. You now have six congruent scalene triangles. Use one triangle as a template and trace the triangle onto what's left of the top half-sheet. Cut again. You now have twelve congruent scalene triangles. Label the interior angles in each triangle *a*, *b*, and *c*, as shown, making sure that all twelve angles labeled as *a* are congruent, all twelve labeled as *b* are congruent, and all twelve labeled as *c* are congruent. Using your twelve congruent scalene triangles, try to create a tessellation.

Observe the angles about each point. How many times did each angle of the triangle fit about each point? What is the sum of the measures of the three angles of a triangle? Compare your results with the results of others. State your next conjecture.

 C-79 —?— triangle will create a monohedral tessellation.

You have seen squares and rectangles tile the plane. You can probably visualize tiling with parallelograms. Will any quadrilateral tessellate? Let's investigate.

  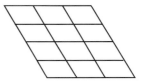 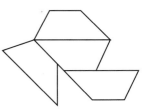

Investigation 8.5.2

Cut out twelve congruent quadrilaterals (not parallelograms). Label the interior angles in each quadrilateral *a*, *b*, *c*, and *d*. Using your twelve congruent quadrilaterals, try to create a tessellation.

Observe the angles about each point. How many times did each angle of your quadrilateral fit at each point? What is the sum of the measures of the angles of a quadrilateral? Compare your results with the results of others. State a conjecture.

C-80   —?— quadrilateral will create a monohedral tessellation.

Because the regular pentagon does not create a tessellation, you know that not every pentagon will tessellate. Is there at least one pentagon that will tessellate? Yes, there

are many. But how many? Prior to 1968, it was thought that all tessellating pentagons could be classified into five types. But in that year R. Kershner found three more and thought that the problem had been solved. No further discoveries were made until 1975, when Martin Gardner wrote about the problem in *Scientific American.* Soon Gardner wrote about another type found by Richard James III. After reading about this new discovery, Marjorie

Marjorie Rice (left) and Dr. Doris Schattschneider

Rice began her own investigations. With no formal training in mathematics beyond the high school level, she discovered within a few months a tenth type of pentagon that tessellates. By 1977, Marjorie Rice had discovered three more types. Mathematics professor Doris Schattschneider of Moravian College brought Rice's research to the attention of the mathematics community and confirmed that Rice had indeed discovered what professional mathematicians had been unable to uncover.

In 1985, a fourteenth type of tessellating pentagon was discovered by Rolf Stein, a German graduate student. Are *all* the types of convex pentagons that tessellate now known? The tessellating pentagon problem remains unsolved.

Two of the pentagonal tessellations discovered by Marjorie Rice

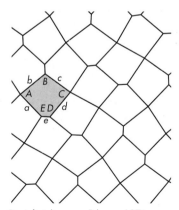

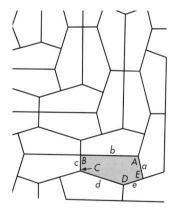

Rice's first discovery, February 1976
$2E + B = 2D + C = 360°$
$a = b = c = d$

Type 13, discovered in December 1977
$B = E = 90°$ $2A + D = 360°$ $2C + D = 360°$
$a = e$ $a + e = d$

Capital letters represent angle measures in the shaded pentagon. Lowercase letters represent lengths of sides.

Exercise Set 8.5

1. The Cairo street tiling shown at right is a very beautiful tessellation that uses equilateral pentagons (the sides are congruent but not the angles). The pentagon is shown below right, with angle measures that will help you draw your own. (Point *M* is the midpoint of the base.) Use a ruler and a protractor to draw an equilateral pentagon on poster board or heavy cardboard. (For an added challenge, you can try to construct the pentagon by using just a compass and a straightedge, as Egyptian artisans likely would have.) Cut out the pentagon and tessellate with it. Color your design.

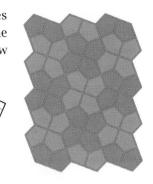

2. Another way to produce a very similar pentagonal tessellation is to make the dual of the semiregular tessellation shown in Lesson 8.4, Exercise 5. Try it.

3. Create a full-page color tessellation of triangles or quadrilaterals. Make the sides of your tile 4 cm to 6 cm long and leave a 3 cm to 4 cm border along the edges of your paper.

4. A nonconvex (concave) quadrilateral is shown at right. Can any nonconvex quadrilateral tile the plane? Try it. Create your own nonconvex quadrilateral and try to create a tessellation with it. Decorate your drawing.

In Exercises 5–7, you'll explore special rectangle tessellations found in many Japanese homes. Traditionally, mats called *tatami* are used as a floor covering in Japan. The original *tatami* were composed of a thick rice straw inner core with a soft reed cover. *Tatami* measure about three feet by six feet by about two inches thick. The size of rooms in traditional Japanese homes are often given in *tatami* numbers (for example, a 6-mat room or an 8-mat room). The arrangement of *tatami* in a room is an art in itself. It is quite common for the seams of *tatami* to form T-shapes. However, arranging four *tatami* so that they come together at one vertex to form a cross is avoided because it is difficult to get a good fit within a room this way.

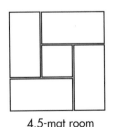

4.5-mat room

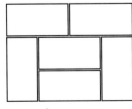

6-mat room

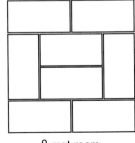

8-mat room

5. Use graph paper to design an arrangement of *tatami* for a 10-mat room. In how many different ways can you arrange the mats so that there are no places where four mats meet at a point (no cross patterns)? Assume that the mats measure 3′ × 6′ and that each room must measure at least 9′ wide. Show all your solutions.

6. The roof in a traditional Japanese wood-frame home is supported on pillars rather than by walls. Thus the home's interior partitions are not load-bearing and can be easily moved. In a traditional Japanese home, one room can serve many functions, and because the partitions are movable, the size of the rooms change according to their functions.

What is the combined area (total *tatami* number) of the two central rooms (living room and family room) of the Japanese home shown at right? On graph paper, draw the border of the two rooms (shown in blue in the diagram), then arrange the *tatami* into a different configuration within the same border. Draw bold lines along *tatami* edges to indicate where partitions could go in this new configuration to create two or more rooms. Give names to or describe functions of the new rooms. What are some advantages of being able to redesign interior rooms?

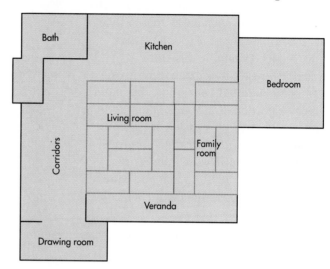

7. It is possible to arrange *tatami* into a rectangular pattern in such a way that there are no fault lines. A fault line is a straight line between the edges of *tatami*, passing all the way through a rectangular arrangement. Below left is an example of an arrangement that has a fault line. The smallest possible rectangle with no fault lines requires 15 *tatami*. There are at least two ways to arrange a *15-tatami* rectangle—one is shown below right. Can you find the other? (This puzzle actually originated as a domino puzzle. Dominos, like *tatami*, have one side that is half the length of the longer side. A set of dominos is helpful for this exercise.)

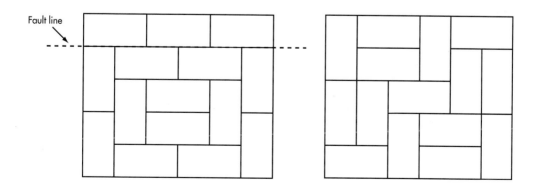

8. When British scientist Sir Roger Penrose of the
   University of Oxford is not at work on quantum
   mechanics or relativity theory, he's inventing
   mathematical games. Penrose came up with a
   special tiling that uses two shapes, a kite and a
   dart. (The dart is a concave kite.) Together the
   kite and the dart can create very interesting tiling
   patterns if Penrose's extra rules for the tiling
   procedure are followed. He placed two small dots
   at two vertices of each tile, as shown below. The
   tiles must be placed in such a way that each
   vertex with a dot always touches only other
   vertices with dots. By adding this extra

Penrose tiling at the Center for Mathematics and Computing, Carleton
College, Northfield, Minnesota

   requirement, Penrose's tiles make what is called a nonperiodic tiling. That is, as
   you continue to build out your tiling, the pattern does not repeat by
   translations. A portion of a Penrose tiling is shown at right.

   Try it. Copy the two tiles shown at right—
   the kite and the dart with their dots—onto
   patty paper. Create your own unique Penrose
   tiling. Draw in pencil at first so that you can
   erase the dots when you're finished. Then color
   your design.

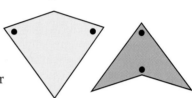

9. **Group Activity** Which pattern blocks can be used to create monohedral tilings? Sketch
   the tilings.

10. **Group Activity** Use a combination of two different types of pattern blocks to create a
    tiling that is 1-uniform.

11. **Group Activity** Use a combination of two or more types of pattern blocks to create a
    tiling that is 2-uniform.

Improving Visual Thinking Skills—*Picture Patterns IV*

Draw what comes next
in each picture pattern.

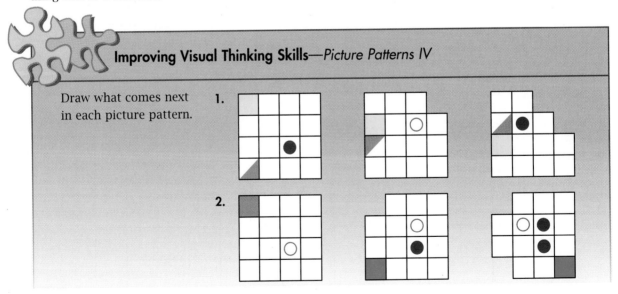

Lesson 8.6

Tessellations Using Only Translations

Islamic artists are the masters of tessellation. Islamic artists were not only trained in geometry, but held the philosophy that mathematics is essential to understanding the universe. They expressed this philosophy in their art. The Alhambra, a thirteenth-century Moorish palace in Granada, Spain, is one of today's finest examples of the precise mathematical art of Islam. Tessellations abound in the exquisite tile-work throughout the palace.

Sketches of majolica tile, M. C. Escher, 1936

In 1936, M. C. Escher traveled to Spain and became fascinated with the tile patterns of the Alhambra. Escher spent days in the Alhambra, sketching the tessellations on the walls and the ceilings. One of his sketches is shown above. Escher wrote of tessellations: "This is the richest source of inspiration that I have ever tapped." But Escher did not limit himself to pure geometric tessellations as did many Islamic artists. He wrote:

> What a pity it was that Islam forbade the making of "images." . . . I find this restriction all the more unacceptable because it is the recognizability of the components of my own patterns that is the reason for my never-ceasing interest in this domain.

Escher spent many years learning how to use translations, rotations, and glide reflections on grids of equilateral triangles and parallelograms to create tessellations of birds, fish, reptiles, and humans. One striking example of Escher's work is the tessellation he used in a tile mural for the Liberal Christian Lyceum in The Hague, the Netherlands. In this lesson you will learn to create your own tessellations by translation of recognizable shapes.

Symmetry drawing E105, M. C. Escher, 1960

The following four steps demonstrate how Escher may have created his "Pegasus" tessellation by replacing two sides of a square with the partial outlines of Pegasus and translating them to complete the motif.

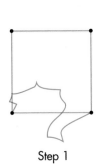

Step 1

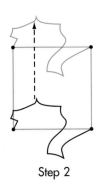

Step 2

Step 3

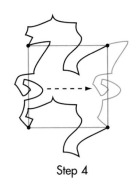

Step 4

One simple way to create a non-polygonal tessellation is by changing the opposite sides of a square or parallelogram tessellation. The steps below show how geometry student Robert Canete created *Leap Frog*.

Leap Frog, Robert Canete, geometry student

Step 1　Robert started with one square from a tessellation of squares (although any parallelogram will work with this method, and he could have used square dot paper). He connected vertices *A* and *B* of the square with a curve. Let's call the curve $\overset{\frown}{AB}$ (curve *AB*).

Step 2　Then he placed tracing paper or clear plastic over $\overset{\frown}{AB}$ and copied the curve with a felt tip pen onto the tracing paper or clear plastic. Next he placed the copy beneath the original and slid it so that the endpoints of $\overset{\frown}{AB}$ lined up with the endpoints of $\overline{CD}$. Then he retraced the curve onto the original so that it connected the endpoints of $\overline{CD}$.

Step 3　He repeated Step 1 with a curve that connected points *A* and *D*. Call the curve $\overset{\frown}{AD}$.

Step 4　He then copied $\overset{\frown}{AD}$ onto tracing paper or clear plastic and transferred the curve across to the opposite side, $\overline{BC}$.

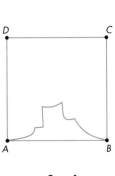

Step 1

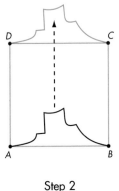

Step 2

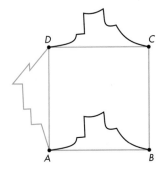

Step 3

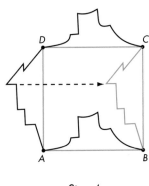

Step 4

Step 5 When completed, he traced the entire figure onto the tracing paper or the clear plastic and moved the figure to the next square. He then traced the entire figure onto the next square. He finally filled the grid of squares with his figure, creating a translation tessellation.

Dog Prints, the illustration below right, is another translation tessellation that uses squares as its basic structure.

The translation technique outlined on the previous page can be used with regular hexagons. The only difference is that with hexagons there are three sets of opposite sides. Therefore you need to draw three sets of curves and translate them to opposite edges. The six steps below demonstrate the process used to create the *Monster Mix* tessellation below.

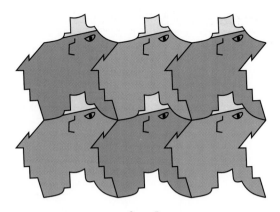

Step 5

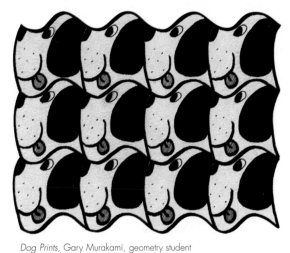

Dog Prints, Gary Murakami, geometry student

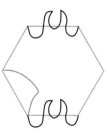

Monster Mix
Mark Purcell
geometry student

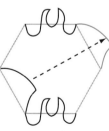

Step 1

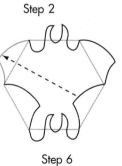

Step 2

Step 3

Step 4

Step 5

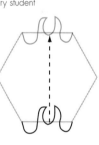

Step 6

Exercise Set 8.6

In Exercises 1–3, copy each tessellating shape and fill it in so that it becomes a recognizable figure. The Escher designs and the student tessellations featured in this lesson took a great deal of time and practice. When you create your own tessellating designs of recognizable shapes, you'll appreciate the need for this practice!

1.

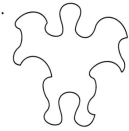

2.

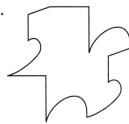

3.

In Exercises 4–6, identify the basic tessellation grid (squares, parallelograms, or regular hexagons) used to create each translation tessellation.

4.

Cat Pack, Renee Chan, geometry student

5.

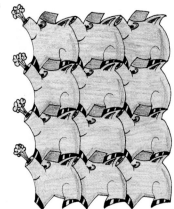

Snorty the Pig, Jonathan Benton
geometry student

6.

Old Wise One, Serene Tam
geometry student

7. Copy the figure and the grid below onto a patty paper. Show how you can use other patty papers to create a tessellation on a grid of parallelograms with the figure.

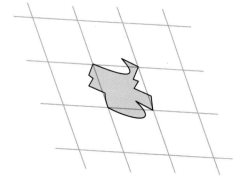

8. Copy the figure and the grid below onto a patty paper. Show how you can use other patty papers to create a tessellation on a grid of regular hexagons with the figure.

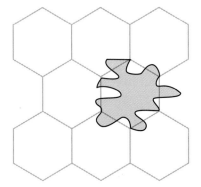

9.* Now it's your turn to create a tessellation of recognizable shapes, using squares as the basic structure and using the translation method described in this lesson. At first, you will probably end up with shapes that look like amoebas or spilled milk, but with practice you will see recognizable images within those amoeba-like tessellating shapes. Decorate your design. Give it a title. You will need the following materials.

- Tracing paper, patty paper, or clear plastic
- Square dot paper or paper with a grid of large squares
- Colored pencils, felt tip markers, and ballpoint pens

10. Try this exercise with a grid of regular hexagons as the basic structure. Create a tessellation of recognizable shapes by using the translation method. Decorate your design. In addition to the materials listed in Exercise 9, you will need the following.

- Isometric dot paper or paper with a grid of large hexagons

In Exercises 11 and 12, reflect the line over the line of reflection indicated, then determine the equation of the new image line.

11. Reflect $y = \frac{2}{3}x - 3$ over the y-axis.

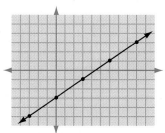

12. Reflect $y = \frac{-5}{2}x + 2$ over the y-axis.

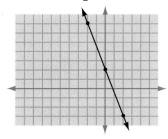

In Exercises 13 and 14, give the numerical symbol for each tessellation.

13.

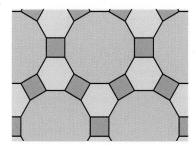

14.

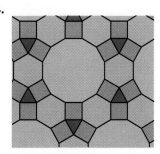

15. A new interstate is about to be built near the two rural towns of Perry and Mason. The two towns will have to pay a portion of the costs to build roads to one junction point on the freeway. Where should the junction point be located so that the total length of the two straight roads is as short as possible? (Of course, one of the roads is going to be named Della Street. Any suggestion for the other road?) Copy the figure at right onto a patty paper and locate the junction point.

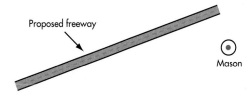

16. The route of a rancher takes him from the house at point *A* to the south fence, then over to the east fence, then to the corral at point *B*. Copy the figure at right onto a patty paper and locate the points on the south and east fences that minimize the rancher's route.

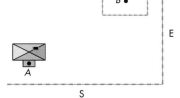

Improving Reasoning Skills—*Crossnumber Puzzle I*

Copy the crossnumber grid at right. In this puzzle, the answers you know will help you answer the problems you don't know. Did Napoleon really study geometry? You'll soon find out. Each clue consists of a statement followed by two numbers. If the statement is true, enter the first number of the pair. If the statement is false, enter the second number.

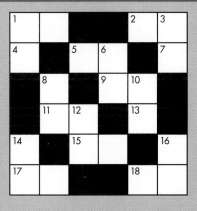

Across

1. Every quadrilateral is a trapezoid. (47, 14)

2. If two angles are vertical angles, then they are congruent. (91, 67)

4. The sum of the measures of the angles of a pentagon is 360°. (5, 6)

5. The diagonals of a rectangle are congruent. (25, 48)

7. Each angle of a regular octagon has a measure of 160°. (7, 3)

8. Euclid wrote the *Elements*. (6, 5)

9. Pythagoras believed that numbers ruled the universe. (49, 27)

11. Lewis Carroll was a mathematician who wrote *Alice's Adventures in Wonderland*. (64, 24)

13. The midsegment in a trapezoid is parallel to the bases. (8, 9)

14. The symbol *AB* stands for the line through points *A* and *B*. (5, 1)

15. An isosceles right triangle is impossible. (00, 10)

16. An obtuse right triangle is impossible. (0, 4)

17. Each base angle of an isosceles right triangle measures 45°. (12, 87)

18. Much of Victor Vasarely's art is geometric in form. (11, 28)

Down

1. Leonardo da Vinci invented calculus. (45, 16)

3. Napoleon studied geometry and found a new proof for the Pythagorean Theorem. (13, 77)

6. Sofia Kovalevskaya was the first European woman since the Renaissance to receive a doctorate in mathematics. (54, 89)

8. The angle bisectors in a trapezoid are concurrent. (52, 66)

10. The diagonals of a rhombus are congruent. (79, 98)

12. The incenter of a triangle is also the center of mass. (40, 41)

14. According to legend, a story about Archimedes inspired Sophie Germain to study mathematics. (11, 58)

16. Benjamin Banneker liked to invent and solve mathematical puzzles. (01, 38)

Lesson 8.7

Tessellations That Use Rotations

In Lesson 8.6, you created recognizable shapes by translating curves from opposite sides of a regular hexagon or square. In tessellations using only translations, all the figures face in the same direction. In this lesson you will use rotations of curves on a grid of parallelograms, equilateral triangles, or regular hexagons. The designs formed will have rotational symmetry about points in the tiling.

The steps below describe how you might create a tessellating reptile similar to that created by Escher in the drawing shown at right. Each reptile is made by rotating three different curves about three alternating vertices of a regular hexagon.

Step 1   Connect points S and I with a curve.

Step 2   Rotate $\widetilde{SI}$ about point I so that point S rotates to coincide with point X.

Step 3   Connect points G and X with a curve.

Step 4   Rotate $\widetilde{GX}$ about point G so that point X rotates to coincide with point O.

Step 5   Create $\widetilde{NO}$.

Step 6   Rotate $\widetilde{NO}$ about point N so that point O rotates to coincide with point S.

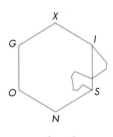

Step 1

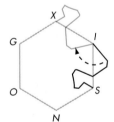

Step 2

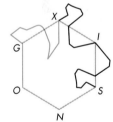

Step 3

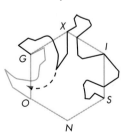

Step 4

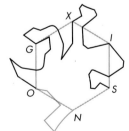

Step 5

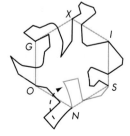

Step 6

Remember that Escher worked long and hard to perfect each of the three curves in his drawing. He adjusted each curve until he got what he recognized as a reptile. When you are working on your own design, keep in mind that you may redraw your curves many times until something you recognize appears.

Escher used his reptile drawing for a number of his works. It appeared in *Metamorphosis II*—a woodcut that is 13 feet long—and in *Metamorphosis III*, which is $22\frac{1}{3}$ feet long! The reptile study was probably best used in his famous lithograph titled *Reptiles*. In *Reptiles*, the little creatures become three-dimensional and leave the two-dimensional drawing, crawl over numerous objects, then re-enter the flat page. Escher loved to play with our perceptions of reality!

Reptiles, M. C. Escher, 1943
©1996 M. C. Escher / Cordon Art – Baarn – Holland
All rights reserved.

Another method used by Escher utilizes rotations on an equilateral triangle grid. Two sides of each equilateral triangle have the same curve, rotated about their common point. The third side is a curve with point symmetry. The following steps demonstrate how you might create a tessellating flying fish like that created by Escher.

Step 1   Connect points F and I with a curve. Then rotate the curve 60° clockwise about point I so that it becomes $\widetilde{IH}$.

Step 2   Find the midpoint S of $\overline{FH}$ and draw $\widetilde{SH}$.

Step 3   Rotate $\widetilde{SH}$ 180° to produce $\widetilde{FS}$. Together $\widetilde{FS}$ and $\widetilde{SH}$ become the point-symmetric $\widetilde{FH}$.

Step 4   With a little added detail, the design becomes a flying fish.

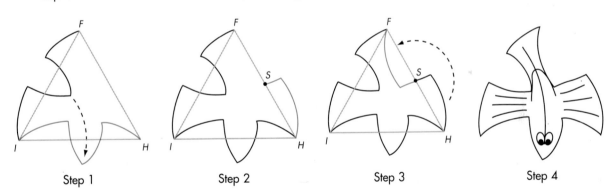

With just a slight variation in the curves, the resulting shape will appear more like a bird than a flying fish. The steps are outlined below.

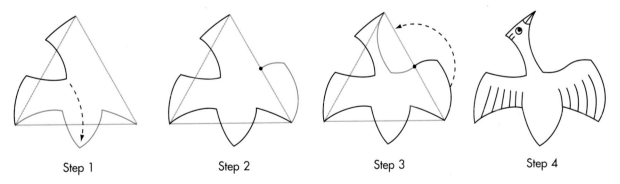

Exercise Set 8.7

In Exercises 1 and 2, identify the basic tessellation grid (equilateral triangles or regular hexagons) used to create the tessellation.

1.

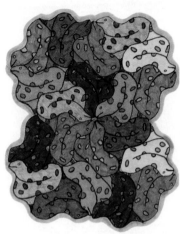

Jack Chow
geometry student

2.

Aimee Plourdes
geometry student

3. Copy the figure and the grid below onto a patty paper. Show how you can use other patty papers to create a tessellation with the figure on a grid of equilateral triangles.

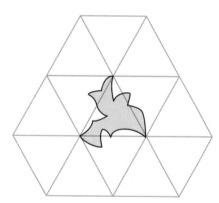

4. Copy the figure and the grid below onto a patty paper. Show how you can use other patty papers to create a tessellation with the figure on a grid of regular hexagons.

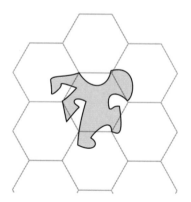

5. Copy onto a patty paper the finished bird (Step 4) from the rotation method described in this lesson. Use the bird to make a tessellating design.

In Exercises 6 and 7, create tessellation designs by using rotations. You will need patty paper, tracing paper, or clear plastic, and a grid of regular hexagons or equilateral triangles or isometric dot paper.

6. Create a tessellating design of recognizable shapes by using a grid of regular hexagons. Decorate and color your art.

7. Create a tessellating design of recognizable shapes by using a grid of equilateral triangles. Decorate and color your art.

8. Geometry student Garret Lum started with a grid of regular hexagons and subdivided each hexagon into three rhombuses to create the design below. Copy a portion of the design and locate the vertices of the regular hexagonal grid. Then divide one hexagon into its three rhombuses.

Garret Lum, geometry student

9. Try Garret Lum's method. Start with a grid of regular hexagons and divide them into congruent rhombuses. Create your own rotation tessellation with your rhombus grid.

10. There is quite a variety of ways in which you can create tessellations by using rotations. The figures below demonstrate how you can create rotation tessellations by using a point-symmetric curve on each side of a parallelogram.

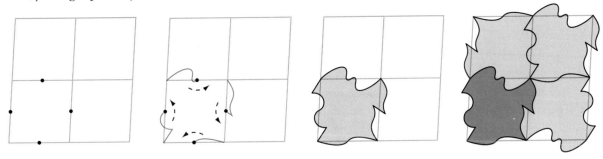

Try it. Start with a grid of parallelograms and create a point-symmetric curve on each side of one of the parallelograms to create your rotation tessellation.

11. The figures at right demonstrate how you can create rotation tessellations by using a combination of translations and rotations on a grid of rectangles.

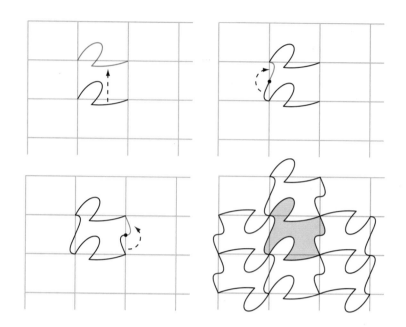

Try it. Start with a grid of rectangles and create a curve on one side. Then translate it to the opposite side. Create point-symmetric curves on the other two sides. Use translations and rotations to fill your grid with copies of this tile. Identify six points on your original tile that are centers of 180° rotational symmetry for the tessellation.

12. The figures below demonstrate how you can create rotation tessellations by rotating on a grid of kites (or on squares and on 60°-120° rhombuses).

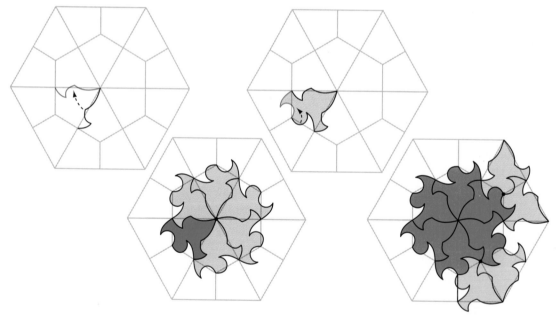

Try it. Start with a grid of regular hexagons and divide them into kites, as shown. Create a curve on each of one pair of noncongruent sides of one kite tile, then rotate each curve to its congruent neighbor. Use rotations to fill your grid with this tile. Notice that the tiling of kites shown above is 3-uniform. It has six kites surrounding the 60° vertex, four kites surrounding the 90° vertex, and three kites surrounding the 120° vertex. Is the 90° vertex a center of 90° rotational symmetry? Explain.

Tessellating with the Conway Criterion

You've discovered that you can tessellate with any triangle, any quadrilateral, some nonregular pentagons, and regular hexagons. The Conway criterion, named for English mathematician John Horton Conway, describes rules for tiles that will always tessellate. The sides of a tile satisfying the Conway criterion need not be straight, but they must have certain characteristics that you'll discover in this project. The rules are surprisingly loose, and it's clear from Escher's work that he knew them long before Conway described them mathematically. In the following investigation, you'll create a tessellation of hexagons that satisfy the Conway criterion. You'll see that tessellating hexagons can be far from regular.

The Conway criterion applied to a hexagon states simply that any hexagon with a pair of opposite sides that are parallel and equal in length will tessellate. First you'll tessellate with a Conway-criterion hexagon, then you'll experiment with special cases and a more general case of the criterion.

Step 1   Construct segment *AB* and point *C*.

Step 2   Select points *A* and *C* and choose Mark Vector in the Transform menu.

Step 3   Select $\overline{AB}$ and point *B* and choose Translate in the Transform menu. You now have your pair of parallel, congruent sides.

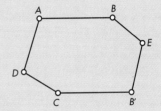

Step 4   Place points *D* and *E* anywhere except on $\overline{AB}$ or $\overline{CB'}$ and construct $\overline{AD}$, $\overline{DC}$, $\overline{BE}$, and $\overline{EB'}$.

You now have a tile that meets the Conway criterion, and believe it or not, this tile will tessellate no matter how you distort it. Here's how.

Step 5   Construct the midpoint *F* of $\overline{AD}$ and the midpoint *G* of $\overline{BE}$.

Step 6   Construct the polygon interior of *ADCB'EB*.

Step 7   Use the Transform menu to mark point *G* as center. Select the polygon interior and point *F* and rotate them 180° about point *G*.

Step 8   Give the image polygon a different color or shade.

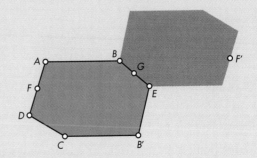

Step 9   Mark vector *FF′*. Select the two polygon interiors
          and translate them by this marked vector.

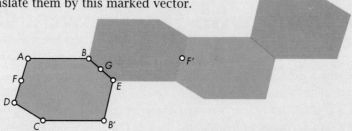

You now have a row of polygons. Can you see how they fit together? It's worth pausing in the construction for a moment to investigate and make a conjecture. Call the polgyons from left to right p_1, p_2, p_3, and p_4. Polygon p_3 is the translation of polygon p_1 by the vector *FF′*. How is polygon p_3 related to polygon p_2? How is polygon p_2 related to polygon p_1? Copy and complete the conjecture.

Conjecture: The compostition of two 180° rotations about two different centers is
                  equivalent to —?—.

Now continue sketching.

Step 10   Mark vector *AC*. Select the row of polygon interiors and translate by this marked
          vector. Translate again one or two more times.

Step 11   Color and/or shade the polygons so that you
          can tell them apart.

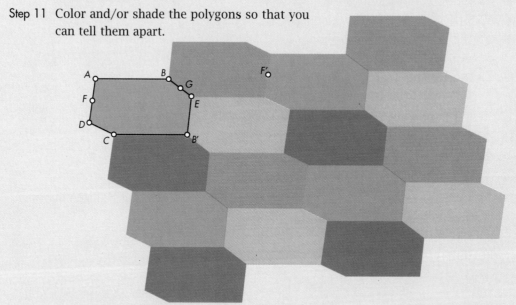

Drag any of the vertices of the hexagon. Are you surprised by the variety of hexagons that will tessellate?

Investigate the following questions.

1.   Drag point *B* on top of point *A*. Now what kind of tessellating shape do you have? This shape is a special case that meets the Conway criterion. What is it about the angles of this shape that guarantee it will tessellate?

2. With point *B* still on top of point *A*, drag point *E* on top of points *B* and *A*. Now what shape do you have? What is it about the angles of this shape that guarantee it will tessellate?

3. Undo so that you have a hexagon again. How can you make the hexagon into a pentagon that will tessellate? Try it. Explain what you did.

4. Undo so that you have a hexagon again. As in any tessellation, the angle measures at each vertex of this tessellation add up to 360°. Explain why they add up to 360°. Hint: Draw lines *AC* and *BB′* and construct a line parallel to *BB′* through point *E*.

5. If you want, you can construct a more general Conway-criterion tile and its tessellation. In a new sketch, start with a Conway-criterion hexagon like that shown below with dashed lines.

Step 1   Construct a jagged edge by drawing segments from point *A* to point *B*. Translate this edge (including points on it) by vector *AC*.

Step 2   Construct the midpoints of the other four sides. From point *B* to midpoint *G*, construct a jagged edge. (You probably don't want to get any messier than two segments.) Rotate this jagged edge by 180° about point *G*.

Step 3   Construct jagged edges on the other three sides by the same method (180° rotations about midpoints).

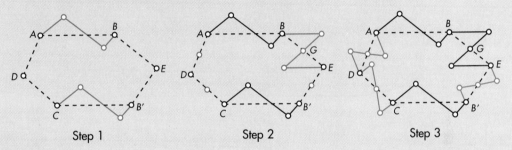

Step 1   Step 2   Step 3

Step 4   Construct the polygon interior and tessellate as you did with the hexagon.

Reference: "Will it Tile? Try the Conway Criterion," D. Schattschneider. *Mathematics Magazine*, vol 53 (Sept 1980): 223–233.

Improving Algebra Skills—*Fantasy Functions II*

If $a \circ b = a^b$ then $3 \circ 2 = 3^2 = 9$
and if $a \triangle b = a^2 + b^2$ then $5 \triangle 2 = 5^2 + 2^2 = 29$.

If $8 \triangle x = 17 \circ 2$, find *x*.

Tessellations That Use Glide Reflections

In this lesson you will use glide reflections to create tessellations. Two methods used by Escher for creating glide-reflection tessellations are demonstrated below. The first method uses a grid of kites; the second uses a grid of parallelograms.

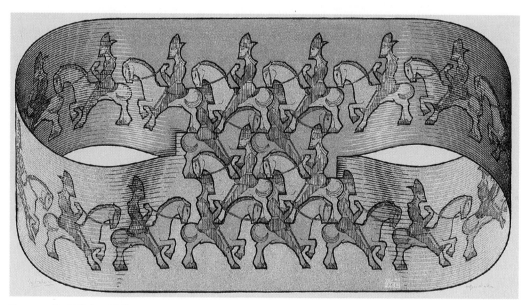

Horseman, M. C. Escher, 1946

In Lesson 8.6, you saw Escher's translation tessellation of the winged horse Pegasus. In the Pegasus tiling all the horses were facing in the same direction. In the drawing *Horseman*, Escher used glide reflections on a grid of glide-reflected kites to get his horsemen facing in opposite directions. The steps below demonstrate how you might make a tessellating design similar to Escher's *Horseman*. (The symbol ↟ indicates a glide reflection.)

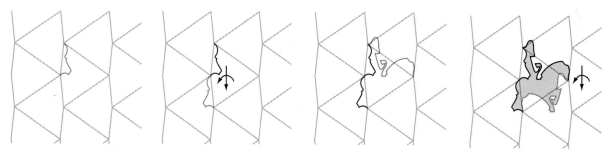

| Step 1 | Step 2 | Step 3 | Step 4 |

In the tessellation of birds shown at right, Escher might have used a grid of glide-reflected parallelograms to create the drawing. The steps below suggest how you might create a glide-reflected tessellation of birds. Start with a grid of glide-reflected parallelograms. Create a curve on the bottom edge of one parallelogram and glide-reflect it to the opposite side. Create a curve on the right edge of the parallelogram and translate it to the opposite side.

Symmetry drawing E108, M. C. Escher, 1967
©1996 M. C. Escher / Cordon Art – Baarn – Holland
All rights reserved.

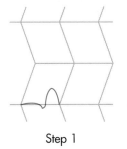

Step 1

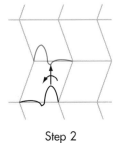

Step 2

Step 3

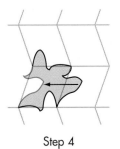

Step 4

Exercise Set 8.8

In Exercises 1 and 2, identify the basic tessellation grid (kites or parallelograms) used to create the tessellation.

1.

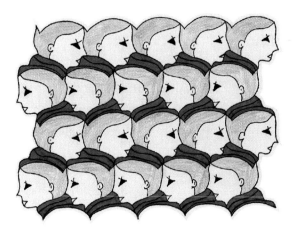

A Boy with a Red Scarf
Elina Uzin
geometry student

2.

Glide Reflection
Alice Chan
geometry student

3. Copy the figure and the grid below onto a patty paper. Show how you can use other patty papers to create a tessellation on a grid of parallelograms with the figure.

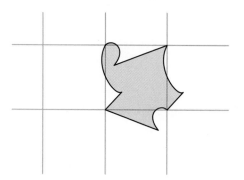

4. Copy the figure and the grid below onto a patty paper. Show how you can use other patty papers to create a tessellation on a grid of kites with the figure.

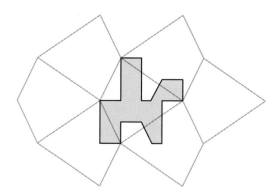

5.* Create a glide-reflection tiling design of recognizable shapes by using a grid of kites. Decorate and color your art.

6.* Create a glide-reflection tiling design of recognizable shapes by using a grid of parallelograms. Decorate and color your art.

There is a variety of ways in which you can create glide-reflection tessellations. The figures below demonstrate how you can create a tessellation by using two different glide reflections. This example uses glide reflections on both pairs of opposite sides of a rectangle.

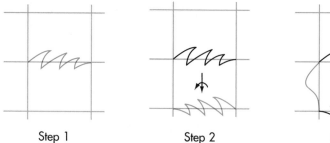

| Step 1 | Step 2 | Step 3 | Step 4 |

7. Try it. Start with a grid of rectangles and create a curve on one of the sides, then glide-reflect it to the opposite side. Create another curve on one of the remaining sides and glide-reflect the curve to the opposite side.

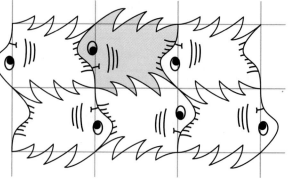

8.* The figures below demonstrate how you can create tessellations by using glide reflections and a point-symmetric curve on a grid of equilateral triangles. The steps below show how you can make a tessellating penguin-like bird similar to that designed by Garret Lum in his *South Pole or Bust* tessellation.

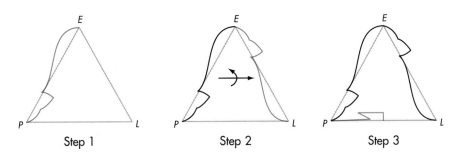

Step 1   Step 2   Step 3   Step 4

Try it. Start with a grid of equilateral triangles and draw a curve on one side of one of the triangles. Glide-reflect this curve to one of the other sides of the triangle. On the third side, draw a point-symmetric curve.

9. The design below left is a tessellation that uses block capital T's. Four different types of shading have been used. Notice that no T is touching another T with the same type of shading. Could this have been done with only three types of shading? The design below right is a tessellation that uses block capital E's. Three different types of shading have been used, and still no E is touching another E with the same type of shading. Which of the block letters in the group at right can be used to create a tessellation? Can any of these block-letter tessellations be shaded with fewer than three types of shading so that no two letters with the same shading touch? What other block letters do you think would tile the plane? What transformations would be used in these tilings? Use graph paper to investigate these questions and make a list that describes your findings.

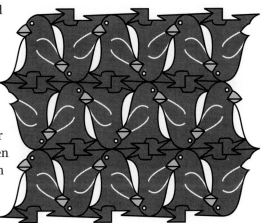

South Pole or Bust, Garret Lum, geometry student

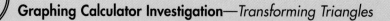

Throughout this text you've done exercises in which you're given a figure on the coordinate plane and you apply (or discover) a rule that transforms the original figure to an image figure. Sometimes the image figure is congruent to the original figure, meaning the transformation rule you used was an isometry.

You can use your calculator to draw figures and transform them. This investigation assumes you have a calculator with statistics graphing capabilities that allow you to draw lines by using data points stored in lists. If you're not already familiar with these functions, you may need to get some extra help from your teacher.

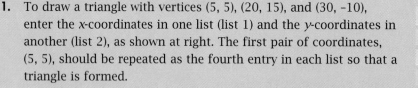

1. To draw a triangle with vertices (5, 5), (20, 15), and (30, –10), enter the *x*-coordinates in one list (list 1) and the *y*-coordinates in another (list 2), as shown at right. The first pair of coordinates, (5, 5), should be repeated as the fourth entry in each list so that a triangle is formed.

L_1	L_2
5	5
20	15
30	–10
5	5

    To make a line graph, turn on a statistics plot and choose the *x*-coordinates from list 1 and the *y*-coordinates from list 2. When you graph, you should get a triangle. Make sure to set the range so that you can see the whole triangle and plenty of space around it. You may want to turn on the grid for this investigation.

2. Make a third list, setting the entries in this list equal to the opposites of the entries in list 1. Turn on a second statistics plot and graph a new triangle by choosing the *x*-coordinates from list 3 and the *y*-coordinates from list 2. How does the position or location of this new triangle compare to the first triangle? What transformation resulted from using the opposites of the original *x*-coordinates?

3. Make a fourth list by adding 5 to each of the *x*-coordinates in list 1 and a fifth list by subtracting 10 from each *y*-coordinate in list 2. Turn on a third statistics plot and graph another triangle by choosing the *x*-coordinates from list 4 and the *y*-coordinates from list 5. How does the position or location of this new triangle compare to the first triangle? What transformation resulted from adding a number to the *x*-coordinate and subtracting a number from the *y*-coordinate?

4. Try other operations on list 1 and list 2 that transform your original triangle. For example, try to perform transformations that will rotate the triangle 180° about the origin or that will reflect the triangle over the *x*-axis. Try creating a transformation that is not an isometry. Explain what you did.

Lesson 8.9

Chapter Review

In this chapter you learned about special transformations in the plane—called isometries—and you revisited principles of symmetry that you first encountered in Chapter 0. Then you applied these concepts to create tessellations. Can you name the three isometries? Can you describe how isometries combine to make other isometries? How can you use reflections to improve your miniature-golf game? What types of symmetry do regular polygons have? What types of polygons will tile the plane? Review this chapter to be sure you can answer these questions.

Exercise Set 8.9

For Exercises 1–12, identify each statement as true or false. For each false statement, sketch a counterexample or explain why it is false.

1. A regular polygon of n sides has n reflectional symmetries and n rotational symmetries.

2. The only three regular polygons that create monohedral tessellations are equilateral triangles, squares, and regular pentagons.

3. Every triangle will create a monohedral tessellation.

4. Every quadrilateral will create a monohedral tessellation.

5. No pentagon will create a monohedral tessellation.

6. No hexagon will create a monohedral tessellation.

7. Two isometries in which the image has the same orientation as the original (the same order of points as you move clockwise) are the translation and rotation transformations.

8. Two isometries in which the image has the opposite orientation as the original are the reflection and glide-reflection transformations.

9. A translation of (5, 12) followed by a translation of (-8, -6) is equivalent to a single translation of (-3, 6).

10. A rotation of 140° followed by a rotation of 260° about the same point is equivalent to the single rotation of 40° about that point.

11. A reflection over a line followed by a second reflection over a parallel line that is 12 cm from the first is equivalent to a translation of 24 cm.

12. There are at least three times as many true statements as false statements in Exercises 1–12.

13. The facade of Chartres Cathedral in France does not have bilateral symmetry. Why not? Sketch the portion of the facade that does have bilateral symmetry.

In Exercises 14–16, identify the type or types of symmetry, including the number of symmetries, in each design. For Exercise 16, describe how you can move candles on the Chanukah menorah to make the colors symmetrical, too.

14.

Mandala, Gary Chen, geometry student

15.

16.

17. Your friendly librarian or the local newsstand may help with this one! Find a logo for a sports team, a car company, or some other organization whose logo has reflectional symmetry. Sketch the logo and its line or lines of reflection symmetry.

18. Find and sketch a logo that has rotational symmetry but not reflectional symmetry.

19. The figures below show the symmetric patterns of a dance called the pavane. The ▭ indicates the male dancer and the ⬭ indicates the female dancer. The arrows indicate the direction of the dancers' movement. The black half of each symbol indicates the front of the dancer. Try your hand at choreography. Design for a group of couples a dance pattern that exhibits some form of symmetry.

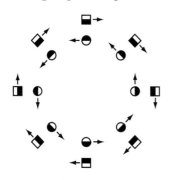  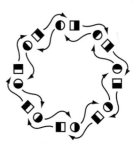

20. Do you need a mirror that is as long as you are tall to be able to see your full height? Actually, no. All you need is . . . ah, that is for you to figure out! Find a full-length mirror and experiment. Use newspapers to cover up the mirror's top or bottom portion (or both) until the smallest portion in which you can still see your full height (y) remains exposed. How does y compare to your height (x)? How should the mirror be positioned? Now, can you explain, with the help of a diagram and what you know about reflections, why a "full-length" mirror need not be as tall as you?

21. Miniature golf pro Sandy Trapp wishes to impress her gallery of fans with a hole in one on the very first hole. How do you think she should hit the ball at T to achieve this feat? Explain.

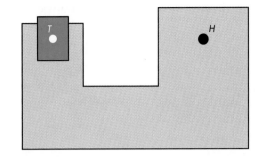

22. Your geometry class has challenged another geometry class at your school to an egg relay. Each team is divided into three groups of runners, designated Runners 1, Runners 2, and Runners 3. Runners 1 from Team A line up at the starting point, S_A, on the football field. A Runner 1 carries an egg balanced in a spoon to a Runner 2 lined up on the west sideline of the field, and Runner 2 carries the egg to a Runner 3 on the east sideline of the field. The Runner 3 carries the egg to the finishing point, F_A, and deposits it into a carton there. Then Runner 3 runs back to the starting point and taps the next Runner 1, who starts the next egg to its destination. Team B does the same thing, except Runners 1 start at point S_B, Runners 2 line up on the east sideline, Runners 3 line up on the west sideline, and they finish at point F_B. The object is to get all your team's eggs to their destination before the other team does.

Copy the figure below. Find the points at which Runners 2 and Runners 3 for each team should line up to maximize their chance for victory (by minimizing their travel distances). Write a short explanation of how you found those points.

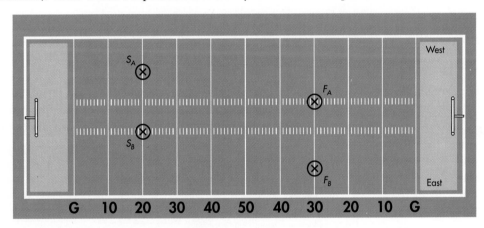

In Exercises 23 and 24, identify the tessellation by giving its numerical symbol.

23.

24.

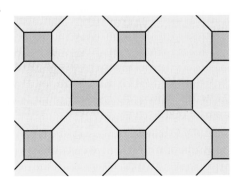

In Exercises 25–27, identify a possible method used to create each tessellation.

25.

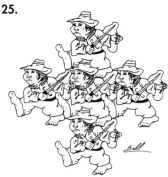

Robert Bell
geometry student

26.

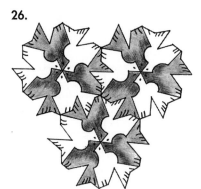

Serene Tam
geometry student

27.

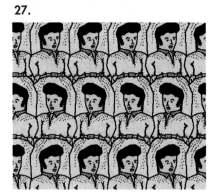

Sightings, Peter Chua and Monica Grant
Created using TesselMania!™ software

In Exercises 28 and 29, copy the figure and grid onto a patty paper. Determine whether or not the figure can be used to create a tessellation on the grid. Explain your reasoning.

28.

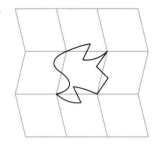

29.

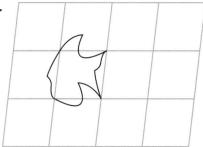

30. In his woodcut *Day and Night,* Escher gradually changes the shape of the patches of farmland into black and white birds. The birds are flying in opposite directions, and, because of this, the birds appear to be glide reflections of each other. But at second glance Escher's method appears to be even more complicated than it first appeared. Notice that the tails of the white birds curve down, while the tails of the black birds curve up. With even closer inspection, it's clear that this is not a glide-reflection tiling at all! When two birds are taken together as one tile (a 2-motif tile), they create a translation tessellation. Use patty papers to find the one tile (made up of the two birds) that Escher used to create his original translation tessellation.

Day and Night, M. C. Escher, 1938
©1996 M. C. Escher / Cordon Art – Baarn – Holland

Assessing What You've Learned

Try one or more of the following assessment suggestions.

Update Your Portfolio

- Choose one of the tessellations you did in this chapter and add it to your portfolio. Describe why you chose it and explain the transformations you used and the types of symmetry it has.

Organize Your Notebook

- Review your notebook to be sure it's complete and well organized. Are all this chapter's isometries and symmetry types included in your definition list or conjecture list? Write a one-page chapter summary.

Write in Your Journal

- This chapter is something of a departure from what you've been doing. For one thing, it lays a heavy emphasis on applying geometry to create art. Write about connections you see between geometry and art. Does creating geometric art give you a greater appreciation for art and for geometry? Explain.

Performance Assessment

- While a classmate, a friend, a family member, or a teacher observes, carry out one of the investigations from this chapter. Explain what you're doing at each step, including how you arrive at the conjecture.

Give a Presentation

- Give a presentation about one of the investigations or projects you did or about one of the tessellations you created.

Improving Reasoning Skills—*Logical Liars*

Five students have just completed a logic contest. To confuse the school's reporter, Lois Lang, each student agreed to make one true and one false statement to her when she interviewed them. Lois was clever enough to figure out the winner. Are you? See if you can determine the results of the contest from the students' statements.

Frances:	Kai was second. I was fourth.
Leyton:	I was third. Charles was last.
Denise:	Kai won. I was second.
Kai:	Leyton had the best score. I came in last.
Charles:	I came in second. Kai was third.

Cooperative Problem Solving

Games in Space

It is the year 2066. The board of directors of the recreational space station *Empyrean* has just sent a directive to the station's education and research council, asking that the council develop some new games. The purpose of the games will be to entertain and possibly educate students visiting from Earth.

You are part of a committee on the council. Your committee's task is to design and build a game. After designing and testing your game, collect its materials, assemble them, print its rules, replay and retest it, and, finally, package it. Develop your game with the following criteria in mind.

• The game is intended for eighth-grade students and older.

• The game must be educational and entertaining. (It might give drill and practice in learning basic geometric facts, or it might develop reasoning skills, problem-solving skills, or communication skills.)

• The game must have a clear set of rules.

• It is recommended, but not required, that the game have a set of playing pieces and a playing board larger than the size of a standard chess board.

• The game's average playing time should be less than one hour.

Area

Square Limit, M. C. Escher, 1964

In this chapter you will use your geometry tools to discover formulas that are used to find the areas of rectangles, parallelograms, triangles, trapezoids, kites, regular polygons, circles, and combinations of these. You may already know some of the formulas. Your investigations will be physical proofs of these formulas. You will use these area formulas to solve problems in this chapter and in the chapters to follow.

Lesson 9.1

Areas of Rectangles and Parallelograms

The **area** of a plane figure is the measure of the region enclosed by the figure.

People in many occupations work with areas. Carpenters calculate the areas of walls, floors, and roofs to order materials for construction. Painters calculate the area of surfaces to be painted so that they know how much paint will be needed for a job. Decorators need to know the areas of carpeting and drapery materials to install in homes. In this chapter you will

Tile layers need to find the area of a floor to determine how many tiles to buy.

discover formulas for finding the areas of the regions within triangles, parallelograms, trapezoids, kites, regular polygons, and circles.

The area of a figure is measured by the number of squares of a unit length (square units) that can be arranged to completely fill that figure. If two polygons are congruent, their areas are equal.

For some figures, like the figure at right, the squares may have to be cut up and rearranged. The area of this figure is 14 square units. There are 12 unit squares that fit completely inside the figure. The remainder of the figure can be filled by cutting up and rearranging two additional squares.

You probably already know many area formulas. Think of the investigations leading to area formulas in this chapter as physical demonstrations of these formulas. Physical demonstrations may help you remember the formulas more easily.

Length: 1 unit

Area: 1 square unit

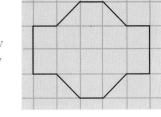

Area: 14 square units

Investigation 9.1.1

It's easy to find the area of rectangles. Find the area of each rectangle in square units.

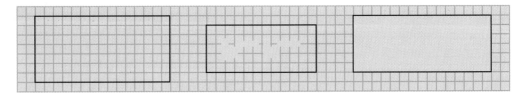

To find the area of the first rectangle you can simply count squares. You can do the same for the second, but it's a little harder because portions of some squares are blanked out. To find the area of the third rectangle, you could copy the rectangle, draw in the lines, and count the squares, but there's an easier method.

Any side of a rectangle (or parallelogram) may be called a **base**. A segment perpendicular to the base with one endpoint on the base and the other endpoint on the opposite side is the **altitude** to that base. The length of the altitude is the **height**. Sometimes the term *altitude* is also used to refer to the height.

Because the length of the base indicates the number of squares in each row and the height indicates the number of rows, you can use these terms to state a formula for the area. Do you see how? State your next conjecture.

C-81 The area of a rectangle is given by the formula —?—, where A is the area, b is the length of the base, and h is the height of the rectangle (***Rectangle Area Conjecture***).

Investigation 9.1.2

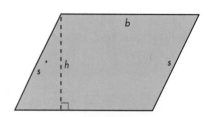

Draw a parallelogram onto a piece of heavy paper or cardboard. Label the parallelogram as shown at right. Construct an altitude from the vertex of the upper obtuse angle to the lower base. Label it as shown. Cut out the parallelogram and then cut along the altitude. You will have two pieces, a triangle and a trapezoid. Try arranging the two pieces into other shapes without overlapping them. Is the area of each of these new shapes the same as the area of the original parallelogram? Can you form a rectangle as one of your new shapes? What is the area of this rectangle? What is the area of the original parallelogram? State your next conjecture.

C-82 The area of a parallelogram is given by the formula —?—, where A is the area, b is the length of the base, and h is the height of the parallelogram (***Parallelogram Area Conjecture***).

How do you use the Parallelogram Area Conjecture? Here are some tips. Write down the formula. Substitute the given or known values into the formula. Solve for the remaining variable. If the dimensions are measured in inches, feet, or yards, the area is measured in square inches (sq in.), square feet (sq ft), or square yards (sq yd). If the dimensions are measured in centimeters or meters, the area is in square centimeters (cm^2) or square meters (m^2). Let's look at an example.

Example

What is the height of a parallelogram that has an area of 7.13 m^2 and a base 2.3 m long?

$$A = bh$$
$$7.13 = (2.3)h$$
$$\frac{7.13}{2.3} = h$$
$$h = 3.1$$

The height measures 3.1 m.

Take Another Look 9.1

 Use a geometry computer program to construct a parallelogram whose perimeter can vary but whose area stays constant.

Exercise Set 9.1

In Exercises 1–4, estimate the area of each figure. The area of each square is one square unit.

1.
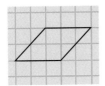

2.
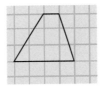

3.
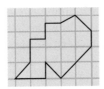

4.
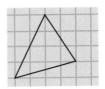

In Exercises 5–12, each quadrilateral is a rectangle. In Exercises 13–20, each quadrilateral is a parallelogram. Use the appropriate unit in each answer. Area is represented by A and P stands for perimeter.

5. $A = -?-$

12 m
19 m

6. $A = -?-$

$2\,1/2''$
$7\,1/4''$

7. $A = -?-$

4.5 cm
9.3 cm

8. $A = 96$ sq yd
$b = -?-$

12 yd
b

9. $A = 273$ cm^2
$h = -?-$
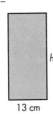
13 cm

10. $A = 375$ sq ft
$h = -?-$
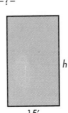
15'

11. $P = 40$ ft
$A = -?-$

7'

12. $A = 264$ sq ft
$P = -?-$

24'

13. $A = -?-$
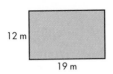
9″   8″
12″

14. $A = -?-$
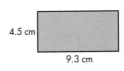
10 cm   9 cm
13 cm

15. $A = -?-$

34 m   36 m
39 m

16. $A = -?-$

3.5 cm   5 cm

17. $A = 176$ sq yd
$h = -?-$
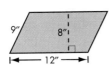
16 yd
13 yd h

18. $A = 48x^2$ sq in.
$b = -?-$
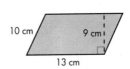
b
$6x$

19. $A = 2508$ cm^2
$P = -?-$
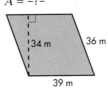
44 cm   48 cm

20. Find the shaded area.
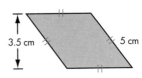
9'   7'
12'

both. (The examples below may give you ideas.) Color and tape the blocks together. Put a border around your paper quilt and sign your work. Title your quilt.

Create a block design
and its reflection.

Make multiple copies.

Experiment with translations, reflections, and rotations.

Translations

Rotations and reflections

How was this done?

Project

Area Problem Solving with Quilts

If you completed the project Quilt Making, you used geometry to create a quilt block design. In this project, you and your group members play the role of quilt manufacturers, and you need to calculate the total area of each of the different types of material used to create different traditional quilt blocks.

Many traditional quilt block designs use geometric shapes. They can range from a simple nine patch, which is based on 9 squares, to more complicated designs like Jacob's Ladder or Underground Railroad, which are based on 16, 25, or even 36 squares. Traditional quilt patterns have names like Snail's Trail, Boy's Nonsense, Monkey Wrench, and Log Cabin. Below are some examples of other traditional patterns. Solve the problems related to area and traditional quilt block designs that follow.

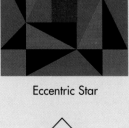

Eccentric Star

Ohio Star

Pine Tree

Bachelor's Puzzle

1. In the Eccentric Star pattern, the size of the block is 10 inches by 10 inches. Calculate the sum of the area of all the black triangles, the sum of the areas of all the dark blue triangles, the sum of the areas of all the purple triangles, and the area of the dark purple square in the middle. You are going to produce a queen-sized quilt for your customer. The standard queen-sized mattress measures 60″ × 80″. How many Eccentric Star blocks will you need to cover that area? (Your quilt's border will cover the sides, which is called the drop, and the pillow tuck of the bed.) From past experience you know that you will need an additional 20% of each fabric to allow for seams and unavoidable mistakes. How much of each color fabric will you need, not counting material for the border?

2. The Ohio Star block measures 12 inches by 12 inches. Calculate the sum of the areas of all the red patches, the sum of the areas of all the blue patches, and the area of the yellow patch. You are going to produce a king-sized quilt for your customer. The king-sized mattress measures 72″ × 84″. How many Ohio Star blocks will you need to cover that area? Again, assume you will need an additional 20% of each fabric to allow for seams and errors. How much of each color fabric will you need? How much fabric will you need for a 15″ border to cover the drop and pillow tuck?

3. The Pine Tree block measures 12 inches by 12 inches. Calculate the sum of the areas of all of the green patches. You are going to produce a twin-sized quilt for your customer. The standard twin-sized mattress measures 39″ × 75″. How many Pine Tree blocks

Exercise Set 9.6

In Exercises 1–8, find the shaded area. The radius of each circle is r. If two circles are shown, r is the radius of the smaller and R is the radius of the larger. All given measurements are in centimeters.

1. $r = 6$
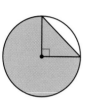

2. $r = 8$
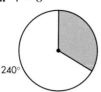

3.* $r = 16$
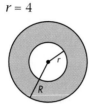

4. $r = 2$

5. $r = 8$
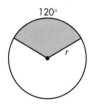

6.* $R = 7$
$r = 4$
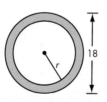

7. $r = 2$
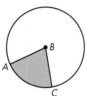

8. $R = 12$
$r = 9$

In Exercises 9 and 10, find the radius. In Exercises 11 and 12 find $m \angle ABC$.

9.* The shaded area is 12π cm².
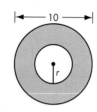

10. The area of the annulus is 32π cm².

11. The shaded area is 120π cm².
$r = 24$ cm

12. The shaded area is 10π cm².
$R = 10$, $r = 8$

13. Suppose the pizza slice in the photo at the beginning of this lesson is a sector with a 36° angle in a circle with a radius of 20 ft. If a can of tomato sauce will cover 3 ft² of pizza, how many cans would be required to cover this slice?

In Exercises 14–17, what is the shaded area in each figure? In Exercises 15–17, the circles are externally tangent. The area of the circle or circles in each figure is what percentage of the area of the square? All given measurements are in centimeters.

14.

18. In Investigation 9.1.2, you cut a right triangle from one end of a parallelogram and reassembled the two parts into a rectangle. You can also cut a triangular region into three parts and reassemble them into a rectangular region, as shown at right. In doing this cutting and reassembling, you **rectify** (make into a rectangle) the parallelogram and the triangle. You can also cut any trapezoidal region into three parts and reassemble them into a rectangular region. Try it.

19. The fascination with dissecting geometric shapes and reassembling them into rectangles can be traced back to early Greek mathematicians. A famous "impossible problem" from ancient Greece is the squaring of the circle problem: Construct with a compass and a straightedge a square having an area equal to that of a given circle.

It wasn't until the nineteenth century that this problem was proved impossible to solve with only a compass and a straightedge. In the time it took to prove this, many geometric discoveries were made while attempting to "square the circle." Even today, professional mathematicians still get mail from geometric explorers who think they have squared the circle.

A fifteenth-century geometric investigator who was fascinated with this problem was Leonardo da Vinci (1452–1519). In attempting to solve the quadrature problem, Leonardo and others were successful in rectifying some special shapes made up of parts of circles. The illustrations below demonstrate how to rectify the pendulum.

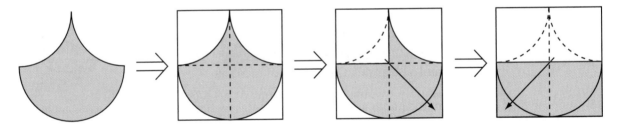

In a series of diagrams, demonstrate how to rectify each figure.

a.   b.   c.   d.

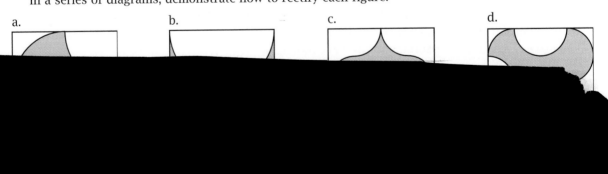

Lesson 9.7

Surface Area

In this lesson you will find the surface area of prisms, pyramids, cylinders, and cones. The **surface area** of each of these solids is the sum of the areas of all the faces or surfaces that enclose the solid. The faces include the solid's top and bottom (**bases**) and its remaining surfaces (**lateral faces** or **surfaces**).

Prism

Pyramid

Cylinder

Cone

To find the surface areas of prisms and pyramids, follow these steps.

Step 1   Draw a diagram of each face of the solid as if the solid were cut apart at the edges and laid flat. Label the dimensions.

Step 2   Calculate the area of each face. If some faces are identical, you need only calculate the area of one and multiply by the number of identical faces.

Step 3   Find the total area of all the faces (bases and lateral faces).

Example A

Find the surface area of the prism below right. Each face is a rectangle.

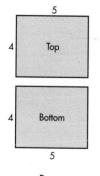

Bases

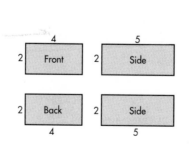

Lateral faces

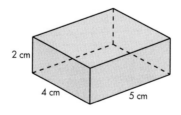

Surface area = $(2)(4)(5) + (2)(2)(4) + (2)(2)(5)$
$= 40 + 16 + 20$
$= 76$

The surface area of the prism is 76 cm^2.

Example B

Find the surface area of the square-based pyramid below right. The height of each triangular lateral face is called the **slant height**.

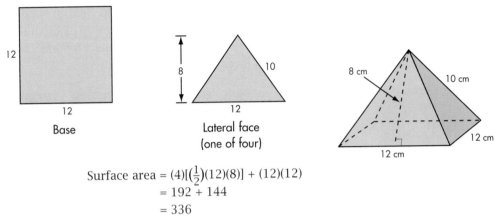

Base

Lateral face
(one of four)

$$\text{Surface area} = (4)\left[\left(\tfrac{1}{2}\right)(12)(8)\right] + (12)(12)$$
$$= 192 + 144$$
$$= 336$$

The surface area of the square-based pyramid is 336 cm^2.

The total surface area of a cylinder is the sum of the lateral surface area and the areas of the bases. The lateral surface is the curved surface of a cylinder. You can think of the lateral surface as a wrapper. You can slice the wrapper and lay it flat to get a rectangular region. The height of the rectangle is the height of the cylinder. The base of the rectangle is the circumference of the circular base of the cylinder. The lateral surface area is the area of the rectangular region.

Example C

Find the surface area of the cylinder below right to the nearest square inch.

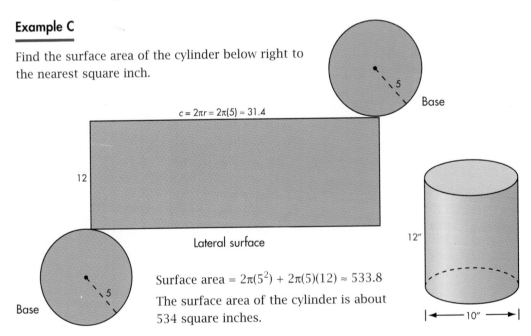

Base

$c = 2\pi r = 2\pi(5) \approx 31.4$

12

Lateral surface

Base

$$\text{Surface area} = 2\pi(5^2) + 2\pi(5)(12) \approx 533.8$$

The surface area of the cylinder is about 534 square inches.

Finding the surface area of a right cone is related to how you found the surface area of a pyramid. To find the lateral surface of an octagonal pyramid, sum the areas of the eight triangles forming the lateral surface. To avoid confusing slant height with the height of the pyramid, use the variable l rather than h for slant height.

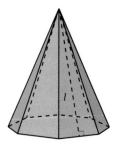 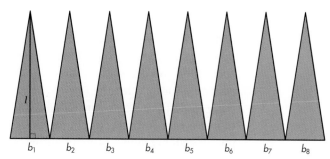

$$\text{Lateral surface area} = \tfrac{1}{2}b_1 l + \tfrac{1}{2}b_2 l + \tfrac{1}{2}b_3 l + \tfrac{1}{2}b_4 l + \tfrac{1}{2}b_5 l + \tfrac{1}{2}b_6 l + \tfrac{1}{2}b_7 l + \tfrac{1}{2}b_8 l$$
$$= \tfrac{1}{2}l(b_1 + b_2 + b_3 + b_4 + b_5 + b_6 + b_7 + b_8)$$
$$= \tfrac{1}{2}l \times (\text{perimeter of base})$$

Now imagine a pyramid whose polygonal base has more than eight sides. What does a regular polygon with 100 sides look like? With 1000 sides? What would a pyramid with 1000 lateral faces look like? As the number of sides of a regular polygon increases, the polygon approaches a circle. Its perimeter approaches the circumference of a circle. As the number of faces of a pyramid increases, it begins to look like a cone. Therefore we can take the formula for the lateral surface area of a pyramid and

Is the tower of this building in Kashan, Iran, a cone or a pyramid? What makes it hard to tell?

substitute circumference (C) for perimeter, which gives us a formula for the lateral surface area of a cone: lateral surface area $= \tfrac{1}{2}Cl = \tfrac{1}{2}(2\pi r)l = \pi r l$.

Example D

Find the total surface area of the right cone with slant height 10 cm and radius 5 cm to the nearest centimeter.

Total surface area = lateral surface area + base area
$$= \pi r l + \pi r^2$$
$$= (\pi)(5)(10) + \pi(5)^2$$
$$= 75\pi \approx 235.6$$

The surface area of the cone is about 236 cm².

Take Another Look 9.7

1. Use algebra to show that the total surface area of a cone is given by the expression $\pi r(l + r)$, where l represents slant height and r represents radius.

2. In Examples A–C in this lesson, you found the surface area of a prism, a pyramid, and a cylinder without using surface area formulas. Instead you used formulas for polygons and circles to find the areas of the different surfaces that made up each solid, then you added those areas to find the total surface area. For the cone, you derived a formula for the lateral surface area. Look back at Examples A–C to see if you can derive a single formula for the total surface area of each solid. Use algebra to make your formulas as simple as possible. (In other words, use as few variables as possible and combine as many terms as possible.) Try each formula on an example of your own making.

Exercise Set 9.7

In Exercises 1–10, find the surface area for each solid. All quadrilaterals are rectangles. All given measurements are in centimeters.

1.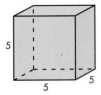
5
5 5

2.
37
37 9

3.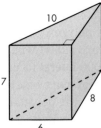
10
7
8
6

4. Round your answer to the nearest cm².

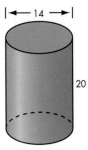
|← 14 →|
20

5. The base is a square.

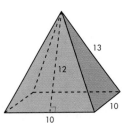
13
12
10
10

6.* Round your answer to the nearest cm².

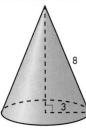
8
3

7.* The base is a regular hexagon with apothem $a \approx 12.1$ and side $s \approx 14$. Each lateral face is a rectangle with height $h = 7$.

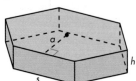
a
h
s

8.* The base is a regular pentagon with apothem $a \approx 11$ and side $s \approx 16$. Each lateral edge $t \approx 17$, and the height of a face $l \approx 15$. Give your answer to the nearest cm².

t
a
s

9.* $D = 8$, $d = 4$, $h = 9$

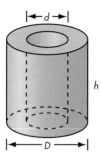

10.* Round your answer to the nearest cm².

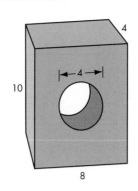

11. Explain how you would find the surface area of an obelisk.

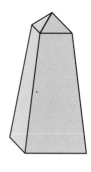

12.* Claudette and Marie are about to paint the exterior walls of their country farm home (all vertical surfaces) and to put new cedar shingles on the roof. Their home has a gambrel roof typical of the houses of rural gentry back in their home country, France. The paint selected costs $25 per gallon and covers 250 square feet per gallon.

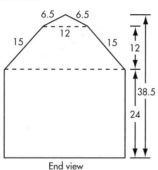

End view

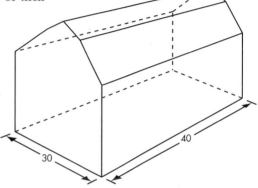

The wood shingles cost $65 per bundle, and each bundle covers 100 square feet. How much will this home improvement cost Claudette and Marie? All measurements are in feet.

13. The shape of the spinning dishes this Sri Lankan dancer is balancing are called **frustrums** of cones. Think of them as cones with their tops cut off. Use your compass to draw onto a sheet of paper pieces that you can cut out and tape together to form a frustrum of a cone.

14. A circular oil spill at 6 a.m. has a radius of 1.0 kilometer. By 7 a.m., the radius of the spill is 1.5 kilometers. By 8 a.m., the radius of the spill is 2.0 kilometers. At this rate, what will be the area of the oil spill if it's not contained by noon?

15.* Cycle City occupies a circular region 8 km in diameter. Two civil defense sirens are to be installed—one 2 km to the east of the center of the city and the other 2 km to the west of the center. If the sound from each siren will travel up to 2 km, what percent of the city will be covered by the sound of the sirens?

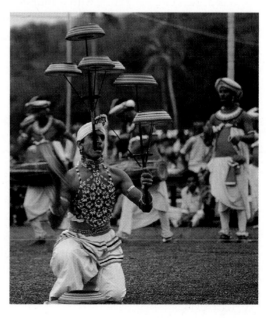

16.* The bull's-eye of the dart board shown has a diameter of 8 cm. The width of each ring is 4 cm. If a random toss hits the target, what is the probability that it hits the bull's-eye?

4 4 4 4 4 4 4 4 4 4
40 cm

17. Hector is a very cost-conscious buyer of produce. When he buys asparagus from his neighborhood grocer, Victoria Vegitale, he buys large bundles that are each 44 cm in circumference. However, today Victoria has only small bundles that are 22 cm around. So Ms. Vegitale offers to sell Hector two 22-cm bundles for the same price as one 44-cm bundle. Is this a good deal or bad deal? Why?

18. In 1792, visiting Europeans presented horses and cattle to Hawaii's King Kamehameha I. The king placed a *kapu*, or taboo, on the animals so that they could not be eaten. In a short thirty years, they were so plentiful that Kamehameha III lifted the *kapu*. Cattle ranching soon developed when Mexican *vaqueros* came to Hawaii to teach Hawaiians how to ride horses for ranching. The Spanish word *vaquero* was replaced with the Hawaiian word *paniolo*. Today, Hawaiian cattle ranching is big business. (Parker Ranch on the island of Hawaii is the largest privately-owned ranch in the United States.) An interesting "grazing geometry" is being used on Hawaii's Kahua Ranch. Ranchers divide the grazing area into sectors. Because the grazing area is not a perfect circle, sectors with a smaller radius have a greater central angle measure. All the sectors have approximately the same area. The cows are rotated through each sector in turn. By the time they return to the first sector, the grass has grown back and is ready for the cycle to repeat.

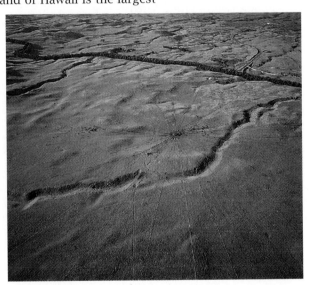

Suppose a circular ranch with a radius of 3 km was divided into 16 congruent sectors. In a 1-year cycle, how long would the cattle graze in each sector? What would be the area of each sector?

Improving Visual Thinking Skills—*Moving Coins*

Create a triangle of coins similar to that at near right. How can you move exactly three coins so that the triangle is pointing down rather than up? When you have found a solution, draw two triangles of circles like those shown in the diagram. Write letters in each of the empty circles of your drawing to show your solution.

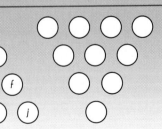

10 Pythagorean Theorem

Waterfall, M. C. Escher, 1961

In this chapter you will discover a property of right triangles known as the Pythagorean Theorem. The Pythagorean Theorem is one of the most important concepts in all of mathematics: It allows you to calculate the distance between two points. You will encounter the theorem of Pythagoras and its applications in many other math classes, such as calculus. In this chapter you will also discover a number of conjectures related to the Pythagorean Theorem that you will use to solve problems.

Lesson 10.1

The Theorem of Pythagoras

FUNKY WINKERBEAN by Batiuk. Reprinted with special permission of North America Syndicate.

There is a surprising relationship between the lengths of the three sides of any right triangle. This property of right triangles is probably the most useful in all high school mathematics because it helps you calculate the distance between two points. Nobody knows at what point in history this relationship was first discovered. The ancient Babylonians and Chinese recognized this relationship, and some math historians believe that the ancient Egyptians also used a special case of this property of right triangles.

In a right triangle, the side opposite the right angle is called the **hypotenuse**. The other two sides are called **legs**. In the figure at right, a and b represent the lengths of the legs, and c represents the length of the hypotenuse. (And no, a hypotenuse is not a large animal that hangs out around watering holes.)

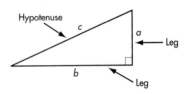

Investigation 10.1

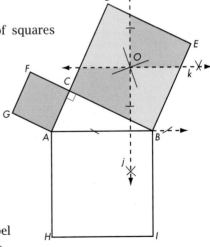

There is a special relationship among the areas of squares constructed on the three sides of a right triangle. The dissection puzzle in this investigation is intended to get you thinking about this special relationship. On a full sheet of paper, perform Steps 1–6. Note that the arcs and the segment extensions necessary to complete Steps 1 and 2 are not indicated in the figure.

Step 1   Construct a scalene right triangle in the middle of your paper (hypotenuse down). Label it so that the hypotenuse is $\overline{AB}$ and the longer leg is $\overline{BC}$.

Step 2   Construct a square on each side of the triangle. Label the square on the longer leg *BCDE*. Label the square on the shorter leg *AGFC*. Label the square on the hypotenuse *ABIH*.

Step 3   Locate the center of *BCDE* (intersection of the two diagonals). Label the point *O*.

Lesson 10.3

Word Problems

You must do things you think you cannot do.
— *Eleanor Roosevelt*

FUNKY WINKERBEAN by Batiuk. Reprinted with special permission of North America Syndicate.

Exercise Set 10.3

In this exercise set, round answers to the nearest 0.1 unit.

1. What is the length of the diagonal of a square whose sides measure 8 cm?

2.* The lengths of the three sides of a right triangle are consecutive integers. Find them.

3.* The lengths of the three sides of a right triangle are consecutive even integers. Find them.

4. Find the area of a right triangle with a hypotenuse that measures 17 cm and one leg that measures 15 cm.

5.* The diagonal of a square measures 32 meters. What is the area of the square?

6. The legs of an isosceles triangle measure 6 cm, and the base measures 8 cm. Find the area.

7. A rectangular garden 6 meters wide has a diagonal measuring 10 meters. Find the perimeter of the garden.

8. How high up on a building will a 15-foot ladder reach if the foot of the ladder is placed five feet from the building?

9. A baseball infield is a square, each side measuring 90 feet. To the nearest foot, what is the distance from home plate to second base?

10. A rectangular closet is 2 feet deep, 3 feet wide, and 8 feet high. What is the length (to the nearest inch) of the longest pole that can fit within the closet? In other words, find the length of the diagonal.

11.* A flagpole has cracked 9 feet from the ground and has fallen as if hinged. The top of the flagpole hit the ground 12 feet from the base. How tall was the flagpole before it fell?

12. To find the distance between two points A and B on the opposite ends of a lake, a surveyor sets a stake at point C so that angle ABC is a right angle. By measuring, she finds AC to be 169 meters and BC to be 65 meters. How far across the lake is point A from point B?

13. Dr. Rhonda Bend is exploring the Martian landscape. She is standing at point C, 288 meters from the base of a vertical cliff (point B). To find the height of the cliff, she focuses a sonic beam at a rock on the top of the cliff (point A), as shown in the figure below. The beam bounces off the rock and returns. She records the time it takes for the sonic beam to return and calculates the distance from point A to point C to be 480 meters. To the nearest meter, what is the height of the cliff?

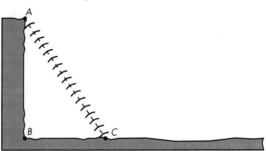

Improving Visual Thinking Skills—*Folding Cubes II*

In each of the problems below, the figure at left represents a folded cube. When the cube is unfolded, which figure at right will it become?

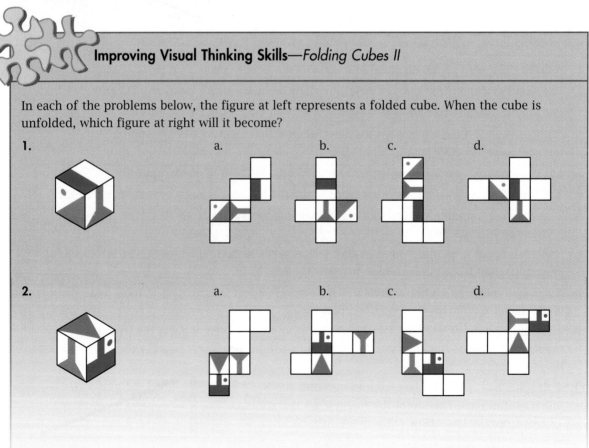

Graphing Calculator Investigation—*The Height Reached by a Ladder*

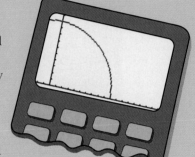

Suppose a house painter rests a 20-foot ladder against a building, then decides the ladder needs to rest against the building 1 foot higher. Will moving the ladder 1 foot toward the building do the job? If the ladder needs to rest against the building 2 feet lower, will moving the ladder 2 feet away from the building do the trick? Investigate the ladder problem on paper and with your graphing calculator by following Steps 1–3.

Step 1   Draw a picture of a ladder leaning against a vertical wall, with the foot of the ladder resting on a horizontal floor. Label the height reached by the ladder y and the distance from the base of the wall to the foot of the ladder x. Use the Pythagorean Theorem to write an equation relating x, y, and the length of the ladder.

Step 2   Solve the equation from Step 1 for y. You now have a function for the height reached by the ladder in terms of the distance from the wall to the foot of the ladder. Enter this equation in your calculator.

Step 3   Before you graph the equation, think about the range you'll want for your graph window. What are the greatest and least values possible for x and y? Enter a reasonable range for the graph window, then graph the equation.

Now answer these questions about your graph.

1. Describe the curve given by the equation you graphed. (Keep in mind that depending on the range for your graph window, the curve may be stretched.)

2. Trace along the graph, starting with $x = 0$. Write down values (rounded to the nearest 0.1 unit) for the height reached by the ladder when $x \approx 3$, 6, 9, and 12. If you were to move the foot of the ladder away from the wall 3 feet at a time, would each move result in the same change in the height reached by the ladder? Explain.

3. Find the value for x that gives a y-value approximately equal to x. How is this value related to the length of the ladder? Draw a picture of the ladder in this position. What angle does the ladder make with the floor?

4. Would you lean a ladder against a wall in such a way that x were greater than y? Explain. How does your graph support your explanation?

5. Model the leaning ladder with a piece of paper and a ruler. Call the left edge of the paper the wall and the bottom of the paper the floor. "Lean" a ruler so that the 20-cm mark meets the wall and the end of the ruler (0 cm) is on the floor. Draw a point at the 10-cm mark to represent a paint can at the midpoint of the ladder. Find five or six more such midpoints for different positions of the ladder. Connect the points with a curve. What kind of curve is this? What's its equation? Graph the equation on your calculator.

Lesson 10.4

Two Special Right Triangles

In an isosceles triangle, the sum of the square roots of the two equal sides is equal to the square root of the third side.

— *The Scarecrow in* The Wizard of Oz
 by L. Frank Baum

In this lesson you will use the Pythagorean Theorem to discover some relationships between the sides of two special right triangles. Problems involving these right triangles are often found on college entrance exams and achievement tests and are often used in trigonometry. Because working with this relationship involves squares and square roots, you will need to review some operations on square roots.

Simplifying square roots

$\sqrt{50} = \sqrt{25 \times 2} = \sqrt{25} \times \sqrt{2} = 5\sqrt{2}$
$\sqrt{84} = \sqrt{4 \times 21} = \sqrt{4} \times \sqrt{21} = 2\sqrt{21}$
$\sqrt{126} = \sqrt{9 \times 14} = \sqrt{9} \times \sqrt{14} = 3\sqrt{14}$

Multiplying square roots

$(\sqrt{3})(\sqrt{2}) = (\sqrt{6})$
$(\sqrt{5})^2 = (\sqrt{5})(\sqrt{5}) = 5$
$(2\sqrt{3})^2 = (2\sqrt{3})(2\sqrt{3}) = 4 \times 3 = 12$

Investigation 10.4.1

This first investigation is an opportunity to practice simplifying square roots. Work with a partner on the square root practice below before moving on to Investigation 10.4.2. Express each square root in its simplest form. Check your answers in the Hints section.

1.* $\sqrt{12}$ **2.*** $\sqrt{18}$ **3.*** $\sqrt{24}$ **4.*** $\sqrt{32}$ **5.*** $\sqrt{40}$

6.* $\sqrt{48}$ **7.*** $\sqrt{60}$ **8.*** $\sqrt{75}$ **9.*** $\sqrt{83}$ **10.*** $\sqrt{85}$

Express each product in its simplest form.

11.* $(3\sqrt{2})^2$ **12.*** $(4\sqrt{3})^2$ **13.*** $(2\sqrt{3})(\sqrt{2})$ **14.*** $(3\sqrt{6})(2\sqrt{3})$ **15.*** $(7\sqrt{3})^2$

In Investigation 10.4.2, you will discover a relationship between the lengths of the legs and the hypotenuse of an isosceles right triangle. This triangle is also referred to as a 45-45 right triangle because each of its acute angles measures 45°. If you bring the opposite corners of a square piece of paper together and fold, you see half the square. The edges of this half square form an isosceles right triangle.

Isosceles right triangle

Investigation 10.4.2

Find the length of the hypotenuse of each isosceles right triangle. Simplify the square root each time to reveal a pattern.

1.* $a =$ –?–

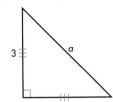

2. $b =$ –?–

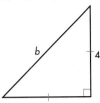

3. $c =$ –?–

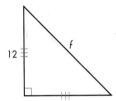

4. $d =$ –?–

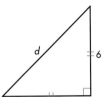

5. $e =$ –?–

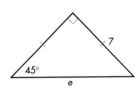

6. $f =$ –?–

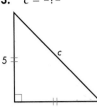

Did you notice something interesting about the relationship between the length of the hypotenuse and the length of the legs in each problem of this investigation? State your observations as your next conjecture.

C-90 In an isosceles right triangle, if the legs have length x, then the hypotenuse has length —?— (***Isosceles Right Triangle Conjecture***).

You can use algebra to verify the Isosceles Right Triangle Conjecture.

$$c^2 = x^2 + x^2$$
$$c^2 = 2x^2$$
$$c = x\sqrt{2}$$

This property can also be demonstrated physically on a geoboard or square dot paper. A right triangle with each leg of length 1 unit has a hypotenuse of $\sqrt{2}$, as shown at right. A right triangle with each leg of length 2 units has a hypotenuse of $2\sqrt{2}$. A right triangle with each leg of length 3 units has a hypotenuse of $3\sqrt{2}$.

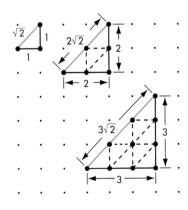

The second special right triangle is the 30-60 right triangle. Imagine folding an equilateral triangle along one of its lines of symmetry. The right triangle you get is a 30-60 right triangle. A 30-60 right triangle is half of an equilateral triangle.

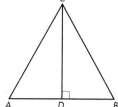

30-60 right triangle

Investigation 10.4.3

Let's start by using a little deductive thinking to reveal a useful relationship in 30-60 right triangles. Triangle *ABC* is equilateral, and $\overline{CD}$ is an altitude.

1. What are *m∠A* and *m∠B*?
2. What are *m∠ACD* and *m∠BCD*?
3. What are *m∠ADC* and *m∠BDC*?
4. Is △*ADC* ≅ △*BDC*? Why?
5. Is $\overline{AD}$ ≅ $\overline{BD}$? Why?

Notice that altitude $\overline{CD}$ divides the equilateral triangle into two right triangles with acute angles that measure 30° and 60°. Look at just one of the 30-60 right triangles, say △*ADC*. How do *AC* and *AD* compare? State your findings as a conjecture.

C-91 In a 30-60 right triangle, if the side opposite the 30° angle has length *x*, then the hypotenuse has length —?—.

Investigation 10.4.4

Let's see what else you can discover about 30-60 right triangles. Find the length of the indicated side in each 30-60 right triangle by using the conjecture you just made. All measurements are in centimeters.

1. *a* = –?–

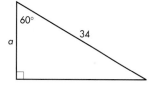

2. *b* = –?–

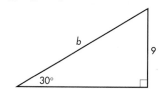

3. *c* = –?–

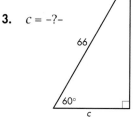

4. *d* = –?–

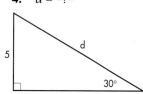

5. *e* = –?–

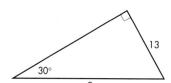

6. *f* = –?–

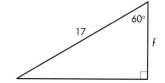

Now use the conjecture you made in Investigation 10.4.3 and the Pythagorean Theorem to find the length of each indicated side.

7. $j = $ –?–

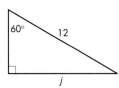

8. $k = $ –?–

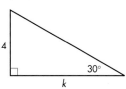

9. $m = $ –?–

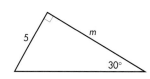

10. $n = $ –?–

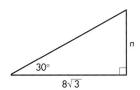

11. $p = $ –?–

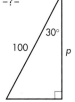

12. $s = $ –?–

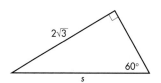

You should have noticed a pattern in your answers. Combine your observations with your latest conjecture and state your next conjecture.

C-92 In a 30-60 right triangle, if the shorter leg has length x, then the longer leg has length —?— and the hypotenuse has length —?— (***30-60 Right Triangle Conjecture***).

This property can also be demonstrated physically on isometric dot paper. The 30-60 right triangle shown at right with a hypotenuse of length 2 units has a longer leg with a length of $\sqrt{3}$. A second 30-60 right triangle with a hypotenuse of length 4 units has a longer leg with a length of $2\sqrt{3}$. A third 30-60 right triangle with a hypotenuse of length 6 units has a longer leg with a length of $3\sqrt{3}$.

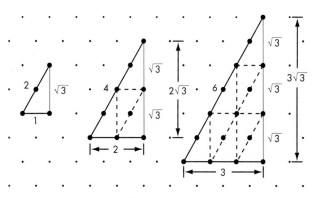

So far you are familiar with the special right triangles shown below.

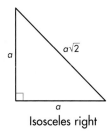

Isosceles right

30-60 right

Take Another Look 10.4

1. Use algebra to show how the 30-60 Right Triangle Conjecture follows from the Pythagorean Theorem.

2. Use a geometry computer program or a compass and a straightedge to construct a square root spiral out to at least √8. The first part of the spiral is shown at right.

3. Use the SSS Congruence Conjecture to verify the converse of the 30-60 Right Triangle Conjecture. That is, show that if a right triangle has sides with length x, $x\sqrt{3}$, and $2x$, then it is a 30-60 right triangle.

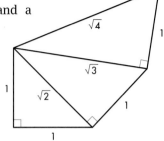

Exercise Set 10.4

Express each square root in its simplest form.

1. $\sqrt{90}$ 2. $\sqrt{96}$ 3. $\sqrt{120}$ 4.* $\sqrt{185}$ 5.* $\sqrt{490}$

6. $\sqrt{576}$ 7.* $\sqrt{720}$ 8. $\sqrt{722}$ 9. $\sqrt{784}$ 10. $\sqrt{828}$

Express each product in its simplest form.

11. $(2\sqrt{2})^2$ 12.* $(4\sqrt{3})^2$ 13. $(5\sqrt{5})(\sqrt{3})$ 14.* $(2\sqrt{6})(\sqrt{12})$ 15.* $(6\sqrt{8})^2$

Solve Exercises 16–30 by using your new conjectures. In most of the exercises, you don't need to use the Pythagorean Theorem. All measurements are in centimeters.

16. $a = $ -?-

17.* $b = $ -?-

18. What is the perimeter of square $SQRE$?

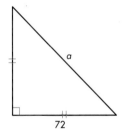

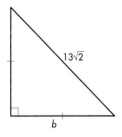

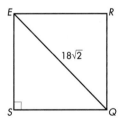

19.* What is the area of the triangle?

20. $c = $ -?-

21. $d = $ -?-

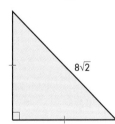

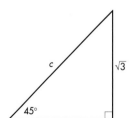

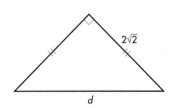

22.* *a* = –?–   *b* = –?–

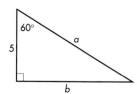

23. *c* = –?–   *d* = –?–

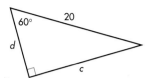

24. *e* = –?–   *f* = –?–

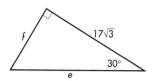

25. *g* = –?–   *h* = –?–

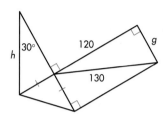

26.* *k* = –?–   *m* = –?–

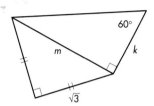

27. *n* = –?–   *p* = –?–

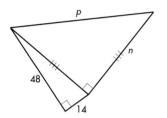

28. Find the coordinates of *P*.

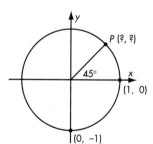

29. Find the coordinates of *A*.

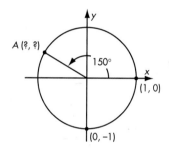

30. Find the coordinates of *T*.

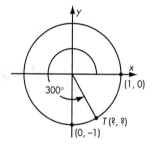

31. What's wrong with this picture?

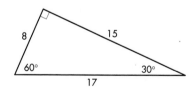

32. Which triangle has the greater area? Explain.

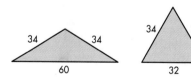

33.* Find the dimensions of two different rectangles with a diagonal of 25 cm.

34.* Sketch and label a figure that demonstrates $\sqrt{32} = 4\sqrt{2}$. (Use dot or graph paper.)

35. In equilateral $\triangle ABC$, segments $\overline{AE}$, $\overline{BF}$, and $\overline{CD}$ are each an angle bisector, a median, and an altitude simultaneously. These three segments divide the equilateral triangle into six overlapping 30-60 right triangles and six smaller, nonoverlapping 30-60 right triangles. One of the six overlapping 30-60 right triangles is $\triangle CDB$. Find the other five. One of the six nonoverlapping 30-60 right triangles is $\triangle ADM$. Find the other five.

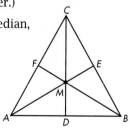

In Exercises 36–38, choose either patty papers or a compass and a straightedge and perform the constructions.

36.* Given the segment with length a below, construct segments with lengths $a\sqrt{2}$, $a\sqrt{3}$, and $a\sqrt{5}$.

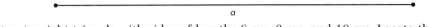

a

37. Construct a right triangle with sides of lengths 6 cm, 8 cm, and 10 cm. Locate the midpoint of each side. Construct a semicircle on each side with the midpoints of the sides as centers. Find the area of each semicircle. What do you notice?

38. Construct a right triangle with sides of lengths 6 cm, 8 cm, and 10 cm. Construct an equilateral triangle on each side of the right triangle, using each side of the right triangle as a side of an equilateral triangle. Find the area of each equilateral triangle. What do you notice?

39. The *Jiuzhang suanshu* is an ancient Chinese mathematics text containing 246 problems. The solutions to some problems require the use of the *gou gu* (Pythagorean Theorem). The *gou gu* would be $(gou)^2 + (gu)^2 = (xian)^2$. Here is a *gou gu* problem translated from the ninth chapter of *Jiuzhang*.

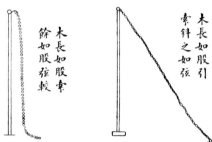

There is a rope hanging from the top of a pole with three chih of it lying on the ground. When it is tightly stretched so that its end just touches the ground, it reaches eight chih from the base of the pole. How long is the rope?

40.* A = {3, 13, 15}; B = {4, 8, 12}; C = {5, 9, 17}

Sets A, B, and C contain lengths of segments. If one length is randomly selected from set A, one length from set B, and one length from set C, what is the probability that segments with the selected lengths can form a right triangle?

Improving Reasoning Skills—*Mudville Monsters*

The starting eleven members of the Mudville Monsters football team and their coach Osgood Gipper have been invited to compete in the Smallville Punt, Pass, and Kick Competition. Upon arriving at the outskirts of the town, they find they must get across the deep Smallville River. The only available way across is with a small boat owned by two very small Smallville football players. The boat holds just one Monster visitor or the two Smallville players. The Smallville players agree to help the Mudville players across if the visitors agree to pay $5 each time the boat crosses the river. If the Monsters have a total of $100 among them, do they have enough money to get all players and the coach to the other side of the river?

Lesson 10.5

Multiples of Right Triangles

He is educated who knows how to find out what he doesn't know.
— *George Simmel*

What happens if you add the same length to each side of a right triangle?

A quick check shows that adding the same number to each of the numbers in a Pythagorean triple does not create a new Pythagorean triple. Adding 2 to each number in 3-4-5 gives 5-6-7, but $5^2 + 6^2 > 7^2$. Likewise, subtracting 1 from each number in 3-4-5 gives 2-3-4, but $2^2 + 3^2 < 4^2$.

What happens if you double the lengths of the sides of a right triangle? Will the new lengths form a right triangle? What if all three lengths are tripled? Let's find out.

Investigation 10.5.1

Step 1   Select one of the common right triangles shown below. Double the length of each side of your chosen triangle.

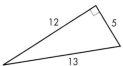
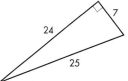

Step 2   Substitute these new lengths into the Pythagorean formula. If they work in the equation, then the new lengths will form a right triangle.

Step 3   Triple the length of each side of another triangle from the group above.

Step 4   Substitute these new lengths into the Pythagorean formula. Will they form a right triangle?

Step 5   Now multiply the length of each side of another triangle by another number. (It doesn't have to be a whole number.)

Step 6   Substitute these new lengths into the Pythagorean formula. Will they form a right triangle?

You should be ready to state a conjecture about multiples of right triangles.

C-93   If you multiply the lengths of all three sides of any right triangle by the same number, the resulting triangle will be a —?— (***Pythagorean Multiples Conjecture***).

In other words, if $a^2 + b^2 = c^2$, then $(an)^2 + (bn)^2 = (cn)^2$. This is a very useful conjecture. If you are familiar with the Pythagorean triple 3-4-5, then you will also be able to solve problems that use multiples of this triple, such as 6-8-10 or 9-12-15.

The lengths of the sides of the four right triangles shown at the beginning of this investigation are examples of Pythagorean triples called **primitives**. The three numbers have no common integer factors. From these primitives, you can create new right triangles by multiplying the primitives by the same number. These new Pythagorean triples are called **multiples**.

In this investigation, you discovered that if you multiply the lengths of all three sides of a right triangle by the same number, the new lengths will form a right triangle. The Pythagorean Multiples Conjecture raises the question: If two sides of a right triangle have a common factor, must the third side have the same factor? Let's find out.

Investigation 10.5.2

Step 1   Select two integers that have a common factor (for example, 6 and 8). They can represent the lengths of two sides of a right triangle.

Step 2   Sketch two different right triangles. Label the two legs of one of the triangles with the two numbers you selected in Step 1. Label the hypotenuse r. On the second triangle, label the hypotenuse with the greater number and label one leg with the other number. Label the third side s.

Step 3   Use the Pythagorean Theorem to solve for r and s. (If 6 and 8 are the lengths of the two legs, then $r = 10$. When 6 is the length of a leg and 8 the length of the hypotenuse, $s = \sqrt{28} = 2\sqrt{7}$. Both 6 and 8 have 2 as a common factor, and both 10 and $2\sqrt{7}$ have 2 as a common factor.) Do your r and s have the same factor as the two integers you chose?

Repeat this investigation with a second pair of integers that have a common factor. Do your r and s have the same factor as your second pair of integers? Compare your results with the results of others near you. You should be ready to state a conjecture.

    C-94 If the lengths of two sides of a right triangle have a common factor, then —?—.

If you recognize that the lengths of two sides of a right triangle have a common factor, you can use this conjecture to conclude that the third side must also have that same factor. Let's look at an example of how you can use this conjecture.

Example

Find the length of one leg of a right triangle with a hypotenuse of 35 cm and a leg of 28 cm.

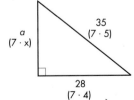

Step 1   Because the lengths of two of the sides are multiples of 7, the length of the third side must be a multiple of 7.

Step 2   When the three sides are written without the common factor 7, the other factors (x, 4, and 5) are revealed as part of the familiar 3-4-5 triple.

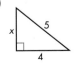

Step 3   If the triangle is a multiple of a 3-4-5 right triangle, then $x = 3$.
Therefore the length of the third side in the original triangle
is $(3)(7)$, or 21.

The leg is 21 cm long.

Take Another Look 10.5

1. Use a geometry computer program's dilation tool to demonstrate the Pythagorean Multiples Conjecture.

2. Use algebra to demonstrate the Pythagorean Multiples Conjecture. That is, if $a^2 + b^2 = c^2$, then show that $(an)^2 + (bn)^2 = (cn)^2$ follows from it.

3. Conjecture 94 can also be quickly verified with algebra. There are two cases. The two legs could have a common factor (Case 1), or the hypotenuse and a leg could have a common factor (Case 2). The proof of Case 1 is shown below.

   Let ar and br be the lengths of the legs of a right triangle (two sides having a common factor r) and let x represent the length of the third side (in this case, the hypotenuse). Because the triangle is right, the three values work in the Pythagorean formula.

   Therefore x (the length of the hypotenuse) also has r as a factor.

   Prove Case 2. That is, show that if the hypotenuse and a leg of a right triangle have a common factor, then the other leg must have the same factor.

$$(ar)^2 + (br)^2 = x^2$$
$$a^2 r^2 + b^2 r^2 = x^2$$
$$r^2(a^2 + b^2) = x^2$$
$$r \sqrt{a^2 + b^2} = x$$

Exercise Set 10.5

Each of Exercises 1–9 (except one) involves one of the four most common Pythagorean primitives. Recognize them and you can save yourself a lot of work! All measurements are in centimeters unless otherwise indicated.

1. $a = \text{–?–}$

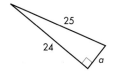

2. $b = \text{–?–}$

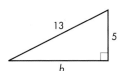

3. What is the perimeter?

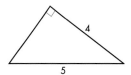

4. $c = \text{–?–}$

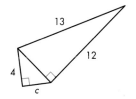

5.* The area of the rectangle is 168 sq ft.
$d = \text{–?–}$

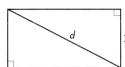

6. What is the area of the shaded rectangle?

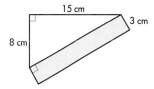

7. What is the shaded area?

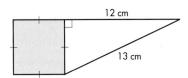

12 cm

13 cm

8.* The arc is a semicircle. What is the shaded area?

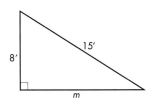

7"

25"

9. $m = -?-$

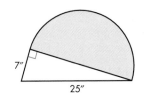

15'

8'

m

10.* Copy the table below. Then complete it by creating multiples of the most common Pythagorean primitives.

Pythagorean triples

Primitives	Doubles	Triples	4 times	10 times
3-4-5	6-8-10	a.* –?–	b. –?–	c. –?–
5-12-13	d. –?–	15-36-39	e. –?–	f. –?–
8-15-17	g. –?–	h. –?–	32-60-68	i. –?–
7-24-25	j. –?–	k. –?–	l. –?–	70-240-250

In Exercises 11–20, each right triangle has sides whose lengths are multiples of a Pythagorean primitive. All measurements are in centimeters. Check your completed table of multiples of Pythagorean triples (Exercise 10) or use your new conjectures to solve for the indicated value.

11.* $a = -?-$

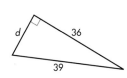

34

16

a

12. $b = -?-$

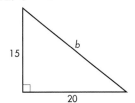

15

b

20

13. $c = -?-$

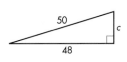

50

48

c

14. $d = -?-$

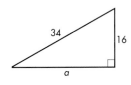

d

36

39

15.* $e = -?-$

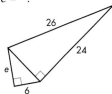

26

24

e

6

16. $f = -?-$

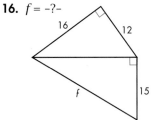

16

12

f

15

17. $r = -?-$ $s = -?-$

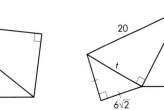

s

30°

r

24

32

18. $t = -?-$ $v = -?-$

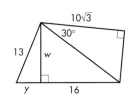

20

30°

v

t

$6\sqrt{2}$

19.* $w = -?-$ $y = -?-$

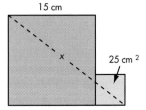

$10\sqrt{3}$

30°

13

w

y

16

20. $x = -?-$

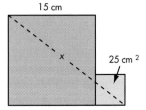

15 cm

x

25 cm ^{2}

21. Which triangle has the greater area? Explain.

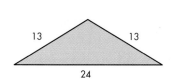

22. Draw and label two different right triangles, each with a 65-cm hypotenuse. Be sure to indicate the lengths of the two legs.

23.* Sketch and label a figure that demonstrates $\sqrt{27} = 3\sqrt{3}$. (Use isometric dot paper or graph paper to aid your sketch.)

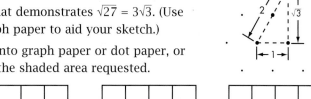

24.* Copy the two blank grids onto graph paper or dot paper, or use a geoboard. Then find the shaded area requested.

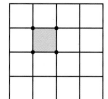   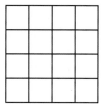

The area of the shaded square is 1.

The area of the shaded square is 2.

Shade in a square with an area of 5.

Shade in a square with an area of 10.

25. Find the coordinates of point *T*.

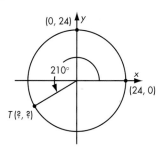

26. Find the coordinates of point *S*.

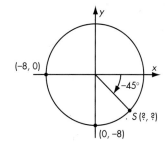

27. Annette is building a redwood gate 4 ft wide by 6 ft tall. She needs to determine the length of the diagonal brace. Find the length of the diagonal to the nearest inch.

28. The area of a rhombus is 96 square centimeters, and the longer diagonal measures 16 cm. Find the perimeter of the rhombus.

29.* At Martian high noon, Dr. Rhonda Bend leaves the Martian U.S. Research Station, traveling east at 60 km/hr. One hour later, Professor I. M. Bryte takes off from the station, heading north straight for the polar icecap at 50 km/hr. How far apart will the doctor and the professor be 4 hr after Rhonda's departure? Express your answer to the nearest kilometer.

30.* Kim travels 1 mile north, then 2 miles east, then 3 miles north again, and then once more east for 4 miles. How far is she from her starting point? (It is less than $5 + \sqrt{5}$ miles.)

A Pythagorean Fractal

If you wanted to draw a picture to state the Pythagorean Theorem without words, you'd probably draw a right triangle with squares on each of the three sides, like the figure shown to the right.

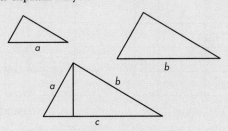

An equivalent statement of the theorem has an even simpler picture: a right triangle divided into two right triangles. In the figure shown below right, a right triangle with hypotenuse c is divided into two smaller triangles, the smaller with hypotenuse a and the larger with hypotenuse b. Obviously the areas of the two smaller triangles add up to the area of the whole triangle. What's surprising is that all three triangles have the same angle measures. Can you explain why? Though different in size, the three triangles all have the same shape. We call figures that have the same shape similar figures. You'll learn more about similarity in Chapter 12. The fact that any right triangle can be divided into two right triangles that are similar to the original is at the core of why the Pythagorean Theorem is true. You'll use similar triangles to prove the theorem in Chapter 16.

For this project, you'll use similar triangles to combine the simple Pythagorean Theorem pictures above into a beautifully complex fractal picture like that shown below. The fractal starts with a right triangle with squares on the sides, then similar triangles are built onto these squares, and squares are built onto those triangles, and so on.

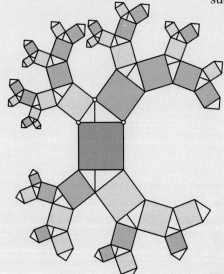

You'll need to follow the steps very carefully to successfully create the figure. In particular, selection order is very important in these steps. If your first attempts don't work out the way you expect, be patient. Try again. You'll find the beauty and the mystery of the final product to be worth it.

Before you begin, you'll need a script tool for constructing a square given an edge. Such a script can be found in the sample scripts that come with Sketchpad. Open a new sketch and experiment with the script tool to see whether the tool you're using constructs a square above the starting edge or below the starting edge when you drag from left to right. The steps in this project assume that your square tool constructs a square below the edge when you drag from left to right, or above the edge when you drag from right to left, as shown in the figure at right. Knowing how your script behaves becomes important near the end of the project.

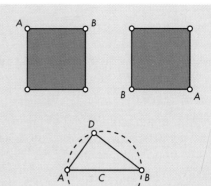

In these first few steps you'll construct a right triangle. Start with a new sketch.

Step 1 Construct $\overline{AB}$ and midpoint C.

Step 2 Construct a circle from point C to point B.

Step 3 Construct $\overline{AD}$ and $\overline{BD}$, where point D is on the circle.

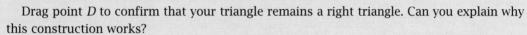

Step 4 Hide point C and the circle.

Drag point D to confirm that your triangle remains a right triangle. Can you explain why this construction works?

You're now ready to begin constructing squares on the sides and similar triangles on those squares.

Step 5 Construct a square on the hypotenuse AB of right triangle ABC. (Use a script tool.)

Step 6 Construct the midpoints of two sides and a segment connecting those midpoints, as shown.

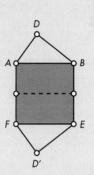

Step 7 Use the Transform menu to mark this segment as a mirror, then reflect point D over the segment.

Step 8 Construct $\triangle FD'E$.

Triangle $FD'E$ is congruent to $\triangle ABD$. In the next few steps, you'll construct squares and similar triangles on the legs of $\triangle ABD$. You'll record a script for these steps so that you can continue the process and create a fractal.

Step 9 Use the File menu to open a new script. Click record.

Step 10 Construct a line through point D, perpendicular to $\overline{AB}$.

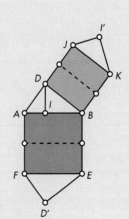

Step 11 Construct the point of intersection I, hide the line, and construct $\overline{DI}$.

Step 12 Construct a square on $\overline{BD}$.

Step 13 Construct the midpoints of two sides and a segment connecting those midpoints.

Step 14 Mark the segment as a mirror and reflect point I over it.

Step 15 Construct right triangle $JI'K$. How does this right triangle compare to the original right triangle ABD?

Step 16 Repeat Steps 12–15 on side *DA:* Construct a square, construct its midsegment, reflect point *I* over the midsegment, and construct a right triangle on the square.

Step 17 Hide the six midpoints and the three midsegments.

Step 18 Color or shade the two squares on the legs differently from each other and from the square on the hypotenuse.

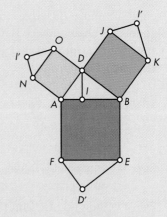

You're not done yet, but you're getting there! Your script should still be recording. So far you've got a right triangle with squares on the sides. The right triangle is divided into similar triangles, and those similar triangles are constructed on the squares. In the following steps you'll add steps to your script that will allow you to repeat this process indefinitely.

Step 19 Scroll to the beginning of your script to confirm that the givens are a point, a straight object, and two other points.

Step 20 In triangle *I′KJ*, select, in order, point *I′*, $\overline{KJ}$, point *K*, and point *J*. Click Loop in the script.

Step 21 In triangle *I′ON*, select, in order, point *I′*, $\overline{ON}$, point *O*, and point *N*. Click Loop in the script.

Step 22 Stop the script.

You're finally ready to make your fractal grow. To play the script on each branch of your figure, follow these steps.

Step 23 Just as you did in Step 20, select, in order, point *I′*, $\overline{KJ}$, point *K*, and point *J*. Click Fast in the script. Choose recursion depth 1. (If you think your computer is fast enough and your screen is big enough, you can try recursion depth 2.)

Step 24 Just as you did in Step 21, select, in order, point *I′*, $\overline{ON}$, point *O*, and point *N*. Click Fast in the script. Choose recursion depth 1 (or 2).

Step 25 In △*D′FE*, select, in order, point *D′*, $\overline{FE}$, point *F*, and point *E*. Click Fast in the script. Choose recursion depth 1 (or 2).

If all went well, you should now have a branching Pythagorean fractal. In a construction as complicated as this, it's easy for something to go wrong. If your script folds squares in on themselves instead of branching out, chances are the selection order got mixed up someplace, either while you were recording the script or when you played it. Don't despair! Undo back to the beginning, then redo until you have your right triangles, squares on the sides, and a right triangle on each square. Try your script again, experimenting with different selection orders.

If you successfully made the Pythagorean fractal, you're ready to investigate its fascinating patterns. First try dragging a vertex of the original triangle. Be patient. Depending on your machine, it may take time for the figure to change, especially if you played your script to depth 2. Explore the following questions.

1. Does the Pythagorean Theorem still apply to the branches of this figure? That is, does the sum of the areas of the branches on the legs equal the area of the branch on the hypotenuse? See if you can answer the question without actually measuring all the areas.

2. a. Consider your original sketch of a right triangle with a square built on each side. This simple sketch is stage 0 of your fractal. At stage 1, you add three triangles and six squares to your construction. On a piece of paper, draw a rough sketch of stage 1. Use the sketch on your screen to help you.

   b. How much area do you add to this fractal between stage 0 and stage 1? (Don't measure any areas to answer this.)

   c. Draw a rough sketch of stage 2. How much area do you add between stage 1 and stage 2?

   d. How much area is added at any new stage?

   e. Determine the stage of the construction on your screen. Also determine how its area compares to its area at stage 0.

   f. A true fractal exists only after an infinite number of stages. If you could build a true fractal based on the construction in this activity, what would be its area?

3. Do you see some squares in the figure that are identical to each other? Give the same color and shade to sets of squares that have the same area. What do you notice about these sets of squares other than their equal area? Describe any patterns you find in sets of equal-area squares.

4. Describe any other patterns you can find in the Pythagorean fractal.

5. Experiment using Sketchpad to create other types of fractals.

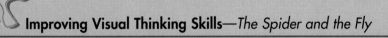

Improving Visual Thinking Skills—*The Spider and the Fly*

Henry E. Dudeney (1847–1930) was one of the greatest inventors of puzzles in the English-speaking world. Many of the puzzles that challenge people today can be traced back to him. His spider-and-fly problem first appeared in an English newspaper. It has challenged puzzle enthusiasts around the world for over three quarters of a century.

Inside a rectangular room, measuring 30′ in length and 12′ in width and height, a spider is at a point on the middle of one of the end walls, 1′ from the ceiling, as shown at point *A*; and a fly is on the center of the opposite wall, 1′ from the floor, as shown at point *B*. What is the shortest distance that the spider must crawl to reach the fly, which remains stationary? Of course, the spider never drops or uses its web, but crawls fairly.

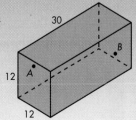

Lesson 10.6

Return of the Word Problems

You may be disappointed if you fail,
but you are doomed if you don't try.
— Beverly Sills

Exercise Set 10.6

In Exercises 1–9, watch for special right triangles and multiples of familiar Pythagorean triples.

1. Find the area of a right triangle whose leg measures 6 feet and whose hypotenuse measures 10 feet.

2. Find the length of the hypotenuse of an isosceles right triangle whose legs measure 3.5 meters. Express your answer accurate to the nearest tenth of a meter.

3.* The area of an isosceles right triangle is 98 square inches. What is the length of the hypotenuse? Express your answer accurate to the nearest inch.

4.* Meteorologist Paul Windward and his fiancée, geologist Raina Stone, are rushing to Lost Wages, Nevada, to wed at the Lost Wages Wedding Emporium. Paul lifts off in his balloon at noon from Pecos Gulch, heading east for Lost Wages. With the prevailing wind blowing west to east, he will average a land speed of 30 km/hr. This will allow him to arrive in Lost Wages in 4 hours. Meanwhile, Raina is 160 km north of Pecos Gulch. At the moment of Paul's lift off, Raina hops into her Jeep™ and heads directly for Lost Wages. At what average speed must she travel to arrive at Lost Wages at the same time Paul does?

5.* A giant California redwood tree 36 meters tall cracked in a violent earthquake and fell as if hinged. The tip of the once beautiful tree hit the ground 24 meters from the base. Researcher Red Woods wishes to investigate the crack. How many meters up from the base of the tree does he have to climb?

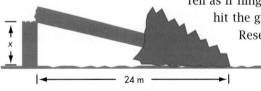

24 m

6.* What is the longest stick that can be placed within a box whose inside dimensions are 24 inches, 30 inches, and 18 inches?

7.* A 25-foot ladder is placed against a building. The bottom of the ladder is 7 feet from the building. If the top of the ladder slips down 4 feet, how many feet will the bottom slide out? No, it is not 4 feet. This is a two-step problem, so draw two right triangles.

8.* The front and back walls of an A-frame cabin are shaped like isosceles triangles, each with a base measuring 10 m. The equal sides of each isosceles triangle measure 13 m. The entire front of the cabin is made of double-pane insulated glass 1 cm thick. What is the area of one isosceles triangle? If the glass was purchased for $120/m², what did the glass for the front of the cabin cost?

9. It all began one afternoon when Romeo's mother would not let him out of the house. This caused his girlfriend, Julie, to wonder, "Where is he now?" So Julie decided to visit Romeo and talk to him from beneath his balcony. As Julie was calling out, "Romeo, oh Romeo, where the heck are you, Romeo?" she slipped and fell into the moat that was directly beneath the balcony. Then and there she decided that the two of them should elope! After drying herself off, she went to the local hardware store and purchased a rope so that Romeo could slide down it and escape. If Romeo's balcony is 6 meters up from the moat, what is the shortest length of rope needed to reach from the balcony to the ground on the opposite side of the 4.5-meter-wide moat?

Romeo and Julie

10.* The inclined plane or ramp makes it possible to lift heavy objects with less force than is necessary to lift them straight up. An object's weight is a measure of the force of gravity being exerted on that object. For example, a force of 100 pounds is required to keep a 100-pound object from falling to the ground. **Work** is a measure of force applied over some distance. The work required to lift a 100-pound object (like the cylinder shown) 1 foot above the ground is 100 foot-pounds. To lift it 2 feet above the ground requires 200 foot-pounds of work. With the help of an inclined plane, the force required to do the work is greatly reduced. If you build a 4-foot-long ramp to roll the cylinder up to a height of 1 foot, you'll do 100 foot-pounds of work over a 4-foot distance, meaning you only need to apply 25 pounds of force at any given moment.

4 feet   1 foot

How much work (foot-pounds) is necessary to lift 80 pounds straight up 2 feet? If a ramp 8 feet long is built to raise the 80 pounds up 2 feet, how much force (measured in pounds) will it take?

11. Hector can only exert 70 pounds of force and he needs to lift a 160-pound steel ball up 2 feet. What is the minimum length of ramp he should build?

12. According to the Americans with Disabilities Act, the slope of a wheelchair ramp must have a slope no greater than $1/12$. What would be the length of ramp necessary to gain a vertical distance of 4 feet? How much force would be required while going up the ramp if a person and his chair together weighed 200 pounds?

13. Can you express the rule of the inclined plane algebraically? Complete the formulas: The work (w) necessary to lift p pounds up a height h feet is given by the formula $w = (–?–)(–?–)$. The work done applying a force f along a ramp of length l is given by the formula $w = (–?–)(–?–)$. Set these equations equal to write a formula for f in terms of p, h, and l.

14. **Group Activity** The tangram puzzle is believed to be over 4000 years old. According to Chinese folklore, Tan was taking a porcelain tile to the emperor when he dropped and broke it into seven pieces, as shown.

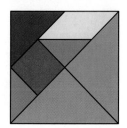

Tan spent the rest of his life trying to put the pieces back into a square. During these efforts, he was successful in creating hundreds of other shapes (a few are shown below). The seven pieces include five isosceles right triangles, a square, and a parallelogram. If the area of the square piece is 4 cm^2, find the dimensions of the other six pieces.

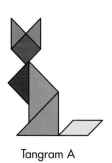

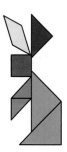

| Tangram A | Tangram B | Tangram C | Tangram D |

15. **Group Activity** Make a set of your own seven tangram pieces and create the swan (C) and the horse with rider (D).

16. What's wrong with this picture?

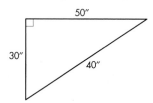

17. What's wrong with this picture?

Lesson 10.7

Distance in Coordinate Geometry

Scott is located at the intersection of Second Street and 3rd Avenue, and his sister Viki is located at the intersection of Seventh Street and 8th Avenue. If Viki were to walk on the sidewalks along the streets and avenues between her and Scott, her shortest route would be ten blocks. It is easy to find distances along the horizontal or the vertical: You simply count blocks. However, if Scott were able to fly straight to Viki, you would need the Pythagorean Theorem to calculate the distance he traveled. To the nearest meter, what is the distance from Scott to Viki if each block is approximately 50 meters long?

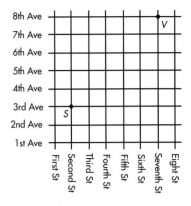

A grid of streets is like a coordinate plane. A coordinate grid is made of two sets of parallel lines, one set running perpendicular to the other set. Thus every segment in the plane (nonvertical and nonhorizontal segments) is the hypotenuse of some right triangle. You can use the Pythagorean Theorem to find the distance between any two points on a coordinate plane.

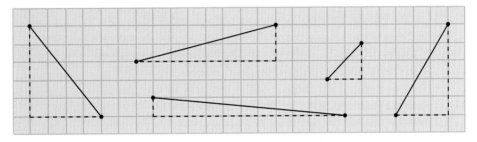

Investigation 10.7

Step 1   Copy each graph onto your own graph paper. Using the grid lines as a guide, turn each given segment into the hypotenuse of a right triangle.

1.

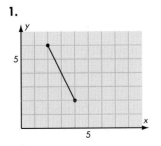

2.

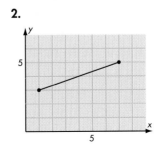

3.

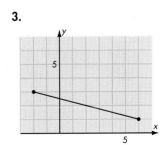

Step 2   Using the grid lines as a guide, turn each given segment into the hypotenuse of a right triangle, then find the length of each segment.

4.* 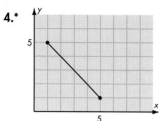   **5.** 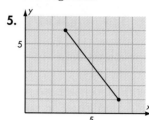   **6.**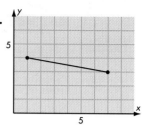

Step 3   Graph the points, then find the distances between them.

7.* (1, 2), (13, 7)   **8.** (–5, –8), (3, 7)   **9.** (–9, –6), (3, 10)

10. What is the distance from Scott to Viki to the nearest meter? (You did read the introduction, didn't you?)

   In the previous ten problems, you found the length of a segment by connecting it to a right triangle on the graph paper and then applying the Pythagorean Theorem. What if the points are so far apart that your graph paper isn't large enough to plot them? For example, what is the distance between the points (15, 37) and (42, 73)? Clearly the coordinates are too large to graph on ordinary graph paper. You need a rule or a formula that uses the coordinates of the two given points to calculate the distance between those points.

   To find the distance between points A and B shown at right, you can simply count the squares on side $\overline{AC}$ and the squares on side $\overline{BC}$, then use the Pythagorean Theorem to determine AB. However, when the distances are too great to count physically, there is still a nice way to find the lengths. You can find the vertical distance BC by subtracting the y-coordinates of points A and B. Because distance is never negative, subtract the lesser coordinate from the greater. The vertical distance $BC = 8 - 1 = 7$. You can find the horizontal distance, AC, by subtracting the x-coordinates of points A and B. The distance $AC = 7 - 2 = 5$. Now you can find the length of $\overline{AB}$: $AB^2 = (7 - 2)^2 + (8 - 1)^2$, and therefore $AB = \sqrt{5^2 + 7^2} = \sqrt{74}$.

   Can you generalize this result and come up with a formula for the distance between any two points in the coordinate plane? Compare your observations with those of others. Copy and complete the conjecture below and add it to your conjecture list.

C-95   If the coordinates of points A and B are (x_1, y_1) and (x_2, y_2), respectively, then
$AB^2 = (-?-)^2 + (-?-)^2$ and $AB = \sqrt{(-?-)^2 + (-?-)^2}$ (***Distance Formula***).

   Let's look at a few examples to see how you can use the Distance Formula to both find the distance between two points and find the equation of a circle.

Example A

Find AB if the coordinates of point A are (8, 15) and the coordinates of point B are (–7, 23).

$$AB^2 = (8 - -7)^2 + (23 - 15)^2$$
$$= 15^2 + 8^2$$
$$= 289$$
$$AB = \sqrt{289} = 17$$

Example B

The distance from any point (x, y) on a circle to the center (5, 4) is 7 units. Use the Distance Formula to write an equation for the circle.

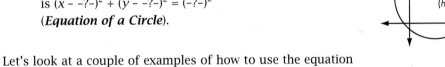

$$\sqrt{(x - 5)^2 + (y - 4)^2} = 7$$

Square both sides of the equation.

$$(x - 5)^2 + (y - 4)^2 = 49$$

Can you generalize Example B for a circle with radius r and center (h, k)? Copy and complete the conjecture below and add it to your conjecture list as an extension of the Distance Formula.

C-96 The equation for a circle with radius r and center (h, k) is $(x - -?-)^2 + (y - -?-)^2 = (-?-)^2$ (***Equation of a Circle***).

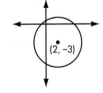

Let's look at a couple of examples of how to use the equation of a circle.

Example C

Find the equation of the circle with center (2, –3) and radius 4.

$$(x - h)^2 + (y - k)^2 = r^2$$

Substitute 2 for h, –3 for k, and 4 for r.

$$(x - 2)^2 + (y - -3)^2 = 4^2$$
$$(x - 2)^2 + (y + 3)^2 = 16$$

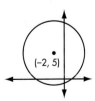

Example D

Find the center and radius of the circle $(x + 2)^2 + (y - 5)^2 = 36$.

$$(x - -2)^2 + (y - 5)^2 = 6^2$$

So the center is (–2, 5) and the radius is 6.

Exercise Set 10.7

In Exercises 1–3, find the distance between each pair of points.

1.* (10, 20), (13, 16)   **2.** (15, 37), (42, 73)   **3.** (–19, –16), (–3, 14)

4. Find the perimeter of $\triangle ABC$ with vertices $A(2, 4)$, $B(8, 12)$, and $C(24, 0)$.

5. Determine whether $\triangle DEF$ with vertices $D(6, -6)$, $E(39, -12)$, and $F(24, 18)$ is scalene, isosceles, or equilateral.

6.* Determine whether $\triangle GHI$ with vertices $G(2, 6)$, $H(18, 2)$, and $I(12, 12)$ is isosceles, right, isosceles right, or equilateral.

For Exercises 7–9 use $\triangle ABC$ with vertices $A(-2, -2)$, $B(4, 0)$, and $C(0, 6)$.

7. Find midpoints M, N, and P of $\overline{AC}$, $\overline{CB}$, and $\overline{AB}$, respectively.

8. Find the slopes of $\overline{MN}$ and $\overline{AB}$, the slopes of $\overline{MP}$ and $\overline{BC}$, and the slopes of $\overline{NP}$ and $\overline{AC}$. How do they compare?

9. Find the lengths of $\overline{MN}$ and $\overline{AB}$, the lengths of $\overline{MP}$ and $\overline{BC}$, and the lengths of $\overline{NP}$ and $\overline{AC}$. How do they compare?

For Exercises 10–12, find the equation of the circle.

10. Center (0, 0); $r = 4$ **11.** Center (2, 0); $r = 5$ **12.** Center (3, 3); through (0, –1)

13.* Guido Palumbo is a Venetian gondolier. Guido needs to know how deep it is in front of his pier. He notices a water lily sticking straight up from the water, whose blossom is 8 cm above the water's surface. Guido pulls the lily to one side, keeping the stem straight, until the blossom touches the water at a spot 40 cm from where the stem first broke the water's surface. From this data Guido is able to calculate the depth of the water. Can you? What is the depth of the water?

14.* A circle of radius 6 has a chord AB of length 6. If point C is selected randomly on the circle, what is the probability that $\triangle ABC$ is obtuse?

Lesson 10.8

Circles and the Pythagorean Theorem

In Chapter 7, you discovered a number of properties that involved right angles in and around circles. In this lesson you will use the conjectures you made, along with the Pythagorean Theorem, to solve some challenging problems. Let's review. Two of the most useful conjectures are listed below.

A tangent to a circle is perpendicular to the radius drawn to the point of tangency (*Tangent Conjecture*).

Angles inscribed in a semicircle are right angles (C-70).

Let's look at a few examples of how you can use these conjectures with the Pythagorean Theorem to solve problems.

Example A

$AN = 12$ cm. $\overrightarrow{TA}$ is tangent to circle N. Find the shaded area.

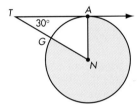

Because $\overrightarrow{TA}$ is a tangent, $\triangle TAN$ is a 30-60 right triangle. The area of the circle is $\pi 12^2$, or 144π cm^2. Because $\frac{60}{360}$ is $\frac{1}{6}$, the shaded area is $\frac{5}{6}$ of the area of the circle. Therefore the shaded area is $\frac{5}{6}(144\pi)$, or 120π cm^2.

Example B

$AB = 6$ cm and $BC = 8$ cm. Find the shaded area.

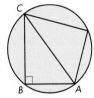

Because $\angle ABC$ is a right angle, $\overset{\frown}{ABC}$ is a semicircle and $\overline{AC}$ is a diameter. If $AB = 6$ cm and $BC = 8$ cm, then $AC = 10$ cm (by the Pythagorean Theorem) and the radius is 5 cm. If the radius is 5 cm, the area of the circle is $\pi 5^2$, or 25π cm^2.

Exercise Set 10.8

In Exercises 1–6, use your circle conjectures and the Pythagorean Theorem to find the shaded area in each figure. Assume lines and rays that appear tangent are tangent. All measurements are given in centimeters.

1.* $OD = 24$

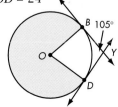

2. $RC = 9$

3.* $BT = 6\sqrt{3}$

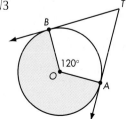

4.* $HA = 8\sqrt{3}$

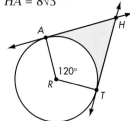

5. $HT = 8\sqrt{3}$

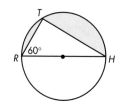

6.* $HO = 8\sqrt{3}$

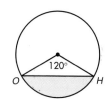

7. Find the coordinates of point *M*.

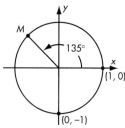

8. Find the coordinates of point *K*.

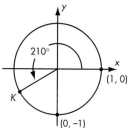

9. Find the coordinates of point *E*.

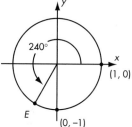

10.* A 3-meter-wide circular track is shown at right. The radius of the inner circle is 12 meters. What is the longest straight path that stays on the track? (In other words, find *AB*.)

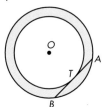

11.* An annulus has a 36-cm chord of the outer circle that is also tangent to the inner concentric circle. Find the area of the annulus.

12.* Sector *ARC* has a radius of 9 cm and an angle that measures 80°. When sector *ARC* is cut out and $\overline{AR}$ and $\overline{RC}$ are taped together, they form a cone. The length of $\overset{\frown}{AC}$ becomes the circumference of the base of the cone. What is the height of the cone?

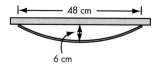

13. In her latest expedition, Ertha Diggs has uncovered a portion of terra-cotta pipe. This relic is believed to be part of an early water drainage system. To determine the diameter of the original pipe, she lays a meter stick across the portion and measures the length of the chord at 48 cm. The depth of the portion from the midpoint of the chord is 6 cm. From these two measurements, she is able to determine the diameter of the original pipe. Can you? What was the pipe's original diameter?

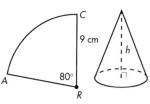

14.* A belt on a piece of machinery needs to be replaced. The belt runs around two wheels, crossing between them so that the larger wheel turns the smaller in the opposite direction. The diameter of the larger wheel is 36 cm, and the diameter of the smaller is 24 cm. The distance between the centers of the two wheels is 60 cm. The belt crosses 24 cm from the center of the smaller wheel. What is the length of the belt?

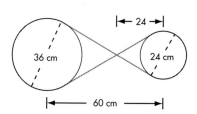

15.* Will Derness is a ranger for the National Forestry Service. His job is to measure the circumference of certain marked trees in his section of forest. The marked trees have

been selected for cutting to "thin" the forest, which will provide room for the growth of unmarked trees. The largest marked tree has a circumference of 336 cm at a point 5 m up from its base. To the nearest centimeter, what is the side length of the largest square piece that can be cut from a cross section of this tree? (In other words, calculate the length of the side of a square that can be inscribed in a circle with a circumference of 336 cm.) Round your answer to the nearest centimeter.

16.* Each of three circles of radius 6 cm is tangent to the other two, and they are inscribed in a rectangle, as shown. What is the height of the rectangle?

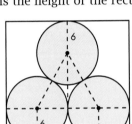

17.* The Gothic arch is based on the equilateral triangle. If the base of the arch measures 80 cm, what is the area of the Gothic arch (shaded region)?

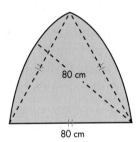

Improving Reasoning Skills—*Reasonable 'rithmetic I*

Each letter in these problems represents a different digit.

1. What is the value of *B*?

2. What is the value of *J*?

```
    3 7 2
    3 8 4
  + 9 B 4
  ─────────
  C 7 C A
```

```
        E F 6
     x  D 7
    ─────────
    D D F D
    J E D
    ─────────
    H G E D
```

Lesson 10.9

Pythagorean Potpourri

Any time you see someone more successful than you are, they are doing something you aren't.

— Malcolm X

Exercise Set 10.9

For Exercises 1–8, use patty papers or a compass and a straightedge.

1. Construct right $\triangle DEF$ with $\angle D$ as an acute angle and $\overline{DE}$ the hypotenuse. Next, construct a circle that circumscribes $\triangle DEF$. What is $\overline{DE}$ in relation to the circle?

2.* Given $\angle A$ and radius $\overline{OR}$, construct circle O so that it is tangent to both sides of $\angle A$.

3.* Construct a tangent to circle O from a given point X outside the circle.

4.* Construct the right $\triangle TOM$ given the median $\overline{OA}$ to the hypotenuse $\overline{TM}$, and one leg $\overline{OT}$.

5. Construct the right $\triangle MAT$ given the median $\overline{AE}$ to the hypotenuse $\overline{MT}$, and the altitude $\overline{AG}$ to the hypotenuse.

In Exercises 6–8, use $\overline{AB}$.

6.* Construct a 30-60 right triangle given AB, the length of the hypotenuse.

7.* Construct a 30-60 right triangle given AB, the length of the shorter leg.

8.* Construct a 30-60 right triangle given AB, the length of the longer leg.

In Exercises 9–11, each large triangle is equilateral. Use the properties of equilateral triangles, 30-60 right triangles, and medians to find x and y in each figure.

9.

10.

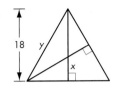

11.

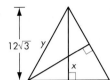

In Exercises 12–14, find the area of each equilateral triangle. Find the area of the inscribed circle. Find the area of the circumscribed circle. The area of the circumscribed circle is how many times as great as the area of the inscribed circle?

12.* $AB = 6$ cm

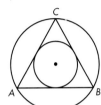

13. $DE = 2\sqrt{3}$ cm

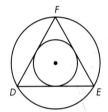

14.* $IJ = 9\sqrt{3}$ cm

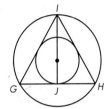

15.* The circle-packing efficiency of a square was calculated in Chapter 9 to be about 78.5%. What is the circle-packing efficiency of an equilateral triangle? That is, the area of a circle inscribed within an equilateral triangle is what percentage of the area of the equilateral triangle?

16.* The area of equilateral $\triangle REG$ is $25\sqrt{3}\,\mathrm{m}^2$. Find the perimeter of $\triangle REG$.

17. Find the area of a regular hexagon with each side measuring 12 cm.

18.* Find the area of a regular hexagon circumscribed about a circle with a radius of 9 inches.

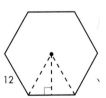

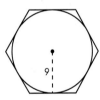

In Exercises 19 and 20, each triangle is equilateral. Find the shaded area in each figure to the nearest square centimeter.

19.* $NP = 12$ cm

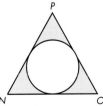

20. $QS = 12$ cm

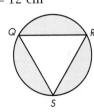

21. What is the area of a circle that circumscribes an isosceles right triangle whose legs measure 8 cm?

22. What is the area of a circle inscribed within an isosceles right triangle whose legs measure 8 cm?

23. What is the area of the square that can be placed within an isosceles right triangle whose legs measure 8 cm? Two sides of the square rest on the legs of the triangle, and a vertex of the square touches the hypotenuse of the triangle.

24. The regular octagon shown below has one vertex at (10, 0) and center at the origin (0, 0). Find the coordinates of the other seven vertices.

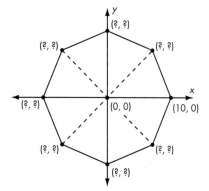

25. The regular hexagon shown below has one vertex at (10, 0). The center of the inscribed and circumscribed circles is the origin (0, 0). Find the coordinates of the other five vertices.

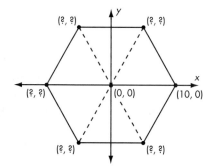

26.* Captain Fordie La George has just launched his shuttle craft *Exitprise* from the orbiting space platform 250 km above the earth. The trajectory of the craft is tangent to the space platform's orbital path and is traveling at 250 km/hr. Approximately how far is the shuttle craft from the earth's surface 4 hr after launch? (The diameter of the earth is approximately 12,750 km.)

27.* **Group Activity** The Gothic arch at right was constructed from arcs of circles, as shown. The base of the arch measures 360 cm. The two small Gothic arches and the circle are reserved for stained glass windows. The remaining regions within the large Gothic arch are masonry. What percentage of the large Gothic arch is reserved for windows?

28.* **Group Activity** Compare the packing arrangements of the cans shown below. The diameter of each can is 2.5 inches. For each arrangement, what is the total area of the box? Are there any reasons why one arrangement may be better than the other? Why is the arrangement at left used?

Lesson 10.10

Chapter Review

School is where you go to learn to communicate,
but teachers say, "No talking!"
— Gallagher

If in fifty years you've forgotten everything else you learned in geometry, you'll probably still remember the Pythagorean Theorem. (Though let's hope you don't really forget everything else!) That's because it has so many applications and you'll encounter it in so many math and science classes throughout your schooling.

It's one thing to remember $a^2 + b^2 = c^2$. It's another to know what it means and to be able to apply it. Review your work from this chapter to be sure you understand how to find a missing length (leg or hypotenuse) in a right triangle. How can you quickly find the length of the hypotenuse in an isosceles right triangle? How are the sides related in a 30-60 right triangle? How is the Pythagorean Theorem used to find the distance between two points in the coordinate plane? How can the theorem be extended to three dimensions? Practice applying the Pythagorean Theorem in the following exercise set.

Exercise Set 10.10

For Exercises 1–12, identify each statement as true or false. For each false statement, sketch a counterexample or explain why the statement is false.

1. The hypotenuse is always the side opposite the right angle in a right triangle and is always the longest side.

2. Any three positive integers that work in the Pythagorean formula are called right triangle numbers.

3. In a right triangle, if a and b are the lengths of the legs and c is the length of the hypotenuse, then $(a + b)^2 = c^2$.

4. If the lengths of the three sides of a triangle work in the Pythagorean formula, then the triangle must be a right triangle.

5. If you multiply the lengths of all three sides of any right triangle by the same number, the resulting triangle will be a larger right triangle.

6. If you add the same length to each side of a right triangle, the resulting triangle will be a right triangle.

7. In an isosceles right triangle, if the legs are of length x, then the hypotenuse is of length $x\sqrt{3}$.

8. In a 30-60 right triangle, if the shorter leg is of length x, then the longer leg is of length $2x$ and the hypotenuse is of length $x\sqrt{3}$.

9. If the coordinates of points A and B are (x_1, y_1) and (x_2, y_2), respectively, then $AB^2 = (x_1 - y_1)^2 + (x_2 - y_2)^2$ and $AB = \sqrt{(x_1 - y_1)^2 + (x_2 - y_2)^2}$.

10. * The midpoint of the hypotenuse of a right triangle is equidistant from all three vertices of the triangle.

11. If a right rectangular prism has dimensions a, b, and c, then the length d of its space diagonal is found by the formula $d = \sqrt{a^2 + b^2 + c^2}$.

12. There are more than four true statements in Exercises 1–12.

13. Simplify $3\sqrt{320}$.

For Exercises 14–21, measurements are given in centimeters.

14. $x = \text{-?-}$

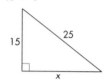

15. Is $\triangle ABC$ a right triangle?

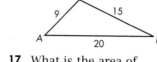

16. The area of the square is 144 cm². What is d?

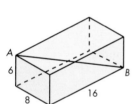

17. What is the area of the triangle?

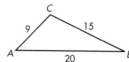

18. $AB = \text{-?-}$

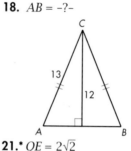

19. $AB = \text{-?-}$

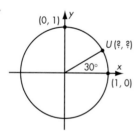

20. What is the area of trapezoid $ABCD$?

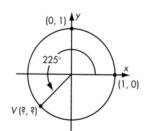

21. * $QE = 2\sqrt{2}$

What is the shaded area?

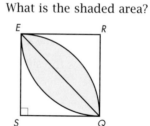

For Exercises 22–24, find the coordinates of the given points.

22.

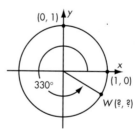

23.

24.

25. What is the length of the hypotenuse of a right triangle whose legs measure 300 feet and 400 feet?

26. What are the lengths of the two legs of a 30-60 right triangle if the length of the hypotenuse is $12\sqrt{3}$?

27. Determine whether $\triangle ABC$ with vertices $A(3, 5)$, $B(11, 3)$, and $C(8, 8)$ is isosceles, isosceles right, or equilateral.

28.* Two cars leave point *A* at noon. One car travels north at 45 mi/hr; the other travels east at 60 mi/hr. How far apart are the two cars after two hours?

29.* To the nearest foot, find the original height of a fallen flagpole that cracked and fell as if hinged, forming an angle of 45 degrees with the ground. The tip of the pole hit the ground 12 feet from its base.

30.* Flora Fluty is away at camp and wishes to mail her wooden flute back home to her family. The flute is 24″ long. Will it fit diagonally within a box whose inside dimensions are 12″ × 16″ × 14″? (Ignore the thickness of the flute.)

31. Find the area of an equilateral triangle whose sides measure 6 meters.

32.* Find the circumference of a circle inscribed within an equilateral triangle whose height measures $12\sqrt{3}$.

33. Find the perimeter of an equilateral triangle if its height measures $7\sqrt{3}$.

34. After an argument, Paul and Paula walk away from each other at what appears to be a right angle. Paul is walking 2 km/hr, and Paula is walking 3 km/hr. After 1 hr Paula stops walking. After 2 hr Paul stops. If at that time they are 5 km away from each other, did they walk away from each other at right angles? Now that they have decided to make up, how long will it take them to reach each other if they continue to walk at their same speeds straight toward each other?

35.* Desert prospector Sagebrush Sally hops onto her dirt bike and, with a full tank of gas, leaves camp traveling 60 km/hr east. After 2 hr she stops and does a little prospecting—with no luck. So she hops back onto her bike and heads north for 2 hr at 45 km/hr. She stops again, does a little more prospecting, and this time hits pay dirt.

Because the greatest distance Sagebrush Sally has ever traveled on one tank of gas is 350 km, she assumes that that is the maximum distance she will be able to travel on this trip. Does she have enough fuel to get back to camp? If not, what is the closest she can come to camp? What should she do?

36. A circle has a central angle *AOB* that measures 80°. If point *C* is selected randomly on the circle, what is the probability that △*ABC* is obtuse?

37. A block of wood is attached to a wire through a pulley 1.5 meters above the path of the block. The distance from the pulley to the point on the block at which the wire is attached is 3.9 meters. How far will the block move if the wire is pulled in 1.4 meters?

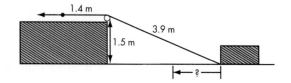

38.* How far will the wire have to be pulled in if the block from Exercise 37 is to be moved a second time, horizontally and to the left, a distance of 1.2 meters?

39. Al baked a square pan of brownies for himself and his two sisters. To divide the pan into three parts, he measured three 30° angles from one of the right-angled corners and cut the brownies into three large pieces. Two pieces form right triangles and the middle piece forms a kite. Did Al divide the pan of brownies equally? Explain your reasoning.

40. You are standing 12 feet from a cylindrical corn-syrup storage tank. The distance to a point of tangency on the tank is 35 feet. What is the radius of the tank?

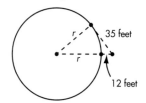

Assessing What You've Learned

Try one or more of the following assessment suggestions.

Update Your Portfolio

• Choose a project, a Take Another Look activity, or one of the more challenging problems you did in this chapter and add it to your portfolio. Document it according to your teacher's instructions.

Organize Your Notebook

• Review your notebook to be sure it's complete and well organized. Be sure you have included all this chapter's conjectures in your conjecture list. Write a one-page chapter summary.

Write in Your Journal

• Why do you think the Pythagorean Theorem is considered one of the most important theorems in mathematics?

Performance Assessment

• While a classmate, a friend, a family member, or a teacher observes, do an investigation, a Take Another Look activity, or a problem from this chapter. Explain what you're doing at each step, including how you arrive at your conjecture or solution.

Write Test Items

• Work with group members to write test items for this chapter. Include simple exercises and complex applications problems. Try to demonstrate more than one approach in your solutions.

Give a Presentation

• Create a poster, a model, or visual aids for the overhead projector and give a presentation about the Pythagorean Theorem.

Cooperative Problem Solving

Pythagoras in Space

1. Astrophysicists at the lunar research facility Galileo II are planning a very large construction on the back side of the lunar surface. They want to make an obvious mark on the moon's surface to indicate the existence of an intelligent civilization in case an intelligent life form from another galaxy sends robotic space probes into our solar system. The astrophysicists have agreed that the Pythagorean Theorem is an important fact all intelligent life forms should know, and so will construct a large right triangle with a square on each side. The sides of the right triangle will measure 30 km, 40 km, and 50 km. Bulldozers will smooth away the regions within the three squares and the right triangle, and the borders of the squares will be made of one-meter-wide reflective material. What is the total area that must be smoothed away? What is the total length of meter-wide reflective material needed for this monumental task? What do you think of the idea?

2. The first robot built by the Galileo II Artificial Intelligence Lab is a very simple one. The experimental robot's first program instructs it to travel in a rectangular spiral: 1 m east, 2 m north, 3 m west, 4 m south, 5 m east again, and so on, until it runs into something. If the robot travels uninterrupted at the rate of 1 m/min, how far away from its starting point will it be after 2 hr?

3.* On the lunar space port Galileo, three large water pipes run from the hydroponics and water recycling plant to the dormitory-recreation facility. Engineers plan to enclose these water pipes within a ceramic-coated triangular casing to protect them from radiation. Each of the water pipes has an outside diameter of 60 cm. If the cross section of the casing is an equilateral triangle, what is the smallest length possible for a side?

4.* Machinists at the space port's machine shop have fabricated a pipe that is 8 m long and 3 cm in diameter. They must transport it down a corridor 3 m wide by 3 m high that has a right-angled corner. The pipe must not be bent or twisted. Will it fit around the corner? What is the longest pipe that will fit around the corner?

5.* The space port's research team is planning an anniversary celebration. They plan to decorate the facility by wrapping a wreath of olive branches around the microwave receiver tower. The tower is a cylindrical column 70 m tall and 8 m in circumference. Because the space port is celebrating its thirtieth anniversary, the team plans to wrap the wreath uniformly about the column exactly 30 times. How long will the spiral wreath be?

Volume

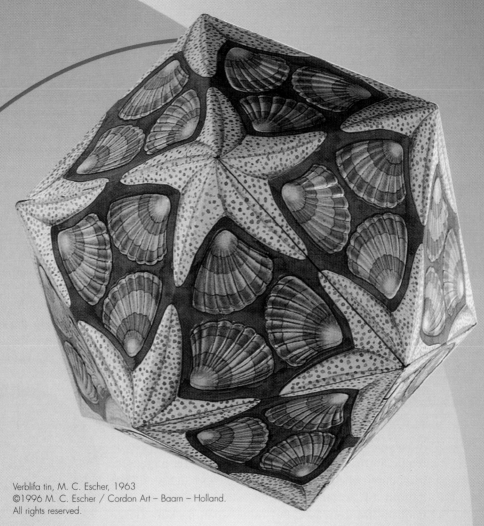

Verblifa tin, M. C. Escher, 1963
©1996 M. C. Escher / Cordon Art – Baarn – Holland.

Volume is the amount of space contained within an object. In this chapter you will discover the formulas for finding the volumes of prisms, pyramids, cylinders, cones, and spheres. You will also discover the formula for the surface area of a sphere. For these discoveries, you will use what you have already learned about the area formulas and the Pythagorean Theorem—by now you should know them by heart. Knowing how to calculate volume is important in many occupations. Of all the geometric topics, area, volume, the Pythagorean Theorem, and similarity (the subject of the next chapter) have the most practical applications.

Lesson 11.1
Polyhedrons, Prisms, and Pyramids

In geometry many of the figures you work with are flat (plane figures). Plane figures have two dimensions. In this chapter you will work with solids. Solids have three dimensions: length, width, and height. You see and touch solids all the time. Some solids, like rocks and plants, are very irregular, but many other solid objects have shapes that can be easily described using common geometric terms. Some of these geometric solids occur in nature: viruses, oranges, crystals, the earth itself. Others

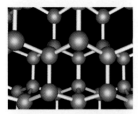

Amethyst crystal

The Ramat Polin housing complex in Jerusalem, Israel, resembles naturally occurring crystals.

are manufactured: books, buildings, baseballs, soup cans, ice cream cones.

At the molecular level, three-dimensional geometry plays a very important role in a number of common substances. The three-dimensional geometry of carbon is an interesting example. Carbon atoms arranged in a very rigid tetrahedral lattice result in diamonds, one of the hardest materials around! But carbon bonded in planes of hexagonal rings results in graphite, a soft material used as a lubricant. The results are totally different with the same chemical element. The difference is in the geometry.

The geometry of diamonds

The geometry of graphite

Scientists recently discovered that carbon can also bond into very large, symmetrical molecules. Named fullerenes, after U.S. engineer Buckminster Fuller (1895–1983), who studied the properties of dome structures, these carbon molecules are popularly called buckyballs. The molecule shown below has the same symmetry as a soccer ball.

Three-dimensional geometry plays one of its most important roles at the molecular level with simple H_2O, or water. Most substances contract in volume (become more dense) when they decrease in temperature. Water is a rare exception. Water increases in volume when frozen, becoming less dense. The motion of water molecules slows down as the temperature decreases. But rather than packing closer together, as in most substances, the molecules in water move apart into a nice, rigid three-dimensional lattice. Water's exceptional behavior is very important for life on this planet! Because ice is less dense than water, ice floats on water,

The buckyball

staying closer to the sun and heat. If ice were heavier (denser) than water, it would sink to the bottom of the ocean, away from heat sources. Eventually the oceans would fill from the bottom up with ice, and we would have an ice planet. What a cold thought.

The problems on the previous page should lead you to a formula for the volume of any right rectangular prism. For example, if a right rectangular prism holds six layers of cubes (1 in. on each edge) and each layer is four cubes by five cubes, then the volume would be (4)(5)(6), or 120 cubic inches. Notice that the number of cubes resting on the base equals the number of square units in the area of the base. The number of layers of cubes equals the number of units in the height of the prism. Therefore you can multiply the area of the base by the height of the prism to calculate the volume. Complete the statement below.

Conjecture A If B is the area of the base of a right rectangular prism and H is the height of the solid, then the formula for the volume is $V = \underline{\quad?\quad}$.

In Chapter 9, you discovered that you can change parallelograms, triangles, trapezoids, and circles into rectangles to find their area. You can use the same method to find the volume of any right prism with bases of these shapes by finding the area of the base and multiplying by the height of the prism. For example, to find the volume of a right triangular prism, find the area of the triangular base (the number of cubes resting on the base) and multiply by the height (the number of layers of cubes).

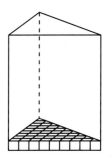

Therefore the formula for finding the volume of right rectangular prisms can be extended to cover all right prisms and right cylinders. Complete the statement below.

Conjecture B If B is the area of the base of a right prism (or cylinder) and H is the height of the solid, then the formula for the volume is $V = \underline{\quad?\quad}$.

What about the volume of a prism or a cylinder that is not a right prism or a right cylinder? Suppose you wanted to find the volume of the oblique rectangular prism shown below, with a base 8.5 inches by 11 inches and a height of 6 inches. The shape of the oblique rectangular prism can be approximated by a slanted stack of three reams of 8.5″ × 11″ paper. The shape can be even better approximated by the individual pieces of paper in a slanted stack.

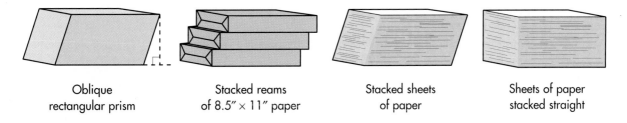

| Oblique rectangular prism | Stacked reams of 8.5″ × 11″ paper | Stacked sheets of paper | Sheets of paper stacked straight |

Rearranging the paper formed into an oblique rectangular solid back into a right rectangular prism changes the shape, but certainly the volume of paper hasn't changed (the area of the base, 8.5″ × 11″, didn't change and the height, 6″, didn't change).

By a similar argument, you can use coffee filters, coins, candies, or filter papers from chemistry to show that an oblique cylinder has the same volume as a right cylinder with the same base and height.

Let's use the stacking model to extend your formula for the volume of right prisms and cylinders to oblique prisms and cylinders. Complete the statement below.

Conjecture C   The volume of an oblique prism or cylinder is the same as the volume of a right prism that has the same base —?— and the same —?—.

Finally, let's combine the three conjectures (A, B, and C) into a single conjecture for finding the volume of any prism or cylinder, whether it's right or oblique.

C-97   If B is the area of the base of a prism or a cylinder and H is the height of the solid, then the formula for the volume is $V = $ —?— (***Prism-Cylinder Volume Conjecture***).

Notice that the same volume formula applies to all prisms, regardless of the type of base they have. To calculate the volume of a prism, first identify the type of base the prism has and use the appropriate area formula to calculate its area. Then multiply the area of the base by the height of the prism. Notice that in oblique prisms, the lateral edges are no longer at right angles to the bases. Thus you do *not* use the length of the lateral edge as the height.

Example A

Find the volume of a trapezoidal prism that has a height of 15 cm. The trapezoidal base has a height of 5 cm and the two bases of the trapezoid measure 4 cm and 8 cm.

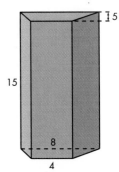

$$V = BH$$
$$= [\tfrac{1}{2}(5)(4 + 8)](15)$$
$$= (30)(15)$$
$$= 450$$

The volume is 450 cm^3.

Example B

Find the volume of a cylinder that has a base with a radius of 6 inches and a height of 7 inches.

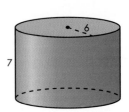

$$V = BH$$
$$= (\pi 6^2)(7)$$
$$= 36\pi(7)$$
$$= 252\pi$$

The volume is 252π in^3

Take Another Look 11.3

You may be familiar with the area model of the expression $(a + b)^2$, shown below. Draw or build a volume model of the expression $(a + b)^3$. How many pieces does your model have? What's the volume of each type of piece? Use your model to write the expression $(a + b)^3$ in expanded form.

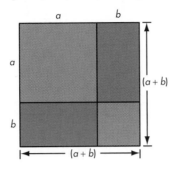
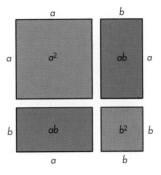

Exercise Set 11.3

Find the volume of each right prism or right cylinder named in Exercises 1-6. All measurements are given in centimeters. Round answers to two decimal places.

1. Rectangular prism   **2.*** Right triangular prism   **3.*** Trapezoidal prism

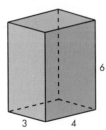
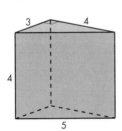
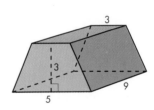

4. Cylinder   **5.*** Semicircular cylinder   **6.*** Cylinder with a 90° slice removed

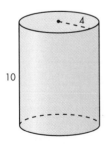
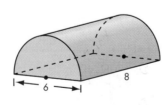
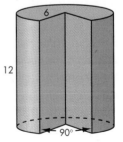

7. Draw and label two different rectangular prisms with volumes of 288 cm³.

8. Draw and label two different circular cylinders with volumes of 1152π cm³.

9. Use the information about the base and height of each solid to find the volume. All measurements are given in centimeters.

Triangular prism

Rectangular prism

Trapezoidal prism
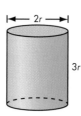

Cylinder

Information about base of solid	Height of solid	Use $b, h, H.$	Use $b, h, H.$	Use $b, b_2, h, H.$	Use $r, H.$
$b = 6, b_2 = 7, h = 8, r = 3$	$H = 20$	a.* $V = -?-$	d. $V = -?-$	g. $V = -?-$	j. $V = -?-$
$b = 9, b_2 = 12, h = 12, r = 6$	$H = 20$	b. $V = -?-$	e. $V = -?-$	h. $V = -?-$	k. $V = -?-$
$b = 8, b_2 = 19, h = 18, r = 8$	$H = 23$	c. $V = -?-$	f. $V = -?-$	i. $V = -?-$	l. $V = -?-$

In Exercises 10–12, express the volume of each solid with the help of algebra. All measurements are given in centimeters. Each quadrilateral is a rectangle.

10.
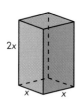

11.
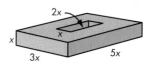

12.
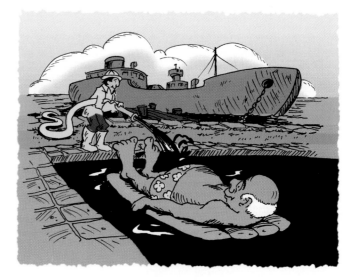

For Exercises 13–17, sketch and label each solid described, then find the volume.

13. A right rectangular prism. The base measures 12 cm by 16 cm. The height is 4 cm.

14. A right trapezoidal prism. The trapezoidal base has a height of 4″ and bases that measure 8″ and 12″. The height of the prism is 24″.

15.* A right circular cylinder with a height of T. The radius of the base is $\sqrt{Q}$.

16. A chocolate cake of diameter 24 cm and height of 14 cm that has a slice missing whose vertex angle measures 45°.

17.* A right triangular prism with height $K + 7$. The base is determined by an isosceles right triangle with a hypotenuse of $K\sqrt{2}$.

18. Although the Exxon *Valdez* oil spill (11 million gallons of oil) is perhaps one of the most notorious oil spills to foul the North American continent, it was minor compared to oil spills during the 1991 Persian Gulf War, in which about 250 million gallons of crude were spilled. A gallon is 0.13368 cubic feet. How many swimming pools 20′ × 30′ × 5′ deep could be filled with 250 million gallons of crude oil?

19. The NAMES Project AIDS Memorial Quilt memorializes persons all around the world who have died of AIDS. As of February 1996, the Quilt was composed of over 32,000 3- by 6-foot panels, which represented less than 10% of the AIDS deaths in the United States alone. At that time, it could cover about 19 football fields with walkways between groups of panels. The 3- by 6-foot panels are assembled into 12-foot squares. As the AIDS epidemic continues, the rapidly increasing number of panels creates a storage challenge for the NAMES Project. (In the beginning of 1996, the NAMES Project was receiving about 50 new panels per week.) When folded, each 12-foot square of panels requires about 1 cubic foot of storage. What was the volume of the Quilt in February 1996? If the storage facility had a floor area of 5,000 square feet, how high were the Quilt panels stacked?

Improving Reasoning Skills—*Murder at the Socratic Liars' Club*

Hemlock Stones, the not-yet-famous consulting detective, has just been called in to solve the mysterious murder at the Socratic Liars' Club. There are five suspects, each of whom has sworn by the club oath to make two true statements and one false one whenever speaking to someone on the club's premises. From their recorded statements below, Hemlock Stones was able to determine "who dunit." Can you?

Professor: I did not kill Henley. I never owned a knife. Lance did it.

Ethel: I didn't kill him. I don't own a knife. The others are drunk.

Phoebe: I'm innocent. Lance is the killer. I don't even know Dutch.

Lance: I'm innocent. Dutch is guilty. The Professor lied when he said I did it.

Dutch: I didn't kill him. Ethel is the murderer. Phoebe and I are old friends.

Lesson 11.4

Volume of Pyramids and Cones

Say what you know, do what you must, come what may.
— *Sofia Kovalevskaya*

In this lesson you will compare the volumes of pyramid-prism pairs and cone-cylinder pairs that have congruent bases and the same height.

If a prism and a pyramid have congruent bases and are of the same height, will they have the same volume? The diagram below suggests that the pyramid will have less volume. How much less? How many times bigger is the prism than the pyramid? Is there a relationship? Does this relationship hold for cylinders and cones?

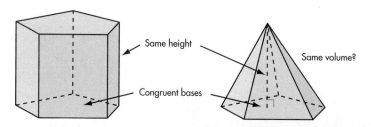

There is a simple relationship between the volumes of prisms and pyramids with congruent bases and the same height and between cylinders and cones with congruent bases and the same height. To discover this relationship you will need the items listed below.

- Hollow pairs of prisms and pyramids with congruent bases and the same height

- Hollow pairs of cylinders and cones with congruent bases and the same height

- Sand, rice, birdseed, or water (if your models are watertight)

- A container for the sand, rice, birdseed, or water

This investigation works best if the tasks are shared. Work in groups of four or five.

Investigation 11.4

Step 1   Choose a prism-pyramid pair with congruent bases and the same height. (Try to use a pair with bases different from those of other groups.)

Step 2   Working over your container, fill your pyramid with sand or water, then pour the contents into your prism. About what fraction of the prism is filled by the volume of one pyramid?

Constructing the Platonic Solids

In the Project: The Five Platonic Solids you discovered what each of the five Platonic solids would look like when unfolded and laid flat. Now use those patterns to construct and assemble the five Platonic solids. You will need the materials listed below.

• Poster board, heavy cardboard, or manila folders

• A compass and a straightedge

• Scissors

• Glue, paste, or cellophane tape

• Colored pens or pencils for decorating the solids

Before you begin, read through the construction tips below.

Construction Tips

1. To construct the icosahedron, octahedron, and tetrahedron, you need to construct networks of equilateral triangles. Build these three solids at the same time.

2. To construct the hexahedron or cube, construct perpendiculars to guarantee that each side is a square.

3. To construct the dodecahedron, first construct regular pentagons. To construct a regular pentagon, follow the six steps outlined below.

Step 1

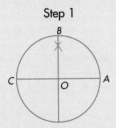

Construct a circle. Construct two perpendicular diameters.

Step 2

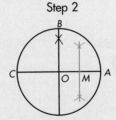

Find *M*, the midpoint of $\overline{OA}$.

Step 3

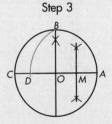

Swing an arc with radius *BM* intersecting $\overline{OC}$ at point *D*.

Step 4

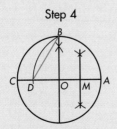

BD is the length of each side of the pentagon.

Step 5

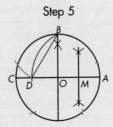

Starting at point *B*, mark off *BD* on the circumference five times.

Step 6

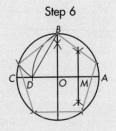

Connect the points to form a pentagon.

4. Construct the dodecahedron in two parts. Each part is composed of five regular pentagons around a regular pentagon. Follow the steps below to construct one part.

Step 1   Construct a large regular pentagon (see Tip 3).

Step 2   Lightly draw all the diagonals in the pentagon. This gives you a smaller regular pentagon, which will be one of the twelve pentagonal faces of your dodecahedron.

Step 3   Through each of the vertices of this central pentagon, draw the diagonals and extend them all the way to the sides of the larger pentagon. Find the five pentagons that encircle the central pentagon.

Step 4   Erase the unnecessary line segments.

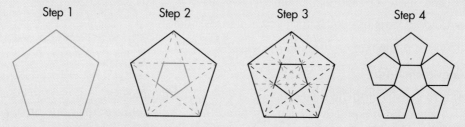

Step 1   Step 2   Step 3   Step 4

5. Decorate each Platonic solid *before* you cut it out.

6. Leave tabs on some edges for gluing your solid together.

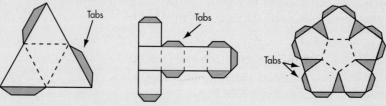

Tabs   Tabs   Tabs

7. Score each edge to be folded by heavily running a ballpoint pen or a compass point over the fold lines. Score on both sides of the paper.

Improving Algebra Skills—*Polynomial Geometry I*

1. Perimeter = –?–

2. Area = –?–

x + 2
x
3x – 1
2x
2x
4x – 3

3. Perimeter = –?–

4. Area = –?–

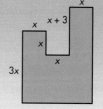

x
x   x + 3
x
x
3x

Lesson 11.6

Displacement and Density

What will happen if you step into a bathtub that is filled to the brim? Right. The water will overflow. What will happen if you fill a glass to the brim with root beer and then add a scoop of ice cream? Right. You'll have a mess! The volume that overflows in each of these situations equals the volume of the solid below the liquid level. This volume is called an object's **displacement**. You can determine the volume of an irregularly shaped object by measuring its displacement.

Example A

Geologist Crystal Stone wishes to calculate the volume of an irregularly shaped rock. She places it into a rectangular prism containing water. The base of the container measures 10 cm by 15 cm. When the rock is put into the container, Crystal notices that the water level rises 2 cm because the rock

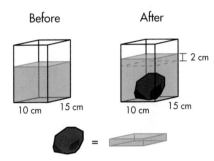

displaces its volume of water. This new "slice" of water has a volume of (2)(10)(15), or 300 cm³. Therefore the volume of the rock is 300 cm³.

An important property of a material is its density. **Density** is the mass of matter in a given volume. Density is calculated by dividing the mass in grams by the volume in cubic centimeters (density = $\frac{mass}{volume}$). A chemist wishing to identify an unknown clump of metal could weigh the clump to determine its mass, determine its volume by displacement, calculate its density, and, finally, look in a chemical handbook in hopes of identifying the compound by its density.

Metal	Density	Metal	Density
Aluminum	2.81 g/cm³	Nickel	8.89 g/cm³
Copper	8.97 g/cm³	Platinum	21.40 g/cm³
Gold	19.30 g/cm³	Potassium	0.86 g/cm³
Lead	11.30 g/cm³	Silver	10.50 g/cm³
Lithium	0.54 g/cm³	Sodium	0.97 g/cm³

Example B

A clump of metal weighing 351.4 g is dropped into a cylindrical container, causing the water level to rise 1.1 cm. The radius of the base of the container is 3.0 cm. What is the density of the metal? Assuming the metal is pure, what is the metal?

$$\text{Volume} = \pi(3.0)^2(1.1)$$
$$= (\pi)(9)(1.1)$$
$$\approx 31.1$$

$$\text{Density} \approx \frac{351.4}{31.1}$$
$$\approx 11.3$$

The density is 11.3 g/cm³. Therefore the metal is lead.

Exercise Set 11.6

1. A rock is added to a container of water, and it raises the water level 3 cm. If the container is a rectangular prism whose base measures 15 cm by 15 cm, what is the volume of the rock?

2. A solid glass ball is dropped into a cylinder with a radius of 6 cm, raising the water level 1 cm. What is the volume of the glass ball?

3. A fish tank 10 inches by 14 inches by 12 inches high is the home of a fat goldfish named Columbus. When he is taken out for some fresh air, the water level in the tank drops $\frac{1}{3}$ inch. What is the volume of Columbus to the nearest cubic inch?

4.* A block of ice is placed into an ice chest containing water. The ice causes the water level to rise 4 cm. The ice chest measures 35 cm by 50 cm by 30 cm high. When ice floats in water, one eighth of its volume floats above the water level and seven eighths floats beneath the water level. What is the volume of the block of ice?

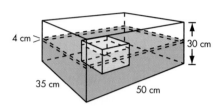

5. A piece of wood placed in a cylindrical container causes the container's water level to rise 3 cm. This type of wood floats half out of the water, and the radius of the container is 5 cm. What is the volume of the piece of wood?

6. Sandy Pyle has found a clump of metal, that appears to be either silver or lead. Hoping it is silver, she drops it into a rectangular container filled halfway with water. She observes that the water level rises 2 cm. Because the base of the container is a square measuring 5 cm on an edge, she is able to calculate the volume of the clump. She weighs the metal and finds that it weighs 525 g. What is the density of the metal? Is it lead or silver?

7.* Chemist Donna Dalton is given a clump of metal and is told that it is sodium. She weighs the metal and finds that it weighs 145.5 grams. To test if it is indeed sodium, she places it into a square prism whose base measures 10 cm on each edge and that is partially filled with a nonreactive liquid. If the metal is sodium, how many centimeters will the liquid level rise?

8. The round bales shown at right are 4 feet in diameter and 4 feet long. Calculate the volume of each bale. Bales this size weigh approximately 900 lbs. What is the density of the packed hay? Square balers make bales whose dimensions are $4' \times 4' \times 6'$ and that weigh approximately 1400 lbs. Is this hay packed more or less densely than the hay in a round bale? Explain.

9. How much does a solid block of aluminum weigh if its dimensions are 4 cm by 8 cm by 20 cm?

10. Which will weigh more: a solid cylinder of gold with a height of 5 cm and a diameter of 6 cm or a solid cone of platinum with a height of 21 cm and a diameter of 8 cm? Which is worth more?

11. Each edge of this alunite crystal pyramid measures 6 cm. You want to drop it into a shallow cylindrical container of crystal-cleansing fluid. If the diameter of the container is 10 cm and the height is 5 cm, will the crystal fit? How much cleansing fluid should you put into the container to cover the crystal without letting the fluid overflow? Explain your reasoning.

12. Sherlock Holmes has just returned home excited. He rushes to his chemical lab, takes a mysterious medallion out of his carrying case, and weighs it. "It weighs 3088 grams," Mr. Holmes says in anticipation. "Now, let's check its volume." He pours water into a glass graduated container with a square base that measures 10 centimeters on each side, then records the water level, which is 53.0 centimeters. He places the medallion into the container and reads the new water level, 54.6 centimeters. After a few minutes of the mental calculation that Holmes enjoys so much, he turns to Dr. Watson. "This confirms my theory about Colonel Banderson. Quick, Watson! The game is afoot. We are off to the train station."

Poor Dr. Watson is still standing with the *London Daily* in his hands. "Holmes, you amaze me. Is it gold?" questions the good doctor.

"If it has a density of 19.3 grams per cubic centimeter, it is gold," smiles Mr. Holmes. "If it is gold, then the Colonel is who he says he is. If it is a fake, then so is Colonel Banderson."

Watson is still waiting for the answer. "Well?" he queries.

Holmes, heading for the door, smiles and says, "It's elementary, my dear Watson. Elementary geometry, that is."

What is the volume of the medallion? Is it gold and is Colonel Banderson who he is?

Sherlock Holmes and the mysterious medallion

Improving Algebra Skills—*An Algebra Mind-Reading Trick II*

Try the following card trick. Use algebra to explain why the trick works.

As the mathemagician, ask your friend to give you any two cards whose sum is less than 10. Tell your friend you will be able to pick a third card that you can use with the other two to make a three-digit number that is divisible by 11. The trick is to add them and find a card equal to that sum.

Suppose you had a 10-inch-square sheet of metal and you wanted to build a small box by cutting out squares from the corners of the sheet and folding up the sides. What size corners should you cut out to get the biggest box possible? In this investigation, you'll explore that question by making models to discover an equation that gives volumes for different boxes. Then you'll graph the equation and use the graph to solve the problem.

1. On graph paper, make four different scale drawings showing the 10-inch-square sheet with corner cuts of 1-inch squares, 2-inch squares, 3-inch squares, and 4-inch squares. What is the volume of each of these boxes? Cut out the box templates and fold and tape them to form boxes. Which has the greatest volume? Do you think you could make a box with greater volume? Try to guess what size cut would give the greatest volume.

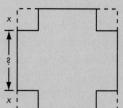

2. Call the length of the corner cut x. What is the length of each side of the box's square base?

3. Write an equation for the volume of the box, y, in terms of x.

4. Before you graph your equation, look back at your answers to problem 1. What do you think would be reasonable ranges for x and y? Enter those values for your calculator graph window, then graph the equation you wrote in problem 3.

5. If you've chosen a good range for your graph window, you should see your graph touch the x-axis in at least two places and reach a maximum somewhere in between. What do the first two x-intercepts represent? What would the boxes look like for x-values near each of those intercepts?

6. Trace the graph to find the value for x that maximizes y. Zoom to find values as precisely as possible.

7. Enter a new equation for the volume of a box formed from a 12-inch-square sheet. What value for x gives a box of maximum volume? Compare your answer for the 12-inch-square sheet with the answer you found in problem 5 for the 10-inch-square sheet. Make a conjecture about what fraction of the side length should be cut from each corner of a square sheet to make a box of maximum volume. Test your conjecture with at least one other sheet size.

Lesson 11.7

Volume of a Sphere

I have a simple philosophy. Fill what's empty. Empty what's full. And scratch where it itches.
— Alice Roosevelt

In this lesson you will develop a formula for the volume of a sphere. (By volume of a sphere, we mean the amount of space contained within the sphere.) In Investigation 11.7 you are going to compare the volume of a cylinder or a cone to the volume of a sphere or a hemisphere. There are a number of commercially available plastic volume measurement kits you can use to do this investigation. You will need a pair of containers (cylinder or cone and sphere or hemisphere) with the same radius. This investigation demonstrates the relationship between the volume of a hemisphere with diameter $2r$ and the volume of a cylinder with the same diameter ($2r$) and height $2r$. This investigation can also be done with a cone-sphere (or hemisphere) pair. In addition to a pair of hollow solids, you will need the items listed below.

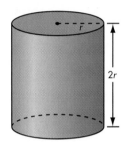

• Sand, rice, birdseed, or water (if your models are watertight)

• Container for the sand, rice, birdseed, or water

Investigation 11.7

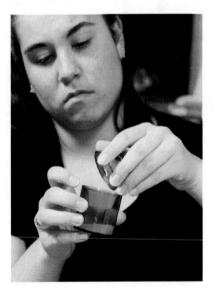

Step 1   Fill the hemisphere with sand, rice, birdseed, or water.

Step 2   Slowly pour the contents of the hemisphere into the cylinder. What fraction of the cylinder does the hemisphere appear to fill?

Step 3   Fill the hemisphere again and pour the contents into the cylinder. What fraction of the cylinder do two hemispheres (one sphere) appear to fill?

    If the radius of the cylinder is r and its height is $2r$, then what is the volume of the cylinder in terms of r? The volume of a sphere is whatever fraction of the cylinder was filled by two hemispheres.

Volume of cylinder	Fraction of cylinder filled by two hemispheres
Volume$_{cylinder}$ = BH	Volume$_{sphere}$ = $(\frac{?}{?})$(Volume$_{cylinder}$)
= $(\pi r^2)(2r)$	= $(\frac{?}{?})(2\pi r^3)$
= $2\pi r^3$	= $(\frac{?}{?})\pi r^3$

What is the formula for the volume of a sphere? State this as your conjecture.

 C-99 The volume of a sphere with radius r is given by the formula —?—
(**Sphere Volume Conjecture**).

Take Another Look 11.7

1. Use algebra and the Sphere Volume Conjecture to show that the volume of a cone with base radius r and height r is half the volume of a hemisphere with radius r. Next show that the ratio of the volume of a cylinder with height r and radius r to the volume of a hemisphere with radius r to the volume of a cone with base radius r and height r is 3 to 2 to 1.

2. Derive the Sphere Volume Conjecture by using a pair of hollow shapes different from those you used in Investigation 11.7. Or use two solids made of the same material and compare weights. Explain what you did and how it demonstrates the conjecture.

Exercise Set 11.7

Find the volume of each solid. All measurements are in centimeters.

1.*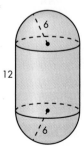

3

2.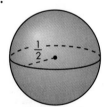

$\frac{1}{2}$

3.

.75

4.*

6

12

6

5.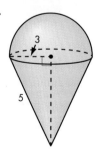

3

5

6.*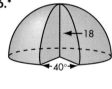

18

40°

7.* A sphere has a volume of 972π cubic inches. Find its radius.

8. A hemisphere has a volume of 18π cm^3. Find its radius.

9. The base of a hemisphere has an area of 256π cm^2. Find its volume.

10. A sphere of ice cream is placed onto your ice cream cone. Both have a diameter of 8 cm. The height of your cone is 12 cm. If you push the ice cream into the cone, will all of it fit?

11.* Lickety Split ice cream comes in a cylindrical container with an inside diameter of 6 inches and a height of 10 inches. The company claims to give the customer 25 scoops of ice cream per container, each scoop being a sphere with a 3-inch diameter. How many scoops will each container really hold? Should Lickety Split be reported to the Better Business Bureau?

12. A cylindrical glass 10 cm tall and 8 cm in diameter is filled to 1 cm from the top with water. If a golf ball 4 cm in diameter is dropped into the glass, will the water overflow?

13. Which is greater, the volume of a hemisphere with radius 2 cm or the total volume of two cones with radius 2 cm and height 2 cm?

14. A can of tennis balls has an inside diameter of 7 cm and a height of 20 cm. If the diameter of a tennis ball is 6 cm, what is the packing efficiency of the tennis ball can? That is, the three balls represent what percentage of the volume of the can?

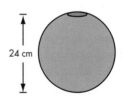

15.* The spherical container shown at right is to be filled with water at a steady rate. It takes 30 seconds to fill the container. Sketch an approximate graph of the height of the water (vertical axis) versus time (horizontal axis).

24 cm

16. What is the volume of the greatest wooden sphere that can be carved out of a wooden cube whose edges measure 12 cm?

Improving Visual Thinking Skills—*Patchwork Cubes*

The large cube on the right is built from 13 double cubes like the one shown plus one single cube. What color must the single cube be and where must it be positioned?

Project

Archimedes' Principle

An important property related to density and displacement is buoyancy. Have you ever been swimming with a friend and noticed how much easier it was to lift or pick up your friend in the water than out of it? Or have you ever played in the ocean or a lake and discovered that you could pick up a large sunken log or boulder and raise it to the surface but couldn't lift it out of the water?

A father teaches his son to swim. The boy's weight in the water is less than it would be on dry land.

The sunken object's apparent loss of weight is called **buoyancy**. The weight lost by an object in water is equal to the weight of the water it displaces. This is called **Archimedes' principle**. Legend has it that Archimedes discovered this principle while trying to solve the problem of how to tell if a crown was made of genuine gold. He realized he could determine the material by weighing the crown under water. The insight came to him while he was bathing. Thrilled by his discovery, Archimedes ran through the streets shouting "Eureka!" wearing just what he'd been wearing in the bathtub.

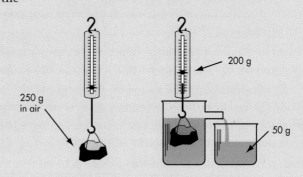

Archimedes' principle states that an object submerged in a liquid is buoyed up by a force equal to the weight of the liquid displaced. For example, if a stone weighing 250 grams in air is placed into water, where it weighs 200 grams, then the water it displaces weighs the missing 50 grams.

Here are some other examples: If you put a block of wood into water, it sinks until it has displaced its own weight of water, then it stops sinking. A boat floats when its weight is less than the weight of the volume of water it displaces. Submarines use Archimedes' principle. When its ballast tanks are filled with air, the submarine's weight is less than the weight of the water it displaces, and it rises to the surface. When the air is blown from the tanks and sea water is allowed into the tanks, the submarine's weight is greater than the weight of the water it now displaces, and it sinks. To raise the submarine, compressed air is used to force the water back out of the tanks. Now the submarine once again weighs less than the water it displaces, and it floats. Pretty neat huh! The same principle also works for hot-air balloons.

Example

A flat-bottomed boat (in the shape of a rectangular prism) has dimensions 4 feet by 6 feet by 3 feet deep. The boat and its three potential passengers weigh 750 pounds. The density of water is 63 pounds per cubic foot. Will the boat be able to hold all three passengers? How high up the boat will the water level reach?

$$\text{weight of water displaced} = \text{weight of object}$$
$$\text{volume displaced} \times \text{density} = 750$$
$$(4)(6)h \times 63 = 750$$
$$h = \frac{750}{(4)(6)(63)}$$
$$h \approx 0.5 \text{ feet}$$

The water will rise only half a foot. Therefore the boat can easily hold all three passengers. Could it hold three more 180-pounders?

Buoyancy Experiment

To get a better understanding of buoyancy, perform this experiment. You will need the following materials.

- An empty half-gallon milk carton with the top cut off, or an open plastic container (for your boat)

- A tub, a large plastic container, or a sink (for your ocean)

- Rice, unpopped popcorn, or dried beans (for your boat's cargo)

- A balance to measure the weight of the boat and its cargo

Step 1   In square centimeters, calculate the area of the base or the bottom of your boat.

Step 2   Add cargo to your boat and weigh the boat with its cargo. If you are using a milk carton, start with about 500 grams of cargo.

Step 3   Calculate how high the water will rise on the sides of your boat. Remember, the boat will displace an equal weight of the water. Recall that 1 cubic centimeter of water weighs 1 g.

Step 4   Try it. Place your cargo-ladened boat into your ocean. Measure the depth that the boat sinks into the water. In cubic centimeters, calculate the volume of the boat beneath the water surface. This volume is also the weight in grams of the displaced water and should equal the weight of the boat plus cargo. How close were your calculations?

Step 5   Finally, calculate the maximum cargo weight your boat can carry without sinking. Then test your calculations. Compete with other groups to see which boat can carry the most cargo without sinking.

Now try the following exercises to reinforce your understanding of Archimedes' principle.

1. Beans are poured into an empty milk carton whose top is cut off. The carton filled with beans weighs 1093 g. The base dimensions of the milk carton are 9.7 cm on each edge, and the carton is 19 cm tall. If the carton of beans is placed into a container of water, will the carton sink or float? If it floats, how high up the side of the carton will the water level rise?

2.* A sheet of aluminum 18 cm by 32 cm by 0.5 cm thick is formed into an open rectangular box by cutting squares 4 cm on a side from each of the corners of the aluminum sheet. When the four sides are folded up and carefully sealed, will the open aluminum box sink or float in water? Explain.

3.* To impress her crew and to help maintain discipline, the fierce pirate Norah S. Grande occasionally performs feats of strength and bravery. Today Norah's ship, *Topaz*, is anchored in the shallows off the Louisiana coast, and a fishing net containing four cannon balls has been dropped over the side of the ship. Norah is going to dive into the water, pick up all four balls, and lift them over her head. Did Norah take on more of a task than she should have? How much does each cannon ball weigh when it is above the surface of the water? How much does a cannon ball weigh when it is in the water? Each cannon ball is a solid lead sphere with a diameter of 16 cm. The density of lead is 11.3 g/cm^3, and the density of salt water is 1.03 g/cm^3.

Lesson 11.8

Surface Area of a Sphere

Nothing in life is to be feared,
it is only to be understood.
— *Marie Curie*

The areas of most small land regions on earth can be found by using area formulas for such plane figures as rectangles, triangles, and circles. However, to find the area of the entire earth's surface, you'd need a formula for the surface area of a sphere. Now that you know how to find the volume of a sphere, you can use that knowledge to arrive at the formula for the surface area of a sphere.

Investigation 11.8

In this investigation you are going to find the surface area of a sphere by first visualizing that the sphere's surface is covered in tiny "sorta polygons." (Thus the surface area, S, of the sphere is the sum of the areas of all the "sorta polygons.") If you picture segments connecting each of the vertices of the "sorta polygons" to the center of the sphere, you can mentally divide the volume of the sphere into many pyramids. Each of the "sorta polygons" is a base for a pyramid, and the radius, R, of the sphere is the height of the pyramid. (Thus the volume, V, of the sphere is the sum of the volumes of all the pyramids.) Now get ready for some algebra.

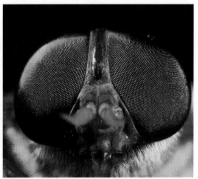

A horse fly's eyes resemble spheres covered by "sorta polygons."

Step 1   Suppose you were to divide the surface of your sphere into 1000 "sorta polygons." If we let B_1, B_2, B_3, . . . , B_{1000} represent the areas of the 1000 "sorta polygons," then the surface area, S, of the sphere can be written as the sum of the 1000 B's. Write this as an equation: $S = B_1 + B_2 + B_3 + \ldots + ?$.

Step 2   Because the volume of the pyramid with base B_1 is $\frac{1}{3}(B_1)(R)$, then the total volume of the sphere, V, is the sum of the volumes of the 1000 pyramids. Written as an equation:
$V = \frac{1}{3}(B_1)(R) + \frac{1}{3}(B_2)(R) + \ldots + \frac{1}{3}(B_{1000})(R)$.
Rewrite this equation by factoring the common expression $\frac{1}{3}R$ from each of the terms on the right: $V = \frac{1}{3}R(? + ? + ? + \ldots + ?)$.

Step 3   But the volume of the sphere is $V = \frac{4}{3}\pi R^3$. Rewrite the equation above by replacing V with $\frac{4}{3}\pi R^3$ (see Step 2) and replacing the sum of the areas of all the "sorta polygons," $B_1 + B_2 + B_3 + \ldots + B_{1000}$, with S (see Step 1).

Step 4   You should now have $\frac{4}{3}\pi R^3 = \frac{1}{3}RS$. Rewrite this equation by solving for the surface area S.

Of course, there's nothing special about 1000 "sorta polygons"; you can use a "zillion" "sorta polygons," and the reasoning is the same. You now have a formula for finding the surface area of a sphere given the radius of the sphere. State this as your next conjecture and add it to your conjecture list.

C-100   The surface area (S) of a sphere with radius r is given by the formula —?— (***Sphere Surface Area Conjecture***).

Take Another Look 11.8

1.* How does the area of the base of a hemisphere (the area of the great circle) compare with the surface area of the hemisphere? Another way to demonstrate the Sphere Surface Area Conjecture is to make the necessary number of copies of the circular region defined by a hemisphere's great circle, cut the copied circular regions into small pieces, and glue them onto the surface of the hemisphere. Try it with half an orange or half a tennis ball. Write a description of what you did and include all the necessary algebra.

2. The Epcot Center in Orlando, Florida, is made of polygonal regions, but the polygonal regions are arranged in little pyramids. Though the overall appearance of the building is spherical, the surface is not smooth. Suppose a perfectly smooth sphere had the same volume as the Epcot Center. Would it have the same surface area? If not, which would be greater, the surface area of the smooth sphere or of the bumpy sphere? Explain.

Exercise Set 11.8

Find the volume and surface area of each solid. All measurements are in centimeters.

1.*
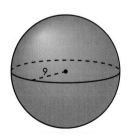

2.
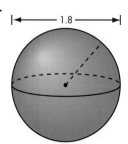

3.
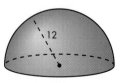

4.* Shaded area = 40π cm². Find the surface area.

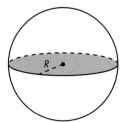

5. Surface area = 64π cm². Find the volume.

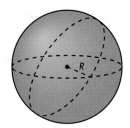

6. Volume = 288π cm³. Find the surface area.

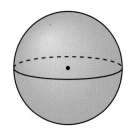

7. If the radius of the base of a hemisphere (the area of the great circle) is *R*, what is the area of the great circle? What is the surface area of the hemisphere? How do they compare? If it takes Jose 4 gallons of wood sealant to cover the hemispherical ceiling of his vacation home, how many gallons of wood sealant are needed to cover the floor?

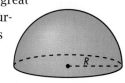

8. A Kickapoo wigwam is a semicylinder with a half-hemisphere on each end. The diameter of both the semicylinder and the half-hemispheres is 3.6 meters. The total length is 7.6 meters. What is the volume of the wigwam and the surface area of its roof?

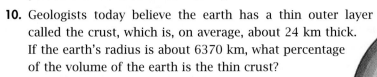

9. Approximately 70% of the earth's surface is covered in water. If the diameter of the earth is approximately 12,750 km, find the approximate area not covered by water. Express your answer to the nearest 100 km².

10. Geologists today believe the earth has a thin outer layer called the crust, which is, on average, about 24 km thick. If the earth's radius is about 6370 km, what percentage of the volume of the earth is the thin crust?

11. For nearly 400 years, the Medici family of Florence, Italy, was successful in farming, banking, and the silk and wool trades. Additionally, the family gave so much support to artists that the Medici name has become synonymous with patronage of the arts.

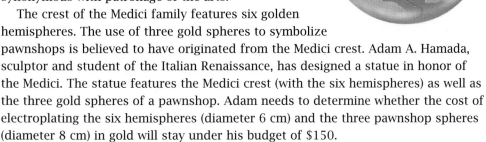

The crest of the Medici family features six golden hemispheres. The use of three gold spheres to symbolize pawnshops is believed to have originated from the Medici crest. Adam A. Hamada, sculptor and student of the Italian Renaissance, has designed a statue in honor of the Medici. The statue features the Medici crest (with the six hemispheres) as well as the three gold spheres of a pawnshop. Adam needs to determine whether the cost of electroplating the six hemispheres (diameter 6 cm) and the three pawnshop spheres (diameter 8 cm) in gold will stay under his budget of $150.

The electroplating shop tells him to expect to pay about 14¢/cm² to electroplate the figures. Will he be able to stay under budget? If not, what diameter pawnshop spheres should he make to stay under budget?

12. A farmer must periodically resurface the interior (wall, floor, and ceiling) of his silo to protect it from the acid created by the silage. The height of the silo to the top of the hemispherical dome is 50 ft and the width is 18 ft. What is the approximate surface area that needs to be treated? If one gallon of resurfacing compound covers about 250 ft², how many gallons are needed for the entire interior of the silo? (Remember to round up!) There is 0.8 of a bushel per ft³. Calculate the number of bushels of grain (to the nearest bushel) this silo will hold.

13. The container shown at right is to be filled with water at a steady rate. It takes 30 seconds to fill the container. Sketch an approximate graph of the height of the water (vertical axis) versus time (horizontal axis).

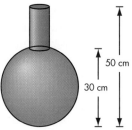

50 cm

30 cm

Improving Reasoning Skills—*Reasonable 'rithmetic II*

Each letter in these problems represents a different digit.

1. What is the value of *C*?
2. What is the value of *D*?
3. What is the value of *K*?
4. What is the value of *N*?

```
    8 7 8 9
    3 B A 7
    4 8 2 A
+   7 A B 5
---------
  2 C 2 8 7
```

```
  D E F F
- E 2 F 6
---------
  1 9 9 7
```

```
        G J
     ┌────────
   7 │ H G K
       2 1
     ───────
       H K
       H K
```

```
            5 2
        ┌────────
    L Q │ N M 2
          N P
        ───────
          M 2
          M 2
```

Lesson 11.9

Revenge of the Volume Problems

Give me a place to stand on and I will move the earth.
— *Archimedes*

Exercise Set 11.9

1. Sylvia Concrete just discovered that someone had opened the valve on her cement truck during the night of March 31 and that all the contents of the truck ran out to form a giant cone of hardened cement. For insurance purposes, Sylvia needs to figure out how much cement is in the cone. The circumference of the base of the cone is 44 feet, and the cone has a height of 6 feet. Calculate the volume of the cement cone to the nearest cubic foot.

2.* An amulet was recently uncovered high in the Andes by Seymour Hills, noted adventurer and archaeologist. Professor Hills must do some calculations to determine if the charm is an authentic Incan relic or a modern tourist trinket. If the volume of the regular hexagonal ring is equal to the volume of the regular hexagonal hole in its center, then it is authentic. If not, then it is a modern imitation. From the dimensions shown in the figure, determine whether the amulet is a modern fake or an authentic Incan relic.

3.* Rosita Aguas is a plumbing contractor. She needs to deliver 200 lengths of steel pipe to a construction site. Each cylindrical steel pipe is 160 cm long, has an outer diameter of 6 cm, and has an inner diameter of 5 cm. Steel has a density of about 7.7 g/cm^3. Rosita needs to know if her quarter-tonne truck can handle the weight of the pipes. (One tonne equals 1000 kg.) To the nearest kilogram, what is the weight of these 200 pipes? How many loads will Rosita have to transport to deliver the 200 lengths of steel pipe?

4. Marco Roni has just made two dozen meatballs. Each meatball has a 2-inch diameter. He wishes to cook them in sauce for a while, but his large pot may not be big enough to hold both them and the sauce. Right now, before the meatballs are added, the sauce is 2 inches from the top of the 14-inch diameter pot. Will the sauce spill over when the meatballs are added to the pot?

5. Inspector Lestrade has sent a small piece of very reactive metal to the crime lab. Lab technician Earl LaMayer weighs the soft metal and finds that it weighs 54.3 g. It appears to be either lithium, sodium, or potassium, all highly reactive with water. Therefore LaMayer places the metal into a glass graduated cylinder of radius 4 cm that contains a nonreactive liquid. The metal causes the liquid level to rise 2.0 cm. Which metal is it?

6.* Can you pick up a solid steel ball of radius 6 inches? Steel has a density of 0.28 pounds per cubic inch. To the nearest pound, what is the weight of the ball?

7. To the nearest pound, find the weight of a hollow steel ball with an outer diameter of 14 inches and a thickness of 2 inches. Steel has a density of 0.28 pounds per cubic inch.

8.* A hollow steel ball has a diameter of 14 inches and weighs 327.36 pounds. Steel has a density of 0.28 pounds per cubic inch. Find the thickness of the ball.

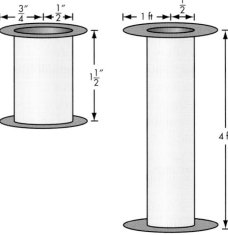

Schulhoff Tam
high school student

9.* Betty Holmes, distant French cousin of the famous Baker Street sleuth, and her friend Professor Hilton Gardens have returned to their flat on Butcher Street with a small Art Deco piece painted deep green. Betty must determine the density of the heavy metal objet d'art to solve the crime she has been working on for weeks. She places it into a regular hexagonal glass graduated prism filled with water and finds that the water level rises 4 cm. Each edge of the hexagonal base measures 5 cm. She carefully places the object on her balance and finds it weighs 5457 g. Professor Gardens calculates the object's volume, then its density. Next, Betty Holmes checks the density table (see Lesson 11.6) to determine its composition. If the object is platinum, it is from the missing Rothschild collection and Inspector Clouseau is the thief. If not, then the Baron is guilty of fraud. What is the composition of the Art Deco objet d'art?

10. This textile worker is inspecting thread being wound onto cylindrical spools. Suppose the small spool shown at right holds 100 yards of thread. If the same thread were wound with the same tightness onto the large spool at far right, how many yards would the spool hold?

Lesson 11.10

Chapter Review

When you stop to think, don't forget to start up again.
— *Anonymous*

In this chapter you discovered a number of formulas for finding volumes and you applied these formulas to solve practical problems. It's as important to remember how you discovered these formulas as it is to remember the formulas themselves. If you remember pouring the contents of a cone into a cylinder with the same height, you might remember how many cones it took and thus be able to derive the formula for a cone. Making connections can help, too. How are a prism and a cylinder alike? What does a cone have in common with a pyramid? Review the chapter to be sure your conjecture list includes formulas for finding the volumes of prisms, cylinders, pyramids, cones, and spheres. You should also be able to find the surface area of a sphere. Why do you think this surface area formula didn't come until the chapter about volume?

Exercise Set 11.10

For Exercises 1–15, identify each statement as true or false. Sketch a counterexample for each false statement or explain why it is false.

1. If K is the area of the base of a prism or a cylinder, and T is the height of the solid, then the formula for the volume is $V = \frac{1}{3}KT$.

2. If C^2 is the area of the base of a pyramid or a cone, and C is the height of the solid, then the formula for the volume is $V = \frac{1}{3}C^3$.

3. The volume V of a sphere with radius r is given by the formula $V = 4\pi r^3$.

4. The surface area S of a sphere with radius R is given by the formula $S = \frac{4}{3}\pi R^3$.

5. Exercise 10 is true.

6. A cylinder has a circular base and a vertex not in the plane of the base.

7. The circle formed by the intersection of a sphere and a plane passing through the center of the sphere is called a great circle.

8. An octahedron is a prism that has an octagonal base.

9. The lateral faces of a prism are never triangles.

10. Exercise 15 is false.

11. The lateral edges of a pyramid are the line segments where the lateral faces intersect.

12. A cylinder is all points in space at a given distance from a given segment.

13. The altitude of a pyramid or a cone is a perpendicular segment from the vertex to the plane of the base.

14. The smallest cylindrical container for a sphere has a volume $\frac{3}{2}$ times the volume of the sphere and a surface area $\frac{3}{2}$ times the surface area of the sphere.

15. There are more true statements than there are false statements in Exercises 1–15.

For Exercises 16–21, find the volume of each solid. Each quadrilateral is a rectangle. All measurements are given in centimeters.

16.

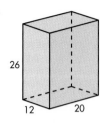

26
12 20

17.

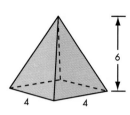

21

|← 14 →|

18.

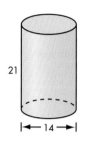

6

4   4

19.

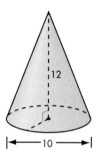

12

|← 10 →|

20.*

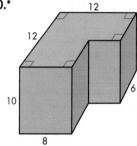

12
12

10

6

8

21.*

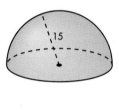

15

For Exercises 22–24, calculate each unknown length given the volume of the solid.

22. Find *h*.
   $V = 896$ cm^3

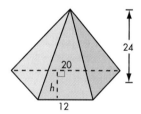

24

20

h

12

23. Find *r*.
   $V = 1728\pi$ cm^3

36

r

24.* Find *r*.
   $V = 256\pi$ cm^3

r

90°

25. A cylinder whose height is 12 cm and whose diameter is 6 cm has sitting on its top base a hemisphere whose diameter is also 6 cm. Find the total surface area of this solid.

26. Find the volume of a triangular pyramid whose base is equilateral with each side 6 meters long. The vertex of the pyramid is 3 meters directly over the incenter of the triangle.

27. Find the volume of a regular hexagonal prism that has a cylinder bored out from the center of its top base to the center of its bottom base. Each edge of the regular hexagon measures 8 cm. The height of the prism is 16 cm. The cylinder has radius 6 cm. Express your answer accurate to the nearest cubic centimeter.

28.* Balloon Boy Bernie pulls out a balloon with a diameter of 8 cm from the back of his balloonmobile. He blows up the balloon to a diameter of 16 cm. How many times has the surface area increased as the diameter has doubled? By how many times has the volume increased?

29. Two rectangular prisms have equal heights but unequal bases. If each of the dimensions of the smaller solid's base is half that of the larger solid's base, then the volume of the larger is how much greater than the volume of the smaller?

30. Two solid cylinders are made of the same material. Cylinder A is six times as tall as cylinder B, but the diameter of cylinder B is four times the diameter of cylinder A. Which cylinder weighs more? How many times more?

31. A ball is placed snugly into the smallest possible box that will completely contain the ball. What percentage of the box is filled by the ball?

32. Which of the following lidless boxes (formed by cutting a square from each corner of a 12″-square piece of paper and folding it into a box) holds the greatest volume: a box made by cutting a 1″ square from each corner, a 2″ square from each corner, or a 3″ square from each corner?

33. The spiral minaret in Samarra, Iraq, shown at right, has an outer walkway that spirals up to the top. Given what you know about calculating the volume of cones and cylinders, describe how you would approximate the volume of a shape like this.

34. Vasarely's op-art painting *Hexa-Tri-C* shows a sphere bulging out of a cube. Counting each dot on a cube edge as one unit, the volume of the cube alone would be (17 units)^3. Suppose the sphere alone has the same volume. What would be its radius? Does your answer seem reasonable based on the painting? Why or why not?

Hexa-Tri-C (1983), Victor Vasarely

35.* Goldie Auric's girlfriend, Silvia Silvers, is just back from a vacation. She has brought back a large, 240-gram gold medallion as a gift for Goldie. The tourist shop she bought the medallion from said it was 22-carat gold ($\frac{22}{24}$, or $\frac{11}{12}$, pure gold and $\frac{1}{12}$ another alloy). Silvia didn't believe this was true, but it was such a great bargain that she couldn't resist. Because the medallion has a rosy tint, the alloy is probably copper. Goldie places the medallion into a graduated cylinder and records the displacement. If it is 22-carat gold, what should the displacement be? If it is only 12-carat gold, what should the displacement be?

36. Draw a right triangle, then sketch the three solids of revolution formed by rotating it around each of its three sides.

Assessing What You've Learned

Try one or more of the following assessment suggestions.

Update Your Portfolio

• Choose a project, a Take Another Look activity, or one of the more challenging problems or puzzles you did in this chapter to add to your portfolio. Be sure to choose a problem for which you can explain your solution. Document it according to your teacher's instructions.

Write in Your Journal

• Describe your own personal problem-solving approach. Are there certain steps you follow when you solve a challenging problem? What are some of your most successful problem-solving strategies?

Organize Your Notebook

• Review your notebook to be sure it's complete and well organized. Be sure you have all the definitions and the conjectures, illustrated with diagrams, in your definition and conjecture lists. Write a one-page summary of Chapter 11.

Performance Assessment

• While a classmate, a friend, a family member, or a teacher observes, demonstrate how to derive one or more of the volume formulas. Explain what you're doing at each step, including how you arrived at the formula.

Write Test Items

• Work with group members to write test items for this chapter. Include simple exercises and complex applications problems. Try to demonstrate more than one approach in your solutions.

Give a Presentation

• Create a poster, a model, or visual aids for the overhead projector, and give a presentation about one or more of the volume conjectures.

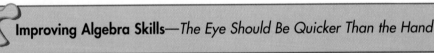

Improving Algebra Skills—*The Eye Should Be Quicker Than the Hand*

How fast can you answer these questions?

1. If $2x + y = 12$ and $3x - 2y = 17$, what is $5x - y$?
2. If $4x - 5y = 19$ and $6x + 7y = 31$, what is $10x + 2y$?
3. If $3x + 2y = 11$ and $2x + y = 7$, what is $x + y$?

Cooperative Problem Solving

Once Upon a Time

You and your CPS team are competing in a school-wide cooperative problem solving event at the lunar colony high school. The high school's geometry teacher, creative writing teacher, and art teacher have created a set of four illustrated geometry problems. The problems are challenging. They are also pure fantasy, and they are intended to set a story-telling mood. Hopefully you will enjoy them and be inspired to create your own word problems.

Each CPS team has three tasks in the event.

Task 1   Work cooperatively in small groups to solve the four problems written by the lunar colony high school teachers. These problems are found on the next few pages.

Task 2   As a team, devise a geometric word problem. Your problem should have a story or setting, should demonstrate a need for an answer, and should use geometry in its solution.

Task 3   Illustrate your problem. While you may have to assign the final illustration to one member of your team, group members should cooperate on all three tasks in the event, including brainstorming about how to illustrate the problem.

Have fun!

1. Dragula, the infamous Transylvanian hemoglobin expert, wishes to calculate the size of the elephant heart he is about to transplant into the body of his ailing friend, Dr. Boris Nogoodnik. The Count places the frozen organ into a graduated cylinder whose diameter measures 18 cm. It contains a 60/40 mix of alcohol and water at 2.5°C. If the liquid level rises 5 cm, what is the volume of the elephant heart? If the elephant heart weighs 2.4 kg, what is the density of the heart (to the nearest 0.1 g/cm^3)? (Boris survived the transplant, but he ran away and joined the circus!)

Count Dragula

2. Secret Agent Carmen Serita Castillo has finally located the fortress of her archenemy, Evil McNasty. She scales a wall, locates the power supply to the fortress, and sets a timer to shut down the fortress's power at midnight. She locates a rooftop door and drops through the door into a darkened room. However, she finds that the villainous McNasty has trapped her at the bottom of a stainless steel vault 2 m × 2 m × 6 m deep. Suddenly, water begins pouring into the vault, filling the small space to a depth of 1.5 m. Almost immediately, a pair of opposite walls start to slowly and inexorably move towards her. After a few minutes of observation, she calculates that the walls are moving at a rate of 5 cm/min. Checking her watch, Carmen Serita determines that the power will shut down in 15 min, stopping the walls dead in their tracks. Will Carmen Serita Castillo be able to float to the top of the rising column of water, reach the door, and escape? Will she be crushed by the walls or doomed to a watery end? What will the water level be when the power goes off?

3.* Young Princess Ali of Dristan had just completed a long and hazardous journey. After years of fighting fire-breathing dragons, she had finally approached her quest: the Mountain of Doom and its cave containing the fabulous treasure of King Caitiff. As Ali was about to enter the fabled cave, there was a tremendous flash of red light and smoke. Out from behind the smoke stepped an old man. "Advance no further! Behold, I am Algebar, the protector of the treasure of Caitiff. You have traveled far, but you must pass one last test before you can lay claim to the treasure." Algebar placed his outstretched hands palms up and shouted, "Pythagoras!" Two puffs of smoke arose. As the smoke cleared, a brilliant gem appeared in each of his hands. One gem, a ruby, was in the shape of a cube; the other, a diamond, was a regular octahedron. "If each edge of this ruby is three fourths as long as each edge of this diamond, which is larger?" questioned Algebar. "Answer correctly and the treasure is yours." The princess answered correctly and claimed the fabulous treasure. If you had been in her spot, would you have been able to figure out which gem had the greater volume? Which gem is larger?

Ali of Dristan and the Treasure of King Caitiff

4. Lord Darman has promised his friend and ally King Bandalf the Elder a priceless gift if he can solve this puzzle.

> I am a shining prism of ivory. Six-eight-twelve my measure be. Pierced through am I by nonintersecting prisms three, the longest being triangular on its face; the shortest being rectangular on its base; the third, isosceles trapezoidal in its place.
>
> The triangle measures three by four by five; the rectangle exactly two by five. The trapezoid is the strangest of them all. Its sides measure two, four, $\sqrt{2}$ twice; quite irrational.
>
> What part of me is missing? For that is our quest: The ratio of what is missing to that of what is left.

Bandalf journeys to the ancient kingdom of Pyria to seek help. There grows the tree where the great Simurgh nests. The Simurgh is a bird of wisdom and, according to ancient tales, has metallic feathers, a peacock's tail, and a small silver head. Rulers, sages, and scholars travel to Simurgh to gain the wisdom of the ages.

Simurgh gives Bandalf a diagram and says, "Now it shall be clear, for none of the piercing prisms intersect. Thus calculate the missing volume and the present volume. Their simple ratio is the answer you seek." What is the prized ratio?

Bandalf the Elder

12 Similarity

Figures that have the same shape are similar. In this chapter you will discover some of the basic properties of similarity and then will use them to solve problems. You will be asked to find the heights of trees, flagpoles, and buildings by measuring shadows and by using similar right triangles. Similarity is important in industry; it is used in the fields of film, photography, optics, architecture, and integrated circuits. Similarity helps explain why elephants have big ears and why movie monsters such as King Kong cannot exist. Solving similarity proportions is a useful skill for working in the fields of chemistry, physics, and medicine.

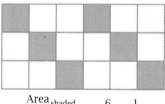

Lesson 12.1

Ratio and Proportion

The amount a person uses his [or her] imagination is inversely proportional to the amount of punishment he [or she] will receive for using it.
— *Anonymous*

Similarity plays a significant role in human history. For example, accurate maps of regions of China have been found dating back to the second century B.C. The cartographers who created these maps must have used principles of similarity to be so accurate. Neolithic cave paintings are small-scale drawings of animals people hunted. Giant geo-

glyphs like the monkey shown at left, made by ancient Peru's Nazca people (110 B.C.–800 A.D.), are some of the largest scale drawings ever created. Some of these animal figures measure over 400 feet long, and their shapes can be seen only from the air. Creating drawings of such scale was possible to do on the ground by using the principles of similarity you'll learn about in this chapter.

The study of similar geometric figures involves ratios and proportions. You may be a little rusty working with ratios and proportions, so let's review. What is a ratio?

*A **ratio** is an expression that compares two quantities by division.*

If a and b are two numbers, then the ratio of a to b is written "a/b." The ratio of a to b is also written "$a{:}b$" or "a is to b."

Example A

Find the ratio of the shaded to the unshaded area.

$$\frac{\text{Area}_{\text{shaded}}}{\text{Area}_{\text{unshaded}}} = \frac{6}{12} = \frac{1}{2}$$

Example B

Find the ratio of 🦎 to 〰️.

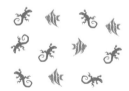

The ratio of 🦎 to 〰️ is $\frac{7}{4}$.

When two ratios are equal, you have a proportion. The expression $\frac{3}{4} = \frac{6}{8}$ is a proportion.

*A **proportion** is a statement of equality between two ratios.*

Proportions are used to solve problems that involve comparing similar objects or situations. You may remember how to solve for a variable in an equation involving fractions. If you have forgotten, one approach is to cross-multiply: if $\frac{a}{b} = \frac{c}{d}$, then $ad = bc$. If one fraction is a multiple of the other, you may use a more direct method. Let's look at a few examples.

Example C

If you work for two weeks and earn $380, how much will you earn in 15 weeks?

$$\frac{380}{2} = \frac{x}{15}$$
$$2x = (380)(15)$$
$$x = \frac{(380)(15)}{2}$$
$$x = 2850 \qquad \text{You will earn \$2850 in 15 weeks.}$$

Example D

Solve for x in $\frac{26}{50} = \frac{x}{75}$.

Before you cross-multiply, ask yourself, Can I reduce fractions? In this case, you can. Rewrite $\frac{26}{50}$ as $\frac{13}{25}$. You then get the equation $\frac{13}{25} = \frac{x}{75}$.

Next, before you cross-multiply, check to see if one numerator (or denominator) is a multiple of the numerator (or denominator) in the other fraction. In this problem, because $25 \cdot 3 = 75$, x must be $13 \cdot 3$, or 39.

Therefore $x = 39$.

Exercise Set 12.1

1. Find the ratio of 🌲 to 🌳.

2.* Find the ratio of the shaded area to the area of the whole figure.

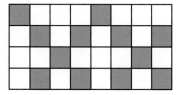

3. Use the figure below to find these ratios: *AC/CD, CD/BD,* and *BD/BC.*

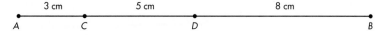

4. Find the ratio of the perimeter of triangle *RSH* to the perimeter of triangle *MFL*.

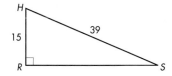

5. Find the ratio of the area of triangle *RSH* to the area of triangle *MFL*.

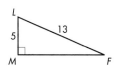

In Exercises 6–11, find the missing number in each proportion.

6. $\frac{7}{21} = \frac{a}{18}$

7. $\frac{10}{b} = \frac{15}{24}$

8. $\frac{20}{13} = \frac{60}{c}$

9. $\frac{4}{5} = \frac{x}{7}$

10.* $\frac{2}{y} = \frac{y}{32}$

11.* $\frac{10}{10 + z} = \frac{35}{56}$

Use a proportion to solve each of Exercises 12–18.

12. A car travels 106 miles on 4 gallons of gas. How far can it be expected to travel on a full tank of 12 gallons?

13. If a 425-pound lunar vehicle weighs 68 pounds on the moon, how much does 150-pound astronaut Luna Luz weigh on the moon? How much would you weigh on the moon?

14.* Pitcher Ernie Runz gave up 34 runs in 106 innings of baseball. What is Ernie Runz's earned run average? In other words, how many runs would he give up in 9 innings? Give your answer accurate to two decimal places.

15. A recipe for 6 dozen cookies calls for $2\frac{1}{2}$ cups of flour. How many cups of flour are needed for 10 dozen cookies?

16. The floor plan of a house is drawn to the scale of $\frac{1}{4}'' = 1'$. The master bedroom measures $3''$ by $3\frac{3}{4}''$ on the blueprints. What is the actual size of the room?

17. The famous consulting detective Hemlock Bones is studying the evidence in the theft of a rare stamp. His client, Sir Osborne Chatsworth III, has an envelope with an 1889 Belgian 6-franc commemorative stamp on it. When Chatsworth bought the stamp, he believed it to be valuable, but an appraiser, Thaddeus O'Malley, claims it is nearly worthless. Hemlock recalls that there were two types of Belgian commemorative stamps made in 1889, slightly different in size. The stamp on Chatsworth's envelope, unfortunately, is indeed the worthless one, measuring 3 cm square. The valuable stamp measures 4 cm square. Prompted by Hemlock's questioning, Chatsworth admits that he let Thaddeus out of his sight during the appraisal. Hemlock suspects that Thaddeus may have removed the valuable stamp from the envelope, replacing it with the worthless stamp! But how can he prove this theory? Fortunately, there's another important piece of evidence: a photograph of the envelope taken for insurance purposes the day before the stamp was appraised. Hemlock must figure out the actual size of the stamp in the photo to determine whether or not Thaddeus made the switch. The length of the actual envelope is 24 cm. In the photo the envelope measures 1.6 cm in length. The stamp in the photo measures 0.2 cm on each edge. Is the stamp on the envelope in the photo rare or worthless? Is Thaddeus O'Malley guilty of the old stamp switch swindle?

The case of the Belgian stamp theft

18. The slope of a roof (pitch) is the ratio of the "rise" of the roof over the "run" of the roof. Say the pitch of a roof is $\frac{8}{15}$. (This ratio is sometimes referred to as an 8-in-15 roof.) If a house 32 feet wide has a 3-in-8 peaked roof what is the total rise of the roof?

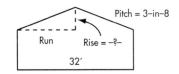

19. Population density is an important ratio. The population density of, say, a city is the ratio of that city's population to its area. Which of the major cities or regions listed in the table has the greatest population density? (Source: *1996 Information Please Almanac*)

Metropolitan area (1995)	Population	Area (mi²)
Tokyo, Japan	28,447,000	1089
Mexico City, Mexico	23,913,000	522
São Paulo, Brazil	21,539,000	451
Bombay, India	13,532,000	95
Hong Kong	5,841,000	23
New York City, United States	14,638,000	1274

Improving Reasoning Skills—*Abbreviated Equations II*

Each equation below contains the first letters of words that will make the equation correct. For example, 12 = M. in a Y. is an abbreviation of the equation 12 = Months in a Year. Find the missing words in each equation.

1. 360 = D. in the I.A. of a Q.
2. 60 = D. in each A. of an E.T.
3. 180 = D. in a S.
4. π = R. of C. to D. in every C.

Improving Visual Thinking Skills—*Build a Two-Piece Puzzle*

Construct two copies of Figure A, shown at right. Here's how to construct the figure.

- Construct a regular hexagon.
- Construct an equilateral triangle on two alternating edges, as shown.
- Construct a square on the edge between the two equilateral triangles, as shown.

Cut out each copy and fold them along the dashed lines into two identical solids, as shown in Figure B. Tape the edges. Now arrange your two solids to form a regular tetrahedron.

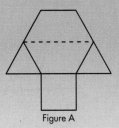

Figure A

Figure B

Lesson 12.2
Similarity

He that lets the small things bind him
Leaves the great undone behind him.
— Piet Hein

Many toys are scale models of real objects, but scale models have many applications besides toys.

Similarity plays an important part in the construction of such large objects as cars and trucks and such small objects as integrated circuits. Before building a car, engineers design scale drawings of the car, use the scale drawings to build scale models, then run tests with the scale models. To fabricate integrated circuits, electrical engineers use a computer to create a large-scale map of the integrated circuit. They then reduce the circuit design and transfer it onto minute silicon chips.

Movies are comprised of scenes scaled down to small images on a piece of film. These images are then scaled way up again to a large screen. Similarity is used not only in the movie-making process, but also as a theme of many movies. The movie *King Kong* was about a giant 30-foot gorilla similar in shape to real gorillas. In the movie *Fantastic Voyage,* people in a submarine were shrunk proportionally until they were small enough to be injected into the bloodstream of another person. Shrinking people may be a far-fetched idea, but current research in nanotechnology is making the possibility of tiny machines taking such a voyage more real than fantastic.

Figures that have the same shape and the same size are congruent figures. Figures that have the same shape but not necessarily the same size are **similar figures**. This, however, is not a precise definition for similarity. What does it mean to be the same shape?

A person looking at reflections in different fun-house mirrors sees different images. Do we want to say that the images are similar to the original? They certainly have a lot of features in common, but they are not similar in a mathematical sense. Similar shapes can be thought of as enlargements or reductions with no irregular distortions. If you can place figure A on a photocopy machine and enlarge or reduce it to fit perfectly over figure B, then figure B and the original figure A are similar. Are all rectangles similar? They certainly have many common characteristics, but they are not all similar because some cannot be enlarged or reduced to fit perfectly over others. What about other geometric figures? All squares are similar to one another and all circles are similar to one another, but all triangles are not similar to one another. In this lesson you will arrive at a mathematical definition for similar polygons.

The two illustrations in Investigation 12.2.1 represent the enlargement and the reduction of an impossible three-dimensional solid.

Let's use the enlarged and reduced figures to see what makes polygons similar. Even though the impossible figures appear to be three-dimensional, please try to see each figure as just a two-dimensional figure.

Investigation 12.2.1

Use patty papers to compare all corresponding angles. Measure the corresponding segments in both hexagons. Find the given ratios of the lengths of corresponding sides.

Is $\angle ABC \cong \angle PQR$?   How do the other corresponding angles compare?

Is $\angle BCD \cong \angle QRS$?

$AB \approx$ –?–	$BC \approx$ –?–	$CD \approx$ –?–
$EF \approx$ –?–	$PQ \approx$ –?–	$QR \approx$ –?–
$RS \approx$ –?–	$TU \approx$ –?–	$\frac{AB}{PQ} \approx$ –?–
$\frac{BC}{QR} \approx$ –?–	$\frac{CD}{RS} \approx$ –?–	$\frac{EF}{TU} \approx$ –?–

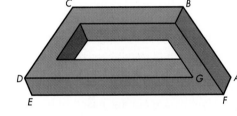

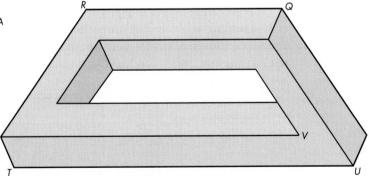

How do the corresponding sides of similar polygons compare? From your observations, you should be able to state that if two polygons are similar, then their corresponding angles are congruent and their corresponding sides are proportional. This statement is reversible. That is, if you construct two polygons that have corresponding angles congruent and corresponding sides proportional, then one polygon is an enlargement or a reduction of the other. In other words, one polygon is similar to the other. Based on these observations, let's state a more mathematical definition for similar polygons.

*Two polygons are **similar polygons** if and only if the corresponding angles are congruent and the corresponding sides are proportional.*

The symbol for the words *is similar to* is ~. You use this symbol in the same way you use the symbol for congruence. If the two quadrilaterals *CORN* and *MAIZ* are similar, you write "*CORN ~ MAIZ*." Just as in statements of congruence, the order of the letters tells you which segments and which angles in the two polygons correspond.

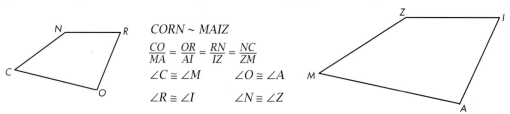

CORN ~ MAIZ

$$\frac{CO}{MA} = \frac{OR}{AI} = \frac{RN}{IZ} = \frac{NC}{ZM}$$

$\angle C \cong \angle M \qquad \angle O \cong \angle A$

$\angle R \cong \angle I \qquad \angle N \cong \angle Z$

In this investigation you discovered that two polygons are similar if and only if their corresponding angles are congruent and the corresponding sides are proportional. Do we need both conditions to guarantee that the two polygons are similar? In other words, if you know only that the corresponding angles of two polygons are congruent, can you conclude that the polygons have to be similar? Or, if corresponding sides of two polygons are proportional, are the polygons necessarily similar?

Polygons with corresponding angles congruent

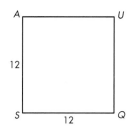

$\angle S \cong \angle R$
$\angle Q \cong \angle E$
$\angle U \cong \angle C$
$\angle A \cong \angle T$
However, $\frac{12}{10} \neq \frac{12}{18}$.

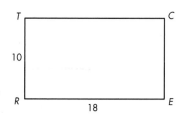

Corresponding angles of square *SQUA* and rectangle *RECT* are congruent, but their corresponding sides are not proportional. Clearly the two polygons are not similar. Therefore you cannot determine whether or not two polygons are similar based only on the fact that the corresponding angles of the two polygons are congruent.

Polygons with corresponding sides proportional

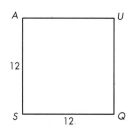

$$\frac{12}{18} = \frac{12}{18}$$
However, $\angle S \not\cong \angle R$.

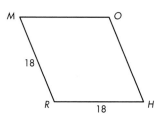

Corresponding sides of square *SQUA* and rhombus *RHOM* are proportional, but their corresponding angles are not congruent. Clearly the two polygons are not similar. Therefore you cannot determine whether or not two polygons are similar based only on the fact that the corresponding sides of the two polygons are proportional.

You have just discovered from the two counterexamples above that to determine whether or not two polygons are similar, you must know both that the corresponding sides of the two polygons are proportional *and* that the corresponding angles are congruent.

You can use the definition of similar polygons to find missing measures in similar polygons.

Example A

Find the measure of the side labeled x and the measure of the angle labeled y in the similar polygons below.

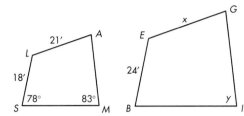

$SMAL \sim BIGE$

$$\frac{18}{24} = \frac{21}{x}$$

$$\frac{3}{4} = \frac{21}{x}$$

$$3x = (4)(21)$$

$$x = 28$$

So the measure of the side labeled x is 28′.

$$\angle M \cong \angle I$$

So the measure of the angle labeled y is 83°.

You can also use the definition of similar polygons to determine whether two polygons are similar if you know the measures of their angles and the lengths of their sides. Look at Example B.

Example B

Determine whether or not the polygons below are similar.

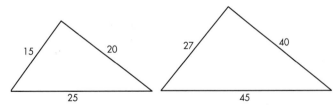

$\frac{15}{27} = \frac{5}{9}$ and $\frac{25}{45} = \frac{5}{9}$,
but $\frac{20}{40} = \frac{1}{2}$.

Therefore the triangles are not similar.

An image on movie film and the image projected onto the screen are similar (as long as the movie is projected at right angles to the screen). For best projection results, a projector is placed a fixed distance away from a screen. If the projector is moved half this distance to the screen, each dimension of the usual image is cut in half. If the projector is moved three times this fixed distance away from the screen, the dimensions of the new image are all three times its usual size.

Earlier in this text you worked with translations, rotations, and reflections. These rigid transformations, or isometries, preserved both size and shape. In rigid transformations, the images are congruent to the original figures. In this lesson you will investigate nonrigid transformations called **dilations**. Is an image in a dilation transformation similar to its original figure? Let's see.

Investigation 12.2.2

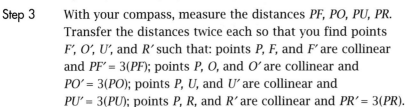

Complete Steps 1–5 to construct a dilation of a quadrilateral.

Step 1   Construct a quadrilateral *FOUR* in the upper third of your paper and place a point *P* above it, as shown at right. (The quadrilateral is like an image on movie film. Point *P* is like the projector's light source and is called the **center of dilation**.)

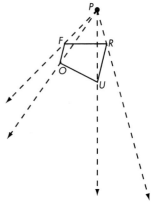

Step 2   From point *P* construct four rays: $\overrightarrow{PF}$, $\overrightarrow{PO}$, $\overrightarrow{PU}$, $\overrightarrow{PR}$. Extend the rays to the end of your paper. (The rays are like the light rays of the projector lamp.)

Step 3   With your compass, measure the distances *PF, PO, PU, PR.* Transfer the distances twice each so that you find points *F′, O′, U′,* and *R′* such that: points *P, F,* and *F′* are collinear and *PF′* = 3(*PF*); points *P, O,* and *O′* are collinear and *PO′* = 3(*PO*); points *P, U,* and *U′* are collinear and *PU′* = 3(*PU*); points *P, R,* and *R′* are collinear and *PR′* = 3(*PR*).

Step 4   You have now located four points, *F′, O′, U′,* and *R′*, that are each three times as far from point *P* as the original four points of the quadrilateral. Construct quadrilateral *F′O′U′R′*. Quadrilateral *F′O′U′R′* is the image of *FOUR* under a dilation with center *P* and a scale factor of 3. (Like the images on the film and the screen, quadrilaterals *FOUR* and *F′O′U′R′* appear to be similar. Are they?)

Step 5   Copy the original quadrilateral onto a patty paper. Compare the corresponding angles of the two quadrilaterals. Are they congruent? Compare the corresponding sides with a compass or a patty paper. Each side of the new quadrilateral is how many times larger than the corresponding side of the original?

Before you make a conjecture about dilation transformations and similar figures, let's look at dilations on the coordinate plane.

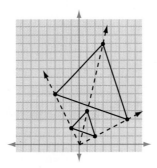   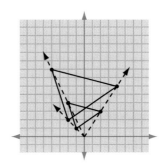

In earlier exercises you worked with translations, rotations, and reflections on the coordinate plane. You discovered that a rule such as $(x, y) \rightarrow (x + 3, y - 2)$ gave you a translation and that a rule such as $(x, y) \rightarrow (x, -y)$ gave you a reflection. You also discovered that a rule such as $(x, y) \rightarrow (2x, 5y)$ gave you a shape that was larger than (but not similar to) the original, while $(x, y) \rightarrow (\frac{2}{3}x, \frac{2}{3}y)$ gave you a shape that was smaller than (and appeared to be similar to) the original. In Investigation 12.2.3, you will perform a dilation centered at the origin of the coordinate plane to see if it creates a similar figure.

Investigation 12.2.3

Complete the steps below to dilate a pentagon on the coordinate plane.

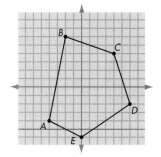

Step 1   Determine the coordinates of the vertices of the pentagon at right and copy it onto your graph paper.

Step 2   Have each member of your group multiply the coordinates of the vertices by one of the following numbers: $\frac{1}{2}$, $\frac{3}{4}$, 2, or 3. (Each member should choose a different number.) Each of these factors is called the scale factor.

Step 3   Locate these new coordinates on your graph paper and draw the new pentagon.

Step 4   Copy the original pentagon onto a patty paper. Compare the corresponding angles of the two pentagons. Are they congruent? Compare the corresponding sides with a compass or a patty paper. The length of each side of the new pentagon is what fraction of the length of the corresponding side of the original?

Compare your results with the results of others near you. You constructed polygons by dilation. Were the images created in the dilations similar to the originals? You should be ready to state a conjecture.

C-101   If one polygon is the image of another polygon under a dilation, then —?— (***Dilation Similarity Conjecture***).

Take Another Look 12.2

Try one or more of these follow-up activities.

1. In Investigation 12.2.3, you dilated figures in the coordinate plane, using the origin as the center of dilation. What if another point in the plane is the center of dilation instead? Copy the polygon at right onto graph paper. Draw the polygon's image under a dilation with a scale factor of 2 and with point A the center of dilation. Repeat using a scale factor of 2/3. Explain how you found the image points. How does dilating about point A differ from dilating about the origin?

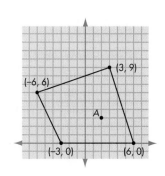

2. You've learned that an ordered pair rule such as $(x, y) \rightarrow (x + b, y + c)$ created a translation. You discovered in this lesson that an ordered pair rule such as $(x, y) \rightarrow (kx, ky)$ created a dilation in the coordinate plane, centered at the origin. What will the rule $(x, y) \rightarrow (kx + b, ky + c)$ yield? Investigate.

3. Use a geometry computer program to investigate dilations. What happens if you dilate by a scale factor less than 0?

Exercise Set 12.2

For Exercises 1 and 2, identify any figure at right that is a reduction or an enlargement of a figure at left.

1.

a.
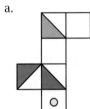

b.
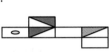

c.

2.

a.

b.

c.

For Exercises 3–5, sketch on graph paper a figure similar, but not congruent to, each figure shown.

3.
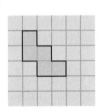

4.

5.

For Exercises 6–8, use algebra to answer each proportion question.

6. If $\frac{15}{a} = \frac{20}{a + 12}$,
then $a = $ –?–.

7. If $\frac{a}{b} = \frac{c}{d}$,
then $ad = $ –?–.

8. If $\frac{a}{b} = \frac{c}{d}$,
then $\frac{b}{a} = $ –?–.

9. Complete the statement: If figure A is similar to figure B and figure B is similar to figure C, then —?—. Draw and label figures to illustrate the statement.

10. Jade and Omar each chipped in $1000 to buy an old boat to fix up. Jade spent $825 on materials, and Omar spent $1650 for parts. They worked an equal number of hours on the boat and eventually sold it for $6,800. How should they fairly divide the $6,800? Explain your reasoning.

11. Altar and Zenor are ambassadors from Titan, the largest moon of Saturn. The atmosphere on Titan is so dense that the Titans have evolved multiple antennae to pick out sound waves. Altar has revealed to the Biological Research Division that the sum of the lengths of a Titan's antennae is a direct measure of that Titan's age. Altar has antennae with lengths 8 cm, 10 cm, 13 cm, 16 cm, 14 cm, and 12 cm. Zenor is 130 years old, and her seven antennae have an average length of 17 cm. How old is Ambassador Altar?

Altar and Zenor

Use the definition of similar polygons to solve Exercises 12–19. All measurements are in centimeters.

12. * *THINK ~ LARGE*
Find *AL, RA, RG, KN.*

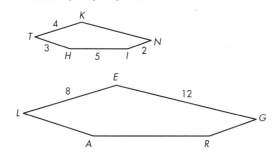

13. Are the polygons similar?

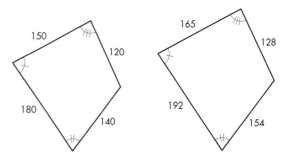

14. *SPIDER ~ HNYCMB*
Find *NY, YC, CM, MB.*

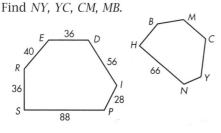

15. * Are the polygons similar?

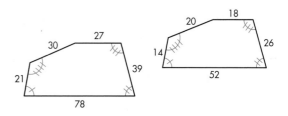

16. △*ACE* ~ △*IKS*
$x = $ –?–
$y = $ –?–

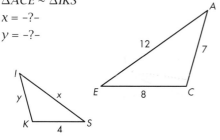

17. △*RAM* ~ △*XAE*
$z = $ –?–

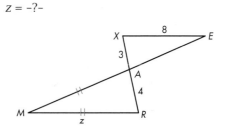

18.* $\overline{DE} \parallel \overline{BC}$

Are corresponding angles congruent?
Are corresponding sides proportional?
Is $\triangle AED \sim \triangle ABC$?

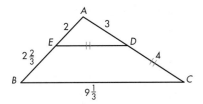

19. $\triangle ABC \sim \triangle DBA$

$m = -?-$
$n = -?-$

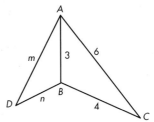

20. Use patty papers or a compass and a straightedge to construct a pentagon similar to the pentagon shown below. Make each side of your pentagon three times as large as its corresponding side in the original pentagon.

21. Copy the quadrilateral shown below onto your graph paper. Draw a similar quadrilateral with each side half the length of its corresponding side in the original quadrilateral.

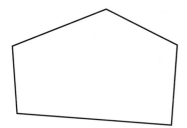

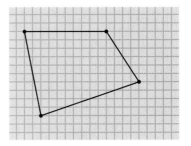

22. The world's largest sculpture carved out of a granite cliff face is that of four United States presidents at Mount Rushmore, located in the Black Hills of South Dakota. Each face is about 60 feet high. If an arm had been carved, how long would it be? Explain how you get your answer. By the way, who are the four presidents?

Making a Mural

*Every child is an artist. The problem is how
to remain an artist after he [or she] grows up.*
— *Pablo Picasso*

Mural artists use similarity to help them create their large artwork. Muralists begin creating a mural by drawing a small picture with a grid of squares drawn over it. They then divide the surface on which the mural will be painted into a similar but larger grid of squares. Proceeding square by square, they draw the lines and shapes of the original drawing into the corresponding positions of the mural surface's large squares. Finally, they paint in the regions to complete the mural.

The design in the small grid of squares below left is similar to the design in the large grid of squares below right. The enlargement was made by matching points in the original drawing to the corresponding points in the larger grid. For example, point *A* in the grid at left is in the same position as point *A'* is in the grid at right.

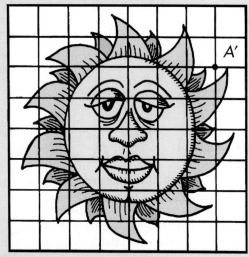

You will need the following materials to create your own mural.

• A cartoon, a photograph, or a small drawing to enlarge

• A large surface for your mural, such as a large sheet of butcher paper

• Drawing equipment, such as a ruler, a meter stick, marking pens, and colored pencils

Begin by constructing lightly in pencil a grid of squares on a photocopy of your cartoon, photograph, or drawing. The more squares you draw, the more accurately your mural will depict your original image. Again in pencil, divide the surface on which you are creating the mural into a similar grid of larger squares. To create your mural, carefully draw the lines and curves of the drawing in the small squares into their corresponding large squares. Finally, color the appropriate regions.

Lesson 12.3

Similar Triangles

In Chapter 5, you found four shortcuts (SSS, SAS, ASA, and SAA) for determining that two triangles are congruent. In Lesson 12.2, we were unable to find simple shortcuts for determining that quadrilaterals are similar. How about for triangles? Are there shortcuts to use for determining that two triangles are similar, like those used to determine that triangles are congruent?

If there are shortcuts, the two examples at left illustrate some limitations. If there is a shortcut using only angles, you need to investigate at least two angles in one triangle that are congruent to the two corresponding angles in the other triangle to see if they force the two triangles to be similar. If there is a shortcut using just sides, the example below left demonstrates that two sides are not sufficient; you need to investigate triangles in which all three pairs of corresponding sides are proportional to see if they force the two triangles to be similar.

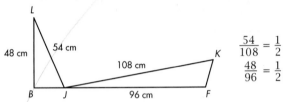

$\angle A \cong \angle D$, but $\triangle ABC$ is not similar to $\triangle DEF$.

$$\frac{54}{108} = \frac{1}{2}$$

$$\frac{48}{96} = \frac{1}{2}$$

$\frac{LJ}{JK} = \frac{LB}{JF}$, but $\triangle LBJ$ is not similar to $\triangle JFK$.

The three investigations in this lesson work best if you work in groups of four or five. Each group member should start with a different triangle and complete the steps outlined for the investigation. Share your results with the other members of your group, and together make a conjecture based on the investigation.

Investigation 12.3.1

If three sides of one triangle are proportional to the three sides of another triangle, must the two triangles be similar? On a sheet of paper, complete Steps 1–3.

Step 1   Draw a triangle.

Step 2   Construct a second triangle whose side lengths are some multiple of the original triangle. (Your second triangle can be larger, with side lengths twice, three times, or four times the original side lengths. Or, it can be smaller, with side lengths one-half, one-third, or one-fourth as long as the original side lengths.)

Step 3   With a protractor, compare the corresponding angles of the two triangles.

Compare your results with the results of others near you. You constructed two triangles whose three pairs of sides are in the same ratio. If this forced the angles to be congruent, then both conditions for similar polygons hold and the two triangles are similar. You should be ready to state a conjecture.

C-102   If the three sides of one triangle are proportional to the three sides of another triangle, then the two triangles are —?— (***SSS Similarity Conjecture***).

Investigation 12.3.2

If two angles of one triangle are congruent to two angles of another triangle, which forces the third pair of angles to be congruent, must the two triangles be similar? On a sheet of paper, complete Steps 1–3.

Step 1 Draw a triangle *ABC*.

Step 2 Construct a second triangle, *DEF*, with $\angle D \cong \angle A$ and $\angle E \cong \angle B$, which forces $\angle F \cong \angle C$ because the measures of three angles must add to 180°.

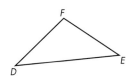

Step 3 Carefully measure the lengths of the sides of both triangles. With the aid of a calculator, calculate and compare the ratios of the corresponding sides. Is $\frac{AB}{DE} \approx \frac{AC}{DF} \approx \frac{BC}{EF}$?

Compare your results with the results of others near you. You constructed two triangles with two pairs of angles congruent, which forced the third pair of angles to also be congruent. If this forced the ratios of the corresponding sides of the two triangles to be equal, then both conditions for similar polygons hold, and it follows that the two triangles are similar. You should be ready to state a conjecture.

C-103 If —?— angles of one triangle are congruent to —?— angles of another triangle, then —?— (***AA Similarity Conjecture***).

In Investigations 12.3.1 and 12.3.2, you discovered two shortcuts for demonstrating similarity between triangles. There remain two possible cases to investigate. Because AA is a shortcut, then ASA, AAS, and AAA are automatically shortcuts. Because SSS is a shortcut, this leaves only SAS and SSA as possible shortcuts that use only three parts of a triangle.

Investigation 12.3.3

For each case, demonstrate that it does or does not force two triangles to be similar.

Case 1 Is SAS a shortcut for similarity? Try to construct two different triangles that are not similar but that have two pairs of sides proportional and that have the pair of included angles equal in measure.

Case 2 Is SSA a shortcut for similarity? Try to construct two different triangles that are not similar but that have two pairs of sides proportional and that have a pair of corresponding nonincluded angles equal in measure. (See Lesson 5.4 on SSA congruence.)

Share your results with others near you. You should be ready to state a conjecture.

C-104 If two sides of one triangle are proportional to two sides of another triangle and —?—, then the two triangles are similar (—?— ***Similarity Conjecture***).

Take Another Look 12.3

Try one or more of these follow-up activities.

1. Are all isosceles triangles similar? Explain why or give a counterexample to show why not. Are all isosceles triangles with base angles that measure 50° similar? Explain.

2. A friend tells you, "My science teacher says that we get a total eclipse of the sun because the ratio of the moon's diameter to its distance from the earth is about the same as the ratio of the sun's diameter to its distance to the earth. But I don't understand how this works. Can you explain?" Draw a diagram and use similar triangles to explain how this works.

3. It is possible for the three angles and two of the sides of one triangle to be congruent to the three angles and two of the sides of another triangle, and yet the two triangles won't be congruent. Two such triangles are shown at right. Use a geometry computer program or patty papers to find another pair of similar (but not congruent) triangles in which five parts of one are congruent to five parts of another.

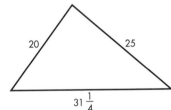

Exercise Set 12.3

Use your new conjectures to solve Exercises 1–14. All measurements are given in centimeters.

1. $g = $ –?–

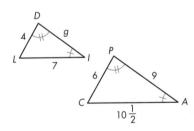

2. $h = $ –?–
   $k = $ –?–

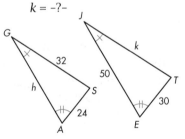

3.* $m = $ –?–

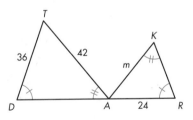

4. $n = $ –?–
   $s = $ –?–

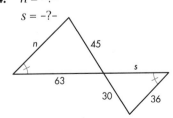

5. Is $\triangle AUL \sim \triangle MST$? Explain.

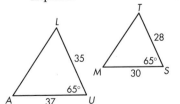

6. Is $\triangle PHY \sim \triangle YHT$? Is $\triangle PTY$ a right triangle? Explain.

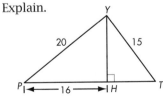

7.* Is $\triangle MOY \sim \triangle NOT$? Explain.

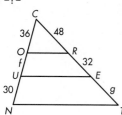

8.* Why is $\triangle TMR \sim \triangle THM \sim \triangle MHR$?
Is $\frac{x}{32} = \frac{32}{60}$? Is $\frac{y}{60} = \frac{60}{68}$?
Find x, y, and h.

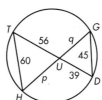

9. $\overline{TA} \parallel \overline{UR}$
Is $\angle QTA \cong \angle TUR$?
Is $\angle QAT \cong \angle ARU$?
Why is $\triangle QTA \sim \triangle QUR$?
$e = -?-$

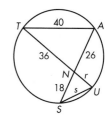

10. $\overline{OR} \parallel \overline{UE} \parallel \overline{NT}$
$f = -?-$
$g = -?-$

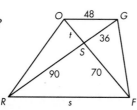

11.* Is $\angle THU \cong \angle GDU$?
Is $\angle HTU \cong \angle DGU$?
Why is $\triangle HUT \sim \triangle DUG$?
$p = -?-$
$q = -?-$

12.* Why is $\triangle SUN \sim \triangle TAN$?
$r = -?-$
$s = -?-$

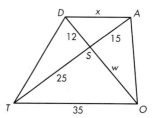

13. *FROG* is a trapezoid.
Is $\angle RGO \cong \angle FRG$?
Is $\angle GOF \cong \angle RFO$?
Why is $\triangle GOS \sim \triangle RFS$?
$t = -?-$
$s = -?-$

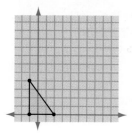

14. *TOAD* is a trapezoid.
$w = -?-$
$x = -?-$

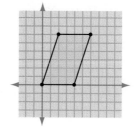

15. Sketch and label two rectangles that are not similar.

16. Sketch and label two isosceles trapezoids that are similar.

For Exercises 17 and 18, use the ordered pair rule to relocate the coordinates of the vertices of each polygon. Is the new figure similar to the original? If they are similar, what is the ratio of the perimeter of the original polygon to the perimeter of the new polygon? What is the ratio of the area of the original polygon to the area of the new polygon?

17. $(x, y) \rightarrow (3x, 3y)$

18. $(x, y) \rightarrow (\frac{1}{2}x, \frac{1}{2}y)$

19. For the past year Phuong has been volunteering at a local SPCA. During that time she's noticed that she goes through seven 35-pound bags of dry dog food every two months and that the animal shelter always houses (at full capacity) eight dogs. Phuong has just transferred to a new, larger SPCA facility that recently opened. This facility can care for 20 dogs at a time. She has been asked to estimate how many pounds of dry dog food the facility should order every three months. Help her out. Explain your reasoning.

20. Oceanographers Taisuke and Sabina have arrived at a marine reserve to study the Hawaiian fish, the Humuhumunukunukuapua'a. They are going to use the

capture-tag-recapture method to determine the size of the Humuhumunukunukuapua'a population. They begin by capturing and tagging 84 Humuhumunukunukuapua'as, which they then release back to the bay. After waiting a week for the fish to swim around the bay and distribute themselves randomly among the general fish population, Taisuke and Sabina catch another 64 Humuhumunukunukuapua'a. Some of these fish are untagged, but twelve are from the tagged group.

From this information they (and you) can estimate the number of Humuhumunuku-nukuapua'a in the reserve. Try it. About how many Humuhumunukunukuapua'a are there in the reserve? As an added challenge, can you say *humu-humu-nuku-nuku-apua'a*?

21.* The Tibetan mandala shown at right is a complex design with a square inscribed within a circle and tangent circles inscribed within the corners of a larger square. In the figure below, find the radius, *r*, of one of the small circles in terms of *R*, the radius of the large circle.

Constructing a Dilation Design

Take a closer look at *Path of Life I*, the M. C. Escher woodcut that begins this chapter. You'll see that all the red fish-like creatures are similar to one another, as are all the white creatures. (The fish around the outside border are congruent to one another, but they're not similar to the other fish.) In each sector of the picture, a dilation transforms the shapes, shrinking them again and again as they approach the picture's center. In this project, you'll make a design using the same transformations—dilations and rotations—that appear in *Path of Life I*.

In Steps 1-9 below, you'll construct two circles and a 45° sector. In a portion of that sector, you'll create three simple shapes that you'll later use to fill in a design by dilating and rotating.

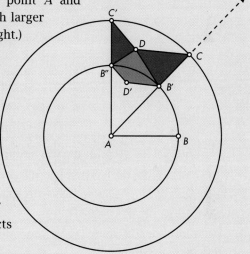

Step 1   Construct a large circle with center point *A* and radius point *B*. (Make your circle much larger than that in the figure shown at right.) Construct $\overline{AB}$.

Step 2   In the Transform menu, mark point *A* as center and rotate point *B* by 45°.

Step 3   Construct $\overrightarrow{AB'}$.

Step 4   Construct a larger circle with center point *A* and radius point *C*, where point *C* is on $\overrightarrow{AB'}$. Hide $\overrightarrow{AB'}$ and construct $\overline{AC}$.

Step 5   Rotate $\overline{AC}$, point *B'*, and point *C* by 45° about center point *A*. This constructs point *B''*, point *C'*, and $\overline{AC'}$.

Step 6   Construct $\overline{C'D}$ and $\overline{DC}$, where *D* is any point in the region between the circles and between $\overline{B''C'}$ and $\overline{B'C}$.

Step 7   Select, in order, $\overline{AB}$ and $\overline{AC}$. Choose Mark Ratio in the Transform menu. (You've marked a ratio of a shorter segment to a longer segment. Because this ratio is less than 1, dilating by this ratio will shrink objects.)

Step 8   Select $\overline{C'D}$, $\overline{DC}$, and point *D* and use the Transform menu to dilate by the marked ratio. (Segments *B''D'* and *D'B'* are the dilated images of $\overline{C'D}$ and $\overline{DC}$.

Step 9   Construct three polygon interiors—two triangles and a quadrilateral—as shown above.

Now you're ready to complete your design in Steps 10 and 11 by dilating and rotating the three polygon interiors you've made so far.

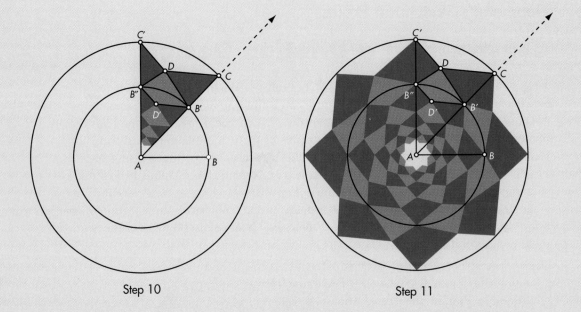

Step 10   Step 11

Step 10 Select the three polygon interiors and dilate them by the marked ratio. (Can you explain why the dilated set of polygons fits next to the original set without any gap or overlap?) Repeat two or three times until the sector is nearly filled.

Step 11 Select all the polygon interiors in the sector and rotate them by 45°. Repeat until you've gone all the way around the circle.

You've now got a design that has the same basic mathematical properties as Escher's *Path of Life I*. Experiment with changing the design by moving different points. Answer the questions below.

1. What locations of point *D* give you a design that has both rotational and reflectional symmetry?

2. You started with two triangles and a quadrilateral. Rotations combined the triangles (shown purple in the figures above) to make more quadrilaterals. Move point *D* anywhere to see how that changes the design. What locations for point *D* give you concave quadrilaterals? What locations for point *D* give you only triangles? What locations for point *D* cause the polygons to overlap?

3. Drag point *C* away from point *A*. What does this do to the dilation ratio? What effect does that have on the figure?

4. Drag point *C* toward point *A*. What happens to the figure when the circle defined by point *C* becomes smaller than the circle defined by point *B*? Explain why this happens.

5. Mathematician Doris Schattschneider of Moravian College is an expert on M. C. Escher and the mathematics he explored. She made this sketch based on *Path of Life I.* Try it yourself or experiment with other dilation-rotation designs of your own. (See activity 6 below.)

6. If you make your own design, here are some ideas you might experiment with.

- Try different angles of rotation. (What's the angle of rotation in the design below left?)

- Can you use polygons whose sides go outside the sector but that still fit together without gaps or overlap when you rotate? See the example below left.

- Make a design by using a two-step transformation: a dilation followed by a rotation. See the example below right. Each arm of this design has what's called **spiral similarity**. (Escher's *Path of Life I* and the design you made in the first part of this project also have spiral similarity. You can color or shade your original design to emphasize the spiral similarity.)

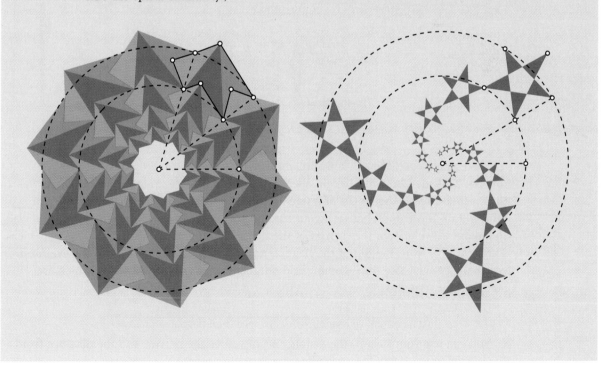

Lesson 12.4

Indirect Measurement with Similar Triangles

Similar triangles can be used to calculate the height of objects that you are unable to measure directly. Suppose that on a sunny day you and a friend wanted to find the height of a lamppost. The lamppost is probably too tall to measure directly. However, the light rays and the vertical lamppost and its shadow form a right triangle that is similar to the right triangle formed by the light rays, your vertical friend, and your friend's shadow. How tall is the lamppost? Because the triangles are similar,

$$\frac{\text{your friend's height}}{\text{your friend's shadow}} = \frac{\text{lamppost's height}}{\text{lamppost's shadow}}.$$

You can easily measure the shadows and your friend. The height of the lamppost is the only value in the proportion that you don't know. Calculate it.

Example A

If a person 5 feet tall casts a 6-foot shadow at the same time that a lamppost casts an 18-foot shadow, what is the height of the lamppost?

$$\frac{5}{6} = \frac{x}{18}$$

$$6x = (5)(18)$$

$$x = \frac{(5)(18)}{6}$$

$$x = 15$$

The lamppost is about 15 feet tall.

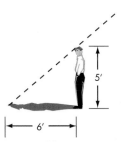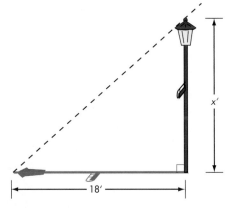

You can also use mirrors and similar triangles to measure indirectly the height of objects. This method will work on overcast days when there are no shadows. Suppose you needed to find the height of a flagpole. Place a mirror with cross hairs (an X) drawn on it flat on the ground between yourself and the flagpole. Look into the mirror and walk to a point at which you see the top of the flagpole lining up with the mirror's cross hairs. The light rays from the top of the flagpole to the mirror and back up to your eye form equal angles (just like the incoming and outgoing angles on a pool table), which are similar right triangles.

$$\frac{\text{your height}}{\text{distance from you to mirror}} = \frac{\text{flagpole's height}}{\text{distance from flagpole to mirror}}$$

You can use your height, the distance from you to the mirror, and the distance from the mirror to the flagpole to calculate the flagpole's height.

10. If 8 oz of dough are needed to make a 10-inch diameter pizza, how many ounces of dough are needed to make a 16-inch diameter pizza of the same thickness?

11. The pentagonal pyramids are similar.

Volume of small pyramid = 320 cm³

$\dfrac{h}{H} = \dfrac{4}{7}$

Volume of large pyramid = –?–

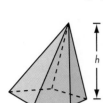 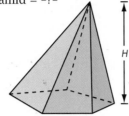

12.* The right cones are similar.

H = –?–   Volume of large cone = –?–

h = –?–   Volume of small cone = –?–

$\dfrac{\text{Volume of large cone}}{\text{Volume of small cone}}$ = –?–

 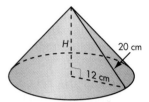

13.* The right trapezoidal prisms are similar.

Volume of small prism = 324 cm³

$\dfrac{\text{Area of base of small prism}}{\text{Area of base of large prism}} = \dfrac{9}{25}$

$\dfrac{h}{H}$ = –?–

$\dfrac{\text{Volume of large prism}}{\text{Volume of small prism}}$ = –?–

Volume of large prism = –?–

 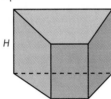

14.* The right cylinders are similar.

Volume of large cylinder = 4608π cu ft

Volume of small cylinder = –?–

$\dfrac{\text{Volume of large cylinder}}{\text{Volume of small cylinder}}$ = –?–

$\left(\dfrac{H}{24}\right)^3$ = –?– H = –?–

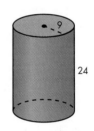 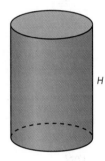

15.* The lengths of corresponding edges of two similar triangular prisms are in the ratio of 5:3. What is the ratio of their volumes?

16.* The volumes of two similar pentagonal prisms are in the ratio of 8:125. What is the ratio of their heights?

17.* The ratio of the weights of two spherical steel balls is 8:27. What is the ratio of their diameters?

18.* ZAP Electronics has just installed an air-conditioning unit in the small warehouse section of their VCR plant. The energy needed to operate an air-conditioning unit is proportional to the volume or space that is being air-conditioned. The energy used to operate the air-conditioning system costs ZAP Electronics about $125 per day. The company is considering installing a similar unit in its main storage warehouse. Each dimension of this warehouse is two and a half times as large as its corresponding dimension in the VCR warehouse. What would be a good estimate of the daily operating cost for the larger warehouse's air-conditioning system?

19. Make four copies of the trapezoid at right. Arrange them into a similar but larger trapezoid.

20. *Mert the metal sculptor has just created a small, solid steel statue of his singing idol, Melvis Bresley. The statue weighs 38 lbs. He plans to make a full-scale version of the statue, which will be four times as large in each dimension. How much will this larger statue weigh if Mert makes it out of solid steel?

Mert the metal sculptor

21. *Similar shapes are often used to display data graphically. However, we must be on guard for graphics that visually misrepresent data. For example, the graph below right shows one-dimensional data—the percentage of doctors in a family practice—displayed in a two-dimensional display. An observer may perceive that the area of each doctor graphic represents the data, but in fact, the heights of the graphics represent the percentage of doctors who are in a family practice. To see how much this graphic distorts the perception of the data, measure the height of each doctor to the nearest 0.1 cm, then find the following ratios.

$$\frac{\% \text{ doctors in '75}}{\% \text{ doctors in '64}} = \frac{16\%}{27\%} \approx 0.59 \qquad \frac{\text{height of graphic in '75}}{\text{height of graphic in '64}} \approx \frac{2.1}{3.6} \approx -?- \qquad \frac{\text{area of graphic in '75}}{\text{area of graphic in '64}} \approx \left(\frac{2.1}{3.6}\right)^2 \approx -?-$$

$$\frac{\% \text{ doctors in '90}}{\% \text{ doctors in '75}} = -?- \approx -?- \qquad \frac{\text{height of graphic in '90}}{\text{height of graphic in '75}} \approx -?- \approx -?- \qquad \frac{\text{area of graphic in '90}}{\text{area of graphic in '75}} \approx -?- \approx -?-$$

How do the ratios of percentages compare with the ratios of heights? Are they close? How do the ratios of the graphics' areas compare with the ratios of percentages of family doctors? Close? Or not so close? Which ratio, heights or areas, should be closer to the ratio of percentages if the graphic is not to be misleading? Write a paragraph explaining why you do or do not find the graphic misleading.

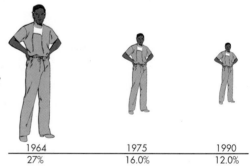

1964	1975	1990
27%	16.0%	12.0%

Improving Visual Thinking Skills—*Painted Faces II*

Unit cubes are assembled to form a larger cube, and then some of the faces of this larger cube are painted. After the paint dries, the larger cube is disassembled into the unit cubes and it is found that 60 of these have no paint on any of their faces. How many faces of the larger cube were painted?

Project

Why Elephants Have Big Ears

But yet it is easy to show that a hare could not be as large as a hippopotamus, or a whale as small as a herring. For every type of animal there is a most convenient size, and a large change in size inevitably carries with it a change of form.
— *J. B. S. Haldane*

In Lesson 12.6, you discovered a very important relationship of similar objects. If the linear dimensions of two solid objects are in the ratio of a to b or $\frac{a}{b}$, then their surface areas are in the ratio of a^2 to b^2 or $(\frac{a}{b})^2$, and their volumes are in the ratio of a^3 to b^3, or $(\frac{a}{b})^3$. The ratio of surface area to volume is of critical importance to all living things. This concept explains why elephants have big ears, why hippos and rhinos have short, thick legs and must spend a lot of time in water, and why movie monsters like King Kong or Godzilla can't exist. Let's take a closer look.

Body temperature All living things process nutrients to produce energy to stay alive. This energy production creates heat that must be radiated from each living thing's surface. Picture two similar creatures, the larger whose each dimension is 3 times larger than that of the smaller creature. The surface area of the larger is 9 times greater and thus has 9 times more surface area with which to radiate heat. That's good. However, the larger creature's volume (and thus the heat it produces) is 27 times greater. This means that each square centimeter of the larger creature's surface must radiate 3 times the heat per square centimeter of the smaller creature's surface. This may not be good.

Bone and muscle strength The strength of a bone or a muscle is proportional not to its mass or volume but to its cross-sectional area. If a youngster 42 inches tall were similar in shape to a 7-foot tall basketball player, then the youngster's bones and muscles would be half the height, half the width, and half the thickness of the ball player's corresponding bones and muscles. The cross-sectional areas of the basketball player's bones and muscles would be 4 times as great and thus could support 4 times the weight. That would be good. However, the basketball player would weigh 8 times the weight of the youngster. This means that each square inch of cross section of bone would have to support twice the weight. This might not be good. It is estimated that the human thigh bone breaks under 10 times the human body weight. Fortunately for basketball players, they are not typically similar in shape to youngsters. Because he would tend to be thin for his height, a 7-foot tall basketball player would be unlikely to weigh 8 times that of a youngster half his height.

Gravity and air resistance Objects falling in a vacuum fall at the same rate. However, an object in air is buoyed up by air resistance. This air resistance is proportional to the surface area of the falling object. Suppose a small mouse is one eighth the length of a large rat. If the two similar creatures fell a great distance (say from a twelve story building), the forces at which they hit bottom would be very different. Because the mass of the mouse is so small relative to air resistance caused by its cross sectional area, the mouse lands and survives. The mass of the rat is so much larger relative to its cross-sectional area that, unfortunately it does not survive. Stated formally, the forces at which the creatures hit bottom are proportional to their masses (volume) and inversely proportional to their surface areas. In other words, because the mass of the large rat is 8^3, or 512, times greater than the small mouse, it would strike the ground with 512 times the force in a vacuum, but because it has 64 times the surface area, the air resistance slows it down by a factor of 64. Therefore the large rat still hits with 8 times the force.

Now it is your turn. Based on what you learned from Lesson 12.6, the reasoning above, and your own research, answer the following questions. Try to be as detailed as possible. Discuss size, surface area, volume (mass), and any necessary biology or physics.

1. Why do large objects cool more slowly than similar small objects?

2. Why are the largest living mammals, the whales, confined to the sea?

3. Why is a beached whale more likely than a beached dolphin to die by overheating?

4. Why are the larger mammals found nearer the poles than at the equator?

5. If a mother and young (smaller) daughter fall into a cold lake, why is the daughter in greater danger of hypothermia?

6. Cold-blooded reptiles are also affected by temperature according to their size. When the weather is cold, iguanas hardly move. When it warms up, they become active. If a small iguana and a large iguana are sunning themselves in the morning sun, which will become active first? Why? Which iguana will remain active longer after sunset? Why?

7.*Why do elephants have big ears?

8. Why do hippos and rhinos have large, thick legs?

9. Why are champion weight lifters seldom able to lift more than twice their weight?

10. Thoroughbred race horses are fast runners but break their legs easily, while draft horses are slow moving and rarely break their legs. Why is this?

11. Assume that a male gorilla can weigh as much as 450 pounds and can reach about 6 feet tall and that King Kong was about 30 feet tall. Could a King Kong really exist? Explain.

12. Very tall professional basketball players are not similar in shape to, say, professional football players. Pro basketball players are very thin compared to pro football players. Why is this? Discuss the advantages and disadvantages of each body type in each sport.

13. An ant can fall 100 times its height and scamper away unhurt, but if a man were to fall 100 times his height, his fall would be fatal. Why?

14. Salvador Dali designed the cover for a program of the ballet *As You Like It,* shown at right. The cover depicts two elephants on long spindly legs. What is the effect created by these two creatures? Could these animals really exist? Explain. Where else in the visual arts, movies, or in literature have you seen or read of impossible creatures? You might wish to go to the art section of your library and find an art book about Salvador Dali and other artists. Did they create other impossible figures?

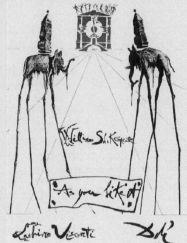

Improving Reasoning Skills—*The Harmonic Triangle*

The pattern at right is called the harmonic triangle. What is the next row in the harmonic triangle?

$$1$$

$$\frac{1}{2} \qquad \frac{1}{2}$$

$$\frac{1}{3} \qquad \frac{1}{6} \qquad \frac{1}{3}$$

$$\frac{1}{4} \qquad \frac{1}{12} \qquad \frac{1}{12} \qquad \frac{1}{4}$$

$$\frac{1}{5} \qquad \frac{1}{20} \qquad \frac{1}{30} \qquad \frac{1}{20} \qquad \frac{1}{5}$$

$$\frac{1}{6} \qquad \frac{1}{30} \qquad \frac{1}{60} \qquad \frac{1}{60} \qquad \frac{1}{30} \qquad \frac{1}{6}$$

$$-?- \qquad -?- \qquad -?- \qquad -?- \qquad -?- \qquad -?- \qquad -?-$$

Lesson 12.7

Proportional Segments by Parallel Lines

In the figure below, $\overline{MT} \parallel \overline{LU}$. Does $\triangle LUV$ appear to be similar to $\triangle MTV$? It does. Let's see if we can support this observation with a paragraph proof.

Given: $\triangle LUV$ with $\overleftrightarrow{MT} \parallel \overline{LU}$

Show: $\triangle LUV \sim \triangle MTV$

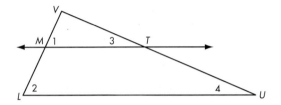

Paragraph Proof

If $\overleftrightarrow{MT} \parallel \overline{LU}$, then $\angle 1 \cong \angle 2$ and $\angle 3 \cong \angle 4$ by the Corresponding Angle Conjecture.
If $\angle 1 \cong \angle 2$ and $\angle 3 \cong \angle 4$, then $\triangle LUV \sim \triangle MTV$ by the AA Similarity Conjecture.
If the two triangles are similar, then the corresponding sides are proportional.
In the figure above, $\frac{LV}{MV} = \frac{VU}{VT} = \frac{LU}{MT}$.

Let's look at two examples of how you can use this discovery to solve problems.

Example A

$\overline{KA} \parallel \overline{BL}$
What is the length of $\overline{CA}$?

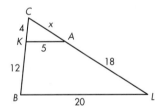

Because $\triangle CKA \sim \triangle CBL$, then

$$\frac{4}{4+12} = \frac{x}{x+18}$$

$$\frac{1}{4} = \frac{x}{x+18}$$

$$4x = x + 18$$

$$3x = 18$$

$$x = 6.$$

Did you notice that the ratio of $\frac{CK}{KB} = \frac{1}{3}$ and that the ratio of $\frac{CA}{AL} = \frac{1}{3}$?

Example B

$\overline{HT} \parallel \overline{WE}$
What is the length of $\overline{WH}$?

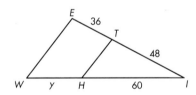

Because $\triangle HIT \sim \triangle WIE$, then

$$\frac{48}{48+36} = \frac{60}{60+y}$$

$$\frac{4}{7} = \frac{60}{60+y}$$

$$240 + 4y = 420$$

$$4y = 180$$

$$y = 45.$$

Did you notice that the ratio of $\frac{WH}{HI} = \frac{3}{4}$ and that the ratio of $\frac{ET}{TI} = \frac{3}{4}$?

Investigation 12.7.1

In each of the problems find x, then find the ratios indicated.

1. $\overleftrightarrow{EC} \parallel \overline{AB}$

2. $\overleftrightarrow{KH} \parallel \overline{FG}$

3. $\overleftrightarrow{QN} \parallel \overline{LM}$

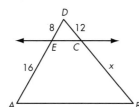

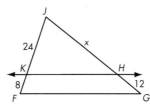

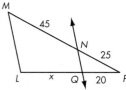

$x = \text{--?--}$

$\dfrac{DE}{AE} = \text{--?--}$, $\dfrac{DC}{BC} = \text{--?--}$

$x = \text{--?--}$

$\dfrac{JK}{KF} = \text{--?--}$, $\dfrac{JH}{HG} = \text{--?--}$

$x = \text{--?--}$

$\dfrac{PQ}{QL} = \text{--?--}$, $\dfrac{PN}{MN} = \text{--?--}$

What do you notice about the ratios of the lengths of the segments that have been cut by the parallel lines? Complete this conjecture: If a line parallel to one side of a triangle passes through the other two sides, then it divides the other two sides —?—.

Is the converse true? That is, if a line divides two sides of a triangle proportionally, is it parallel to the third side?

Investigation 12.7.2

On a sheet of paper, complete Steps 1–5.

Step 1 Draw an acute angle P.

Step 2 Beginning at point P, mark off lengths of 8 cm and 10 cm on one ray. Label the points A and B.

Step 3 Mark off lengths of 12 cm and 15 cm on the other ray. Label the points C and D. Notice that $\frac{8}{10} = \frac{12}{15}$.

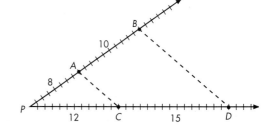

Step 4 Draw $\overline{AC}$ and $\overline{BD}$. Notice that $\overline{AC}$ divides $\overline{PB}$ and $\overline{PD}$ into segments such that $PA/AB = PC/CD$.

Step 5 With a protractor, measure $\angle PAC$ and $\angle PBD$. Are $\overline{AC}$ and $\overline{BD}$ parallel?

Repeat the steps, but this time mark off lengths of 6 cm and 8 cm on one ray and 12 cm and 16 cm on the other ray. Notice that $\frac{6}{8} = \frac{12}{16}$.

Compare your results with the results of others near you. If $\angle PAC \cong \angle PBD$, then $\overline{AC} \parallel \overline{BD}$. Complete this conjecture: If a line cuts two sides of a triangle proportionally, then it is —?— to the third side.

You should be ready to combine the conjectures from the last two investigations into one conjecture.

C-109 If a line parallel to one side of a triangle passes through the other two sides, then it divides them —?—. Conversely, if a line cuts two sides of a triangle proportionally, then it is —?— to the third side (***Parallel Proportionality Conjecture***).

What if more than one line passes through the two sides of a triangle parallel to the third side? Are all the ratios of lengths of corresponding segments equal?

Example

Find x. Find y.

Because $\overline{IT} \parallel \overline{NO}$ in $\triangle NOE$, $\frac{6}{8} = \frac{9}{x}$. Therefore $x = 12$.

Because $\overline{NO} \parallel \overline{PR}$ in $\triangle PRE$, $\frac{(6 + 8)}{10} = \frac{(9 + 12)}{y}$.

Therefore $\frac{14}{10} = \frac{21}{y}$ and $y = 15$.

When you substitute the values for x and y,

$\frac{EI}{NI} = \frac{3}{4}$ and $\frac{ET}{TO} = \frac{3}{4}$, $\frac{NI}{NP} = \frac{4}{5}$ and $\frac{TO}{RO} = \frac{4}{5}$, $\frac{EI}{NP} = \frac{3}{5}$ and $\frac{ET}{RO} = \frac{3}{5}$.

Not only does $\frac{EI}{NI} = \frac{ET}{TO}$, but it is also true that $\frac{NI}{NP} = \frac{TO}{RO}$ and $\frac{EI}{NP} = \frac{ET}{RO}$.

It appears to be true, at least from this example, that if two lines passing through two sides of a triangle are parallel to the third side of the triangle, then the corresponding segments on those two sides are proportional. Let's investigate a few more problems like the example above.

Investigation 12.7.3

Use the Parallel Proportionality Conjecture to answer each question.

1. $\overline{FT} \parallel \overline{LA} \parallel \overline{GR}$

$x = $ —?— $y = $ —?—

Is $\frac{FL}{LG} = \frac{TA}{AR}$?

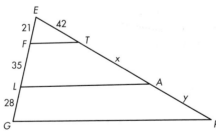

2. $\overline{ZE} \parallel \overline{OP} \parallel \overline{IA} \parallel \overline{DR}$

$a = $ —?— $b = $ —?— $c = $ —?—

Is $\frac{DI}{IO} = \frac{RA}{AP}$? Is $\frac{IO}{OZ} = \frac{AP}{PE}$?

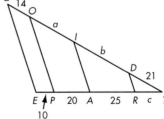

Compare your results from the two problems with the results of others near you. Does it appear that your last conjecture can be extended to hold even if two or more lines pass through the two sides of a triangle parallel to the third side? Complete the conjecture below.

C-110 If two or more lines pass through two sides of a triangle parallel
to the third side, then they divide the two sides —?—
(***Extended Parallel Proportionality Conjecture***).

Your new conjecture is very useful. You already know how to use the perpendicular
bisector construction to divide a segment into two, four, or eight equal parts. Now you
can use your new conjecture to divide a segment into *any* number of equal parts.

Example

Draw segment *AB*. Then divide it into three equal parts.

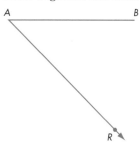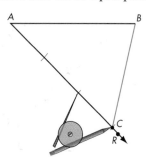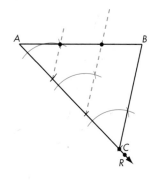

Step 1   Draw $\overline{AB}$. From one endpoint of $\overline{AB}$, say point *A*, draw $\overrightarrow{AR}$ to form $\angle BAR$ with measure
about 45°.

Step 2   On $\overrightarrow{AR}$, mark off three equal lengths with your compass. From the endpoint *C* of the
third segment, draw $\overline{BC}$ to endpoint *B*. You now have $\triangle BAC$ with side $\overline{AC}$ divided into
three equal lengths.

Step 3   Next, through the two points on $\overrightarrow{AC}$, construct rays parallel to side $\overline{BC}$ so that the parallel
rays pass through $\overline{AB}$. The two parallel rays intersect $\overline{AB}$ at two points, dividing the
segment into three equal parts.

Take Another Look 12.7

Try one or more of these follow-up activities.

1.* Complete the algebraic argument showing that if a line parallel to one side of a
triangle passes through the other two sides, then it divides the other two sides
proportionally. In other words, complete the reasoning for the argument given below.

If $\ell \parallel \overline{LU}$, then $\triangle LUV \sim \triangle MTV$.

If $\triangle LUV \sim \triangle MTV$, then $\frac{a+b}{a} = \frac{c+d}{c}$.

If $\frac{a+b}{a} = \frac{c+d}{c}$, then $\frac{a}{b} = \frac{c}{d}$.

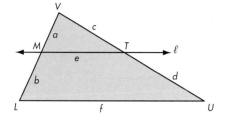

2. Refer to the same diagram of $\triangle LUV$ cut
by line ℓ, parallel to $\overline{LU}$. The Parallel
Proportionality Conjecture tells us that
$\frac{a}{b} = \frac{c}{d}$. Is it also true that $\frac{a}{b} = \frac{e}{f}$? If it's
true, explain why. If not, draw a counterexample.

3.* The Extended Parallel Proportionality Conjecture can be proven with the help of an additional line, as shown below.

The conjecture states: If two or more lines pass through two sides of a triangle parallel to the third side, then they divide the two sides proportionally.

Conjecture: If $\overleftrightarrow{EH} \parallel \overleftrightarrow{LC} \parallel \overline{MI}$, then $\frac{b}{d} = \frac{s}{t}$.

Given: $\overleftrightarrow{EH} \parallel \overleftrightarrow{LC} \parallel \overline{MI}$

Show: $\frac{b}{d} = \frac{s}{t}$

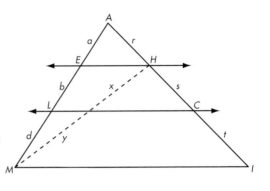

Because you must connect b, d, s, and t into one equation, you construct $\overline{MH}$. This gives you $\triangle HEM$ with $\overline{LC} \parallel \overline{EH}$ and $\triangle MIH$ with $\overline{LC} \parallel \overline{MI}$. Write a proof for the Extended Parallel Proportionality Conjecture.

4.* Is the converse of the Extended Parallel Proportionality Conjecture true? That is, if two lines intersect two sides of a triangle, dividing the two sides proportionally, must the two lines be parallel to the third side? Prove it is true or find a counterexample proving it is not true.

5. Sketch and label a plane intersecting a pyramid parallel to the base of the pyramid. State a conjecture about planes intersecting pyramids parallel to the base of the pyramid. Test your conjecture. Can you explain why you think your conjecture is true?

6. If the three sides of one triangle are parallel to the three sides of another triangle, what might be true about the two triangles? Use a geometry computer program to investigate. Make a conjecture and explain why you think your conjecture is true.

Exercise Set 12.7

All measurements are in centimeters.

1.* $\ell \parallel \overline{WE}$
$a = -?-$

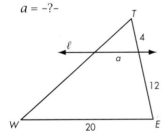

2. $m \parallel \overline{DR}$
$b = -?-$

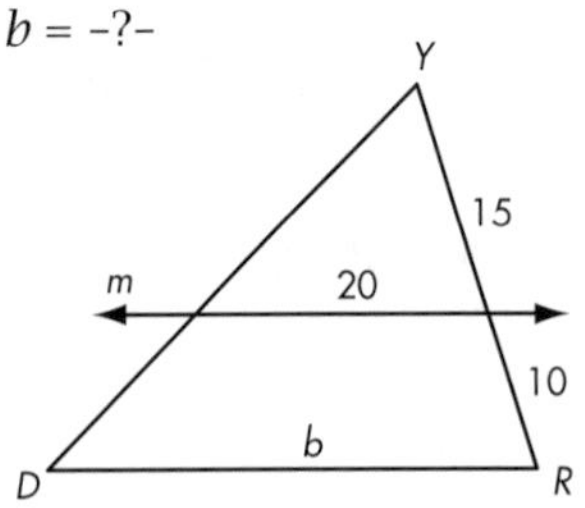

3.* $n \parallel \overline{SN}$
$c = -?-$

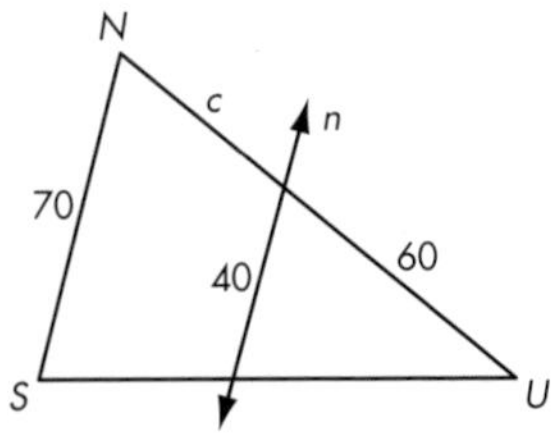

4.* $\ell \parallel \overline{RA}$

$d = $ -?-

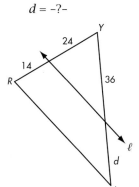

5. $m \parallel \overline{BA}$

$e = $ -?-

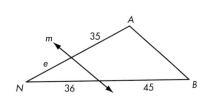

6.* Is $r \parallel \overline{AN}$?

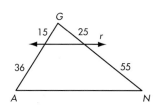

7. Is $m \parallel \overline{FL}$?

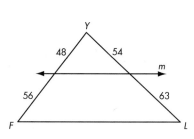

8. $r \parallel s \parallel \overline{OU}$

$m = $ -?- $n = $ -?-

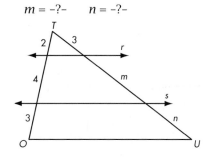

9. $\overline{MR} \parallel p \parallel q$

$w = $ -?- $x = $ -?-

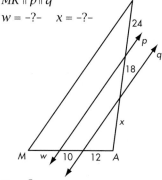

10. Is $m \parallel \overline{EA}$?

Is $n \parallel \overline{EA}$?

Is $m \parallel n$?

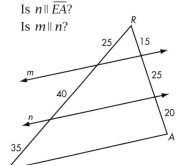

11. Is $\overline{XY} \parallel \overline{OG}$?

Is $\overline{XY} \parallel \overline{FR}$?

Is $FROG$ a trapezoid?

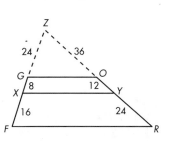

12.* $a = $ -?-

$b = $ -?-

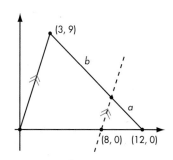

In Exercises 13 and 14, use your compass and straightedge to complete each task.

13. Draw $\overline{EF}$. Then divide it into five equal parts.

14.* Draw $\overline{IJ}$. Then construct a regular hexagon with IJ as the perimeter.

15. The ratio of the volumes of two similar cylinders is 27:64. What is the ratio of the diameters of their similar bases?

16.* The surface areas of two cubes are in the ratio of 49:81. What is the ratio of their volumes?

17. *A sheet of lined paper or graph paper can be used to divide a segment into equal parts. Draw a segment on a piece of patty paper and divide the segment into five equal parts by positioning it onto lined paper or graph paper. What conjecture explains why this works?

18. In Lesson 12.5, you saw how it is possible to divide a triangle into four similar triangles, a parallelogram into four similar parallelograms, and a special trapezoid into four similar trapezoids. These figures are called **rep-tiles**. A rep-tile is a figure that can be repeated to form a larger similar figure (a replica). A rep-tile can also be divided into figures smaller than but similar to itself. Copy the figure at right onto your paper and show how it can be divided into four similar figures.

19. *The truncated cone shown at right was formed by cutting off the top of a cone with a slice parallel to the base of the cone. What is the volume of the truncated cone?

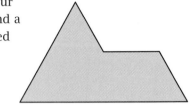

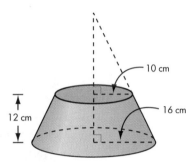

20. Romunda is preparing her specialty, *les cannonballs chocolates,* for this evening's guests of honor, Brian and Maggie. In the original recipe, her special cookie dough is rolled into 36 cannonballs (spheres with 4-cm diameters), then covered in finely ground hazelnuts. However, Romunda has decided to double the recipe. With twice the amount of dough, she reasons that she can still make 36 spheres but they will now be spheres with 8-cm diameters. Is she correct? If not, how many 8-cm diameter spheres can she make by doubling the recipe?

21. *Galileo Galilei (1564–1642) used the drafting instrument shown at right. Called a sector compass, it consists of two sides, each of which displays equal scale. The sector compass is used to construct segments that are some fraction of a given segment. If you wish to construct a segment that is three fourths (or 75%) of a given segment, you adjust the sector compass so that the segment fits between the 100-marks. You then draw the segment connecting the two —?— on the two scales. What points on the compass should you connect? Explain why this works.

22. *Another drafting instrument used to construct segments is the pair of proportional dividers, shown at right. This instrument consists of two styluses of equal length connected by a set screw. The dividers are adjusted for different proportions by loosening and sliding the set screw along grooves in the dividers. Where should the set screw be positioned so that these dividers can make segments that are three fourths of given segments? Explain.

23. The graph on the following page shows one-dimensional data—the price per barrel of light crude oil leaving Saudi Arabia each January from 1973 to 1979—but this data is represented by a three-dimensional display. The heights are supposed to represent the price per barrel, but instead the volumes of the barrels seem to represent the data, which is misleading because the barrels are growing in height, width, and depth. It's even difficult to correctly perceive the changes in height because the barrels are drawn in perspective.

  The ratio of the light crude prices in 1979 and 1975 is $\frac{\$13.34}{\$10.48}$, which is about 1.275. This indicates that the price increased by about 27.5% from 1975 to 1979. Measure the heights of the 1979 and 1975 barrels in millimeters. What's the ratio of their heights? Use the ratio of their heights to find the ratio of their volumes. How does the ratio of their volumes compare to the ratio of their prices? Did the volume of the barrels increase by anything close to 27.5%?

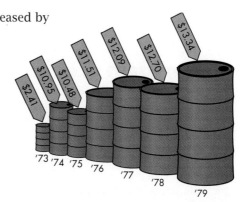

'73 '74 '75 '76 '77 '78 '79

Improving Reasoning Skills—*Crossnumber Puzzle II*

Copy the crossnumber grid at right. In this puzzle, the answers you know will help you answer the problems you don't know. Did Blaise Pascal really invent the one-wheeled wheelbarrow? You'll soon find out. Each clue consists of a statement followed by two numbers. If the statement is true, enter the first number of the pair. If the statement is false, enter the second number.

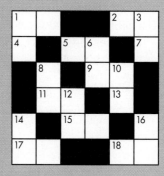

Across

1. The hypotenuse is always the longest side in a right triangle. (76, 91)
2. The consecutive angles of a rhombus are supplementary. (76, 91)
4. There are exactly nine diagonals in a hexagon. (5, 9)
5. Twenty lines in a plane intersect in a maximum of 210 points. (68, 43)
7. Pythagoras taught his followers never to poke a fire with iron because a flame was the symbol of truth. (2, 5)
8. The diagonals of a rectangle are perpendicular to each other. (7, 1)
9. The exact angle formed by the hands of a clock at 1:20 measures 80°. (35, 54)
11. The acute angles of a right triangle are supplementary. (96, 79)
13. SSS and SAS are two ways to show two triangles congruent. (2, 5)
14. Over the entrance to Plato's Academy was inscribed: "Let no one unversed in geometry enter here." (4, 8)
15. The area of a rhombus is equal to half the product of the two diagonals. (68, 72)
16. The formula $A = bh$ is used to find the area of a triangle. (4, 9)
17. If two angles of a triangle are congruent, then the triangle is equilateral. (87, 75)

18. Mathematician Blaise Pascal is credited with the invention of the one-wheeled wheelbarrow. (42, 57)

Down

1. Abraham Lincoln learned logic by studying Euclid's *Elements*. (95, 79)
3. The Sioux culture has traditionally considered 4 the perfect number, which is often associated with their sacred symbol, the circle. (62, 15)
6. The perpendicular bisector of a chord passes through the center of the circle. (33, 85)
8. In a 30-60 right triangle, the side opposite the 30° angle is half the length of the hypotenuse. (17, 79)
10. The measure of each exterior angle of a regular octagon is 45°. (52, 45)
12. The *z* in the abbreviation *oz* for ounce comes from the old Spanish coin *onza de oro*, which contained one ounce of gold. (96, 67)
14. Hypatia of Alexandria is considered the first known woman mathematician. (47, 88)
16. If the corresponding angles of two triangles are congruent, then the two polygons are congruent. (47, 92)

The Golden Ratio

Geometry has two great treasures: one is the theorem of Pythagoras; the other is the division of a line into extreme and mean ratio. The first we may compare to a measure of gold; the second we may name a precious jewel.
— *Johannes Kepler*

The precious jewel that Johannes Kepler (1571–1630) spoke of is called the golden ratio. The **golden ratio** is a number, usually represented with the Greek letter phi (ø), that satisfies the special proportion

$$\frac{1}{ø} = \frac{ø}{1 + ø}.$$

The golden ratio is often represented geometrically by a golden rectangle. A **golden rectangle** is a rectangle whose length and width satisfy the proportion.

$$\frac{w}{l} = \frac{l}{w + l}$$

A golden rectangle whose width is 1 unit has a length of ø units. In other words, the ratio of the length of a golden rectangle to its width is ø.

Mathematicians have been fascinated with the golden ratio since long before Kepler's time. The ratio was of particular interest to mathematicians of ancient Greece. Some researchers believe Greek artists and architects found the golden rectangle to be the most pleasing of rectangular shapes. These researchers claim that golden rectangles were used many times in the design of the most famous of Greek temples, the Parthenon. Whether or not Greek artists and architects actually used the golden ratio is debated, but there is no dispute that Greek mathematicians took great interest in the mathematical properties of the ratio.

What is the actual value of the golden ratio, and how is it found? Let's use algebra to calculate the golden ratio from its definition,

$$\frac{1}{ø} = \frac{ø}{1 + ø}.$$

Cross-multiply to get $1 + ø = ø^2$.

$$ø^2 - ø - 1 = 0$$

Using the quadratic formula, $ø = \dfrac{-(-1) \pm \sqrt{(-1)^2 - 4(1)(-1)}}{2} = \dfrac{1 \pm \sqrt{5}}{2}$.

The positive root is the golden ratio.

1. Use your calculator to determine an approximate value for ø.

2. Use your calculator and the approximate value you found for ø to explore some of the golden ratio's interesting properties.

   a. Calculate $ø^2$. How is it related to ø? Start with the definition of ø and do the algebra to demonstrate this relationship.

   b. Calculate $\frac{1}{ø}$. How is it related to ø? Explain.

   c. Calculate $ø^3 - 3ø$. What is this difference? Use algebra to support your discovery.

3. The golden ratio can also be approximated with the help of a familiar pattern of numbers. Determine the next three numbers in this pattern:

   1, 1, 2, 3, 5, 8, 13, 21, 34, 55, 89, 144, 233, –?–, –?–, –?–.

In Chapter 1, you worked with this sequence of numbers. It's called the Fibonacci sequence. The numbers in the Fibonacci sequence occur in many branches of mathematics as well as in nature and in art. To determine a decimal approximation of the golden ratio, you can calculate the ratio of pairs of consecutive Fibonacci numbers. Use a calculator to find the ratios of the following pairs of consecutive Fibonacci numbers as decimals accurate to six places.

   a. $\frac{34}{21} \approx$ –?–   b. $\frac{144}{89} \approx$ –?–   c. $\frac{377}{233} \approx$ –?–

   d. $\frac{610}{377} \approx$ –?–   e. $\frac{987}{610} \approx$ –?–   f. $\frac{1597}{987} \approx$ –?–

Notice that as the Fibonacci numbers grow larger and larger, the ratio gets closer and closer to the golden ratio.

4. Measure the length across the top of the Parthenon, shown in the photo on the previous page, to the nearest millimeter. If this is the length of a golden rectangle, you can find the width by dividing by ø. Try it. Now measure the photo again and see if the value you found for the width of a golden rectangle corresponds to the height of the Parthenon. Do you think this face of the Parthenon fits into a golden rectangle? Can you see why some researchers dispute the theory that Greek architects used the golden ratio to build this structure?

5. A golden rectangle can be constructed from a square. Use the method described below to construct as large a golden rectangle as possible on a full sheet of paper.

   Step 1   Construct a square. Label it *GOEN*. Extend $\overline{GO}$. Extend $\overline{NE}$.

   Step 2   Bisect $\overline{GO}$. Label the midpoint *M*. With *ME* as your radius and point *M* as center, construct an arc intersecting $\overleftrightarrow{GO}$ at point *L*.

   Step 3   Construct the rectangle *OLDE*. Rectangle *GLDN* is a golden rectangle.

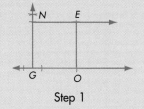

Step 1

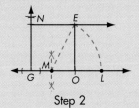

Step 2

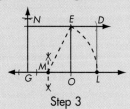

Step 3

6. Use the Pythagorean Theorem to show that the ratio of the length of the longer side to the length of the shorter side is equal to the golden ratio. Let $OE = 2x$, then let $MO = x$. Calculate ME. But $GL = x + ME$. Calculate GL/LD.

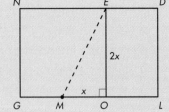

 Show that $\frac{GL}{LD} = \frac{1 + \sqrt{5}}{2}$.

 Did you notice that $\overline{OE}$ divides the golden rectangle into a square and another rectangle? Does the small rectangle appear to be similar to the original? In fact, the smaller rectangle is also a golden rectangle.

7. In problem 5, you constructed a golden rectangle from a square. The small rectangle on the right side of the square was also a golden rectangle. Using the rectangle you constructed in problem 5, divide the small rectangle into a square and an even smaller golden rectangle, as shown at right. Divide this smallest golden rectangle (the one in the upper right corner of the figure) into a square and a golden rectangle.

8. The process you used in problem 7 can also be done in reverse. Construct as small a golden rectangle as possible. Use your construction tools to add a square above it, as shown in the figure, using the long side of the rectangle as the side of the square. The square and the rectangle combine to form a larger golden rectangle. Then add on another square to the right of the new rectangle, creating another larger golden rectangle. Repeat this process two more times.

9. Use the golden rectangles from problem 7 to construct an approximation of the logarithmic spiral. You can do so by constructing a 90° arc in each square. If arcs are drawn in the golden rectangle, as shown, you get a very graceful curve, which is related to the beautiful spiral of a nautilus seashell. The curve approximated by the golden rectangle spiral has many names, due in part to its many different but related properties. René Descartes (1596–1650) called it the equiangular spiral, and Edmond Halley (1656–1743) called it the proportional spiral. Jacob Bernoulli (1654–1705) named it the logarithmic spiral and asked that it be engraved on his tombstone.

10. Use the golden rectangles from problem 8 to construct another approximation of the logarithmic spiral. You can do so by placing a point in the center of each square and then sketching a spiral to connect the center of each square.

11. In his book *Growth and Form,* Sir D'Arcy Thompson calls the type of growth exhibited by the shell of the nautilus gnomonic growth. As an animal grows by gnomonic growth,

the size of the animal's shell increases but its shape remains unchanged. The nautilus shell grows longer and wider to make room for the growing animal within, but it grows at one end only. Each new section increases in size so that the overall shape remains similar. The repeating process of adding on a new square to a golden rectangle to get a new, larger golden rectangle is analogous to the nautilus adding a new chamber to its shell to accommodate its growth.

Gnomonic growth can be modeled with more than just golden rectangles. Start with a square or a rectangle and construct a sequence of squares in which each new square is added onto the new, longer side of the figure. (See the examples below.) You can also start with an isosceles right triangle and construct a sequence of isosceles right triangles in which each new right triangle is added onto the new hypotenuse of the figure. Try it. Start with a square, a rectangle, or an isosceles right triangle and construct a sequence of at least six figures to model the spiraling of gnomonic growth.

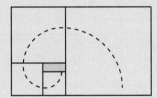
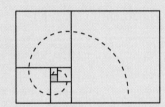
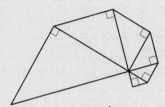

12. A golden triangle is an isosceles triangle in which the ratio of the length of a leg to the length of the base is the golden ratio. Use the construction from problem 5 to construct a golden triangle. Each of the five tips of a pentagram is a golden triangle.

13. As mentioned before, people have been fascinated by the golden ratio for centuries. It is believed by some that the golden ratio is found in the Great Pyramid. Many others believe that artists, including Leonardo da Vinci (1452–1519) and Piet Mondrian (1872–1944), incorporated the golden rectangle into their works because of its aesthetic appeal.

Some researchers believe that classical Greek sculptures of the human body were proportioned so that the ratio of the total height to the height of the navel was the golden ratio. It is claimed that Polykleitus (ca 450 B.C.) used the golden ratio in creating the proportions for his *Doryphoros,* "The Spear Bearer."

What do you think? Measure the height of the statue of Doryphoros, shown in the photo at right, and measure the height of the navel. Is the ratio of total height to navel height close to the golden ratio?

Doryphoros

Lesson 12.8

Chapter Review

Similarity has more applications than just about any geometry topic. Any scale drawing or model, anything that is reduced or enlarged, is governed by properties of similar figures. Think of as many applications as you can in the movie-making industry alone. How many can you list? Similarity is also useful in indirect measurement. Can you describe at least two indirect measurement methods? How are distance ratios in similar figures related to ratios of area and volume? How can similar figures be used to misrepresent data in graphs?

Exercise Set 12.8

For Exercises 1–12, identify each statement as true or false. If possible, sketch a counterexample for each false statement or explain why it is false.

1. If the three sides of one triangle are proportional to the three sides of another triangle, then the two triangles are similar.

2. If two angles of one triangle are congruent to two angles of another triangle, then the two triangles are similar.

3. If two sides of one triangle are proportional to the sides of another triangle, then the two triangles are similar.

4. If the four angles of one quadrilateral are congruent to the four corresponding angles of another quadrilateral, then the two quadrilaterals are similar.

5. An angle bisector in a triangle divides the opposite side into two segments whose lengths are in the same ratio as the corresponding adjacent sides.

6. If two triangles are similar, then their corresponding altitudes, corresponding medians, and corresponding angle bisectors are proportional to their corresponding sides.

7. If two similar polygons (or circles) have corresponding sides (or radii) in the ratio of m/n, then their areas are in the ratio of m/n.

8. If two similar solids have corresponding dimensions in the ratio of m/n, then their volumes are in the ratio of m/n.

9. If a line parallel to one side of a triangle passes through the other two sides, then it divides them proportionally.

10. If a line cuts two sides of a triangle proportionally, then it is parallel to the third side.

11. If two or more lines pass through two sides of a triangle parallel to the third side, then they divide the two sides equally.

12. Six statements in Exercises 1–12 are false.

For Exercises 13–16, solve each proportion.

13. $\frac{x}{15} = \frac{8}{5}$

14. $\frac{4}{11} = \frac{24}{x}$

15. $\frac{4}{x} = \frac{x}{9}$

16. $\frac{x}{x+3} = \frac{34}{40}$

17. The rate at which countries exchange currency is called the exchange rate. On Sunday, January 1, 1995, 1 U.S. dollar bought 4.75 Mexican pesos. However, by the following Sunday, January 8, $1.00 could be exchanged for 5.72 pesos. What was the exchange rate for pesos to dollars on January 1? (One peso bought what decimal fraction of a dollar?) If you owned 10,000 pesos, how much would they have been worth in U.S. dollars on January 1? How much would they have been worth in U.S. dollars on January 8?

In Exercises 18–21, measurements are given in centimeters.

18.* $\triangle ABC \sim \triangle DBA$

    $x = -?-$ $y = -?-$

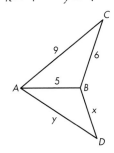

19. $ABCDE \sim FGHIJ$

    $w = -?-$ $x = -?-$ $y = -?-$ $z = -?-$

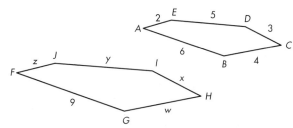

20. $k \parallel \ell \parallel m \parallel n$

    $w = -?-$
    $x = -?-$
    $y = -?-$
    $z = -?-$

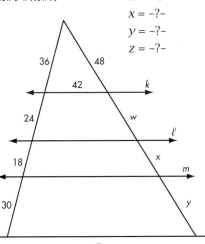

21.* The dimensions of the smaller cylinder are two thirds of the larger. The volume of the larger cylinder is 2160π cm³. Find the volume of the smaller cylinder.

22. Diane is 5′8″ tall and wants to find the height of an oak tree in her front yard. She walks along the shadow of the tree until her head is in a position where the end of her shadow exactly overlaps the end of the treetop's shadow. She is now 11′3″ from the foot of the tree and 8′6″ from the end of the shadow. How tall is her oak tree?

23. After building a rectangular box home for his pet python, Monty, Charlie learns that Lucy has also built a home for Monty but with dimensions twice as great. If it takes one half of a gallon of paint to cover the surface of Charlie's home for Monty, how many gallons of paint would be needed to paint the python home that Lucy built? How many times as much volume will Lucy's box hold as Charlie's?

24.* The Jones family paid \$150 to a painting contractor to stain their 12-ft-by-15-ft back deck. The Smiths, their neighbors, have a similar deck that measures 16 ft by 20 ft. If the Smiths wish to "keep up with the Joneses," what is a proportional price the Smith family should expect to pay to have their deck stained by the contractor?

25. The ratio of the perimeters of two similar parallelograms is 3:7. What is the ratio of their areas?

26. The areas of two circles are in the ratio of 25:16. What is the ratio of their radii?

27. Would 15 pounds of ice cubes that each measure one inch on an edge melt faster than a 15-pound block of ice? Explain.

28.* Construct $\overline{KL}$. Then find a point P that divides KL into the ratio of 2:3.

29.* The Ring-a-Ding Sisters Circus has come to town. P. T. Barnone is the star of the show. She does a juggling act atop a stool that sits on top of a rotating ball that spins at the top of a 20-meter pole. The diameter of the ball is 4 meters, and P. T.'s eye is 2 meters above the ball. The circus manager needs to know the radius of the circular region beneath the ball in which spectators would be unable to see eye to eye with P. T. Find the radius to the nearest tenth meter for the manager so that he can put in the seats for the show. (Use 1.7 for $\sqrt{3}$.)

The Ring-a-Ding Sisters Circus

30. Many fanciful movies, books, and stories have been written about people who are accidentally shrunk to a small fraction of their original height. This change in size creates a variety of changes in their needs. If a person's height were decreased to one twentieth his original size, how would that change the amount of food he'd require, or the amount of material needed to clothe him, or the time he'd need to get to different places? Explain.

13 Trigonometry

Belvedere, M. C. Escher, 1958

In this chapter you will discover some of the properties and the applications of right triangle trigonometry. You will use this branch of mathematics to calculate distances that are difficult or impossible to measure directly. You will probably study trigonometry in greater depth in your next mathematics course—it is a useful subject used in architecture, astronomy, engineering, navigation, and surveying.

Lesson 13.1

Trigonometric Ratios

Research is what I am doing when I don't know what I'm doing.
— *Wernher von Braun*

Right triangle trigonometry is the study of the relationships between the sides and the angles of right triangles. In this lesson you will discover some of these relationships.

In Chapter 12, you used mirrors and shadows to measure heights indirectly. Trigonometry gives you another measuring method. For example, you can use trigonometry to calculate the height of a tree by measuring the angle of elevation and the distance from the vertex of the angle to the tree.

How did right triangle trigonometry develop? When early mathematicians and astronomers calculated the ratios of the sides for different right triangles, they discovered that whenever the ratio of the shorter leg's length to the longer leg's length was close to some specific fraction, the angle opposite the shorter leg was close to some specific measure. This was true no matter how big or small the triangle.

For example, in every right triangle in which the length of the shorter leg divided by the length of the longer leg is close to the fraction $\frac{3}{5}$, the angle opposite the shorter leg measures close to 31°. What is a good approximation for x?

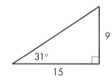

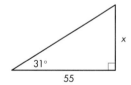

In every right triangle in which the length of the shorter leg divided by the length of the longer leg is close to the fraction $\frac{9}{10}$, the angle opposite the shorter leg measures close to 42°. What is y?

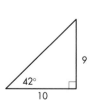
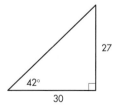

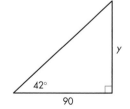

In Exercises 18–23, angle-measuring devices are used to determine a distance.

18. A salvage ship is locating wreckage. The ship's sonar picks up a signal showing wreckage at an angle of depression measuring 12°. The ocean charts for the region list an average depth of 40 meters. If a diver is lowered from the salvage ship at this point, how far can the diver expect to travel along the ocean floor to the wreckage?

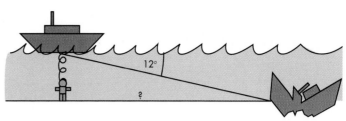

19. According to a Chinese legend from the Han dynasty (206 B.C.–A.D. 220), General Han Xin flew a kite over the palace of his enemy to determine the distance between his troops and the palace. If the general let out 800 meters of string and the kite was flying at a 35° angle of elevation, approximately how far away was the palace from General Han Xin's position?

20. Ben is flying a kite directly over his buddy Franklin, who is 125 meters away. His kite string makes a 39° angle with the level ground. To the nearest meter, how high is his kite?

21. *The angle of elevation from a ship to the top of a 42-meter lighthouse on the shore measures 33°. To the nearest meter, how far is the ship from the shore?

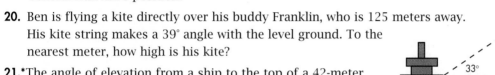

22. *Meteorologist Wendy Storm is using a sextant to determine the height of a weather balloon. When she views the weather balloon through her sextant, which is sighted 1 m above the level ground, she measures a 44° angle up from the horizontal. The radio signal from the balloon tells her that the balloon is 1400 m from her measuring device. To the nearest meter, how high is the balloon? To the nearest meter, how far is it to a position directly beneath the balloon?

23. *A lighthouse is observed by a ship's officer at a 42° angle to the path of the ship. At the next sighting, the lighthouse is observed at a 90° angle to the path of the ship. The distance traveled between sightings is 1800 m. To the nearest meter, what is the distance between the ship and the lighthouse at this second sighting?

Improving Algebra Skills—*Corecting Algebar Misteakes*

What is wrong with each of these algebraic statements? Correct each one.

1. $x^2 + x = x^3$

2. $x^2 + x^2 + x^2 = x^6$

3. $2x + 3y = 5xy$

4. $\frac{2x}{7} + \frac{3x}{5} = \frac{5x}{12}$

5. $8x - x = 8$

6. $x^2 + y^2 = (x + y)^2$

7. $\frac{1}{x} + \frac{1}{x} = \frac{2}{2x}$

8. $\sqrt{x^2} = x$

Project

Indirect Measurement

If I have seen farther than others it is because I have stood on the shoulders of giants.
— Isaac Newton

In this project you will use trigonometric methods and equipment—a scientific calculator and an angle-measuring device—to determine the heights of two different-sized objects that you are unable to measure directly. This project works best in groups of four or five. Divide up the tasks, which include being responsible for taking measurements, recording the measurements, performing the calculations, keeping track of the equipment, and being the person measured. You will need the following equipment.

- Measuring tape or meter sticks.

- Notebook for recording your measurements.

- Scientific calculator.

- Device for measuring angles. (Three sample angle-measuring devices are shown below. Build one of them or make one of your own design.) For more detail about how to make them and how they work, see Project: Making a Clinometer in Chapter 4.

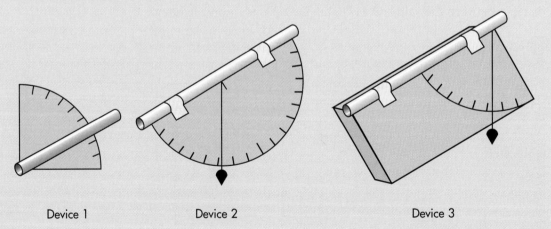

Device 1   Device 2   Device 3

Before you begin, make a table of your measurements, similar to that below.

Name of object to be measured	Angle of elevation	Height of observer's eye	Distance from observer to object	Calculated height of object
1. —?—	–?–	–?–	–?–	–?–
2. —?—	–?–	–?–	–?–	–?–

Follow Steps 1–4 to determine indirectly the heights of the two objects.

Step 1 Begin by measuring one person's eye height. (We'll call this person the observer.) Record the measurement in your table.

Step 2 Locate two tall objects with heights that would be difficult to measure directly, such as a school building, a football goal post, a flagpole, or a tall tree.

Step 3 With your angle-measuring device, measure the observer's viewing angle from the horizontal to the top of each object to be measured. Measure the distance from the observer to the base of each object. Enter the measurements into the table.

Step 4 Perform all the calculations to approximate the heights of the two objects.

Improving Reasoning Skills—*Container Problem III*

You have an unmarked 11-liter container, an unmarked 3-liter container, and an unlimited supply of water. In table, symbol, or paragraph form, describe the process necessary to end up with exactly 10 liters of water in one container.

11 L

3 L

Improving Algebra Skills—*Symbol Juggling*

If $V = \frac{1}{3}BH$, $B = \frac{1}{2}h(a + b)$, $h = 2x$, $a = 2b$, $b = x$, and $Hx = 12$, find the value of V in terms of x.

Lesson 13.3

The Law of Sines

Rules are for the obedience of fools and the guidance of wise men.
— David Ogilvy

So far you have used trigonometry only to solve problems involving right triangles. Trigonometry can also be used with triangles other than right triangles. For example, if you know the measures of two angles and one side of a triangle (either ASA or SAA), you can find the other two sides with the help of a trigonometric property called the law of sines. The law of sines can be discovered while using trigonometry to find areas of triangles. Let's look at an example.

Example A

Find the area of $\triangle ABC$ if $AB = 150$ meters, $BC = 100$ meters, and $m\angle B = 40°$.

$$\sin 40° = \frac{CD}{100}$$

$$CD = (100)(\sin 40°)$$

$$\text{Area}_{\triangle ABC} \approx (0.5)(150)CD$$

$$\text{Area}_{\triangle ABC} \approx (0.5)(150)[(100)(\sin 40°)]$$

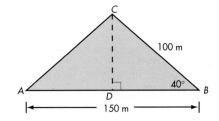

The area is about 4821 m².

In Investigation 13.3.1, you will arrive at a formula for the area of a triangle given the lengths of two sides and the measure of the angle between them.

Investigation 13.3.1

Find the area of each triangle.

1.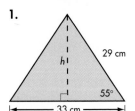

2.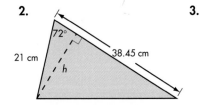

3.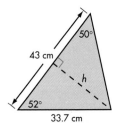

Now do the same with algebra for the triangle shown at right. Use a trigonometric function to find an expression for the height h in terms of side a and angle C. Then use the formula for the area of a triangle to derive an equation for the area of a triangle given two sides and the angle between them. State your equation as your next conjecture.

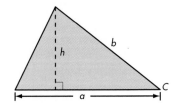

C-111 The area of a triangle is given by the formula $A = \dfrac{?}{}$, where a and b are the lengths of two sides and C is the angle between them.

In Investigation 13.3.2, you will use the formula for the area of a triangle to arrive at the property called the law of sines.

Investigation 13.3.2

Use the figure at right to perform Steps 1-6.

Step 1 Find altitude h in terms of a and the sine of an angle.

Step 2 Find altitude h in terms of b and the sine of an angle.

Step 3 Show $\dfrac{\sin A}{a} = \dfrac{\sin B}{b}$.

Step 4 Find altitude k in terms of c and the sine of an angle.

Step 5 Find altitude k in terms of b and the sine of an angle.

Step 6 Show $\dfrac{\sin B}{b} = \dfrac{\sin C}{c}$.

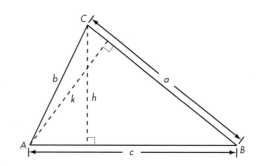

Steps 3 and 6 can be combined to form the law of sines. Complete the conjecture below and add it to your conjecture list.

C-112 For a triangle with angles of measures A, B, and C and sides of lengths a, b, and c (a opposite A, b opposite B, and c opposite C), $\dfrac{\sin A}{?} = \dfrac{\sin B}{?} = \dfrac{\sin C}{?}$.

Now, if you know the right combination of parts (for example, ASA or SAA), you can use the law of sines to find lengths of sides or measures of angles of triangles.

Example B

Find the length b of side $\overline{AC}$ in $\triangle ABC$ with $BC = 350$ cm, $m\angle A = 59°$, and $m\angle B = 38°$.

Start with the law of sines.

$$\frac{\sin A}{a} = \frac{\sin B}{b}$$

Now solve for b.

$$b \sin A = a \sin B$$

$$b = \frac{a \sin B}{\sin A}$$

$$b = \frac{(350)(\sin 38°)}{\sin 59°}$$

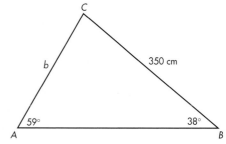

The length b of side $\overline{AC}$ is approximately 251 cm.

You can also use the law of sines to find the measure of a missing angle but only if you know whether the angle is acute or obtuse. For example, if you knew in $\triangle ABC$ that $BC = 160$ cm, $AC = 260$ cm, and $m\angle A = 36°$, you would not be able to use that information to find $m\angle B$ because the given information doesn't determine a triangle. Remember in Chapter 5 when you constructed noncongruent triangles by using SSA? In this case, there are two possible measures for $\angle B$ (see the figures below).

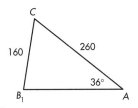 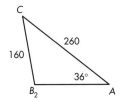

Because in this chapter you've only investigated trigonometric ratios for acute angles, you'll be asked to find only acute angle measures.

Example C

Find the measure of acute $\angle B$ in $\triangle ABC$ with $BC = 250$ cm, $AC = 150$ cm, and $m\angle A = 69°$.

Start with the law of sines.

$$\frac{\sin A}{a} = \frac{\sin B}{b}$$

Now solve for $\sin B$.

$$a \sin B = b \sin A$$

$$\sin B = \frac{b \sin A}{a}$$

$$\sin B = \frac{(150)(\sin 69°)}{250}$$

The measure of $\angle B$ is approximately 34°.

Take Another Look 13.3

Try one or more of these follow-up activities.

1. Use algebra to show that $\dfrac{a}{\sin A} = \dfrac{b}{\sin B} = \dfrac{c}{\sin C}$.

2. In this lesson you were reminded that SSA does not determine a triangle. For that reason, you've been asked to find only acute angles by using the law of sines when the given information about a triangle was SSA. Take another look at a pair of triangles determined by SSA. In the figure at right, how is $\angle D$ related to $\angle CBA$? Use your protractor to find $m\angle D$, then calculate $m\angle CBA$. Use your calculator to find the sine of each angle. Find the sines of another pair of angles that are related in the same way. Complete the conjecture: $\sin \theta = \sin (-?-)$.

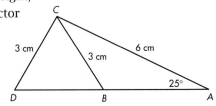

Exercise Set 13.3

In Exercises 1–3, find the area of each figure to the nearest square centimeter.

1. Area = –?–

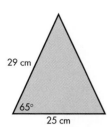

29 cm
65°
25 cm

2. Area = –?–

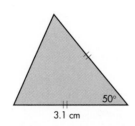
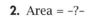

50°
3.1 cm

3.* Area = –?–

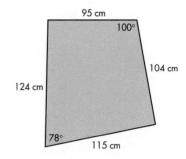

95 cm
100°
104 cm
124 cm
78°
115 cm

4.* Find the area of a regular octagon inscribed within a circle whose radius is 12 cm.

In Exercises 5–7, find the length of each indicated side to the nearest centimeter.

5.* w = –?–

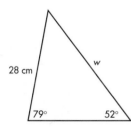

28 cm
w
79°
52°

6. x = –?–

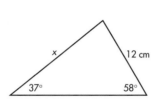

x
12 cm
37°
58°

7. y = –?–

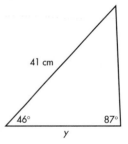

41 cm
46°
87°
y

In Exercises 8–10, find the measure of each indicated angle in each acute triangle to the nearest degree.

8.* m∠A = –?–

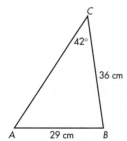

C
42°
36 cm
A   29 cm   B

9. m∠B = –?–

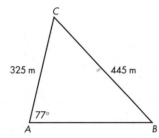

C
325 m   445 m
77°
A   B

10. m∠C = –?–

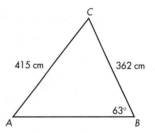

C
415 cm   362 cm
63°
A   B

11.* Igor's pet bat, Natasha, is flying at the end of a 50-foot leash. Using an angle-measuring device, Igor spots Natasha at a 55° angle up from the horizontal. To the nearest foot, how high is Natasha flying if the leash is taut and anchored to the ground?

12. Archaeologist Ertha Diggs is using an angle-measuring device to determine the height of an ancient temple. When she views the top of the temple through her device, she records a 37° angle up from the horizontal. She is standing 130 meters from the center of the temple's base, and her eye is 1.5 meters above the ground. To the nearest tenth of a meter, how tall is the temple?

13.* According to legend, Galileo used the leaning tower of Pisa to conduct his experiments on the laws of gravity. When he dropped objects from the top of the 55-meter tower (measured length, not height, of tower), they landed 4.8 meters from the tower's base. To the nearest degree, what is the angle ϕ that the tower leans off from the vertical?

14. One of the most impressive of the Mayan pyramids is El Castillo pyramid in Chichén Itzá, Mexico. The pyramid has a platform on its top and a flight of 91 steps on each of its four sides. (Four flights of 91 steps equal 364 steps in all. The top platform adds a level, so the pyramid consists of 365 levels, which represent the 365 days of the Mayan year.) To the nearest centimeter, what is the height of the top of the steps if each of the 91 steps is 30 cm deep by 26 cm high? To the nearest degree, what is the angle of ascent?

15. Alphonse is directly over the 2500-meter landing strip in his hot-air balloon (point *A*) and is observed at one end of the strip by Beatrice (point *B*) with an angle of elevation measuring 39°. Meanwhile, at the other end of the strip, Collette (point *C*) observes the balloon with an angle of elevation of 62°. What is the distance between Alphonse and Beatrice? What is the distance between Alphonse and Collette? How high up is Alphonse when he is observed?

16.* A tree is growing vertically on a hillside that is at a 16° angle to the horizontal. The tree casts a shadow 18 m long up from the slope. If the angle of elevation of the sun measures 68°, how tall is the tree?

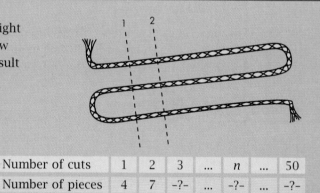

Improving Visual Thinking Skills—*Rope Tricks*

1. If the rope shown at right is cut as indicated, how many pieces would result from 50 cuts?

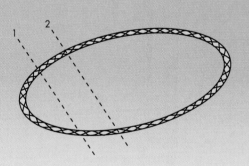

Number of cuts	1	2	3	...	*n*	...	50
Number of pieces	4	7	-?-	...	-?-	...	-?-

2. If the rope shown at right is cut as indicated, how many pieces would result from 50 cuts?

Number of cuts	1	2	3	...	*n*	...	50
Number of pieces	2	4	-?-	...	-?-	...	-?-

Lesson 13.4

The Law of Cosines

A ship in a port is safe, but that is not what ships are built for.
— *Benazir Bhutto*

You've solved a variety of problems with the Pythagorean Theorem ($c^2 = a^2 + b^2$). It is perhaps your most important geometric conjecture. In Chapter 9, you discovered that the Distance Formula was really just the Pythagorean Theorem. You even used the Pythagorean Theorem to derive the Equation of a Circle.

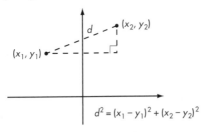

$$d^2 = (x_1 - y_1)^2 + (x_2 - y_2)^2$$

Distance Formula

$$(x - h)^2 + (y - k)^2 = r^2$$

Equation of a Circle

In trigonometry also there are relationships derived from the Pythagorean Theorem. One of them is even called the Pythagorean Identity. Complete the steps in Investigation 13.4 to derive the Pythagorean Identity.

Investigation 13.4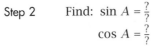

Step 1 Find: $(\sin 27°)^2 + (\cos 27°)^2 = ?$
$(\sin 53°)^2 + (\cos 53°)^2 = ?$
$(\sin 78°)^2 + (\cos 78°)^2 = ?$

Are you ready to make a conjecture? Let's see if you can derive your conjecture. Use the figure at right to perform Steps 2–5.

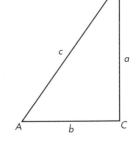

Step 2 Find: $\sin A = \dfrac{?}{?}$

$\cos A = \dfrac{?}{?}$

Step 3 Substitute your results from Step 2 into the equation
$(\sin A)^2 + (\cos A)^2 = \left(\dfrac{?}{?}\right)^2 + \left(\dfrac{?}{?}\right)^2.$

Step 4 Because the two fractions in this equation have the same denominator, combine them into one fraction.
$(\sin A)^2 + (\cos A)^2 = \dfrac{(?)^2 + (?)^2}{c^2}$

Step 5 But $a^2 + b^2 = c^2$ because the triangle is a right triangle.
Therefore substitute c^2 for $a^2 + b^2$ in the numerator of the equation and simplify.

You should now be ready to state the Pythagorean Identity. Complete the conjecture and add it to your conjecture list.

C-113 For any angle A, —?— (***Pythagorean Identity***).

Even though the Pythagorean Theorem is so very powerful, its use is still limited to right triangles. Another trigonometric property, the law of cosines, does for all triangles what the Pythagorean Theorem does for right triangles!

In his *Elements,* Greek mathematician Euclid establishes two propositions that generalize the Pythagorean Theorem. Today we refer to these propositions as the law of cosines. We might picture the law of cosines being developed when a mathematician first asked, "What would happen to the Pythagorean formula if the right angle were instead acute or obtuse?" Huh? Take a look at the figures at right.

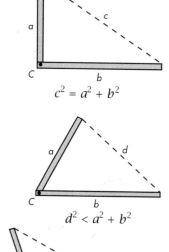

If two sticks with lengths a and b were connected at one pair of endpoints, forming a right angle ($\angle C$), the distance c between the other endpoints could be calculated from $c^2 = a^2 + b^2$.

If the sticks, still attached at the same endpoints, were moved so that the angle between them ($\angle C$) was now an acute angle, the distance d between the endpoints would be less than before.

That is, $d^2 < a^2 + b^2$ or $d^2 = a^2 + b^2 -$ something.

If the sticks were pushed apart so that the angle between them ($\angle C$) was now an obtuse angle, the distance f between the endpoints would be greater than before.

That is, $f^2 > a^2 + b^2$ or $f^2 = a^2 + b^2 +$ something.

This observation led to a mathematical search for that "something." The "something" turned out to be $2ab \cos C$. The Pythagorean Theorem was thus generalized to all triangles. This property is called the law of cosines.

C-114 For any triangle with sides of lengths a, b, and c, and with C the angle opposite the side with length c, $c^2 = a^2 + b^2 - 2ab \cos C$ (***Law of Cosines***).

The steps used to derive the law of cosines are left for you as a Take Another Look activity.

The problems in this lesson will be restricted to finding the sine, cosine, or tangent of angles measuring less than 90°. (The trigonometric ratios for angles measuring greater than 90° must be redefined. This is done with a coordinate geometry approach to trigonometry, which you'll learn about in a later course.)

In Lesson 13.3, you used the law of sines for triangle problems involving two angles and a side (ASA or SAA). The law of cosines can be used when you are given three side lengths or two side lengths and an angle measure (SSS or SAS). Let's look at a few examples of how to use the law of cosines.

Example A

Find the length r of side $\overline{CT}$ in acute $\triangle CRT$ with
$RT = 45$ cm, $CR = 52$ cm, and $m\angle R = 36°$.

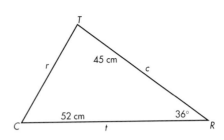

To find r, use the law of cosines:
$c^2 = a^2 + b^2 - 2ab \cos C$.

In this problem, the law of cosines becomes
$$r^2 = c^2 + t^2 - 2ct \cos R.$$

Substitute the known values.
$$r^2 = 45^2 + 52^2 - 2(45)(52)(\cos 36°)$$

To find r, find the value of $2(45)(52)(\cos 36°)$ and store it in the memory of your
calculator. (Your display should show 3786.1995.) Next, calculate the value of $45^2 + 52^2$
and subtract the value in memory. Then find the square root. (You should get
30.705056.) Therefore the length r of side $\overline{CT}$ is about 31 cm.

If you are given all three side lengths of a triangle (SSS), the law of cosines can be used
to find the measure of an angle.

Example B

Find the measure of $\angle Q$ in acute $\triangle QED$ with
$ED = 250$ cm, $QD = 175$ cm, and $QE = 225$ cm.

Use the law of cosines. To find $m\angle Q$, start with the form
$$q^2 = e^2 + d^2 - 2ed \cos Q.$$

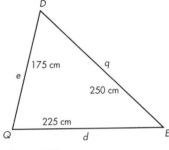

Solve for $\cos Q$.
$$\cos Q = \frac{q^2 - e^2 - d^2}{-2ed}$$

Substitute the known values.
$$\cos Q = \frac{250^2 - 175^2 - 225^2}{-2(175)(225)}$$

When you calculate $\cos^{-1}$ for the number above, your
display should show 76.225853. The measure of angle Q is about 76°.

Take Another Look 13.4

Try one or both of these follow-up activities.

1. The law of cosines for the triangle below left, using angle C, is $c^2 = a^2 + b^2 - 2ab \cos$
   C. State the law of cosines in two different ways, using angle A and angle B.

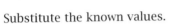

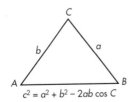

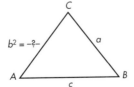

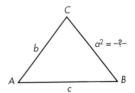

2.* Derive the law of cosines.

Exercise Set 13.4

Find the length of each indicated side to the nearest centimeter.

1. * $w = $ –?–

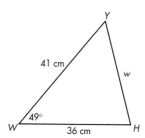

2. $x = $ –?–

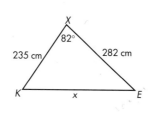

3. $y = $ –?–

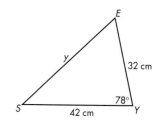

Find the measure of each indicated angle to the nearest degree.

4. * $m\angle A = $ –?–

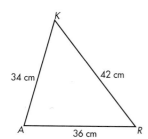

5. $m\angle B = $ –?–

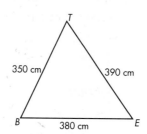

6. $m\angle C = $ –?–

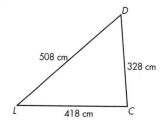

7. * To the nearest degree, find the measure of the smallest angle in a triangle whose side lengths are 4 m, 7 m, and 8 m.

8. Two sides of a parallelogram measure 15 cm and 20 cm, and one of the diagonals measures 30 cm. What are the measures of the angles of the parallelogram to the nearest degree?

9. If two 24-centimeter radii of a circle form a central angle measuring 126°, what is the length of the chord connecting the two radii to the nearest tenth of a centimeter?

10. A cargo company needs to load truck trailers into ship cargo containers. The trucks must drive up a ramp to a loading platform 30 ft off the ground, but they have difficulty driving up a ramp at an angle steeper than 20°. How long does the ramp need to be?

11. Chip Woodman, the foreman at the paper plant, must estimate the volume of a conical woodchip pile. The distance from the tip of the cone to the edge of the base (the slant height) is 304 feet and forms a 54° angle with the ground. What is the height of the cone to the nearest foot? What is the area of the base of the cone to the nearest thousand square feet? What is the volume of the cone to the nearest hundred thousand cubic feet?

12. A lighthouse 55 meters above sea level spots a distress signal from a sailboat. The angle of depression to the sailboat measures 21°. To the nearest meter, how far away is the sailboat from the base of the lighthouse?

13. Housepainters have established a general safety rule for leaning ladders against buildings. What might happen if the angle with the ground is too small? What might happen when a painter climbs up the ladder and the angle is too great? Picasso Painting Company has instructed its painters to always place their ladders at an angle measuring between 55° and 75° from the level ground. If Regina places the foot of her 25′ ladder 6′ from the base of the wall, is the ladder placed safely? What is the angle at which she has placed the ladder? If it is not a safe angle, should she move the ladder in or out? How far?

14.* Ertha Diggs has uncovered the remains of a square-based Egyptian pyramid. The base is intact and measures 130 meters on each side. The top portion of the pyramid has eroded away over the centuries, but what remains of each face of the pyramid forms a 65° angle with the ground (angle of ascent). To the nearest meter, what was the original height of the pyramid?

15.* Captain Ace Malloy is flying commercial flight 1123 out of San Francisco. He is heading east at 720 km/hr when he spots an electrical storm straight ahead. He turns the jet 20° to the north to avoid the storm and continues in this direction for 1 hr. Then he makes a second correction so that he can travel back toward his original flight path. Eighty minutes after his second correction, he enters his original flight path at an acute angle and his craft is back on course. By avoiding the storm, how much time did Ace lose from his original flight plan?

Lesson 13.5

Problem Solving with Trigonometry

One ship drives east and another drives west
With the self-same winds that blow,
'Tis the set of the sails and not the gales
Which tells us the way to go.

— *Ella Wheeler Wilcox*

There are many practical applications of trigonometry. Some of these applications involve vectors. In earlier vector activities, you actually measured with rulers and protractors the size of the resulting vectors, or the angles between vectors. Now, you will be able to calculate the resulting vectors with the law of cosines or the law of sines.

Example

If rowing instructor Upson Downes is in a stream flowing north to south at 3 km/hr and he is rowing northeast at his usual steady rate of 4.5 km/hr, what direction east of north is he actually moving? At what speed is he traveling?

First we draw and label the vector parallelogram as shown. We know the lengths of the two sides and the measure of the included angle, so we can use the law of cosines to find the length (speed) of the resultant vector (r).

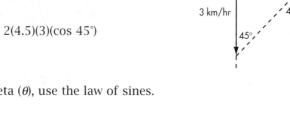

$$r^2 = 4.5^2 + 3^2 - 2(4.5)(3)(\cos 45°)$$
$$r^2 \approx 10.158$$
$$r \approx 3.2 \text{ km/hr}$$

To find the angle theta (θ), use the law of sines.

$$\frac{\sin \theta}{3} = \frac{\sin 45°}{3.2}$$
$$\sin \theta = \frac{(3)(\sin 45°)}{3.2} \approx 0.6629$$
$$\theta = 41.5°.$$

Therefore Upson Downes is moving 86.5° east of north.

Exercise Set 13.5

1.* The steps to the front entrance of a public building rise a total of 1 m. A portion of the steps are to be torn out and replaced by a ramp for wheelchairs. By a city ordinance, the angle of inclination for a wheelchair ramp cannot measure greater than 4.5°. What is the minimum distance from the entrance that the ramp must begin? Express your answer to the nearest tenth of a meter.

2. A lighthouse is east of a Coast Guard patrol boat. The Coast Guard station is 20 km north of the lighthouse. The Coast Guard radar officer aboard ship measures the angle between the lighthouse and the station to be 23°. To the nearest kilometer, how far is the ship from the station?

3. The Archimedean screw (also called the Archimedean snail) is a water-raising device used since antiquity to raise water from a river for irrigation. The device consists of a wooden screw enclosed within a cylinder, and when the cylinder is turned, the screw raises the water. The Archimedean screw is very efficient at an angle measuring 25°. If a screw needs to raise water 2.5 meters, how long should its cylinder be?

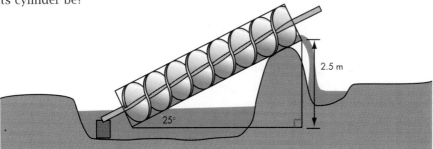

2.5 m

25°

4. A surveyor needs to calculate the distance from the dock on Nomansan Island (point *C*) to the dock on the mainland (point *A*). First, he locates a point *B* up the shoreline so that $\overline{AB}$ is perpendicular to $\overline{AC}$. If *AB* measures 150 meters and $m\angle ABC = 58°$, what is the distance between the two docks to the nearest meter?

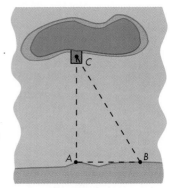

5. A tall evergreen tree has been damaged in a strong wind. The top of the tree is cracked and bent over, touching the ground as if the trunk were hinged. The tip of the tree touches the ground 20 feet 6 inches from the base of the tree and forms a 38° angle with the ground. Forest ranger Willow Green must determine the original height of the damaged tree. Help her. To the nearest foot, what was the tree's original height?

6. A pocket of opal matrix is known to be 24 meters below point *A*. The Outback Mining Company, however, is unable to dig straight down because sacred aboriginal artifacts are buried a few meters below the earth's surface. The company has been given permission to dig at point *B*, 8 meters away from point *A*. At what angle to the level surface must the mining crew dig to reach the opal matrix? What distance must they dig?

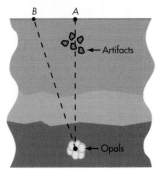

7.* Army Private Olivia Drabb observes her friend Amelia Allheart in her fighter plane. Olivia knows that Amelia is going to be flying over her at a level elevation of 6.3 km. At one point she observes the plane moving directly toward her from the west at an angle of elevation measuring 49°. Forty seconds later she turns and sees Amelia moving away from her, heading east at an angle of elevation measuring 15°. How fast is Amelia's plane flying in kilometers per hour?

8. * Farmer Goldie McDonald is planning to lay a water pipe through a small hill. She wants to determine the length of pipe needed. From a point off to the side of the hill on level ground, she runs one rope to the pipe's entry point and a second rope to the exit point. The first rope measures 14.5 meters, the second measures 11.2 meters, and they meet at an angle measuring 58°. What is the length of pipe needed? At what angle with the first rope should the pipe be laid so that it comes out of the hill at the correct exit point?

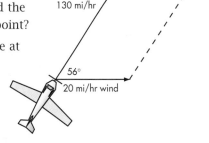
130 mi/hr
56°
20 mi/hr wind

9. * Geovannie is flying his Cessna on a heading shown in the figure at right. His instruments show an air speed of 130 mi/hr. (Air speed is the speed of a plane as it flies in still air, that is, ignoring wind.) There is a 20 mi/hr wind intersecting the path of the plane, as shown. What is the speed of the plane (its ground speed)?

10. * Annie and Sashi are backpacking in the Sierras. They walk 8 km from their base camp at a bearing of 42°. (Recall that a bearing is the angle measured clockwise from north.) They stop for some birdwatching. After lunch and a game of backgammon, they pack up and change direction to a new bearing of 137°. After 5 km they meet up with their friends Elina and Nancy at the designated rendezvous point, the waterfalls. To the nearest tenth of a kilometer, how far are Annie and Sashi from their base camp?

Annie and Sashi backpacking in the Sierras

11. * At what bearing must Sashi and Annie travel from the waterfalls to return to their base camp?

Geometer's Sketchpad Project

Trigonometric Ratios and the Unit Circle

When you defined trigonometric ratios in terms of the sides of a right triangle, you were restrained to talking about angles with measures less then 90°. But in the coordinate plane, it's possible to define trigonometric ratios for angles with measures greater than 90° and less than 0°. In this project you'll discover those definitions.

Step 1 In the Display menu, set Preferences to show distance units in inches and to show angle measures in directed degrees.

Step 2 In the Graph menu, choose Create Axes. You'll get a pair of axes, a point A at the origin, and a point B at $(1, 0)$.

Step 3 Construct a circle with center A and radius point B. This circle with radius 1 unit is called the **unit circle**.

In the coordinate plane, trigonometric ratios are defined in terms of this unit circle. In Steps 4–6, you'll construct a right triangle within the unit circle so that you can see the relationship between the right triangle definitions and the unit circle definitions for trigonometric ratios.

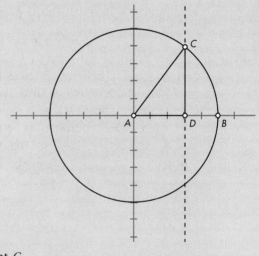

Step 4 Construct radius AC, where C is any point on the circle. Locate point C in the first quadrant for now.

Step 5 Construct a line through point C, perpendicular to the x-axis.

Step 6 Use the Segment tool to construct right $\triangle ADC$, where point D is the point of intersection of the line through point C and the x-axis. Hide the line through point C.

Now that you have a right triangle, with point C in the first quadrant, think about your right triangle definitions of sine and cosine. Answer the following questions before you continue your construction.

1. In right triangle DAC, the cosine of $\angle DAC$ is defined as $\dfrac{\text{length of adjacent side}}{-?-}$. The length of the hypotenuse is –?–. Measure the coordinates of point C. Which coordinate corresponds to the cosine of $\angle DAC$?

2. In right triangle DAC, the sine of $\angle DAC$ is defined as $\dfrac{-?-}{-?-}$. Which coordinate of point C corresponds to the sine of $\angle DAC$?

3. In right triangle DAC, the tangent of $\angle DAC$ is defined as $\dfrac{-?-}{-?-}$. How can you use the coordinates of point C to calculate the tangent of $\angle DAC$?

You can use Sketchpad to check your answers to the questions above. Select point *D*, point *A*, and point *C*, in order, and choose Angle in the Measure menu. (With Preferences set to measure angles in directed degrees, selection order is important. Positive angle measures are measured in a counterclockwise direction. You should get a positive angle measure when point *C* is in the first or second quadrant.) Use Sketchpad's calculator to calculate the sine, cosine, and tangent of ∠*DAC*. Were you right about how the cosine, sine, and tangent compare with the *x*- and *y*-coordinates of point *C*?

You can also construct the tangent of ∠*DAC* as it relates to the unit circle. Continue your construction by following Steps 7–9.

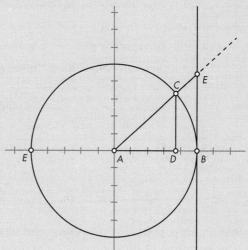

Step 7   Construct a line through point *B*, perpendicular to the *x*-axis. (This line is tangent to the unit circle.)

Step 8   Construct line *AC* and segment *AE*, where *E* is the point at which $\overleftrightarrow{AC}$ intersects the tangent line. Hide $\overleftrightarrow{AC}$.

Step 9   Measure the coordinates of point *E*. Which of these coordinates is related to the tangent of ∠*DAC*?

You should now have point coordinates based on your unit circle that give the sine, cosine, and tangent of angles with measures between –180° and 180°. Drag point *C* around the circle and observe these measures. When point *C* is in the first quadrant, you get the right-triangle trigonometric ratios you've used before now. By dragging point *C* into other quadrants, you can see how these ratios are defined for obtuse angles and negative angle measures. Answer the following questions.

4. How does the sine of ∠*DAC* change as the measure of the angle goes from 0° to 90° to 180°? From –180° to –90° to 0°?

5. How does the cosine of ∠*DAC* change as the measure of the angle goes from 0° to 90° to 180°? From –180° to –90° to 0°?

6. If the sine of one angle is equal to the cosine of another angle, how are the angles related to each other?

7. Explain why the *y*-coordinate of point *E* gives you the tangent of ∠*DAC*. Hint: Use similar triangles to show that *EB = CD/AD*.

8. For what angles is the tangent of ∠*DAC* negative?

9. What happens to the tangent of ∠*DAC* as its measure approaches 90° or –90°? Can you explain why the coordinates of point *E* in your sketch disappear at these angles?

10. Use Sketchpad's Tabulate command in the Measure menu to make your own tables of approximate values of the sine, cosine, and tangent of angle measures from 0° to 90° (use 15° intervals). What do you think your table should say for the tangent of 90°?

With a little more sketching, you can add an interesting animation to your sketch. You can contruct points that will trace curves representing the sine and tangent functions. Move point *C* into the first quadrant, then follow Steps 10–18.

Step 10 Construct $\overline{AF}$, where point *F* is on the *x*-axis.

Step 11 Measure the coordinates of point *F*, then locate it as close to (6.28, 0) as possible. (If that point is off screen, you can scale the axes by dragging point *B* closer to point *A*.)

Step 12 Construct point *G* on $\overline{AF}$ and a line through point *G*, perpendicular to $\overline{AF}$.

Step 13 Construct lines through points *C* and *E*, parallel to the *x*-axis.

Step 14 Where the horizontal line through point *C* intersects the vertical line through point *G*, construct a point and label it *Sin*. Change the color of the point (if you have a color monitor).

Step 15 Where the horizontal line through point *E* intersects the vertical line through point *G*, construct a point and label it *Tan*. Change the color of the point (if you have a color monitor).

Step 16 Choose Trace Point in the Display menu for each of these two points of intersection.

Step 17 Select point *G*, $\overline{AF}$, point *C*, and the circle. Choose Animation in the Action Button submenu of the Edit menu. Choose one way for point *G* on the segment and one way for point *C* on the circle.

Step 18 Move point *G* to the origin and move point *C* to (1, 0). Double-click the animation button to trace curves that show the motion of the sine and tangent functions. These curves show how the sine and tangent functions change as ∠*DAC* changes.

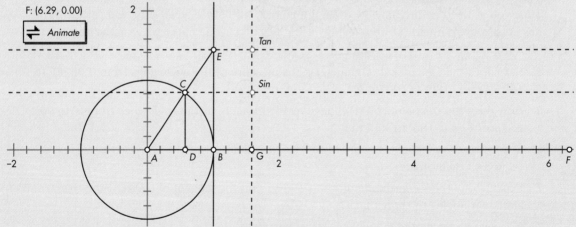

What's special about 6.28 as the *x*-coordinate of point *F*? Experiment with other locations for point *F* to see what happens. For the curves representing sine and tangent, the values on the *x*-axis correspond to angle measures in units called radians: 2π radians = 360°. You'll learn more about radian angle measure in a later course. To explore the graphs of trigonomeric functions using degree measures, try the Graphing Calculator Investigation—Trigonometric Functions.

You've seen many applications where you can use trigonometry to find unknown distances or angle measures in triangles. Another important application of trigonometry is

to model periodic phenomena. Ocean tides and the motion of a pendulum are examples of periodic phenomena because they repeat themselves over time. In this investigation, you'll discover characteristics of the graphs of trigonometric functions, including their periodic nature. The functions you'll graph are defined not only for acute angles but also for angles with measures less than 0° and greater than 90°. To learn more about how these functions are defined you can do (or just read) the Geometer's Sketchpad Project: Trigonometric Ratios and the Unit Circle that precedes this investigation.

1. Set your calculator to degree mode and define a window optimized for trigonometric functions. (This window should have an x range of about –360 to 360 and a y range of at least –2 to 2.) Graph the equation $y = \sin x$. Explain why you think this graph is said to be periodic.

2. Trace along this graph, starting at $x = 0$. At what value for x does the graph start repeating itself? (The horizontal distance from any point on the graph to a point where the graph starts repeating itself is called the period of the graph.)

3. What are the maximum and minimum values for y?

4. Graph the equation $y = \cos x$ on the same graph. How does the graph of $y = \cos x$ compare to the graph of $y = \sin x$? Trace along one of the graphs. For what values of x are the sine and cosine the same?

5.* Clear the sine and cosine graphs and graph $y = \tan x$. (Depending on your calculator, your graph may include vertical lines.) Trace along this graph. What happens to the value of tan x as x approaches 90°? Use what you know about the definition of tangent in right triangles to explain why this is so.

6. Graph $y = \sin x$ again on the same graph as $y = \tan x$. In question 2 above, you discovered where the sine graph starts repeating itself. Compare the period of the tangent function to the period of the sine function.

Lesson 13.6

Chapter Review

Trigonometry was developed by astronomers who wished to map the stars. Obviously, measurements of stars and planets are hard to make by direct methods. In this chapter, you've seen that many indirect measurement problems can be solved by using triangles. Using sine, cosine, and tangent ratios, you can find unknown lengths and angle measures if you know just a few measures in a right triangle—or in any triangle using the law of sines or the law of cosines. That's why triangles are so useful in problem solving.

A common mnemonic for remembering how trigonometric ratios are defined is SOH CAH TOA. Say it out loud. What do you think it stands for? What's the least you need to know about a right triangle in order to find all its measures? What parts of a nonright triangle do you need to know in order to find the other parts? Describe a situation in which an angle of elevation or depression can help you find an unknown height.

Exercise Set 13.6

Use the definitions of the three trigonometric ratios to complete each statement.

1. $\sin A = -?-$
$\cos A = -?-$
$\tan A = -?-$

2. $\sin B = -?-$
$\cos B = -?-$
$\tan B = -?-$

3.* $\sin \phi = -?-$
$\cos \phi = -?-$
$\tan \phi = -?-$

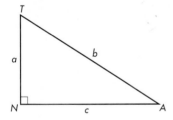

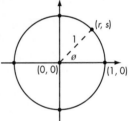

Use a scientific calculator to determine each value accurate to four decimal places.

4. $\sin 57° \approx -?-$

5. $\cos 9° \approx -?-$

6. $\tan 88° \approx -?-$

Find the measure of each angle to the nearest degree.

7. $\sin A = 0.5447$

8. $\cos B = 0.0696$

9. $\tan C = 2.9043$

In Exercises 10–18, express each angle measure to the nearest degree, each length to the nearest centimeter, area to the nearest square centimeter, and volume to the nearest cubic centimeter.

10. Area = -?-

11. $w = -?-$

12. $\triangle ABC$ is acute.
$m\angle A = -?-$

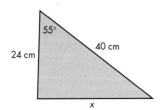

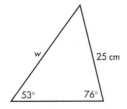

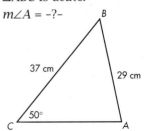

Deductive Reasoning

You used inductive reasoning throughout this text to discover many basic properties of geometry. In this chapter you will learn about deductive reasoning, which is commonly called logical reasoning. You will learn to translate logical arguments into symbolic form, then use rules of logic to determine whether those arguments are valid.

Möbius Strip II, M. C. Escher, 1963
©1996 M. C. Escher / Cordon Art – Baarn – Holland.
All rights reserved.

"That's Logical!"

*I refuse to join any club that would
have me for a member.*
— Groucho Marx

"That's logical!" You've probably heard that expression many times. What do we mean when we say someone is thinking logically? One dictionary defines *logical* as "capable of reasoning or using reason in an orderly fashion that brings out fundamental points."

"Prove it!" That's another expression you've probably heard many times. It is an expression that is used by someone concerned with logical thinking. In daily life, proving something often means you can present some authority or fact to support a point. In Example A, an eyewitness is cited as one person's proof.

Example A

"All right, Butch. Where were you on the night of January 11?" asked Inspector Vida.

"Inspector," replied Butch in an injured tone, "I can prove I was with my dear mom. Call and ask her."

To prove a point, you may sometimes cite accepted rules to support a conclusion. This often occurs in sports.

Example B

Samantha Slugger has two strikes against her. She attempts to bunt and fouls the ball. The umpire calls her out. Samantha complains, "Prove I'm out!" The ump states the official league rule on attempted bunts with two strikes.

Exercise Set 14.1

Use logical reasoning to solve each problem.

1.* Beatrice is older than Catherine, and Michelle is younger than Catherine. Is Beatrice older or younger than Michelle? Explain.

2. Every cheerleader at Washington High School is a junior. Mark is a cheerleader at Washington High School. What year is Mark in school?

3. Every student at Brightmore High School who takes the auto shop class also takes the driver's education class. Roberto and Sharon are both students at Brightmore High School. Roberto is not taking driver's education. Sharon is taking auto shop. Is Roberto taking auto shop? Is Sharon taking driver's education? Explain.

4. Boris and Natasha each own a very unique pet. One owns a monkey and the other owns an alligator. The alligator's owner would like to own a golden retriever, but she is allergic to animal hair. Who owns the monkey? Explain.

5. Assume the following two statements are true: Either parallelogram *PARE* is a rhombus or it is a rectangle. Parallelogram *PARE* is not a rhombus. What can you conclude?

6. David tells Manny, "I am over six feet tall." "No, you aren't, and I can prove it," states Manny. How might Manny prove it?

7. Ann proudly tells Amy, "I can spell *parallelogram* correctly." "Prove it!" replied Amy. How might Ann prove it?

8. Highway Patrol Officer Brandenburg pulled Trucker Tucker over for speeding. "But officer, I wasn't speeding," pleaded Tucker. "The other vehicles were just going very slow." Officer Brandenburg growled, "You were speeding, and I can prove it." In court, how will the officer prove Tucker was speeding?

9. Jock says he can run the 100-yard dash faster than anyone else in the world. How many people would Jock have to beat to prove his statement? How many people would have to beat Jock to disprove his statement?

10. Diana has been asked to explain to her cousin why $\triangle ABC$ is a right triangle. She says, "Because the two sides $\overline{AB}$ and $\overline{AC}$ are marked equal and one angle measures 45°, the triangle must be a right triangle." Is she correct? Is her reasoning correct? Fill in the missing steps in her reasoning.

11. If the diagonals of a quadrilateral are perpendicular, then the quadrilateral is either a kite, a rhombus, or a square. If a quadrilateral is either a rhombus or a square, then its diagonals bisect each other. Suppose you know that the diagonals of a quadrilateral *ABCD* are perpendicular but do not bisect each other. What can you conclude about quadrilateral *ABCD*? Explain.

Lesson 14.2

Sherlock Holmes

There are two types of reasoning in mathematics: inductive and deductive. Thus far in this book, you have been using inductive reasoning by observing patterns and making conjectures about your observations. This is the creative, investigative form of reasoning that mathematicians use most often. In the coming chapters, you will take a look at deductive reasoning to see if your discoveries are logically consistent.

Deductive reasoning, or logical reasoning, is the process of demonstrating that if certain statements are accepted as true, then other statements can be shown to follow from them.

Illustration by Sidney Paget, *The Strand Magazine,* 1892

What better way to begin our study of deductive reasoning than with that deduction expert, Sherlock Holmes. Sit back and read the following excerpt from *The Adventure of the Dancing Men* by Sir Arthur Conan Doyle.

Holmes had been seated for some hours in silence with his long, thin back curved over a chemical vessel in which he was brewing a particularly malodorous product. His head was sunk upon his breast, and he looked from my point of view like a strange, lank bird, with dull gray plumage and a black top-knot.

"So, Watson," said he, suddenly, "you do not propose to invest in South African securities?"

I gave a start of astonishment. Accustomed as I was to Holmes's curious faculties, this sudden intrusion into my most intimate thoughts was utterly inexplicable.

"How on earth do you know that?" I asked.

He wheeled round upon his stool, with a steaming test-tube in his hand, and a gleam of amusement in his deep-set eyes.

"Now, Watson, confess yourself utterly taken aback," said he.

"I am."

"I ought to make you sign a paper to that effect."

"Why?"

"Because in five minutes you will say that it is all so absurdly simple."

"I am sure that I shall say nothing of the kind."

"You see, my dear Watson," he propped his test-tube in the rack, and began to lecture with the air of a professor addressing his class, "it is not really difficult to construct a series of inferences, each dependent upon its predecessor and each simple in itself. If, after doing so, one simply knocks out all the central inferences and presents one's audience with the starting point and the conclusion, one may produce a startling, though possibly a meretricious, effect. Now, it was not really difficult, by inspection of the groove between your left forefinger and thumb, to feel sure that you did not propose to invest your small capital in the gold fields."

"I see no connection."

"Very likely not; but I can quickly show you a connection. Here are the missing links of the very simple chain: 1. You had chalk between your left finger and thumb when you returned from the club last night. 2. You put chalk there when you play billiards, to steady the cue. 3. You never play billiards except with Thurston. 4. You told me four weeks ago that Thurston had an option on some South African property which would expire in a month, and which he desired you to share with him. 5. Your checkbook is locked in my drawer, and you have not asked for the key. 6. You do not propose to invest your money in this manner."

"How absurdly simple!" I cried.

Let's take a closer look at the two forms of logical reasoning used by Sherlock Holmes.

Example A

Watson had chalk between his fingers upon returning from the club.

If Watson had chalk on his fingers, then he had been playing billiards.

Therefore Watson had been playing billiards.

Example B

If Watson wished to invest his money in South African securities with Thurston, then he would have had his checkbook when playing billiards with Thurston.

Watson did not have his checkbook when he played billiards with Thurston.

Therefore Watson does not wish to invest his money in South African securities with Thurston.

Example C

If Watson was playing billiards, then he was playing with Thurston.

Watson was playing billiards.

Therefore —?—.

Exercise Set 14.2

Let's look at another Sherlock Holmes adventure from Sir Arthur Conan Doyle's *The Hound of the Baskervilles.* Pay close attention to the conclusions Holmes arrives at from the facts he gathers observing a walking stick.

Mr. Sherlock Holmes, who was usually very late in the mornings, save upon those not infrequent occasions when he was up all night, was seated at the breakfast table. I stood upon the hearth-rug and picked up the stick which our visitor had left behind him the night before. It was a fine, thick piece of wood, bulbous-headed, of the sort which is known as a "Penang lawyer." Just under the head was a broad silver band, nearly an inch across. "To James Mortimer, M.R.C.S., from his friends of the C.C.H.," was engraved upon it, with the date "1884." It was just such a stick as the old-fashioned family practitioner used to carry—dignified, solid, and reassuring.

"Well, Watson, what do you make of it?"

Holmes was sitting with his back to me, and I had given him no sign of my occupation.

"How did you know what I was doing? I believe you have eyes in the back of your head."

"I have, at least, a well-polished, silver-plated coffee-pot in front of me," said he. "But tell me, Watson, what do you make of our visitor's stick? Since we have been so unfortunate as to miss him and have no notion of his errand, this accidental souvenir becomes of importance. Let me hear you reconstruct the man by an examination of it."

"I think," said I, following as far as I could the methods of my companion, "that Dr. Mortimer is a successful, elderly medical man, well esteemed, since those who know him give him this mark of their appreciation."

"Good!" said Holmes. "Excellent!"

"I think also that the probability is in favour of his being a country practitioner who does a great deal of his visiting on foot."

"Why so?"

"Because this stick, though originally a very handsome one, has been so knocked about that I can hardly imagine a town practitioner carrying it. The thick iron ferrule is worn down, so it is evident that he has done a great amount of walking with it."

"Perfectly sound!" said Holmes.

"And then again, there is the friends of the C.C.H. I should guess that to be the Something Hunt, the local hunt to whose members he has possibly given some surgical assistance, and which has made him a small presentation in return."

"Really, Watson, you excel yourself," said Holmes, pushing back his chair and lighting a cigarette. "I am bound to say that in all the accounts which you have been so good as to give of my own small achievements, you have habitually underrated your own abilities. It may be that you are not yourself luminous, but you are a conductor of light. Some people without possessing genius have a remarkable power of stimulating it. I confess, my dear fellow, that I am very much in your debt."

He had never said as much before, and I must admit that his words gave me a keen pleasure, for I had often been piqued by his indifference to my admiration and to the attempts which I had made to give publicity to his methods. I was proud, too, to think that I had so far mastered his system as to apply it in a way which earned his approval. He now took the stick from my hands and examined it for a few minutes with his naked eyes. Then with an expression of interest he laid down his cigarette, and, carrying the cane to the window, he looked over it again with a convex lens.

"Interesting, though elementary," said he as he returned to his favorite corner of the settee. "There are certainly one or two indications upon the stick. It gives us the basis for several deductions.

"I am afraid, my dear Watson, that most of your conclusions were erroneous. When I said that you stimulated me I meant, to be frank, that in noting your fallacies I was occasionally guided towards the truth. Not that you are entirely wrong in this instance. The man is certainly a country practitioner. And he walks a good deal."

"Then I was right."

"To that extent."

"But that was all."

"No, no, my dear Watson, not all—by no means all. I would suggest, for

Illustration by Sidney Paget,
The Strand Magazine, 1901

example, that a presentation to a doctor is more likely to come from a hospital than from a hunt, and that when initials 'C.C.' are placed before that hospital the words 'Charing Cross' very naturally suggest themselves."

"You may be right."

"The probability lies in that direction. And if we take this as a working hypothesis we have a fresh basis from which to start our construction of this unknown visitor."

"Well, then, supposing the 'C.C.H.' does stand for 'Charing Cross Hospital,' what further inferences may we draw?"

"Do none suggest themselves? You know my methods. Apply them!"

"I can only think of the obvious conclusion that the man has practiced in town before going to the country."

"I think that we might venture a little farther than this. Look at it in this light. On what occasion would it be most probable that such a presentation would be made? When would his friends unite to give him a pledge of their good will? Obviously at the moment when Dr. Mortimer withdrew from the service of the hospital in order to start in practice for himself. We know that there has been a presentation. We believe there has been a change from a town hospital to a country practice. Is it, then, stretching our inference too far to say that the presentation was on the occasion of the change?"

"It certainly seems probable."

"Now, you will observe that he could not have been on the staff of the hospital, since only a man well-established in a London practice could hold such a position, and such a one would not drift into the country. What was he, then? If he was in the hospital and yet not on the staff he could only have been a house-surgeon or a house-physician—little more than a senior student. And he left five years ago— the date is on the stick. So your grave, middle-aged family practitioner vanishes into thin air, my dear Watson, and there emerges a young fellow under thirty, amiable, unambitious, absent-minded, and the possessor of a favorite dog, which I should describe roughly as being larger than a terrier and smaller than a mastiff."

I laughed incredulously as Sherlock Holmes leaned back in his settee and blew little wavering rings of smoke up to the ceiling.

"As to the latter part, I have no means of checking you," said I, "but at least it is not difficult to find out a few particulars about the man's age and professional career." From my small medical shelf I took down the Medical Directory and turned up the name. There were several Mortimers, but only one who could be our visitor. I read his record aloud.

"Mortimer, James, M.R.C.S., 1882, Grimpen, Dartmoor, Devon, House surgeon, from 1882 to 1884, at Charing Cross Hospital. Winner of the Jackson prize for Comparative Pathology, with essay entitled 'Is Disease a Reversion?' Corresponding member of the Swedish Pathological Society. Author of 'Some Freaks of Atavism' (Lancet, 1882). 'Do We Progress?' (Journal of Psychology, March, 1883). Medical Officer for the parishes of Grimpen, Thorsley, and High Barrow."

"No mention of the local hunt, Watson," said Holmes with a mischievous smile, "but a country doctor, as you very astutely observed. I said, if I remember right, amiable, unambitious, and absent-minded. It is my experience that it is only an amiable man in this world who receives testimonials, only an unambitious one who abandons a London career for the country, and only an absent-minded one who leaves his stick and not his visiting-card after waiting an hour in your room."

"And the dog?"

"Has been in the habit of carrying this stick behind his master. Being a heavy stick the dog has held it tightly by the middle, and the marks of his teeth are very plainly visible. The dog's jaw as shown in the space between these marks, is too broad in my opinion for a terrier and not broad enough for a mastiff."

Illustration by Sidney Paget, *The Strand Magazine*, 1891

Holmes studied the stick carefully and came up with many surprising conclusions. He concluded that the owner of the stick was a country practitioner who walked a great deal. Additionally, the stick's owner had left Charing Cross Hospital five years earlier, he was absent-minded, and he owned a dog. Complete the logical argument in each deduction below. Return to the story if necessary for Exercises 1–3.

1. If the stick had been knocked about so much, then the stick must have belonged to a country practitioner.
The stick had been knocked about.
Therefore —?—.

2.* If the iron tip of the walking stick was worn down, then —?—.
The iron tip of the walking stick was worn down.
Therefore —?—.

3. If the date on the stick was 1884, then the stick's owner had left the hospital staff five years earlier.
The date on the stick was 1884.
Therefore —?—.

4. If $\triangle ABC$ is isosceles, then the base angles are congruent.
$\triangle ABC$ is isosceles.
Therefore —?—.

5.* If Carolyn has study sessions, then she does well on her tests.
If Carolyn does well on her tests, then she gets good grades on her report card.
Carolyn has study sessions.
Therefore —?—. (two steps)

6. If the contents in the brightly wrapped gift box are roses,
then the contents in the brightly wrapped gift box are flowers.
The contents in the brightly wrapped gift box are not flowers.
Therefore the contents in the brightly wrapped gift box are —?—.

7.* If △ABC is equilateral, then $m\angle A = 60°$.
$m\angle A \neq 60°$
Therefore —?—.

8. If Colonel Moran is the murderer, then he has powder burns on his shirt.
The colonel does not have powder burns on his shirt.
Therefore —?—.

9. If a cone with a radius of r for the circular base has a height of $2r$, then the volume of the cone will equal $\frac{2}{3}\pi r^3$. If the volume of the cone will equal $\frac{2}{3}\pi r^3$, then the volume of the cone will equal the volume of a hemisphere with a radius of r. A certain cone with a radius of r for the circular base has a height of $2r$.
Therefore —?—.

For Exercises 10–13, use your knowledge of geometry and good logical reasoning.

10. If quadrilateral $ABCD$ is a kite with $\overline{AB} \cong \overline{AD}$, then $\overline{AC}$ is the perpendicular bisector of $\overline{BD}$. Quadrilateral $ABCD$ is a kite with $\overline{AB} \cong \overline{AD}$. What can you conclude?

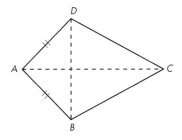

11. If the lengths of the three sides of △RSH are $RS = 14$ cm, $RH = 48$ cm, and $SH = 50$ cm, then △RSH is a right triangle. The lengths of the three sides of △RSH are $RS = 14$ cm, $RH = 48$ cm, and $SH = 50$ cm. If △RSH is a right triangle, then the area of △RSH is 336 cm². What can you conclude?

12. Lin Lian Po claims that the diagonals of every parallelogram bisect the angles of the parallelogram. Can you help her prove her statement true, or can you find a counterexample to prove her conjecture false? Explain.

13. Patch has just made a geometry discovery, and he is trying to convince his fellow group members that it is true. "If circle A has twice the circumference of circle B," explains Patch, "then the area of circle A is twice the area of circle B." Is Patch correct? Present an algebraic argument to help convince his group members that his conjecture is true or provide a counterexample to prove his conjecture false.

14.* When you read the words *therefore, because,* or *since,* a logical argument is often implied. Find two logical arguments within the following paragraph, then write them in this form: If statement A, then statement B. Statement A. Therefore statement B.

Papercutting is a folk art found in cultures around the world, and cut-paper patterns have been given as gifts for centuries. Paper patterns were first used as patterns for pottery or textiles, but they later became an art form themselves. A popular traditional Chinese cut-paper pattern depicts the carp. One ancient Chinese folk tale tells of a carp that, in trying to migrate upstream, leaped so high it cleared the famous Dragon Gate. The carp is therefore thought to represent great leaps in good fortune—good luck! Because the Chinese color of good luck is red, the carp is often cut from red paper.

Lesson 14.3

Forms of Valid Reasoning

Readers are plentiful, thinkers are rare.
— *Harriet Martineau*

In Lesson 14.2, Sherlock Holmes demonstrated his amazing ability to use deductive reasoning, or logic. Holmes used certain forms of valid reasoning to draw his conclusions. In this lesson you will learn to translate logical arguments from English into symbols. Then you, too, will learn to use two of the accepted forms of reasoning found in logical arguments.

A **logical argument** consists of a set of premises and a conclusion. Each given statement is a **premise**, and the statement arrived at through reasoning is called the **conclusion**. An argument is **valid** if the conclusion has been arrived at through accepted forms of reasoning. If all the premises of an argument are true and the argument is valid, then and only then must the conclusion be true.

If you draw a conclusion by using an unaccepted form of reasoning, then the argument is **invalid**. An **unsound** argument is simply an argument in which one or more premises are false. An unsound argument, like an invalid one, may or may not have a true conclusion.

English logician and mathematician Bertrand Russell (1872–1970) was once teaching a class at the City College of New York in which he illustrated the difference between validity and soundness. Russell stated that it is possible to "prove" anything if you have at least one false premise. One of his students challenged him. "All right," the student said, "Suppose that $1 + 1 = 1$. Prove that you are the pope." Russell answered almost immediately. "I am 1 and the pope is 1, and if we are considered together, we are $1 + 1$—which is 1—so we are one person and I am the pope."

When you translate an argument into symbolic form, you use capital letters (*P, Q, R, S*, for example) to stand for simple statements that are either true or false. When you write "IF *P* THEN *Q*," you are writing a **conditional statement**. For example, if *P* stands for the sentence "Watson had chalk between his fingers" and *Q* stands for the sentence "Watson was playing billiards," then the conditional statement IF *P* THEN *Q* translates to "If Watson had chalk between his fingers, then Watson was playing billiards."

Example A

LET *P:* I study.

LET *Q:* I get good grades.

IF *P* THEN *Q:* If I study, then I get good grades.

One of the accepted forms of reasoning used by Sherlock Holmes is called modus ponens.

*According to **modus ponens** (MP), if you accept IF P THEN Q as true and you accept P as true, then you must logically accept Q as true.*

Holmes's use of modus ponens is shown below. The symbolic form of the argument is on the right.

Example B

English argument	Symbolic translation
If Watson had chalk between his fingers, then he had been playing billiards. Watson had chalk between his fingers. Therefore Watson had been playing billiards.	LET *P:* Watson had chalk between his fingers. LET *Q:* Watson had been playing billiards. IF *P* THEN *Q* *P* THEREFORE *Q*

Let's look at another example of modus ponens.

Example C

English argument	Symbolic translation
Sonya had the flu. If Sonya had the flu, then Sonya gave the flu to Melissa. Therefore Sonya gave the flu to Melissa.	LET *S:* Sonya had the flu. LET *M:* Sonya gave the flu to Melissa. *S* IF *S* THEN *M* THEREFORE *M*

The **negation** of a sentence is made by placing the word *not* in the sentence appropriately or by preceding the sentence with the phrase "It is not the case that." For example, the negation of the sentence "Paula will receive her inheritance" can be expressed in two ways: "Paula will not receive her inheritance" or "It is not the case that Paula will receive her inheritance." If *P* stands for the sentence "Paula will receive her inheritance," then NOT *P* stands for the sentence "Paula will not receive her inheritance."

To express the negation of a sentence that already contains a negation, remove the *not* or precede the sentence with "It is not the case that." For example, the negation of "It is not raining" can be expressed in two ways: "It is raining" or "It is not the case that it is not raining." The two expressions are equivalent. This property is called **double negation.** If *P* stands for a sentence that contains a negative, such as "Mike did not win the sweepstakes," then NOT *P* stands for the sentence "It is not the case that Mike did not win the sweepstakes" or "Mike won the sweepstakes."

Another valid form of reasoning used by Sherlock Holmes is called modus tollens.

*According to **modus tollens** (MT), if you accept IF P THEN Q as true and you accept NOT Q as true, then you must logically accept NOT P as true.*

Holmes's use of modus tollens is illustrated in Example D.

Example D

English argument	Symbolic translation
If Watson wished to invest his money in South African securities with Thurston, then he would have had his checkbook with him when playing billiards with Thurston. Watson did not have his checkbook when playing billiards with Thurston. Therefore Watson does not wish to invest his money in South African securities with Thurston.	LET S: Watson wished to invest in South African securities with Thurston. LET T: Watson had his checkbook when playing billiards with Thurston. IF S THEN T NOT T THEREFORE NOT S

Let's look at another example of modus tollens.

Example E

English argument	Translation
I do not get wet when I go outside. If it is raining, then I get wet when I go outside. Therefore it is not raining.	LET W: I get wet when I go outside. LET R: It is raining. NOT W IF R THEN W THEREFORE NOT R

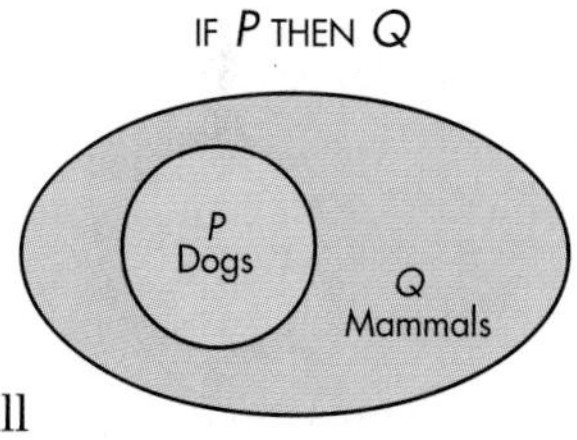

A Venn diagram can be a useful tool for "seeing" logic statements. For example, a conditional statement such as IF P THEN Q can be restated as "If a point is in region P, then it must be in region Q" or "All the points of P are in Q," which can be shown visually by the Venn diagram at right. To illustrate this, suppose region P represents all dogs and region Q represents all mammals. It is logical to make the conditional statement "If Spot is a dog, then Spot is a mammal."

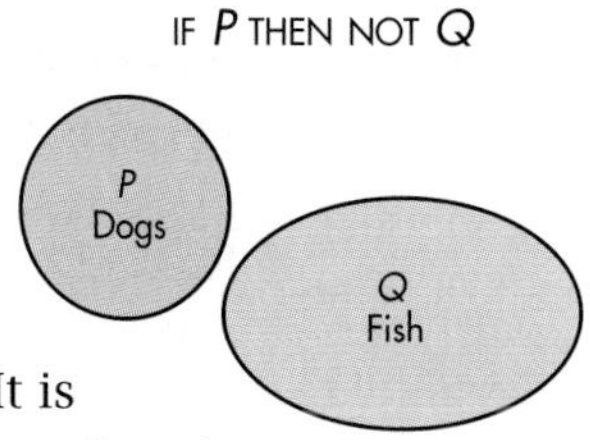

A Venn diagram is also helpful for visualizing logic statements that contain negations. For example, a conditional statement such as IF P THEN NOT Q can be restated as "If a point is in region P, then it must not be in region Q" or "None of the points of P are in Q," which can be shown visually by the Venn diagram at right. To illustrate this, suppose region P represents all dogs and region Q represents all fish. It is logical to make the conditional statement "If Spot is a dog, then Spot is not a fish."

Exercise Set 14.3

1.* Phuong declares, "I live at 2332 Oak Street." Izuru replies, "I don't believe you. Prove it." How might Phuong prove it?

2. Edith, Ernie, and Eva are an economist, an electrician, and an engineer, but not necessarily in that order. The economist does consulting work for Eva's business. Ernie hired the electrician to rewire his new kitchen. Edith earns less than the engineer but more than Ernie. Match the names with the occupations.

In Exercises 3-6, use the legend below to translate from symbols into English.

LET *P:* I get a job.   LET *Q:* I will earn money.

LET *R:* I will go to the movies.   LET *S:* I will spend my money.

3.* IF *P* THEN *Q*   4. IF *Q* THEN *R*   5. IF NOT *P* THEN *R*   6. IF NOT *R* THEN NOT *S*

In Exercises 7-10, use the legend below to translate from English into symbols.

LET *P:* Today is Wednesday.   LET *Q:* Tomorrow is Thursday.

LET *R:* Friday is coming.   LET *S:* Yesterday was Tuesday.

7.* If today is Wednesday, then tomorrow is Thursday.

8. If tomorrow is Thursday, then Friday is coming.

9. If yesterday was not Tuesday, then tomorrow is not Thursday.

10. If yesterday was Tuesday, then Friday is coming.

In Exercises 11-13, create a Venn diagram illustrating each conditional statement.

11. IF *R* THEN *S*   12.* IF NOT *R* THEN *S*   13. IF NOT *R* THEN NOT *S*

In Exercises 14-16, write a conditional statement that fits each Venn diagram.

14.
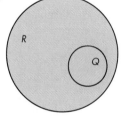

15.
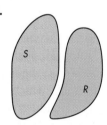

16.
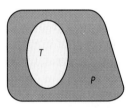

In Exercises 17-31, if the two premises fit the logically valid reasoning pattern of modus ponens or modus tollens, state which reasoning pattern is used and state the conclusion in English. Then translate the complete argument into symbolic form. If the statements do not fit the logically valid reasoning pattern of modus tollens or modus ponens, simply write "No valid conclusion."

17.* If Sherlock Holmes gathers all the clues, then he will be victorious. Sherlock Holmes gathers all the clues.

18. If Golum is carnivorous, then Golum is dangerous. Golum is dangerous.

19. Triangle *ABC* is isosceles. If a triangle is isosceles, then its base angles are congruent.

20. If Chris goes out with David, then she will have fun on Saturday night. Chris will not have fun on Saturday night.

21. If Merlo the Magnificent performs magic on Wednesday, then he will receive his royalty check from the Magician's Guild. Merlo the Magnificent performs magic on Thursday but not on Wednesday.

22. If $\overline{ED}$ is a midsegment in $\triangle ABC$, then $\overline{ED}$ is parallel to a side of $\triangle ABC$. $\overline{ED}$ is a midsegment in $\triangle ABC$.

23. If $2x + 3 = 17$, then $x = 7$. But $x \neq 7$.

24. If the sun rises in the west, then morning shadows point east. Morning shadows do not point east.

25.* If Van takes the bus, then she will be late for her job interview. Van does not take the bus.

26. If Hemlock Bones decodes the secret message, then Agent Otto will not be captured. Agent Otto will not be captured.

27.* If the Egyptian obelisks and the Mayan stele were made by close descendants of the same culture, then they would have many similarities. The Mayan stele is very different in shape and design from the Egyptian obelisks.

28. If $\angle A$ and $\angle B$ are vertical angles, then they are congruent. $\angle A$ and $\angle B$ are not vertical angles.

29. If $ABCD$ is a rectangle, then its diagonals are congruent. The diagonals of $ABCD$ are not congruent.

30. If parallelogram $ABCD$ is not a rhombus, then $\overline{AB} \cong \overline{CB}$. But $\overline{AB} \cong \overline{CB}$.

Obelisks at Karmac, Egypt   Mayan stele at Chichén Itzá, Mexico

31. If the lengths of the three sides of $\triangle ABC$ are 8, 15, and 17, then the triangle is a right triangle. The lengths of the three sides of $\triangle ABC$ are not 8, 15, 17.

32. Aja, Bea, Corrina, and Dottie are on four different sports teams. Each young woman plays on only one team. They play soccer, basketball, baseball, and tennis. Bea's position is goalie. The tallest of the four plays basketball, and the shortest is a tennis player. Corrina is taller than Dottie but shorter than both Bea and Aja. Match each person with her sport. Explain your reasoning.

33. What is one possible conditional statement implied by the advertisement shown at right? What does the company selling the product want the consumer to do? What conclusion does the company want the consumer to make?

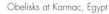

How would you know if your fish had fleas?

Keep your fish fit and frisky with Flea-Free Fine Fish Collars!

. . . because fish can't scratch themselves.

34. Refer to the story about Bertrand Russell at the beginning of this lesson. Is Russell's argument that he is the pope sound or unsound? Explain.

Lesson 14.4

The Symbols of Logic

Mathematicians are very symbol-minded people.
— *Trammell Crow*

Logicians use symbols to shorten logical arguments, making them easier to understand.

IF *P* THEN *Q* is written $P \rightarrow Q$.

NOT *R* is written ~*R*.

The symbol $\therefore$ stands for THEREFORE.

Each statement below has been translated from English to symbolic form.

English statement	Symbolic translation
If the evening sky is red, then tomorrow will be a beautiful day.	LET *P:* The evening sky is red. LET *Q:* Tomorrow will be a beautiful day. $P \rightarrow Q$
If yesterday was Tuesday, then tomorrow will not be Friday.	LET *Y:* Yesterday was Tuesday. LET *T:* Tomorrow will be Friday. $Y \rightarrow \sim T$
It is not true that if I fail the test, then I will not graduate.	LET *F:* I fail the test. LET *G:* I will graduate. $\sim(F \rightarrow \sim G)$

The logical argument below is translated from English to symbolic form.

English statement	Symbolic translation
If the integer is divisible by 6, then it is divisible by 3. The integer is not divisible by 3. Therefore it is not divisible by 6.	LET *P:* The integer is divisible by 6. LET *Q:* The integer is divisible by 3. $P \rightarrow Q$ $\sim Q$ $\therefore \sim P$

Exercise Set 14.4

In Exercises 1-6, use logician's shorthand to shorten each statement.

1.* IF *P* THEN NOT *Q*

2. IF NOT *Q* THEN *R*

3. IF *Q* THEN NOT *R*

4. IF NOT *Q* THEN NOT *P*

5.* IT IS NOT TRUE THAT IF *R* THEN *S*

6. IT IS NOT TRUE THAT IF NOT *P* THEN *R*

In Exercises 7–9, translate each conditional sentence from English to symbolic form. Use the letter substitutions indicated below.

LET *P:* World peace is in jeopardy.
LET *Q:* International tensions are reduced.
LET *R:* The destruction of civilization could result.
LET *S:* The build-up of nuclear weapons continues.

7. If international tensions are not reduced, then world peace is in jeopardy.

8. If the build-up of nuclear weapons continues, then the destruction of civilization could result.

9. It is not true that if the build-up of nuclear weapons does not continue, then international tensions are reduced.

In Exercises 10–15, translate each symbolic sentence into an English sentence. Use the legend you used in Exercises 7–9.

10. $P \rightarrow R$ 11. $S \rightarrow \sim Q$ 12.* $\sim Q \rightarrow R$

13. $\sim P \rightarrow \sim S$ 14.* $\sim(\sim P \rightarrow \sim Q)$ 15. $\sim(\sim S \rightarrow \sim R)$

In Exercise 16–24, identify each symbolic argument as modus ponens (MP) or modus tollens (MT). If the argument is not valid, write "No valid conclusion."

16. $T \rightarrow R$
    T
    $\therefore R$

17.* $S \rightarrow \sim Q$
    Q
    $\therefore \sim S$

18. $\sim Q \rightarrow R$
    $\sim Q$
    $\therefore R$

19.* $T \rightarrow R$
    R
    $\therefore \sim T$

20. $P \rightarrow Q$
    $\sim P$
    $\therefore \sim Q$

21. $\sim R \rightarrow S$
    $\sim S$
    $\therefore R$

22.* $\sim P \rightarrow (R \rightarrow Q)$
    $\sim P$
    $\therefore (R \rightarrow Q)$

23.* $T \rightarrow \sim P$
    $\sim P \rightarrow \sim Q$
    T
    $\therefore \sim Q$

24.* $\sim R \rightarrow P$
    $P \rightarrow \sim Q$
    Q
    $\therefore R$

In Exercises 25–30, if the two premises fit the logically valid reasoning pattern of modus tollens or modus ponens, state the conclusion in English. Set up a legend showing the sentence that each letter represents, then translate the complete argument into symbolic form. If the statements do not fit the logically valid reasoning pattern of MT or MP, simply write "No valid conclusion."

25. If the treasure is discovered, then pirate Rufus T. Ruffian will walk his own plank. The treasure is discovered.

26. If you use Shining Smile toothpaste, then you will be successful. You do not use Shining Smile toothpaste.

27. If Chrissy is an athlete, then she scores high on spatial visualization tests. Chrissy scores high on spatial visualization tests.

28. If $\overline{AC}$ is the longest side in $\triangle ABC$, then $\angle B$ is the largest angle in $\triangle ABC$. $\angle B$ is not the largest angle in $\triangle ABC$.

29. If $\angle B$ is the largest angle in $\triangle ABC$, then the measure of $\angle B$ is greater than 60°. $\angle B$ is the largest angle in $\triangle ABC$.

30. If $\ell_1 \parallel \ell_2$, then $\angle 1 \cong \angle 2$. $\ell_1 \nparallel \ell_2$.

Lesson 14.5

The Law of Syllogism

What is thinking? I should have thought I would have known.
— Karl Gerstner

So far you have learned two forms of valid reasoning, modus ponens and modus tollens. Now you will learn a third form of valid reasoning, the law of syllogism.

What conclusion can you draw from the two conditional statements below?

If I study, then I'll do well on the test tomorrow. If I do well on the test tomorrow, then I'll get a good grade in the class.

If you accept the two statements as true, then you must accept the truth of the conclusion: If I study, then I'll get a good grade in the class.

This valid form of reasoning is called the law of syllogism. The law of syllogism draws a conditional conclusion from two conditional statements.

*According to the **law of syllogism** (LS), if you accept IF P THEN Q as true and if you accept IF Q THEN R as true, then you must logically accept IF P THEN R as true.*

The argument below is an example of the law of syllogism.

English statement	Symbolic translation
If I eat a pizza after midnight, then I will have nightmares. If I have nightmares, then I will get very little sleep. Therefore if I eat pizza after midnight, then I will get very little sleep.	LET *P:* I eat pizza after midnight. LET *N:* I will have nightmares. LET *S:* I will get very little sleep. $P \rightarrow N$ $N \rightarrow S$ $\therefore P \rightarrow S$

Let's look at another example of the law of syllogism.

English statement	Symbolic translation
If I earn money, then I will buy a computer. If I get a job, then I will earn money. Therefore if I get a job, then I will buy a computer.	LET *P:* I will earn money. LET *Q:* I will buy a computer. LET *R:* I get a job. $P \rightarrow Q$ $R \rightarrow P$ $\therefore R \rightarrow Q$

Often a logical argument contains more than one logically valid reasoning pattern. Each of the following examples uses two logically valid reasoning patterns.

English statement	Symbolic translation
If I study all night, then I will miss my nighttime soap opera. If Jeannine comes over to study, then I study all night. Jeannine comes over to study. Therefore I will miss my nighttime soap opera.	LET *P:* I study all night. LET *Q:* I will miss my nighttime soap opera. LET *R:* Jeannine comes over to study. $P \rightarrow Q$ $R \rightarrow P$ R $\therefore Q$

You can show that this argument is valid in two logical steps.

$R \rightarrow P$
R
$\therefore P$ by modus ponens

$P \rightarrow Q$
P
$\therefore Q$ by modus ponens

English statement	Symbolic translation
If the consecutive sides of a parallelogram are congruent, then the parallelogram is a rhombus. If the parallelogram is a rhombus, then its diagonals are perpendicular bisectors of each other. The diagonals are not perpendicular bisectors of each other. Therefore the consecutive sides of the parallelogram are not congruent.	LET *P:* The consecutive sides of a parallelogram are congruent. LET *Q:* The parallelogram is a rhombus. LET *R:* The diagonals are perpendicular bisectors of each other. $P \rightarrow Q$ $Q \rightarrow R$ $\sim R$ $\therefore \sim P$

You can show that this argument is valid in two logical steps.

$P \rightarrow Q$
$Q \rightarrow R$
$\therefore P \rightarrow R$ by the law of syllogism

$P \rightarrow R$
$\sim R$
$\therefore \sim P$ by modus tollens

You can use a Venn diagram to visualize the law of syllogism. The Venn diagram at right illustrates the argument $P \rightarrow Q$ and $Q \rightarrow R$, therefore $P \rightarrow R$. For example, suppose region P contains all squares, region Q contains all rectangles, and region R contains all parallelograms. If quadrilateral *ABCD* is a square, then it is a rectangle. If *ABCD* is a rectangle, then it is a parallelogram. Therefore if quadrilateral *ABCD* is a square, it must be a parallelogram.

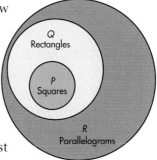

Exercise Set 14.5

In Exercises 1-6, if the premises fit the logically valid reasoning pattern of the law of syllogism, modus tollens, or modus ponens, state the conclusion in English. Set up a legend showing the sentence that each letter represents, then translate the complete argument into symbolic form. If the statements do not fit the logically valid reasoning pattern of LS, MT, or MP, simply write "No valid conclusion."

1. If you use Shining Smile toothpaste, then you will be successful. If you use Shining Smile toothpaste, then you may become a television star.

2. If a young woman is an athlete, then she probably scores high on spatial visualization tests. If she scores high on spatial visualization tests, then she will probably do well in geometry.

3. If Professor Moriarty wrote a paper about the binomial theorem, then he is familiar with Pascal's triangle. Professor Moriarty is not familiar with Pascal's triangle.

4. If you are open-minded, you will listen to both sides of a story. If you listen to both sides of a story, then you will make more intelligent decisions. If you make more intelligent decisions, then you will have a more successful career.

5. If two base angles of a triangle are congruent, then the triangle has two sides congruent. If two sides of a triangle are congruent, then the triangle is isosceles. The triangle is not isosceles.

6. If $\overline{EF}$ is a segment connecting the midpoints of the nonparallel sides $\overline{AD}$ and $\overline{BC}$ in trapezoid $ABCD$, then $\overline{EF}$ is a midsegment of trapezoid $ABCD$. If $\overline{EF}$ is a midsegment of trapezoid $ABCD$, then $\overline{EF}$ is parallel to side $\overline{AB}$. If $\overline{EF}$ is parallel to side $\overline{AB}$, then $ABFE$ is a trapezoid.

In Exercises 7-12, identify each symbolic argument as modus ponens (MP), modus tollens (MT), the law of syllogism (LS), or some combination. If the symbolic argument is not valid, write "No valid conclusion."

7. $P \to R$
    $R \to S$
    $\therefore P \to S$

8. $S \to \sim Q$
    S
    $\therefore \sim Q$

9. $\sim Q \to \sim R$
    Q
    $\therefore R$

10. $T \to R$
    $P \to R$
    $\therefore T \to P$

11. $T \to S$
    T
    $P \to \sim S$
    $\therefore \sim P$

12. $\sim R \to S$
    $S \to P$
    $P \to \sim Q$
    $\therefore \sim R \to \sim Q$

In Exercises 13-17, use clear logical reasoning to answer each question.

13. Tuesday evening, Mai Jiang's mother told her, "If you get an A on your chemistry exam tomorrow, I'll do your Saturday chores for you this coming weekend." Mai did get an A on her chemistry exam Wednesday. If Mai Jiang's mother always keeps her word, what should happen this weekend?

14. Coach Swartz told his starting five Wednesday evening, "If any of you don't come to practice tomorrow, then you won't play in Friday's game." Darlene is one of Coach Swartz's starting five. She came to practice Thursday, but she did not play in Friday's game. Did the coach break his word to Darlene? Explain.

15.* Nelson was asked to spend his afternoons sitting in the principal's office after school each day so he would, as the principal put it, "reconsider his past deviations from the norm of acceptable classroom deportment." While waiting in detention and trying to figure out what the principal said, Nelson scribbled 100 statements into his previously empty diary.

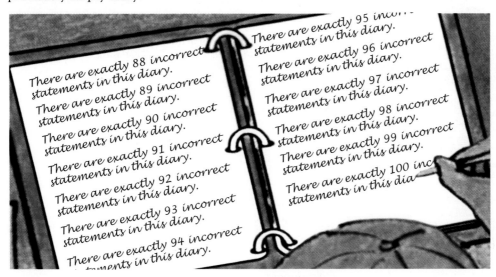

There are exactly 88 incorrect statements in this diary.
There are exactly 89 incorrect statements in this diary.
There are exactly 90 incorrect statements in this diary.
There are exactly 91 incorrect statements in this diary.
There are exactly 92 incorrect statements in this diary.
There are exactly 93 incorrect statements in this diary.
There are exactly 94 incorrect ~~ments~~ in this diary.
There are exactly 95 ~~inco~~ statements in this diary.
There are exactly 96 incorrect statements in this diary.
There are exactly 97 incorrect statements in this diary.
There are exactly 98 incorrect statements in this diary.
There are exactly 99 incorrect statements in this diary.
There are exactly 100 inc statements in this dia

Which of Nelson's 100 statements is (or are) true? Explain.

16. Each day, Jade rides her motorcycle along the coast highway to the university, traveling a distance of 48 miles. She told her boyfriend that she never speeds (the speed limit is 55 miles per hour), but she admitted that yesterday she woke up late and still got to school in a record time of 48 minutes. Can you prove she was speeding at some point during her trip? Explain.

17. Tick-Tack-Toe Logic What are the positions O should *not* play because doing so will allow X to win? Explain.

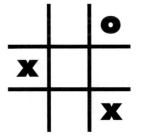

Improving Algebra Skills—*An Algebra Mind-Reading Trick III*

Mind-reading tricks often employ algebra. Try the following mind-reading trick. Use algebra to explain why the trick works.

Pick a number between 1 and 9. Triple it. Now add 7 to your answer. Multiply your results by 5. Subtract 32, then divide your answer by 3. Finally, double your results. Remove the 2 from your answer, and the number left is your original number.

Lesson 14.6

The Law of the Contrapositive

Illustration by John Tenniel

"You should say what you mean," the March Hare went on.

"I do," Alice hastily replied; *"at least—at least I mean what I say—that's the same thing, you know."*

"Not the same thing a bit!" said the Hatter. *"Why, you might just as well say that 'I see what I eat' is the same thing as 'I eat what I see'!"*

— Alice's Adventures in Wonderland *by Lewis Carroll*

Every conditional statement has three other conditionals associated with it. They are the converse, the inverse, and the contrapositive.

Statement:	$P \to Q$
Converse:	$Q \to P$
Inverse:	$\sim P \to \sim Q$
Contrapositive:	$\sim Q \to \sim P$

To create the **converse** of a conditional statement, the two parts of the statement are simply reversed. To create the **inverse**, the two parts are negated. To create the **contrapositive**, the two parts are reversed *and* negated.

In logic, conditional statements are either true or false. If a statement is true, is its converse necessarily true? Earlier in this text you worked with statements and converses, and you discovered that there is no true/false relationship that holds between a statement and its converse. How about its inverse? Contrapositive? Let's look at an example of a true conditional and its converse, its inverse, and its contrapositive.

Example

Statement:	If two angles are vertical angles, then they are congruent. $(P \to Q)$ The statement is *true*.
Converse:	If two angles are congruent, then they are vertical angles. $(Q \to P)$ The converse of the statement is *false*.
Inverse:	If two angles are not vertical angles, then they are not congruent. $(\sim P \to \sim Q)$ The inverse of the statement is *false*.

Contrapositive: If two angles are not congruent, then they are not vertical angles. ($\sim Q \rightarrow \sim P$)
The contrapositive of the statement is *true*.

In this example, the contrapositive has the same truth value as the original conditional, and the converse and the inverse have the same truth value. Is this relationship always true? Let's investigate.

Investigation 14.6

Write the converse, the inverse, and the contrapositive for each of the four conditional statements below. Then identify each statement, converse, inverse, and contrapositive as true or false.

1.* If it is a rose, then it is a flower.

2. If you're out of chocolate cake, you're out of dessert.

3. If the triangle is isosceles, then the triangle's base angles are congruent.

4. If $\triangle ABC$ is congruent to $\triangle DEF$, then $AB = DE$.

For each statement above, did the contrapositive have the same truth value as the original conditional? Did the inverse and the converse have the same truth value? This leads to our fourth form of logical reasoning, the law of the contrapositive.

*The **law of the contrapositive** (LC) says that if a conditional statement is true, then its contrapositive is also true. Conversely, if the contrapositive is true, then the original conditional statement must also be true.*

Law of the contrapositive	Converse of the law of the contrapositive
$P \rightarrow Q$	$\sim Q \rightarrow \sim P$
$\therefore \sim Q \rightarrow \sim P$	$\therefore P \rightarrow Q$

The law of the contrapositive says that any conditional and its contrapositive are logically equivalent. You may replace one with the other.

From the investigation it should be clear that both a statement and its converse may or may not have the same truth value. However, a statement and its contrapositive always have the same truth value (LC). If one is true, then the other is true. If one is false, then the other is false. Look again at the inverse and the converse of one of the statements in the investigation. Notice that each is the contrapositive of the other. The inverse and the converse are logically equivalent. They are either both true or both false.

So far, you have learned four basic forms of valid reasoning.

Four forms of valid reasoning			
$P \rightarrow Q$	$P \rightarrow Q$	$P \rightarrow Q$	$P \rightarrow Q$
P	$\sim Q$	$Q \rightarrow R$	$\therefore \sim Q \rightarrow \sim P$
$\therefore Q$	$\therefore \sim P$	$\therefore P \rightarrow R$	
by MP	by MT	by LS	by LC

Exercise Set 14.6

In Exercises 1–24, determine whether or not each logical argument is valid. If the argument is valid, state what logical reasoning pattern or patterns it follows (MP, MT, LS, LC). If the argument is not valid, write "No valid conclusion."

1. $P \to Q$
   P
   $\therefore Q$

2. $R \to S$
   S
   $\therefore R$

3. $P \to R$
   $\sim R$
   $\therefore \sim P$

4.* $\sim P \to S$
   $\sim P$
   $\therefore S$

5. $\sim R \to \sim S$
   $\sim R$
   $\therefore \sim S$

6. $\sim P \to \sim R$
   R
   $\therefore P$

7. $\sim P \to S$
   $\sim S$
   $\therefore P$

8.* $\sim P \to \sim Q$
   $\sim Q$
   $\therefore P$

9. $P \to Q$
   $\sim P$
   $\therefore \sim Q$

10. $R \to P$
    $P \to S$
    $\therefore R \to S$

11. $S \to P$
    $R \to P$
    $\therefore S \to R$

12.* $\sim(P \to Q)$
    $\therefore \sim P \to \sim Q$

13.* $P \to Q$
    $Q \to S$
    $S \to T$
    $\therefore P \to T$

14. $P \to \sim Q$
    Q
    $\therefore \sim P$

15. $\sim Q \to \sim R$
    Q
    $\therefore R$

16. $\sim S \to P$
    $R \to \sim S$
    $\therefore R \to P$

17. $R \to S$
    $\therefore \sim S \to R$

18. $R \to T$
    $\therefore \sim T \to \sim R$

19. $R \to P$
    $T \to \sim R$
    $\therefore R \to T$

20. $\sim P \to Q$
    $\therefore \sim Q \to P$

21. $P \to Q$
    $\sim R \to \sim Q$
    $\therefore P \to R$

22.* $\sim T \to \sim Q$
    $S \to Q$
    $\therefore S \to T$

23. R
    $R \to (P \to T)$
    $\therefore P \to T$

24. $(P \to Q) \to R$
    $P \to Q$
    $R \to S$
    $\therefore S$

25. Use the Venn diagram to make a conditional statement.

26. Write the converse, the inverse, and the contrapositive of the conditional statement you made in Exercise 25.

27. Show how you can use modus ponens and the law of the contrapositive to make the same logical conclusions as modus tollens.

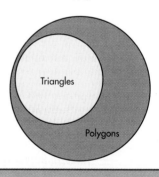

Triangles

Polygons

Improving Reasoning Skills—*The Dealer's Dilemma*

In the middle of dealing 52 cards among 4 players for a bridge game, you must stop for a phone call. When you return, no one can remember where the last card was dealt. (And, of course, no cards have been touched.) Without counting the number of cards in anyone's hand or the number of cards yet to be dealt, how can you rapidly finish dealing, giving each player exactly the same cards she or he would have received if you hadn't been interrupted?

Direct Proofs

If there is an opinion, facts will be found to support it.
— Judy Sproles

There are three basic approaches to proving logical arguments. These approaches consist of direct proofs, conditional proofs, and indirect proofs. In this lesson you will look at direct proofs. In a direct proof of a logical argument, the given information or premises are stated, then valid patterns of reasoning, such as MP, MT, LS, and LC, are used to arrive directly at the conclusion.

Below is the symbolic form of a logical argument. This form is called a two-column proof. In a two-column proof, each statement in the logical argument is written in the left-hand column. The reason for each statement is written directly across in the right-hand column. To show that the argument is valid, use a direct proof.

Example

The following is an example of a direct proof.

Premises: $P \rightarrow Q$

            $R \rightarrow P$

            $\sim Q$

Conclusion: $\sim R$

Proof:

1. $P \rightarrow Q$ 1. Premise
2. $\sim Q$ 2. Premise
3. $\sim P$ 3. From lines 1 and 2, using MT
4. $R \rightarrow P$ 4. Premise
5. $\therefore \sim R$ 5. From lines 3 and 4, using MT

Because the proof arrived at $\sim R$ by using only the premises and logical reasoning, the argument is valid. You should be convinced that if $P \rightarrow Q$, $R \rightarrow P$, and $\sim Q$ are true, then $\sim R$ must also be true. Like many proofs, this proof could have been written in more than one way.

You can also visualize the proof with a flow chart, as shown below. Place in boxes the statements of the argument (or proof), and place beneath the boxes the reasons you are able to make the statements. Use arrows to show the flow of reasoning. The argument flows from the premises to the conclusion.

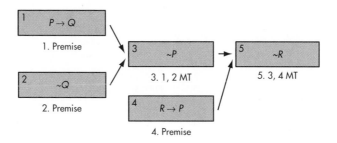

Exercise Set 14.7

For Exercises 1 and 2, copy the proof of the logical argument, including the list of premises and the conclusion. Provide each missing reason.

1.* Premises: P
          $P \rightarrow Q$
          $Q \rightarrow R$
          $R \rightarrow S$

  Conclusion: S

  Proof:
  1. $P \rightarrow Q$ 1. —?—
  2. $Q \rightarrow R$ 2. —?—
  3. $P \rightarrow R$ 3.* —?—
  4. $R \rightarrow S$ 4. —?—
  5. $P \rightarrow S$ 5. —?—
  6. P 6. —?—
  7. $\therefore S$ 7. —?—

2. Premises: $P \rightarrow Q$
          $Q \rightarrow \sim R$
          R

  Conclusion: $\sim P$

  Proof:
  1. $Q \rightarrow \sim R$ 1. —?—
  2. R 2. —?—
  3. $\sim Q$ 3. —?—
  4. $P \rightarrow Q$ 4. —?—
  5. $\therefore \sim P$ 5. —?—

3.* Copy the proof below, including the list of premises and the conclusion. Provide each missing statement and reason.

  Premises: $P \rightarrow R$
          $T \rightarrow S$
          $\sim T \rightarrow P$
          $\sim S$

  Conclusion: R

  Proof:
  1. $T \rightarrow S$ 1. —?—
  2. $\sim S$ 2. —?—
  3. $\sim T$ 3. —?—
  4. $\sim T \rightarrow P$ 4. —?—
  5. P 5.* —?—
  6. —?—   6. —?—
  7. —?—   7. —?—

4.* There can be more than one way to prove an argument. The argument in Exercise 3 can also be proved as shown below. Copy this proof, including the list of premises and the conclusion. Provide each missing statement and reason.

  Premises: $P \rightarrow R$
          $T \rightarrow S$
          $\sim T \rightarrow P$
          $\sim S$

  Conclusion: R

  Proof:
  1. $T \rightarrow S$ 1. —?—
  2. $\sim S \rightarrow \sim T$ 2.* —?—
  3. $\sim T \rightarrow P$ 3. —?—
  4. $\sim S \rightarrow P$ 4. —?—
  5. $P \rightarrow R$ 5. —?—
  6. —?—   6. —?—
  7. —?—   7. —?—
  8. —?—   8. —?—

In Exercises 5–7, provide each step and reason necessary to prove the logical argument.

5. Premises: $S \rightarrow Q$
$\qquad\qquad\;\; P \rightarrow S$
$\qquad\qquad\;\; \sim R \rightarrow P$
$\qquad\qquad\;\; \sim Q$
Conclusion: R

6. Premises: $P \rightarrow Q$
$\qquad\qquad\;\; \sim P \rightarrow R$
$\qquad\qquad\;\; \sim R$
Conclusion: Q

7. Premises: P
$\qquad\qquad\;\; Q \rightarrow R$
$\qquad\qquad\;\; R \rightarrow S$
$\qquad\qquad\;\; P \rightarrow Q$
Conclusion: S

8. The following argument in English has been translated into a symbolic argument. Prove that the argument is valid.

English statement	Symbolic translation
If you listen to both sides of a story, then you will make an intelligent decision. If you are open-minded, then you will listen to both sides of a story. You did not make an intelligent decision. Therefore you are not open-minded.	LET P: You listen to both sides of a story. LET Q: You will make an intelligent decision. LET R: You are open-minded. $P \rightarrow Q$ $R \rightarrow P$ $\sim Q$ $\therefore \sim R$

In Exercises 9–11, translate the English argument into symbolic form, then prove that the argument is valid.

9.* If all wealthy people are happy, then money can buy happiness. If money can buy happiness, then love doesn't exist. But love exists. Therefore not all wealthy people are happy.

10. If 60 out of 104 majors at the local university require calculus, then calculus is an important class to take in college. If calculus is an important class to take in college, then high school students should take all the math necessary to enter a calculus class. In fact, 60 out of 104 majors at the local university require calculus. Therefore high school students should take all the math necessary to enter a calculus class.

11. If $\triangle ABC$ is isosceles, then base angles $\angle A$ and $\angle B$ are congruent. If $\overleftrightarrow{DE}$ is parallel to $\overleftrightarrow{AC}$, then the alternate interior angles $\angle A$ and $\angle D$ are congruent. Line DE is parallel to $\overleftrightarrow{AC}$. Triangle ABC is isosceles. Therefore $\angle A$ and $\angle B$ are congruent and $\angle A$ and $\angle D$ are congruent.

12.* Pirate Captain Coldhart has left us some clues about the location of his latest buried treasure: his footprints in the sand. According to a note found in one of his boots, he moved the very heavy treasure chest from the base of one palm tree and carried it to a number of different palm trees before deciding at which tree to rebury his booty. He thought that because his path returned over and over to the various trees, no one would be able to figure out which tree he buried his treasure beneath, but you can! Where did he start? Where did he end his walk and bury the treasure? Explain.

13.* Upon capture, Captain Coldhart was brought before Judge Bertrand Whitehead Boole—a former logic teacher—for sentencing. Judge Boole, who loves logic games, told Coldhart to stand and make one statement. "If your statement is true, then you will be sentenced to spend the rest of your life on a diet of broccoli. If your statement is false, you will be sentenced to spend your life on a diet of Brussels sprouts." Because Coldhart had never been a fan of vegetables to begin with, a choice between a life on broccoli or Brussels sprouts inspired him to make a clever statement that spared him sentencing. What statement did he make?

Improving Algebra Skills—*How Fat Is the Cat Named Rat?*

Young Carlos and his little sister Christina have a pet cat named Rat. They notice that poor Rat has put on a few extra pounds lately, thus they have decided to weigh their furry feline friend. Unfortunately, their scale does not register any weight under 100 pounds. They very cleverly decide that each of them will step on the scale with Rat and record their total weight. Carlos plus Rat weigh 125 pounds, and Christina plus Rat weigh 118 pounds. However, neither Carlos nor Christina weighs 100 pounds so they can't weigh themselves standing on the scale alone. They do, however, stand on the scale together. Brother and sister together weigh a total of 189 pounds. From this they are able to determine Rat's weight. Can you?

Improving Algebra Skills—*Happy Birthday, Sweet Sixteen*

The five-by-five grid of squares shown at right is a magic birthday card for anyone turning sixteen. Here is what the birthday person does with the card.

Step 1   Circle any number in the five-by-five array and cross out all the other numbers in your chosen number's row and column.

Step 2   Circle a second number from those remaining in the five-by-five array and cross out all the other numbers in your second circled number's row and column.

Step 3   Continue circling numbers, then crossing out all the remaining numbers in the row and the column.

6	1	8	2	4
5	0	7	1	3
3	−2	5	−1	1
7	2	9	3	5
4	−1	6	0	2

This process will leave you with five numbers circled and the others crossed out. The sum of the circled numbers should be 16. Amazingly, the sum will be 16 no matter what numbers you choose! Try it a couple of times. Can you explain why this works?

Lesson 14.8

Conditional Proofs

If you wait for tomorrow, tomorrow comes. If you don't wait for tomorrow, tomorrow comes.

— Traditional saying from Sierra Leone

One of the most common methods of proof in mathematics is the conditional proof. A conditional proof is used to prove that a $P \to Q$ type of statement follows logically from a set of premises. A conditional proof begins when an assumption is made. It is assumed that the first part of the conditional statement, called the **antecedent**, is true. Then logical reasoning is used to demonstrate that the second part of the conditional statement, called the **consequent**, must also be true. If this process is successful, it's demonstrated that IF P IS TRUE, THEN Q MUST BE TRUE. In other words, a conditional proof shows that the antecedent implies the consequent.

Example

The following is an example of a conditional proof.

Premises: $P \to R$
            $S \to \sim R$

Conclusion: $P \to \sim S$

Proof:

1. P	1. Assume the antecedent
2. $P \to R$	2. Premise
3. R	3. From lines 1 and 2, using MP
4. $S \to \sim R$	4. Premise
5. $\sim S$	5. From lines 3 and 4, using MT

Assuming P is true, the truth of $\sim S$ is established.

$\therefore P \to \sim S$

Although the logical argument above was proved using a conditional-proof format, the same argument could also have been proved using a direct-proof format. Many logical arguments can be proved using more than one type of proof. With practice you will be able to determine which method you should use for a particular argument. Here's how the two-column proof above fits into a flow-proof format.

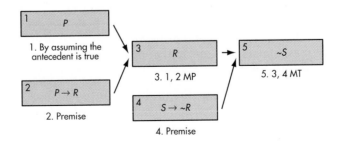

Exercise Set 14.8

For Exercises 1 and 2, copy the proof, including the list of premises and the conclusion. Provide each missing statement or reason.

1.* Premises: $P \rightarrow R$
$S \rightarrow \sim R$
$\sim P \rightarrow Q$

Conclusion: $\sim Q \rightarrow \sim S$

Proof:

1.	$\sim Q$	1.	—?—
2.	$\sim P \rightarrow Q$	2.	—?—
3.	P	3.	From lines 1 and 2, using —?—
4.	$P \rightarrow R$	4.	—?—
5.	R	5.	From lines 3 and 4, using —?—
6.	$S \rightarrow \sim R$	6.	—?—
7.*	$\sim S$	7.*	From lines 5 and 6, using —?—

Assuming —?— is true, the truth of —?— is established.
∴ $\sim Q \rightarrow \sim S$

2. Premises: $\sim R \rightarrow \sim Q$
$T \rightarrow \sim R$
$S \rightarrow T$

Conclusion: $S \rightarrow \sim Q$

Proof:

1.	S	1.	—?—
2.	$S \rightarrow T$	2.	—?—
3.	T	3.	From lines 1 and 2, using —?—
4.	$T \rightarrow \sim R$	4.	—?—
5.	$\sim R$	5.	—?—
6.	—?—	6.	—?—
7.	—?—	7.	—?—

Assuming S is true, the truth of $\sim Q$ is established.
∴ —?—

In Exercises 3–8, provide each step and reason necessary to prove the logical argument.

3. Premises: $\sim T \rightarrow S$
$T \rightarrow (R \rightarrow Q)$
$P \rightarrow \sim S$

Conclusion: $P \rightarrow (R \rightarrow Q)$

4. Premises: P
$Q \rightarrow R$

Conclusion: $(P \rightarrow Q) \rightarrow R$

5. Premises: $P \rightarrow (Q \rightarrow R)$
$R \rightarrow S$
Q

Conclusion: $P \rightarrow S$

6. Premises: $(P \rightarrow Q) \rightarrow R$
$\sim(P \rightarrow Q) \rightarrow T$
$R \rightarrow S$

Conclusion: $\sim S \rightarrow T$

7. Premises: $P \rightarrow Q$
$\sim R \rightarrow S$
$R \rightarrow \sim Q$
$\sim S$

Conclusion: $\sim P$

8. Premises: $P \rightarrow Q$
$\sim P \rightarrow S$
$R \rightarrow \sim S$
$\sim Q$

Conclusion: $\sim R$

In Exercises 9–11, translate the argument into symbolic terms, then prove the argument is valid. Some arguments may contain statements that do not logically support the conclusion. Be on guard for unnecessary statements.

9. If Evette is innocent, then Alfa is telling the truth. If Romeo is telling the truth, then Alfa is not. If Romeo is not telling the truth, then he has something to gain. Romeo has nothing to gain. Therefore if Romeo has nothing to gain, then Evette is not innocent.

10.* If the consecutive sides of a parallelogram are congruent, then the parallelogram is a rhombus. If a parallelogram is a rhombus, then its diagonals are perpendicular bisectors of each other. If a parallelogram is a rhombus, then its diagonals bisect the angles. The diagonals are not perpendicular bisectors of each other in parallelogram *ABCD*. Therefore the consecutive sides of parallelogram *ABCD* are not congruent.

11. If $\triangle ABC \cong \triangle DEF$, then $\angle A \cong \angle D$. If $\angle A \cong \angle D$, then $a = d$. But $a \neq d$. Therefore $\triangle ABC \not\cong \triangle DEF$.

12. How does a conditional proof differ from a direct proof? Give your own example of a logical argument with a conditional conclusion. Show how you can prove your argument with a direct proof and with a conditional proof.

Improving Visual Thinking Skills—*Folding Cubes III*

Figure X, as you have seen in earlier puzzles, can be folded along the dotted lines into a cube. Figure Y can also be folded along the dotted lines into a cube.

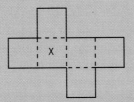

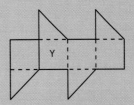

Which of the following figures can be folded along the dotted lines into a cube?

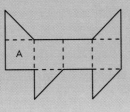

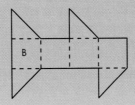

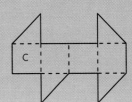

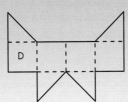

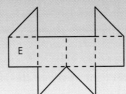

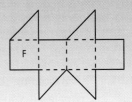

Lesson 14.9

Indirect Proofs

How often have I said to you that when you have eliminated the impossible, whatever remains, however improbable, must be the truth?

— *Sherlock Holmes in* The Sign of Four
by Sir Arthur Conan Doyle

The third type of proof is an indirect proof. An indirect proof is a clever, almost sneaky approach to proving something. In an indirect proof, all the possibilities that can be true are recognized. Next, all but one possibility are eliminated when they are shown to contradict some given fact or accepted idea. Therefore it must be accepted that the one remaining possibility is true. The mystery story below is an example of an indirect proof.

Five people alone on a tropical island have a loud argument after a card game. When four of them return to the scene of the argument later in the day, they discover the fifth person unconscious with an arrow in his back. The hero of the story, Sheerluck, a cousin of Sherlock Holmes, eliminates himself as a suspect because he knows he didn't commit the crime. He eliminates his fiancé as a suspect because she has been with him all day. He eliminates Colonel Moran because the colonel recently injured both arms and therefore would have been unable to shoot with a bow and an arrow. Sheerluck also eliminates the wounded man as a suspect because a wound in the back cannot be self-inflicted. There is only one other person on the island who could have commited the crime. Sheerluck concludes that the guilty party is this fifth person, Sir Charles Mortimer.

You can use indirect reasoning as a clever strategy for answering a multiple-choice question if you are unsure of the correct answer.

Example A

In which year was a United States president born?

a. 1492   b. 1676   c. 1809   d. 1979

Assuming that you cannot select the correct choice immediately, let's look at the four choices. If you can eliminate three of them, the remaining choice must be the correct answer. You can eliminate 1492 and 1676 because the person would be a hundred years old or older in 1776, the year the United States became a country. You can also eliminate

1979 because anyone born in 1979 would not be old enough to be president—the minimum age is thirty-five. Because you have only four choices and you have eliminated three of them, the remaining choice, 1809, must be correct. Therefore a United States president was born in 1809. Do you know who he was?

In mathematics and logic, if you wish to show that a statement is true, you can begin by assuming that it is not true. Then you can demonstrate that this assumption leads to a contradiction. For example, if you are given a set of premises and are asked to show that some conclusion P is true, begin the proof by assuming that the opposite of P, namely $\sim P$, is true. Then show that this assumption leads to a contradiction of an earlier statement. If $\sim P$ leads to a contradiction, it must be false and P must be true.

Let's look at an example of an indirect proof.

Example B

Premises: $R \rightarrow S$
$\sim R \rightarrow \sim P$
P

Conclusion: S

Proof:

1.	$\sim S$	1.	Assume the opposite of the conclusion
2.	$R \rightarrow S$	2.	Premise
3.	$\sim R$	3.	From lines 1 and 2, using MT
4.	$\sim R \rightarrow \sim P$	4.	Premise
5.	$\sim P$	5.	From lines 3 and 4, using MP
6.	P	6.	Premise

But lines 5 and 6 contradict each other. It's impossible for both P and $\sim P$ to be true.

Therefore $\sim S$, the original assumption, is false. If $\sim S$ is false, then S is true.

$\therefore S$

Exercise Set 14.9

For Exercises 1–3, determine the correct answer by indirect reasoning. Explain how you eliminated each incorrect choice.

1. What is the capital of Mali?
   a. Paris
   b. Tucson
   c. London
   d. Bamako

2. Which Italian scientist used a new invention called the telescope to discover the moons of Jupiter?
   a. Sir Edmund Halley
   b. Julius Caesar
   c. Galileo Galilei
   d. Madonna

3. Which person twice won a Nobel prize?
   a. Sherlock Holmes
   b. Marie Curie
   c. Joan of Arc
   d. Leonardo da Vinci

For Exercises 4-6, copy the proof of the logical argument, including the list of premises and the conclusion. Provide each missing statement or reason.

4. Premises: $P \rightarrow Q$

$R \rightarrow P$

$\sim Q$

   Conclusion: $\sim R$

Proof:

1. R 1. Assume the opposite of the —?—
2. $R \rightarrow P$ 2. —?—
3. P 3. From lines 1 and 2, using —?—
4. $P \rightarrow Q$ 4. —?—
5. Q 5. From lines —?— and —?—, using MP
6. $\sim Q$ 6. —?—

But lines 5 and 6 contradict each other. Therefore R, the assumption, is false.

$\therefore \sim R$

5.* Premises: $P \rightarrow (Q \rightarrow R)$

$Q \rightarrow \sim R$

Q

   Conclusion: $\sim P$

Proof:

1. P 1. Assume the —?— of the —?—
2. $P \rightarrow (Q \rightarrow R)$ 2. —?—
3. $Q \rightarrow R$ 3.* —?—
4. Q 4. —?—
5. R 5. —?—
6. —?—   6. —?—
7. —?—   7. From lines —?— and —?—, using —?—

But lines —?— and —?— contradict each other. Therefore P, the assumption, is false.

$\therefore \sim P$

6. Premises: $(P \rightarrow S) \rightarrow \sim Q$

$\sim Q \rightarrow R$

$\sim R$

   Conclusion: $\sim(\sim S \rightarrow \sim P)$

Proof:

1. $\sim S \rightarrow \sim P$ 1. Assume the —?— of the —?—
2. —?—   2. —?—
3. —?—   3. —?—
4. —?—   4. —?—
5. —?—   5. —?—
6. —?—   6. —?—
7. —?—   7. —?—

But lines —?— and —?— contradict each other. Therefore $\sim S \rightarrow \sim P$, the assumption, is false.

$\therefore \sim(\sim S \rightarrow \sim P)$

It's your turn. In Exercises 7-10, provide the steps and the reasons to prove each logical argument. You may not need to use indirect proof.

7. Premises: $(R \rightarrow S) \rightarrow P$

$T \rightarrow Q$

$\sim T \rightarrow \sim P$

$\sim Q$

   Conclusion: $\sim(R \rightarrow S)$

8. Premises: $P \rightarrow (R \rightarrow Q)$

P

$\sim(\sim Q \rightarrow S)$

   Conclusion: $\sim(\sim R \rightarrow S)$

9. Premises: $(P \rightarrow Q) \rightarrow \sim(R \rightarrow Q)$

                 $S \rightarrow \sim R$

                 $\sim S \rightarrow Q$

Conclusion: $\sim(\sim Q \rightarrow \sim P)$

10. Premises: $P \rightarrow Q$

                 $Q \rightarrow \sim R$

                 $T \rightarrow R$

Conclusion: $T \rightarrow \sim P$

In Exercises 11–13, translate the arguments, then prove they are valid.

11. If Clark is a mathemagician, then everyone at the theater has a good time. If everyone at the theater has a good time, then Lois is not his assistant. Lois is his assistant. Therefore Clark is not a mathemagician.

12.* Margarita and Sergio will try out for the school play only if Tiffany and Tuong Cam decide to try out. However, if Dene tries out for the play, then Tiffany and Tuong Cam will not try out. Dene tries out for the play. Therefore Margarita and Sergio will not try out for the school play.

13.* If Boris becomes a pastry chef, then if he gives in to his desire for chocolate mousse, then his waistline will suffer. If his waistline suffers, then his dancing will suffer. Boris gives in to his desire for chocolate mousse. However, his dancing will not suffer. Therefore Boris does not become a pastry chef.

14. Invent a set of premises and a conclusion you can deduce logically from your premises. Then write the proof youself, or give your premises and conclusion to a classmate and have her or him write the proof.

Improving Visual Thinking Skills—*Constructing a Persian Tile Design*

This Persian tile design dates back to the eleventh century. It is a tessellation of kites and squares. You saw this design while discovering the theorem of Pythagoras. Re-create this design with a compass and a straight-edge, then color it.

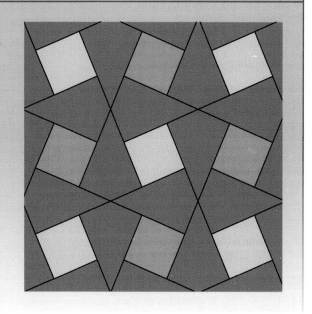

Lesson 14.10

Chapter Review

In Chapter 14 you learned the basic rules of deductive reasoning. Like Sherlock Holmes, you can use your new logical thinking skills to solve logical puzzles and mysteries. You can also use the laws of logic to write valid arguments and proofs.

   Can you recall all the laws of valid reasoning you can use in a logic proof? Can you recall the different types of logic proofs? In Chapters 15 and 16 you'll apply your deductive reasoning skills to geometry proofs.

Exercise Set 14.10

Match.

1. IF–THEN statements
2. P in a $P \rightarrow Q$ statement
3. Q in a $P \rightarrow Q$ statement
4. Converse of $P \rightarrow Q$
5. Inverse of $P \rightarrow Q$
6. Contrapositive of $P \rightarrow Q$
7. If $P \rightarrow Q$ is true and P is true, then Q is true.
8. If $P \rightarrow Q$ is true and $\sim Q$ is true, then $\sim P$ is true.
9. If $P \rightarrow Q$ is true and $Q \rightarrow R$ is true, then $P \rightarrow R$ is true.
10. Given statements in an argument
11. Final statement in an argument
12. Deduction expert

a. Modus ponens
b. Modus tollens
c. Law of syllogism
d. Conditional statements
e. Conclusion
f. Premises
g. $Q \rightarrow P$
h. $\sim P \rightarrow \sim Q$
i. $\sim Q \rightarrow \sim P$
j. Antecedent
k. Consequent
l. Contradiction
m. Sherlock Holmes
n. Dr. Watson
o. Professor Moriarty

In Exercises 13-16, translate each English statement into symbolic notation. Use the following letter substitutions.
LET P: Smoking causes cancer.
LET Q: It is dangerous to smoke.
LET R: Smoking is glamorous.
LET S: Smoking is addictive.
LET T: It is difficult to stop smoking.

13. If smoking causes cancer, then it is dangerous to smoke.
14. If smoking is glamorous, then smoking doesn't cause cancer.
15. If smoking is not addictive, then it is not difficult to stop smoking.
16. It is not true that if smoking is not addictive, then smoking is glamorous.

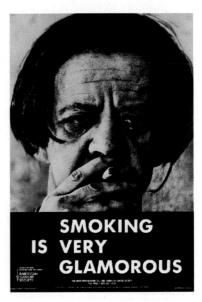

SMOKING IS VERY GLAMOROUS

AMERICAN CANCER SOCIETY

In Exercises 17–20, translate each symbolic statement into English. Use the following letter substitutions.

LET *P:* Students are smart.   LET *Q:* Students will study.
LET *R:* Teachers are human.   LET *S:* They make mistakes.

17. $P \rightarrow Q$ **18.** $\sim S \rightarrow \sim R$

19. Converse of $\sim P \rightarrow S$ **20.** Inverse of $S \rightarrow P$

In Exercises 21–24, provide the steps and the reasons to prove each logical argument.

21. Premises: $(\sim Q \rightarrow P) \rightarrow \sim R$
                   $T \rightarrow S$
                   $\sim R \rightarrow \sim S$
                   T

Conclusion: $\sim(\sim P \rightarrow Q)$

22. Premises: $\sim P \rightarrow (Q \rightarrow \sim W)$
                   $\sim S \rightarrow Q$
                   $\sim T$
                   $P \rightarrow T$

Conclusion: $W \rightarrow S$

23. Premises: $(P \rightarrow Q) \rightarrow (R \rightarrow S)$
                   $(\sim S \rightarrow \sim R) \rightarrow T$

Conclusion: $\sim T \rightarrow \sim(P \rightarrow Q)$

24. Premises: $R \rightarrow (Q \rightarrow S)$
                   $S \rightarrow P$
                   $Q \rightarrow \sim P$
                   Q

Conclusion: $\sim R$

25. In his book *Candide* French satirist Voltaire (1694–1778) included a character named Pangloss, a "philosopher" who presents a number of logical arguments. Many of Pangloss's arguments are valid, but they're often unsound. Here's an example of one of his many attempts to prove that "this is the best of all possible worlds." "It is proved . . . that things cannot be other than they are, for since everything was made for a purpose, it follows that everything is made for the best purpose. Observe: Our noses were made to carry spectacles, so we have spectacles."

The validity of this argument can be analyzed by translating it into symbols.

Disappearing Bust of Voltaire (1941), Salvador Dali
Can you see the bust of Voltaire in the painting?

LET *P:* We have spectacles.
LET *Q:* Our noses were made to carry spectacles.
LET *R:* The best purpose for our noses is to carry spectacles.
IF *P* THEN *Q*
IF *Q* THEN *R*
P
THEREFORE *R*

Is the argument valid? What form of reasoning is used? Is the argument sound? If you think it is unsound, which premise (or premises) do you think is false? Explain.

In Exercises 26 and 27, translate the argument into symbolic form, then prove it.

26. If Sherlock Holmes is successful, then Professor Moriarty will be apprehended. If Dr. Watson doesn't slip up, then Sherlock Holmes will locate the missing clue. If Sherlock Holmes locates the missing clue, then he is successful. Dr. Watson doesn't slip up. Therefore Professor Moriarty will be apprehended.

27. People from Seattle, Washington, like to joke with tourists that it is easy to predict the weather there: "If you can see Mount Rainier, then it is going to rain. If you cannot see Mount Rainier, then it is raining." From these two premises you can conclude that if it is not raining, then it is going to rain.

Assessing What You've Learned

You're probably coming close to the end of the school year now. This is a good time to look back over the entire year and reflect on your accomplishments and the challenges you faced. If you read my note to you at the beginning of this text, you may remember that one of my goals in this book is that you should have fun doing geometry.

Now I'd like to hear from you. This is your chance to assess not only what you've learned, but also how this book has helped you along the way. Write me a letter and let me know what you think of the book. Did you like the approach of discovering concepts for yourself by doing investigations? Do you feel like you've become a better problem solver? What experiences did you enjoy most using the book? What are some things that frustrated you? Overall, did you have fun?

Send your letter to Michael Serra, c/o Key Curriculum Press, P.O. Box 2304, Berkeley, CA 94702. I'm sorry I won't have time to reply to individual letters, but I promise to take the time to read them all.

Improving Visual Thinking Skills—3 × 3 Inductive Reasoning Puzzle III

Sketch the figure missing in the lower right corner of this three-by-three pattern.

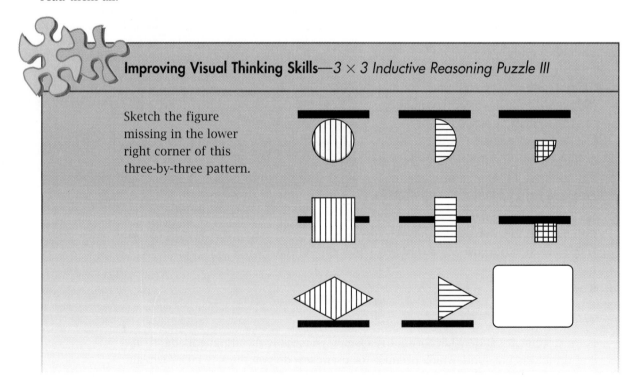

Cooperative Problem Solving

Logic in Space

1. Three mysterious black boxes recently appeared in the cargo bay of the space station Entropy. The three boxes are labeled "widgets," "gadgets," and "widgets and gadgets." A note was attached to one of the boxes, obviously by some prankster on the recreational space station Empyrean, which read, "Each label is incorrect. But it is possible to determine the correct label for each box by selecting only one object from one of the boxes. How?" The note was signed "Space Happy." How indeed?

2. A group of six astronauts has been asked to explore three sectors of the space station to determine which area would be best for building a new housing project. They know that they are to work in teams of two, each team consisting of an engineer and a biologist. Leroy, Malik, and Nancy are biologists. Linda, Michelle, and Nikko are engineers. Unfortunately, a computer glitch has erased some of their instructions. Help them to figure out who the teams will be based on the information they have.

   Linda is new, so she should work with an experienced team leader.

   Nikko is not scheduled to work with Malik this time.

   Because this is Leroy's first mission he should be in charge of collecting soil samples.

   Malik joined the team at the same time as Leroy.

3. The six-astronaut crew of Entropy has just received four small visitors from Titan. The six astronauts and the four Titans are scheduled to attend the Lunar Space Conference at the lunar space port Galileo. At the moment, however, only two of the space station's lunar shuttles are in operation. Each can carry only two Titans or one astronaut on any one trip between the station and the moon surface. How many one-way shuttle trips must be made to convey all ten conventioneers to the conference?

4. When all shuttles are in service, the docking stations of Entropy and Empyrean are quite busy. This active shuttle schedule works out well for the seven close friends and space cadets of the graduating class of 2043. Adrianne is stationed on Entropy, and Brenda, Carl, Dianna, and Eduardo are stationed on Empyrean. Their cadet classmates, Fred and Gianna, are stationed at the lunar research facility Galileo II. Brenda's job brings her to Entropy every second day. Carl's schedule brings him to Entropy every third day. Dianna's work requires her to check in at Entropy every fourth day. Eduardo's duties send him to the station every fifth day. Cadets Fred and Gianna have projects that require trips up to Entropy every sixth and seventh day, respectively. Today, to their surprise, they have all arrived on the same day. They have agreed to celebrate the next time they are all together here at Entropy. How many days from today will this happen?

5. Lunar explorer Eugene O. Regan, based at Galileo II, is planning an 800-km trip across the lunar surface to the U.N. Research Facility. Eugene has a logistics problem because the lunar rover at the base can carry only enough fuel to travel a distance of 500 km. Of course, there are no fuel stations between the two bases, so he must build a series of fueling points along the route to store fuel. The lunar-rover fuel is packed in 100-km units. How many full loads of fuel will Eugene need to complete his trip to the U.N. Research Facility?

Geometric Proof

Circle Limit III, M. C. Escher, 1959

In this chapter you will use deductive reasoning to prove many of your earlier conjectures. You'll use flow charts to aid you in planning your proofs. You will learn a number of other proving strategies such as thinking backward and carefully marking your diagrams with given information. You will prove geometric conjectures by two-column proof, flow-chart proof, or paragraph proof. You will see how some discoveries are logically related to each other.

A deductive system depends on assumptions. In the project Non-Euclidean Geometries, you'll learn how different assumptions lead to new geometries. The Escher woodcut *Circle Limit III,* shown above, illustrates one such non-Euclidean geometry system.

Lesson 15.1

Premises of Geometry

Over thousands of years the Babylonians and Egyptians discovered many geometric principles and developed a collection of "rule-of-thumb" procedures for doing practical geometry. The result of trial and error, these procedures were used to compute simple areas and volumes. The procedures were used in surveying to reestablish land boundaries after floods, and they were practical instructions for building canals and tombs.

By 600 B.C. a prosperous new civilization had begun to grow in the trading towns along the coast of Asia Minor and later in Greece, Sicily, and Italy. People had free time to discuss and debate issues of government and law. This led to an insistence on reasons to support statements made in debate. Mathematicians began to use logical reasoning to deduce mathematical ideas.

Greek mathematician Thales of Miletus made a number of valuable geometric conjectures. Unlike most other mathematicians before him, Thales supported his discoveries with logical reasoning. Over the next 300 years, the process of supporting mathematical conjectures with logical arguments became more and more refined. Other Greek mathematicians, including Thales' most famous student, Pythagoras, began linking together chains of logical reasoning. Later students of mathematics at Plato's Academy linked even longer chains of geometric properties together by deductive reasoning. Euclid, in his famous work about geometry and number theory, *Elements,* established a single chain of deductive arguments for most of the geometry then known. Euclid started from a collection of statements he regarded as obviously true (**postulates**). He then systematically demonstrated that one after another geometric discovery followed logically from his postulates and his previously verified conjectures (**theorems**). In doing this, Euclid created a deductive system. A **deductive system** consists of a set of premises and a set of logical rules.

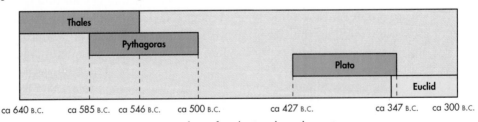

Timeline of early Greek mathematics

Up to now you have been learning geometry inductively, the way early civilizations did. You have observed geometric figures and have made conjectures about them. For example, you discovered that the sum of the measures of the three angles of every triangle is 180°. You did not measure the three angles of every triangle. Instead, you studied enough triangles to become convinced that the conjecture is true.

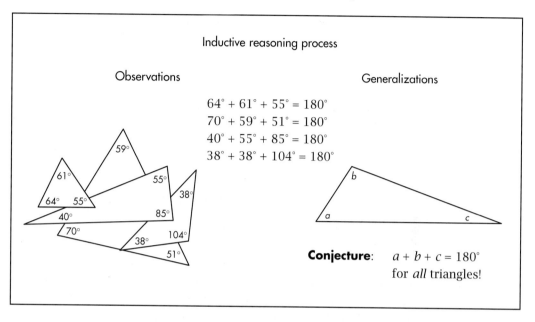

Inductive reasoning process

Observations   Generalizations

$64° + 61° + 55° = 180°$
$70° + 59° + 51° = 180°$
$40° + 55° + 85° = 180°$
$38° + 38° + 104° = 180°$

Conjecture: $a + b + c = 180°$
for *all* triangles!

Deductive reasoning process

Facts accepted as true   Logical consequences

Fact 1 $x + b + y = 180°$ because x, b,   **Conclusion**: $a + b + c = 180°$
and y are measures of angles   for any triangle.
that form a straight line.

Fact 2 $x = a$ and $y = c$ because
alternate interior angles
are congruent.

Fact 3   We can substitute equal
values for equal values.

In this chapter you will look at geometry as Euclid did. You will create a deductive system based on a collection of premises. These premises will include definitions and undefined terms; properties of algebra, equality, and congruence; and postulates. From these premises you will prove some of your earlier conjectures. Proven conjectures will become theorems, and from the theorems you'll prove other conjectures, turning them into theorems.

The process for geometric proofs is the same as for logic proofs except that there are different types of premises.

Premises for geometric arguments

1. Definitions and undefined terms
2. Properties of algebra, equality, and congruence
3. Postulates of geometry
4. Previously accepted or proven geometric conjectures (theorems)

You should already be familiar with the first type of premise in the list: undefined terms (point, line, and plane) and definitions. In fact, you should have a complete list of definitions in your notebook.

Let's take a closer look at the second group of premises: properties of algebra, equality, and congruence.

Properties of Algebra

Here are five important properties of algebra, true for any numbers a, b, and c.

- The **commutative property of addition** says that $a + b = b + a$.
- The **commutative property of multiplication** says that $ab = ba$.
- The **associative property of addition** says that $(a + b) + c = a + (b + c)$.
- The **associative property of multiplication** says that $(ab)c = a(bc)$.
- The **distributive property** says that $a(b + c) = ab + ac$.

Properties of Equality

Whether or not you're familiar with the following names, you've used these properties of equality.

- The **reflexive property of equality** says that $a = a$. In other words, any number is equal to itself. (This property is also called the **identity property**.)

- The **transitive property of equality** says that if $a = b$ and $b = c$, then $a = c$. This property often takes the form of the **substitution property**, which says that if $a = b$, then a may be replaced by b in an algebraic expression. For example, if $x + a = c$ and $a = b$, then $x + b = c$.

- The **symmetric property of equality** says that if $a = b$, then $b = a$.

- The **addition property of equality** says that if $a = b$, then $a + c = b + c$. (Also, if $a = b$ and $c = d$, then $a + c = b + d$.)

- The **subtraction property of equality** says that if $a = b$, then $a - c = b - c$. (Also, if $a = b$ and $c = d$, then $a - c = b - d$.)

- The **multiplication property of equality** says that if $a = b$, then $ac = bc$. (Also, if $a = b$ and $c = d$, then $ac = bd$.)

- The **division property of equality** says that if $a = b$, then $a/c = b/c$ provided $c \neq 0$. (Also, if $a = b$ and $c = d$, then $a/c = b/d$ provided $c \neq 0$ and $d \neq 0$.)

- The **zero product property of equality** says that if $ab = 0$, then $a = 0$ or $b = 0$ or both a and $b = 0$.

- The **square root property of equality** says that if $a^2 = b$, then $a = \pm\sqrt{b}$.

The properties of algebra and equality are used to solve algebraic equations. The solution of an equation is really an algebraic proof. Each step can be supported by a property. The addition property of equality, for example, permits you to add the same number to both sides of an equation to get an equivalent equation.

Example A

Equation: $5x - 12 = 3(x + 2)$

Solution:

$5x - 12 = 3(x + 2)$	Given
$5x - 12 = 3x + 6$	Distributive property
$5x = 3x + 18$	Addition property of equality
$2x = 18$	Subtraction property of equality
$x = 9$	Division property of equality

Example B

Conjecture: If $ax + b = c$, then $x = \dfrac{c - b}{a}$ provided $a \neq 0$.

Proof:

$ax + b = c$	Given (Assume the antecedent is true in a conditional proof.)
$ax = c - b$	Subtraction property of equality
$x = \dfrac{c - b}{a}$	Division property of equality

Why are the properties of algebra and equality important in geometry? Because the lengths of segments and the measures of angles involve numbers, you will occasionally need to use these properties in geometric proofs. Whenever you use a property of algebra or a property of equality in a geometric proof, refer to it by name.

Definition of Congruence and Properties of Congruence

Just as you use equality to express a relation between numbers, you use congruence to express a relation between geometric figures. Recall from the definition of congruent segments and angles that if $AB = CD$, then $\overline{AB} \cong \overline{CD}$ and that if $m\angle A = m\angle B$, then $\angle A \cong \angle B$. You'll extend the first three properties of equality to congruence relationships in Exercise Set 15.1.

Take Another Look 15.1

Try one or more of these follow-up activities.

1. In Chapter 1, you discovered a number of basic properties of odd and even integers. For example, you discovered that the sum of two odd integers is always an even integer. You discovered that $2n$ is a rule that generates even integers and $2n - 1$ is a rule for odd integers. Let $2n - 1$ and $2m - 1$ represent any two odd integers and prove that the sum of two odd integers is always an even integer.

2. Prove that the product of any two odd integers is always an odd integer.

3. Prove that the sum of the first n positive odd numbers—$1 + 3 + 5 + 7 + \ldots 2n - 1$— is the nth square number.

Exercise Set 15.1

1. Which Greek mathematician is credited with being the first to support his geometric discoveries with logical reasoning?

2. Which Greek mathematician is credited with being the first to establish a single chain of deductive arguments for all of geometry?

3. From about 600 B.C. to 300 B.C. geometry changed from the discovered "practical" geometry of early Egyptians and Babylonians to the logically organized geometry of early classical Greeks. What characteristics of classical Greek civilization may have given rise to this change?

4. What is the difference between a postulate and a theorem?

5. Euclid might have stated the addition property of equality (after translating from the Greek) in this way: If equals are added to equals, the results are equal. State the subtraction, multiplication, and division properties of equality as Euclid might have stated them (preferably in English—only use Greek for extra credit!).

6. Write the reflexive property of congruence, the transitive property of congruence, and the symmetric property of congruence. Include a diagram for each property. Illustrate one property with congruent triangles, the second property with congruent segments, and the third property with congruent angles. Remember, these properties may seem ridiculously obvious. This is exactly why they are accepted as premises (which require no proof).

7.* When you state $\overline{AC} \cong \overline{AC}$, what property are you using?

8. State the property that supports the statement: If $\overline{AC} \cong \overline{BD}$ and $\overline{BD} \cong \overline{HK}$, then $\overline{AC} \cong \overline{HK}$.

9. State the property that supports the statement: If $x + 120 = 180$, then $x = 60$.

10. State the property that supports the statement: If $2(x + 14) = 36$, then $x + 14 = 18$.

11. State the property that supports the statement: If $\overline{AC} \cong \overline{BD}$, then $\overline{BD} \cong \overline{AC}$.

In Exercises 12–15, use the properties of equality and algebra to provide each missing reason in the steps to solve the algebraic equation or to prove the algebraic argument.

12.* Equation: $7x - 22 = 4(x + 2)$

Solution:

$7x - 22 = 4(x + 2)$	Given
$7x - 22 = 4x + 8$	—?— property
$3x - 22 = 8$	—?— property of equality
$3x = 30$	—?— property of equality
$x = 10$	—?— property of equality

13. Equation: $\dfrac{5(x - 12)}{4} = 3(2x - 7)$

Solution:

$\dfrac{5(x - 12)}{4} = 3(2x - 7)$	Given
$5(x - 12) = 12(2x - 7)$	—?— property of equality
$5x - 60 = 24x - 84$	—?— property
$5x = 24x - 24$	—?— property of equality
$-19x = -24$	—?— property of equality
$x = \dfrac{24}{19}$	—?— property of equality

14.* Conjecture: If $\dfrac{x}{m} - c = d$, then $x = m(c + d)$ provided $m \neq 0$.

Proof:

$\dfrac{x}{m} - c = d$	—?—
$\dfrac{x}{m} = d + c$	—?—
$x = m(d + c)$	—?—
$x = m(c + d)$	—?—

15. Conjecture: If $a + b = 180$ and $c + d = 180$ and $a = c$, then $b = d$.

Proof:

$a + b = 180$	—?—
$c + d = 180$	—?—
$a + b = c + d$	—?—
$a = c$	—?—
$b = d$	—?—

In Exercises 16 and 17, solve the equation for x and provide a reason for each step of your derivation.

16.* $a(x + b) - c = d$ **17.*** $x^2 + 8x - 3 = 2(4x + 3)$

In Exercises 18–23, copy each statement-reason table. For each reason, state the definition, the property of algebra, or the property of congruence that supports the statement. If the statement is given, then write "given" as your reason.

18.* Given: $ABCD$ is a parallelogram.

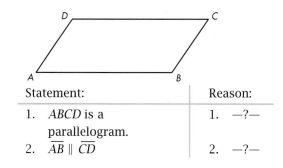

Statement:	Reason:
1. $ABCD$ is a parallelogram.	1. —?—
2. $\overline{AB} \parallel \overline{CD}$	2. —?—

19. Given: $\overline{CD}$ is a median.

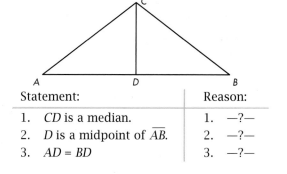

Statement:	Reason:
1. CD is a median.	1. —?—
2. D is a midpoint of $\overline{AB}$.	2. —?—
3. $AD = BD$	3. —?—

20. Given: $\overleftrightarrow{CM}$ is $\perp$ bisector of $\overline{AB}$.

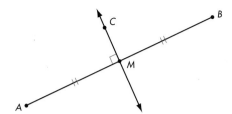

Statement:	Reason:
1. $\overleftrightarrow{CM}$ is $\perp$ bisector of $\overline{AB}$.	1. —?—
2. $AM = MB$	2. —?—
3. $AB \perp \overleftrightarrow{CM}$	3. —?—

21. Given: $AP = PC$
$PC = CD$

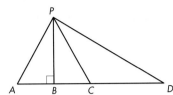

Statement:	Reason:
1. $AP = PC$	1. —?—
2. $PC = CD$	2. —?—
3. $AP = CD$	3. —?—

22. Given: $\overline{AO}$ and $\overline{BO}$ are radii.

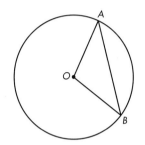

Statement:	Reason:
1. $\overline{AO}$ and $\overline{BO}$ are radii.	1. —?—
2. $AO = OB$	2. —?—
3. $\triangle AOB$ is isosceles.	3. —?—

23. Given: $AB = BE$
B is the midpoint of $\overline{AC}$ and $\overline{DE}$.

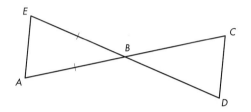

Statement:	Reason:
1. $AB = BE$	1. —?—
2. B is the midpoint of AC and DE.	2. —?—
3. $AB = BC$	3. —?—
4. $BD = BE$	4. —?—
5. $BC = BE$	5. —?—

Improving Reasoning Skills—*Logical Vocabulary*

Here is a logical vocabulary challenge. It is sometimes possible to change one word to another of equal length by changing one letter at a time. Each change, or move, you make gives you a new word. For example, DOG can be changed to CAT in exactly three moves.

$$DOG \Rightarrow DOT \Rightarrow COT \Rightarrow CAT$$

Change MATH to each of the following words in exactly four moves.

1. MATH $\Rightarrow$ -?- $\Rightarrow$ -?- $\Rightarrow$ -?- $\Rightarrow$ ROSE
2. MATH $\Rightarrow$ -?- $\Rightarrow$ -?- $\Rightarrow$ -?- $\Rightarrow$ CORE
3. MATH $\Rightarrow$ -?- $\Rightarrow$ -?- $\Rightarrow$ -?- $\Rightarrow$ HOST
4. MATH $\Rightarrow$ -?- $\Rightarrow$ -?- $\Rightarrow$ -?- $\Rightarrow$ LESS
5. MATH $\Rightarrow$ -?- $\Rightarrow$ -?- $\Rightarrow$ -?- $\Rightarrow$ LIVE

Lesson 15.2

Postulates of Geometry

Geometry is the art of correct reasoning on incorrect figures.
— George Polya

As you build a deductive system, you demonstrate that certain statements are logical consequences of other previously accepted or proven statements. This chain of logical reasoning must begin somewhere, so every deductive system must contain some statements that are never proved. In a geometric system, these statements are traditionally called postulates. They form the third group of premises for geometric arguments. You have already used the first two types of premises—undefined and defined terms and the properties of algebra, equality, and congruence—to write some simple proofs.

You will use some of the geometric conjectures you made earlier this year as postulates. You will prove other conjectures by using your deductive system. After you prove a conjecture, it becomes a theorem.

Postulates should be so basic that we can accept them as true with little debate. Your list of postulates differs somewhat from Euclid's. It includes some statements and assumptions that Euclid made that he didn't call postulates. You will see in the project Non-Euclidean Geometries that

A page from a Latin translation of Euclid's *Elements*. Which of these definitions can you translate?

one of the postulates in both your list and Euclid's list has actually come under debate. As you read through the postulates, think about which one might be most open for dispute.

The postulates of Euclid's geometry are based on the geometric constructions that are possible with a compass and a straightedge. As you've performed basic geometric constructions in this class, you've been assuming some of these "obvious" truths.

P-1 You can construct exactly one line through any two points (***Line Postulate***).

P-2 The intersection of two lines is exactly one point (***Line Intersection Postulate***).

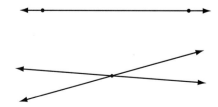

P-3   You can construct exactly one midpoint on any line segment (***Midpoint Postulate***).

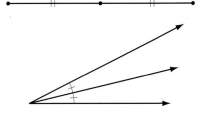

P-4   You can construct exactly one angle bisector in any angle (***Angle Bisector Postulate***).

P-5   Through a point not on a given line, you can construct exactly one line parallel to the given line (***Parallel Postulate***).

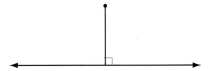

P-6   Through a point not on a given line, you can construct exactly one line perpendicular to the given line (***Perpendicular Postulate***).

P-7   If point *B* is on $\overline{AC}$ and between points *A* and *C*, then $AB + BC = AC$ (***Segment Addition Postulate***).

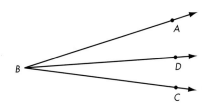

P-8   If point *D* lies in the interior of ∠*ABC*, then $m\angle ABD + m\angle DBC = m\angle ABC$ (***Angle Addition Postulate***).

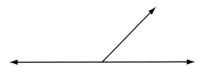

The following postulates are based on some of the most basic and useful of your geometric conjectures.

P-9   If two angles are a linear pair, then they are supplementary (***Linear Pair Conjecture***, now called ***Linear Pair Postulate***).

P-10   If two parallel lines are cut by a transversal, then the alternate interior angles are congruent. Conversely, if two lines are cut by a transversal forming congruent alternate interior angles, then the lines are parallel (***AIA Conjecture***, now called ***AIA Postulate***).

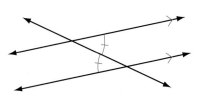

P-11 If the three sides of one triangle are congruent to three sides of another triangle, then the two triangles are congruent (*SSS Congruence Conjecture*, now called *SSS Congruence Postulate*).

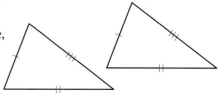

P-12 If two sides and the angle between them in one triangle are congruent to two sides and the angle between them in another triangle, then the two triangles are congruent (*SAS Congruence Postulate*).

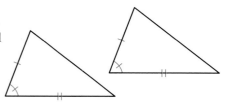

P-13 If two angles and the side between them in one triangle are congruent to two angles and the side between them in another triangle, then the two triangles are congruent (*ASA Congruence Postulate*).

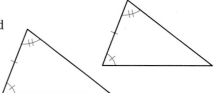

Take Another Look 15.2

You might be asking yourself, Why did we select just the AIA Conjecture as a postulate but not the Corresponding Angles Conjecture (CA) or the Alternate Exterior Angles Conjecture (AEA)? We could have selected all three, but it was not necessary. Given any one as a postulate, you can deduce the other two, making them theorems. Try it. Given the AIA Conjecture as a postulate, deduce the CA Conjecture. Hint: You'll need another postulate in addition to the AIA Postulate.

Exercise Set 15.2

In Exercises 1-8, identify each statement as true or false. Then state which definition, property of algebra, property of congruence, or postulate supports your answer.

1. If M is the midpoint of $\overline{AB}$, then $AM = BM$.

2.* If M is the midpoint of $\overline{CD}$ and N is the midpoint of $\overline{CD}$, then M and N are the same point.

3. If $\overrightarrow{AB}$ bisects $\angle CAD$, then $\angle CAB \cong \angle DAB$.

4. If $\overrightarrow{AB}$ bisects $\angle CAD$ and $\overrightarrow{AF}$ bisects $\angle CAD$, then $\overrightarrow{AB}$ and $\overrightarrow{AF}$ are the same ray.

5. Lines ℓ and m intersect at different points A and B.

6. Line ℓ passes through points A and B and line m passes through points A and B, but lines ℓ and m are not the same line.

7. Point *P* is in the interior of ∠*RAT*, thus *m*∠*RAP* + *m*∠*PAT* = *m*∠*RAT*.

8. If point *M* is on $\overline{AC}$ and between points *A* and *C*, then *AM* + *MC* = *AC*.

In Exercises 9–14, state the definition, property of algebra, property of congruence, or postulate that supports each statement.

9. Given: ∠1 ≅ ∠2

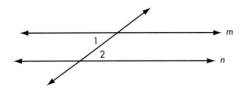

Statement:	Reason:
1. ∠1 ≅ ∠2	1. —?—
2. *m* ∥ *n*	2. —?—

10. Given: △*ABC* ≅ △*DEF*

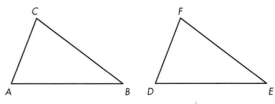

Statement:	Reason:
1. △*ABC* ≅ △*DEF*	1. —?—
2. $\overline{AB}$ ≅ $\overline{DE}$	2. —?—

11. Given: $\overline{AC}$ ≅ $\overline{BD}$, $\overline{AD}$ ≅ $\overline{BC}$

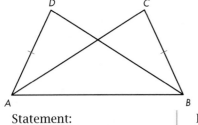

Statement:	Reason:
1. $\overline{AC}$ ≅ $\overline{BD}$ $\overline{AD}$ ≅ $\overline{BC}$	1. —?—
2. $\overline{AB}$ ≅ $\overline{AB}$	2. —?—
3. △*ABC* ≅ △*BAD*	3. —?—
4. ∠*D* ≅ ∠*C*	4. —?—

12.* Given: $\overline{AB}$ ∥ $\overline{CD}$

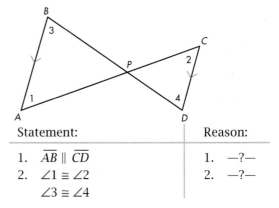

Statement:	Reason:
1. $\overline{AB}$ ∥ $\overline{CD}$	1. —?—
2. ∠1 ≅ ∠2 ∠3 ≅ ∠4	2. —?—

13.* Given: Isosceles △*ABC* ($\overline{AB}$ ≅ $\overline{BC}$)

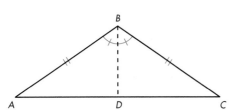

Statement:	Reason:
1. $\overline{AB}$ ≅ $\overline{BC}$	1.* —?—
2. Construct angle bisector $\overline{BD}$.	2.* —?—
3. ∠*ABD* ≅ ∠*CBD*	3. —?—
4. $\overline{BD}$ ≅ $\overline{BD}$	4.* —?—
5. △*BCD* ≅ △*BAD*	5. —?—

14. Given: △*ABC*

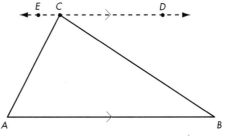

Statement:	Reason:
1. Construct $\overleftrightarrow{CD}$ with $\overleftrightarrow{CD}$ ∥ $\overline{AB}$.	1. —?—
2. ∠*DCB* ≅ ∠*ABC*	2. —?—
3. ∠*ECA* ≅ ∠*CAB*	3. —?—

In Exercises 15 and 16, provide the algebraic proof of the conjecture.

15. Show that the sum of an odd integer and an even integer is always an odd integer.

16. Show that the sum of any three consecutive integers is always divisible by three.

17. Explain why we must have undefined terms in our geometric system.

18. The Declaration of Independence begins with the statement "We hold these truths to be self-evident that . . . ," then goes on to list four postulates of good government. Look up the Declaration of Independence and list the four self-evident truths or postulates that were the original premises of the United States government.

Improving Visual Thinking Skills—*Picture Patterns V*

Sketch the figure that goes in box 12 below.

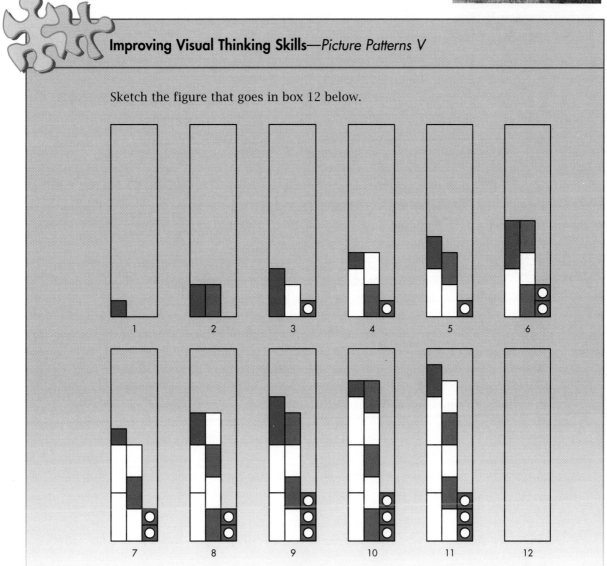

Non-Euclidean Geometries

Anyone who plays games has probably also changed the rules of a game. Sometimes by changing just one rule, a game becomes completely different. You can compare geometry to a game whose rules are the postulates of geometry. If you change even one postulate, you may create a new geometry.

The geometry of Euclid, called Euclidean geometry (the geometry you discover in this text), was based on a number of postulates, or self-evident truths. A postulate, according to the contemporaries of Euclid, was an obvious truth that could not be derived from other postulates. The list below contains the first five of Euclid's postulates.

Postulate 1: You can draw a straight line between any two points.

Postulate 2: You can extend any segment indefinitely.

Postulate 3: You can draw a circle with any given point as center and any given radius.

Postulate 4: All right angles are equal.

Postulate 5: If two straight lines lying in a plane are met by another line and if the sum of the interior angles on one side is less than two right angles, then the straight lines, if extended sufficiently, will meet on the side on which the sum of the angles is less than two right angles.

Euclid's fifth postulate is known as the Parallel Postulate. That's right! There is no mention of parallel in the fifth postulate. It has that name because it is logically equivalent to:

Through a given point not on a given line there passes at most one line that is parallel to the given line.

This version of the Parallel Postulate is called the Playfair axiom, named after Scottish physicist and mathematician John Playfair (1748–1819). Read Postulate 5 again and draw a sketch to illustrate it. Do you see now why it is called the Parallel Postulate?

The first four postulates seem straightforward. The fifth postulate, however, sounds complicated. Because the fifth postulate does not sound like the others, mathematicians tried for centuries to prove that it could be deduced from the others.

Attempting to use indirect proof, mathematicians began by assuming that the fifth postulate was false and then tried to reach a logical contradiction. Two possible assumptions are negations of the Parallel Postulate.

Assumption 1: Through a given point not on a given line there pass *more than one line* parallel to the given line.

Assumption 2: Through a given point not on a given line there pass *no lines* parallel to the given line.

German mathematician Carl Friedrich Gauss (1777–1855), Russian mathematician Nikolai Lobachevsky (1792–1856), and Hungarian mathematician János Bolyai (1802–1860) investigated the first assumption. They found that this assumption led to a new geometry that did not contradict any of Euclid's other postulates or any of the theorems that didn't depend on the parallel postulate. In fact, when they replaced the Parallel Postulate with this assumption, they discovered a new deductive system of geometry. This non-Euclidean geometry is called **hyperbolic geometry**.

German mathematician Georg Friedrich Bernhard Riemann (1826–1866) chose the second assumption. This assumption, together with adjustments to the other four postulates, led to a second non-Euclidean geometry that also did not contradict any of Euclid's other postulates or any of the theorems not dependent on the Parallel Postulate. Riemann's geometry is called **elliptic geometry**.

All the theorems of Euclidean geometry, except those depending directly or indirectly on the Parallel Postulate, are also valid in hyperbolic and elliptic geometries. The most familiar example of a theorem of Euclidean geometry that relies on the Parallel Postulate is the Triangle Sum Theorem: The sum of the measures of the three angles of a triangle is 180°. The corresponding theorem in hyperbolic geometry states: The sum of the measures of the three angles of a triangle is less than 180°. In elliptic geometry the theorem states: The sum of the measures of the three angles of a triangle is greater than 180°.

Elliptic Geometry

You can use the surface of a sphere as a model of elliptic geometry. To establish the model, make two changes to the Euclidean postulates. First, replace the word *line* with the words *great circle.* (A **great circle** is a circle formed on the surface of a sphere by a plane passing through the center of the sphere. For example, the equator is a great circle on the surface of the earth.) Second, agree to call each pair of diametrically opposite points (points on opposite ends of a diameter) pole points.

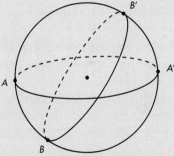

Model of elliptic geometry

In the figure at right, points *A* and *A′* are pole points. So are points *B* and *B′*. More than one great circle can pass through points *A* and *A′*. However, only one great circle can pass through points *A* and *A′* and through points *B* and *B′*.

In Euclidean geometry, lines never end and are infinite in length. In elliptic geometry, great circles never end; however, their length is finite! If you travel the shortest path along a sphere, you will eventually return to your starting point.

That part of a great circle between two points of the great circle is simply an arc of a great circle, which is the spherical equivalent of a line segment.

Riemann assumed that through a given point not on a given line there pass *no* lines parallel to the given line. Simply put, there are no parallel lines in elliptic geometry. Because all great circles intersect, the spherical model of elliptic geometry supports this assumption.

1. Write a set of postulates for elliptic geometry by rewriting Euclid's first four postulates and the Playfair axiom version of the fifth postulate. For example, make sure you replace the word *line* with the words *great circle.*

2. Find a spherical surface to use as a model of elliptic geometry. On your model, show an example of two "lines" perpendicular to the same "line" that are not parallel.

3. On your model, show that two points do not always determine a unique line.

4. Draw an "isosceles triangle" on your model. Does the Isosceles Triangle Theorem hold in elliptic geometry?

5. The sum of the measures of the three angles of a triangle is always greater than 180° in elliptic geometry and always less than 180° in hyperbolic geometry. Draw a "triangle" on your elliptic model and explain why the angle sums for elliptic geometry make sense. Remember to use "lines" for the sides of your triangle!

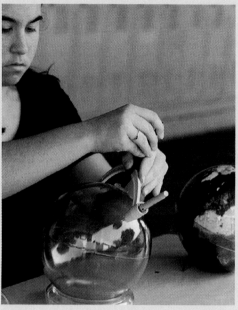

Using a compass to do constructions on a Lénárt Sphere®

Hyperbolic Geometry

To find a model for hyperbolic geometry, we need a geometry in which for every line and a point not on that line, there is more than one parallel line. French mathematician Henri Poincaré (1854–1912) devised such a model. It is called the Poincaré disk, and it consists of all the points in the interior of a circle. The "lines" in this geometry are arcs inside the circle. The arcs must have their endpoints on the circle and must be "perpendicular" to the circle at both endpoints. (The arc endpoints themselves are not part of the lines but represent the lines at infinity.) To measure angles formed by a pair of these "lines," simply measure the angles formed by the tangents to the arcs at their point of intersection.

Distance in this model is more complicated. The "lines" are still defined to have infinite length because you compress distance as you get closer to the edge of the disk. Thus segments that look congruent may have very different lengths, depending on their proximity to the center of the circle. To try to visualize how distance works in this model, look again at Escher's *Circle Limit III* at the beginning of the chapter. Imagine that all the fish of a given color are actually congruent, but that they just appear to get smaller at the edge of the disk because they're far away. The figure at right shows that many "lines" through a common point can be parallel to a given line. "Lines" *p*, *r*, and *s* all pass through point *M*, yet none of them intersect "line" *t*.

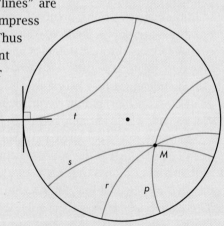

The endpoints of the arcs meet the circle at right angles. This is shown by the tangents drawn at one endpoint of the arc representing line *t*.

12. Can you prove the conjecture in Exercise 7 by using SAS instead of SSS? Explain.

13. Constructing auxiliary lines can be tricky. Suppose you saw this step in a proof: Construct angle bisector *CD* to the midpoint of side *AB* in △*ABC*. What's wrong with that step? Explain.

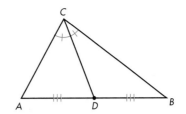

Improving Reasoning Skills—*Seeing Spots*

The arrangement of green and yellow spots at right may appear to be random, but there is a pattern. Each row is generated by the row immediately above it. Find the pattern and add several rows to the arrangement. Do you think a row could ever consist of all yellow spots? All green spots? Could there ever be a row with one green spot? Does a row ever repeat itself?

Improving Algebra Skills—*Polynomial Geometry II*

Find the perimeter and area of each figure below.

1.

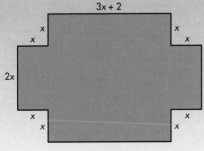

2.

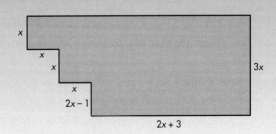

Chapter Review

Mistakes are part of the dues one pays for a full life.
— Sophia Loren

Your list of properties, postulates, and theorems is the beginning of a deductive system for much of geometry. In Chapter 16 you will continue to build on this system by adding more theorems and a few more postulates. Before you start Chapter 16, make sure you really understand the theorems and postulates in your geometric system so far. Also make sure you understand the process of geometric proof. Can you describe some of the different ways you can write a geometric proof?

Exercise Set 15.8

In Exercises 1–10, identify each statement as true or false. For each false statement sketch a counterexample or explain why the statement is false. Consult your theorem list.

1. The angle bisector of the vertex angle of an isosceles triangle is also a median.

2. The median from the vertex angle of an isosceles triangle is also an angle bisector of the vertex angle.

3. If two angles are congruent, then they are vertical angles.

4. Supplements of the same angle or congruent angles are supplementary.

5. A diagonal of a parallelogram divides the parallelogram into two isosceles triangles.

6. The diagonals of a parallelogram are congruent.

7. If two lines are parallel to a third line, then they are parallel to each other.

8. If two lines in the same plane are perpendicular to a third line in the plane, then they are perpendicular to each other.

9. If two parallel lines are intersected by a transversal, then the interior angles on the same side of the transversal are congruent.

10. If the diagonals of a parallelogram are congruent, then the parallelogram is a rectangle.

In Exercises 11–14, consult your postulates list and your list of properties of algebra and equality.

11. When you say $\overline{AC} \cong \overline{AC}$, you are using which property of congruence?

12.*State the postulate that supports the statement: If M is the midpoint of $\overline{CD}$, then $CM + MD = CD$.

13. If you have a line ℓ and a point P, which postulate allows you to construct a line parallel to line ℓ through point P?

14. If $\triangle ABC \cong \triangle DEF$, then $\overline{AC} \cong \overline{DF}$. What is the reason you would give for this conclusion in a proof?

15. How would you state the next sentence as a conditional? The base angles of an isosceles triangle are congruent.

For Exercises 16–19, write a paragraph proof, a two-column proof, or a flow-chart proof of each conjecture. Once you complete the proofs, add the theorems to your list.

16. *Conjecture:* The consecutive angles of a parallelogram are supplementary (C-53).

  Given: Parallelogram *SAND*

  Show: $\angle DSA$ and $\angle NAS$ are supplementary.

17. *Conjecture:* Parallel lines are everywhere equidistant. (If two lines are parallel, then the distances from all points on one line to the other line are equal.)

  Given: $\ell_1 \parallel \ell_2$ with points A and B on ℓ_1 and points C and D on ℓ_2 such that $\overline{AC} \perp \ell_2$ and $\overline{BD} \perp \ell_2$

  Show: $AC = BD$

18. *Conjecture:* If a point is on the perpendicular bisector of a segment, then it is equidistant from the endpoints of the segment (C-1).

  Given: $\overleftrightarrow{CD}$ is $\perp$ bisector of $\overline{AB}$ with point P on $\overleftrightarrow{CD}$.

  Show: $AP = BP$

19. *Conjecture:* The median to the base of an isosceles triangle is also an altitude to the base.

  Given: $\overline{CD}$ is the median to the base $\overline{AB}$ of isosceles $\triangle ABC$.

  Show: $\overline{CD}$ is the altitude to the base.

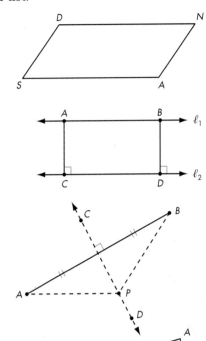

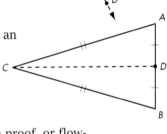

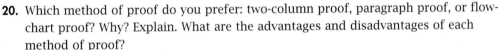

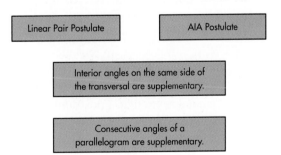

20. Which method of proof do you prefer: two-column proof, paragraph proof, or flow-chart proof? Why? Explain. What are the advantages and disadvantages of each method of proof?

21. When completed, the concept map below shows one way of logically relating the postulates and the theorems used in this chapter to prove that the consecutive angles of a parallelogram are supplementary. Copy and complete the concept map by drawing in the missing arrows.

Logical relationship of postulates and theorems needed to prove that consecutive angles of a parallelogram are supplementary

| Linear Pair Postulate | | AIA Postulate |

Interior angles on the same side of the transversal are supplementary.

Consecutive angles of a parallelogram are supplementary.

Assessing What You've Learned

Update Your Portfolio

- A proof or a short sequence of proofs would make a good addition to your portfolio. Choose a proof that challenged you but on which you think you did your best work. Include a summary of what interested you about that proof and how you planned it.

Write in Your Journal

- Proofs serve a variety of purposes in mathematics. Mathematicians use proofs to verify their findings, to explain why things are true, to communicate their findings to others, to organize mathematical relationships systematically, and as springboards for further discovery. You're just getting started with proofs, so you may not have thought of all those uses for proofs. But from your point of view, what is the most important purpose of proofs in mathematics?

- Write about the difference between knowing that something is true and knowing why that thing is true. What do proofs have to do with that difference? Use a geometric example.

Organize Your Notebook

- With this chapter, you've begun new lists of postulates and theorems. Review your notebook to be sure these lists are complete and well organized. Write a one-page summary of Chapter 15.

Give a Presentation

- Proofs are one way in which mathematicians communicate their findings. Prepare a poster or overhead transparency of a proof. In your presentation, explain how you planned the proof and explain each step of the proof.

Improving Visual Thinking Skills—*Triangle Cards*

Visualize six equilateral triangle-shaped cards whose sides measure 6 cm, 5 cm, 4 cm, and so on down to 1 cm. The triangles with odd perimeters are black; the other triangles are white. Begin by placing the triangle with the 6-cm edge down onto the table. Then place the 5-cm triangle into one corner (A) of the 6-cm triangle. Next, place the 4-cm triangle into corner (B) of the 5-cm triangle, counterclockwise from corner A. Next, place the 3-cm triangle into corner (C) of the 4-cm triangle, counterclockwise from corner B. Continue in this way with the remaining triangles, positioning them so that they rotate inwardly counter clockwise. Draw a top view of what the cards would look like when the process is finished.

CHAPTER 15

Cooperative Problem Solving

The Centauri Challenge

For the past 60 years, astronomers of the lunar colony have been communicating with an intelligent life form in the Centauri star system. Because of the vast distance between the lunar colony and the star system, communication is very slow. It takes 20 years for a message to be sent and a response to be received. The most recent communication included The Centauri Challenge, sent from logic students of the star system to students at the lunar colony.

The Centauri Challenge

Centauri is a formal system of strings of the letters P, Q, R, and S. Four rules govern the strings' behavior. Using some combination of the four rules, it is possible to change one string of letters into a different string.

Rule 1: Any two adjacent letters in a string can change places with each other. (PQ >> QP)

Rule 2: If a string ends in the same two letters, then you may substitute a Q for those two letters. (RSS >> RQ)

Rule 3: If a string begins in the same two letters, then you may add an S in front of those two letters. (PPR >> SPPR)

Rule 4: If a string of letters starts and finishes with the same letter, then you may substitute an R for all the letters between the first and last letters. (PQRSP >> PRP)

   A theorem in this system is a string followed by the symbol >> followed by a string. For example, PQQRSS >> QRQ says that if given string PQQRSS, then you can arrive at QRQ by using the rules. An example of the proof of this theorem is shown on the next page.

Example

Show: PQQRSS >> QRQ

Proof:
PQQRSS	Given
PQQRQ	By Rule 2
QPQRQ	By Rule 1
QRQ	By Rule 4

Therefore PQQRSS >> QRQ.

Challenge 1

Use the rules of Centauri to prove each theorem.

1. PQPRQ >> RQ
2. PQRSSQR >> RQ
3. PSSRS >> RQ
4. PSRQQRSQPSSS >> RQ
5. PQQQQP >> RQ
6. QQQQQQ >> SRQ

Challenge 2

Use the rules of Centauri to answer each question.

1. Can you find a string of five or more letters that cannot be reduced to RQ? If yes, produce it. If not, prove that you cannot.
2. One of the rules of Centauri can be removed without losing any of the first five theorems proved in Challenge 1. Which of the rules can be removed? Why must this rule be used in the sixth theorem? Create another theorem that must use this rule.

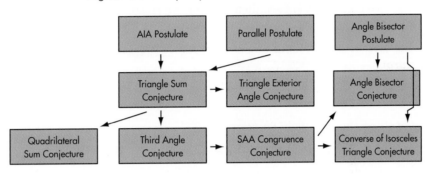

Lesson 16.2

Proving the Triangle Sum Conjecture

Earlier you discovered a number of angle conjectures. The order in which you discovered them is not the order in which one logically follows from another. Some have become postulates; others you will prove in this lesson. The concept map below shows one order in which you can prove the conjectures in this lesson.

Logical relationship of postulates and theorems in this lesson

| AIA Postulate | Parallel Postulate | Angle Bisector Postulate |

| Triangle Sum Conjecture | Triangle Exterior Angle Conjecture | Angle Bisector Conjecture |

| Quadrilateral Sum Conjecture | Third Angle Conjecture | SAA Congruence Conjecture | Converse of Isosceles Triangle Conjecture |

The proof of the Triangle Sum Conjecture is a nice, short proof once you construct a very clever auxiliary line. The conjecture is believed to have been one of the conjectures proved by Greek mathematician Thales. Thales is credited as being the first mathematician to apply deductive reasoning to geometry.

Let's take a look. The first three tasks in the proof of the Triangle Sum Conjecture are shown below. In addition, part of the plan (including the auxiliary line) has been started for you. You get to complete the plan, then perform the proof in Exercise Set 16.2.

Conjecture: The sum of the measures of the three angles of every triangle is 180° (***Triangle Sum Conjecture***).

Given: $\triangle ABC$

Show: $m\angle 1 + m\angle 2 + m\angle 3 = 180°$

Plan: A nice way to prove this conjecture is to construct a line through one vertex, parallel to the opposite side. (The Parallel Postulate allows you to construct it.) You can show $m\angle 4 + m\angle 2 + m\angle 5 = 180°$ with the help of the Linear Pair Postulate and the Angle Addition Postulate. You also can show $m\angle 1 = m\angle 4$ and $m\angle 3 = m\angle 5$ by the AIA Postulate.

Exercise Set 16.2

1. Complete the plan given in this lesson, then create a proof of the Triangle Sum Conjecture. Once finished, the conjecture can be called the Triangle Sum Theorem. Add it to your list.

In Exercises 2–7, perform all five steps of the proof process to prove each conjecture. When you're finished, add the theorems to your list.

2. If two angles of one triangle are congruent to two angles of another triangle, then the remaining two angles are congruent (***Third Angle Conjecture***).

3. The measure of an exterior angle of a triangle equals the sum of the measures of its two remote interior angles (***Triangle Exterior Angle Conjecture***).

4. If two angles and a side that is not between them in one triangle are congruent to two angles and a side that is not between them in another triangle, then the two triangles are congruent (***SAA Congruence Conjecture***).

5.* If a triangle has two congruent angles, then the triangle is isosceles (***Converse of the Isosceles Triangle Conjecture***).

6. The sum of the measures of the four angles of a quadrilateral is 360° (***Quadrilateral Sum Conjecture***).

7.* If a point is on the bisector of an angle, then it is equidistant from the sides of the angle (***Angle Bisector Conjecture***).

Use postulates and theorems to explain Exercises 8–10.

8. Explain why $x = y$.

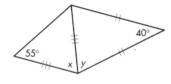

9. Explain why $a + b = 90°$.

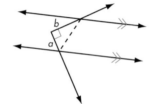

10. Explain why $m + n = 120°$.

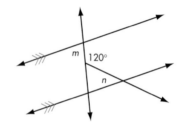

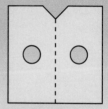

Lesson 16.3

Proving Circle Conjectures

Earlier in the year you discovered a number of circle properties. Now you will prove some of them. To prove circle conjectures, we need to add the Arc Addition Postulate to our list of postulates.

P-14 If point B is on $\widehat{AC}$ and between points A and C, then $m\widehat{AB} + m\widehat{BC} = m\widehat{AC}$ (*Arc Addition Postulate*).

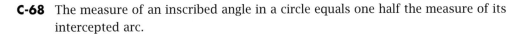

We also need to recall the definitions of the degree measures of arcs.

> The degree measure of a circle is 360°.
>
> The degree measure of a semicircle is 180°.
>
> The degree measure of a minor arc is the measure of its central angle.

Let's start with the proof of the Inscribed Angle Conjecture.

C-68 The measure of an inscribed angle in a circle equals one half the measure of its intercepted arc.

The proof of the Inscribed Angle Conjecture is interesting because you first prove the conjecture for a special condition or case, then you use that special case to extend the conjecture to all cases.

When you inscribe an angle in a circle, exactly one of the cases below must be true.

Case 1	Case 2	Case 3

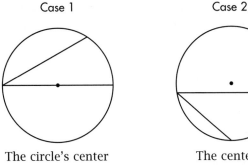 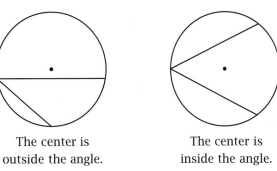

The circle's center is on the angle.	The center is outside the angle.	The center is inside the angle.

To prove the Inscribed Angle Conjecture, we will first prove that it is true when one of the sides of the angle passes through the center of the circle (Case 1). Then we will use this special case to prove the other two cases. When all three cases have been proved, you can rename the conjecture as the Inscribed Angle Theorem.

Conjecture: The measure of an inscribed angle in a circle equals one half the measure of its intercepted arc when a side of the angle passes through the center of the circle (Case 1).

Given: Circle O with $\angle MDR$ on diameter $\overline{DR}$

Show: $m\angle MDR = \frac{1}{2} m\widehat{MR}$

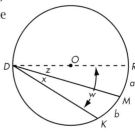

Plan:
- Because all radii are congruent, $\triangle DOM$ is isosceles and $x = z$.
- By the Triangle Exterior Angle Theorem, $x + z = y$.
- So $x + x = y$, which gives $2x = y$. Therefore $x = \frac{1}{2} y$.
- Because $m\angle MDR = x$ and $m\widehat{MR} = y$, $m\angle MDR = \frac{1}{2} m\widehat{MR}$.

Two-column proof of Case 1:

Let $x = m\angle MDR$, $z = m\angle OMD$, and $y = m\angle ROM$.

1. $\overline{DO} \cong \overline{OM}$	1. All radii of a circle are congruent (definition of a circle).
2. $\triangle DOM$ is isosceles.	2. Definition of isosceles triangle
3. $\angle MDR \cong \angle OMD$; $x = z$	3. Isosceles Triangle Theorem
4. $x + z = y$	4. Triangle Exterior Angle Theorem
5. $x + x = y$; $2x = y$	5. Substitution property, addition fact
6. $x = \frac{1}{2} y$	6. Division property of equality
7. $m\angle MDR = \frac{1}{2} m\angle ROM$	7. Substitution property
8. $m\angle ROM = m\widehat{MR}$	8. The measure of a minor arc is equal to the measure of its determined central angle.
9. $m\angle MDR = \frac{1}{2} m\widehat{MR}$	9. Substitution property

A paragraph proof is less structured than a two-column proof. We will use Case 1 to write a paragraph proof for Case 2. You will prove Case 3 in Exercise Set 16.3.

Conjecture: The measure of an inscribed angle in a circle equals one half the measure of its intercepted arc when the center of the circle is outside the angle (Case 2).

Given: Circle O with inscribed angle $\angle MDK$ on one side of diameter DR

Show: $m\angle MDK = \frac{1}{2} m\widehat{MK}$

Paragraph proof of Case 2:

Let $z = m\angle MDR$, $x = m\angle MDK$, and $w = m\angle KDR$. Then, by the Angle Addition Postulate, $w = m\angle KDR = m\angle MDR + m\angle MDK$, or $w = x + z$. Let $a = m\widehat{MR}$ and $b = m\widehat{MK}$. Thus $a + b = m\widehat{MR} + m\widehat{MK} = m\widehat{KR}$ (Arc Addition Postulate).

Stated in terms of x and b, we wish to show that $x = \frac{b}{2}$. From Case 1 we know that $w = \frac{(a + b)}{2}$ and that $z = \frac{a}{2}$.

We know that $w = x + z$, so $x = w - z$ (subtraction property of equality).

Substitute for w and z to get $x = \frac{(a + b)}{2} - \frac{a}{2} = \frac{(a + b) - a}{2} = \frac{b}{2}$. Therefore $m\angle MDK = \frac{1}{2} m\widehat{MK}$.

CHAPTER 16

Exercise Set 16.3

1. Write a proof of Case 3 of the Inscribed Angle Conjecture.

Once you have proved all three cases, you have proved the Inscribed Angle Conjecture. You can now rename it as the Inscribed Angle Theorem and add it to your list. You will use this theorem to prove theorems in the remainder of this exercise set.

When a conjecture can be proved as an immediate consequence of another theorem, the new theorem is a **corollary** of the original theorem. The four conjectures in Exercises 2–5 are corollaries of the Inscribed Angle Theorem. Prove each conjecture and add it to your theorem list.

2. Inscribed angles that intercept the same arc are congruent (C-69).

3. Angles inscribed in a semicircle are right angles (C-70).

4. The opposite angles of a quadrilateral inscribed in a circle are supplementary (C-71).

5. Parallel lines intercept congruent arcs on a circle (C-72).

6. Explain why $\overline{KE} \parallel \overline{YL}$.

7. Explain why $\triangle JIM$ is isosceles.

8. Explain why $\triangle KIM$ is isosceles.

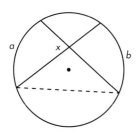

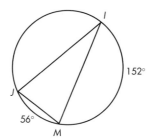

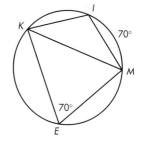

Now for more challenges. Determine whether each conjecture in Exercises 9–12 is true or false. If the conjecture is false, draw a counterexample. If the conjecture is true, prove it by performing all five tasks in the proof process.

9.* The measure of an angle formed by two intersecting chords in a circle is equal to half the sum of the two intercepted arcs.

10.* The measure of an angle formed by two secants intersecting outside a circle is equal to half the difference of the two intercepted arcs.

11.* If circles A and E intersect at points C and K, then $\overline{AE}$ connecting the circles' centers is the perpendicular bisector of the common chord $\overline{CK}$.

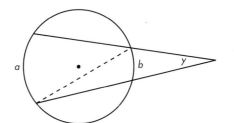

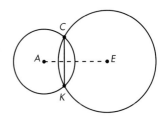

12. If a trapezoid is inscribed within a circle, then the trapezoid is isosceles.

Lesson 16.4

Proving the Pythagorean Theorem

In this lesson you will investigate and discover a similarity conjecture found in right triangles. Once you have proved this new conjecture, you can use it to prove the Pythagorean Theorem!

To prove conjectures involving similarity, you need to add a new postulate to your list. In Chapter 12, you discovered some shortcuts for showing that two triangles are similar. We are going to rename the AA Similarity Conjecture as a postulate so we can use it in our proofs. Here it is.

P-15 If two angles of one triangle are congruent to two
angles of another triangle, then the two triangles are similar
(*AA Similarity Postulate*).

Investigation 16.4

An altitude has been constructed to the hypotenuse of each right triangle below. This construction creates two more right triangles, each within the original right triangle. For cases 1–3 below, calculate the measure of the acute angles in each right triangle.

1. $a = -?-$
  $b = -?-$
  $c = -?-$

2. $a = -?-$
  $b = -?-$
  $c = -?-$

3. $a = -?-$
  $b = -?-$
  $c = -?-$

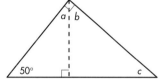 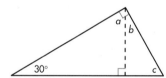 

Do you notice anything special about the two small right triangles in each larger right triangle? How do the two small right triangles compare with each other? How do they compare with the original right triangle? If you have trouble visualizing how the three triangles in each case relate to one another, study the figure below. It shows the two smaller right triangles separated from the original right triangle.

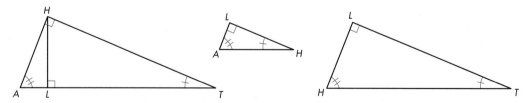

Discuss your observations with your group members. Are all three right triangles similar? You should be ready to state your next conjecture.

C-115 If you drop an altitude from the right angle to the hypotenuse of a right triangle, it divides the right triangle into two right triangles that are —?— to each other and to the original —?—.

Complete the statement of similarity between the three right triangles in the diagram on the previous page: $\triangle HAT \sim \triangle LAH \sim \triangle$ -?-.

Exercise Set 16.4

For Exercises 1–3, write the statement of similarity among the three right triangles.

1. $\triangle ALE \sim \triangle$ -?- $\sim \triangle$ -?-

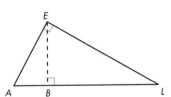

2. $\triangle BEK \sim \triangle$ -?- $\sim \triangle$ -?-

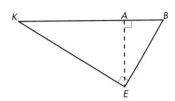

3. $\triangle CHA \sim \triangle$ -?- $\sim \triangle$ -?-

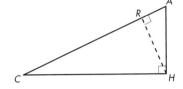

In Exercises 4–6, complete each proportion statement from the similar right triangles. You will use these statements to prove the Pythagorean Theorem.

4. $\dfrac{x}{a} = \dfrac{\text{-?-}}{c}$

5. $\dfrac{y}{b} = \dfrac{b}{\text{-?-}}$

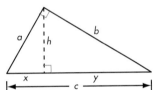

6. $\dfrac{a}{\text{-?-}} = \dfrac{\text{-?-}}{a}$

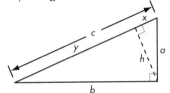

For Exercises 7–9, write a proof for each conjecture. When you finish each proof, add the new theorem to your list.

7. Conjecture: If you drop an altitude from the right angle to the hypotenuse of a right triangle, it divides the right triangle into two right triangles that are similar to each other and to the original right triangle.

8.* Pythagorean Theorem

9.* Converse of the Pythagorean Theorem

10. Explain how the congruence shortcuts stated below and on the following page for right triangles, the HA Congruence Shortcut and the LA Congruence Shortcut, are corollaries of the SAA Congruence Theorem and how the LL Congruence Shortcut for right triangles, stated on the following page, is a corollary of the SAS Congruence Postulate.

• If the hypotenuse and an acute angle of a right triangle are congruent to the hypotenuse and an acute angle of a second right triangle, then the two right triangles are congruent (*HA Congruence Shortcut*).

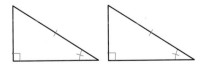

- If one leg and an acute angle of a right triangle are congruent to the corresponding leg and acute angle of a second right triangle, then the two right triangles are congruent (***LA Congruence Shortcut***).

- If the two legs of a right triangle are congruent to two legs of a second right triangle, then the two right triangles are congruent (***LL Congruence Shortcut***).

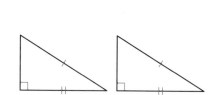

11. *Use the Pythagorean Theorem to prove the HL Congruence Shortcut for right triangles, stated below.

- If the hypotenuse and one leg of a right triangle are congruent to the hypotenuse and leg of a second right triangle, then the two right triangles are congruent (***HL Congruence Shortcut***).

Use your theorems from this lesson to solve Exercises 12–14.

12. * $a = -?-$

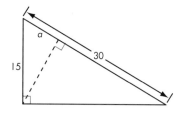

13. $b = -?-$

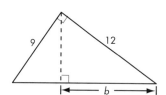

14. $c = -?-$

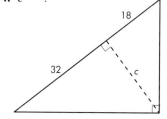

15. *While flying beneath the cloud cover of Saro-Gahtyp Valley, archaeologist and adventurer Dakota Davis was able to photograph mysterious geometric markings on the valley floor. The markings consist of a series of right triangles with a geometric figure on each side of each right triangle. One right triangle has a semicircle on each side. Another right triangle has a regular pentagon on each side. Still another has an equilateral triangle on each side. The ancient valley culture that made the markings left no written language, but the figures hint at a great geometric discovery made by this long-lost civilization. What is that discovery? Prove it for one of the special cases shown.

Dakota Davis and the Saro-Gahtyp Discovery

Lesson 16.5

Indirect Geometric Proofs

We don't need no education.
— "Another Brick in the Wall" by Pink Floyd

Recall the procedure for an indirect proof from Chapter 14, shown below.

1. Begin by assuming the opposite of what you wish to show is true.
2. Then, using logical reasoning, try to reach a contradiction.
3. Once you have a contradiction, point out that because your argument is valid, the assumption must be responsible for the contradiction. Therefore the assumption is false, and the desired conclusion must be true.

In this lesson, you'll apply indirect proof techniques to geometric proofs.

Example A

Conjecture: If $m\angle N \neq m\angle O$ in $\triangle NOT$, then $NT \neq OT$.

   Given: $\triangle NOT$ with $m\angle N \neq m\angle O$

   Show: $NT \neq OT$

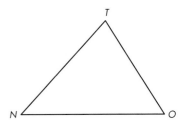

Paragraph proof:

Assume $NT = OT$. If $NT = OT$, then $m\angle N = m\angle O$ by the Isosceles Triangle Conjecture. But this contradicts the given fact that $m\angle N \neq m\angle O$. Therefore the assumption ($NT = OT$) is false and thus the opposite ($NT \neq OT$) is true.

Example B

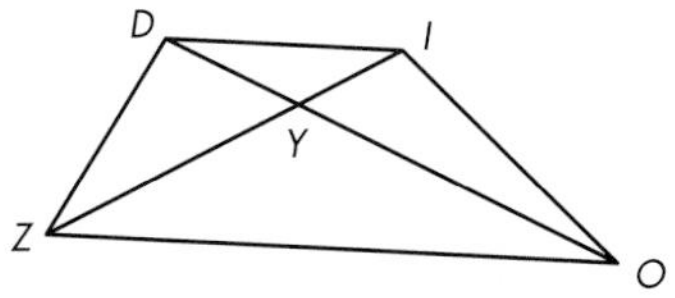

Conjecture: The diagonals of a trapezoid do not bisect each other.

   Given: Trapezoid $ZOID$ with parallel bases $\overline{ZO}$ and $\overline{ID}$ and diagonals $\overline{DO}$ and $\overline{IZ}$ intersecting at point Y

   Show: Diagonals of trapezoid $ZOID$ do not bisect each other. That is, $DY \neq OY$ and $ZY \neq IY$.

Paragraph proof:

Assume the diagonals of trapezoid $ZOID$ *do* bisect each other. In Exercise Set 16.1, you proved that if the diagonals of a quadrilateral bisect each other, the quadrilateral is a parallelogram. So $ZOID$ is a parallelogram. Thus $ZOID$ has two pairs of opposite sides parallel. But because it is a trapezoid, it has exactly one pair of parallel sides. This is contradictory. So the assumption that its diagonals bisect each other is false and the conjecture is true.

Exercise Set 16.5

1. Is the proof in Example A claiming that if two angles of a triangle are not congruent, then the triangle is not isosceles? Explain.

2. Is the proof in Example B claiming that if the diagonals of a quadrilateral bisect each other, then the quadrilateral is not a trapezoid? Explain.

3. Give each missing reason in the indirect proof below.

   Conjecture: No triangle has two right angles.

       Given: $\triangle ABC$

       Show: No two angles are right angles.

   Two-column proof:
   1. Assume $\triangle ABC$ has two right angles.   1. —?—
      (Assume $m\angle A = 90°$ and $m\angle B = 90°$.)
   2. $m\angle A + m\angle B + m\angle C = 180°$ 2. —?—
   3. $90° + 90° + m\angle C = 180°$ 3. —?—
   4. $m\angle C = -?-$ 4. —?—

   But if $m\angle C = 0$, then the two sides $\overline{AC}$ and $\overline{BC}$ overlap. This contradicts the given information. So the assumption is false. Therefore no triangle has two right angles.

4. Answer each question to complete the indirect proof of the Tangent Conjecture.

   Conjecture: A tangent is perpendicular to the radius drawn to the point of tangency.

       Given: Circle O with tangent $\overleftrightarrow{AT}$ and radius $\overline{AO}$

       Show: $\overline{AO} \perp \overleftrightarrow{AT}$

   Paragraph proof:

   a. Assume $\overline{AO}$ is not perpendicular to $\overleftrightarrow{AT}$. Construct a perpendicular from point O to $\overleftrightarrow{AT}$ and label the intersection point B ($\overline{OB} \perp \overleftrightarrow{AT}$). Which postulate tells you this is possible?

   b. Select a point C on $\overleftrightarrow{AT}$ so that B is the midpoint of $\overline{AC}$. Which postulate tells you this is possible?

   c. Next construct $\overline{OC}$. Which postulate tells you this is possible?

   d. $\angle ABO \cong \angle CBO$. What reason (or reasons) tells you this?

   e. $\overline{AB} \cong \overline{BC}$. What definition tells you this?

   f. $\overline{OB} \cong \overline{OB}$. What property of congruence tells you this is possible?

   g. Therefore $\triangle ABO \cong \triangle CBO$. Which congruence conjecture tells you the triangles are congruent?

   h. If $\triangle ABO \cong \triangle CBO$, then $\overline{AO} \cong \overline{CO}$. Why?

   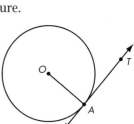

   Thus C must be a point of the circle (because a circle is the set of *all* points in the plane at a given distance from the center and points A and C are both the same distance from the center). Therefore $\overleftrightarrow{AT}$ intersects the circle in *two* points (A and C) and thus $\overleftrightarrow{AT}$ is not a tangent. But this contradicts the given information. So the assumption that $\overline{AO}$ is not perpendicular to $\overleftrightarrow{AT}$ is false, and $\overline{AO} \perp \overleftrightarrow{AT}$.

   Rename the Tangent Conjecture as the Tangent Theorem and add it to your list.

5. Write an indirect proof of the conjecture below.

Conjecture: The bases of a trapezoid have unequal lengths.

Given: Trapezoid *ZOID* with parallel bases $\overline{ZO}$ and $\overline{ID}$

Show: $ZO \neq ID$

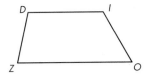

6. Write the given and the show, then plan and write the proof of the following conjecture: Tangent segments from a point to a circle are congruent (***Tangent Segments Conjecture***). This conjecture is a corollary of the Tangent Theorem.

7. Write the given and the show, then plan and write the proof of this circle conjecture: The perpendicular bisector of a chord passes through the center of the circle.

8.* Here is a logic puzzle adapted from *What Is the Name of This Book?* by Raymond Smullyan: Doctors were examining a patient they thought might be delusional. They decided to give the patient a lie-detector test. One of the questions they asked him was "Are you Napoleon?" He replied, "No." The machine showed he was lying. Is the patient delusional? Explain your reasoning.

Improving Visual Thinking Skills—*Visual Analogies*

Which of the designs on the right complete the statements on the left? Explain.

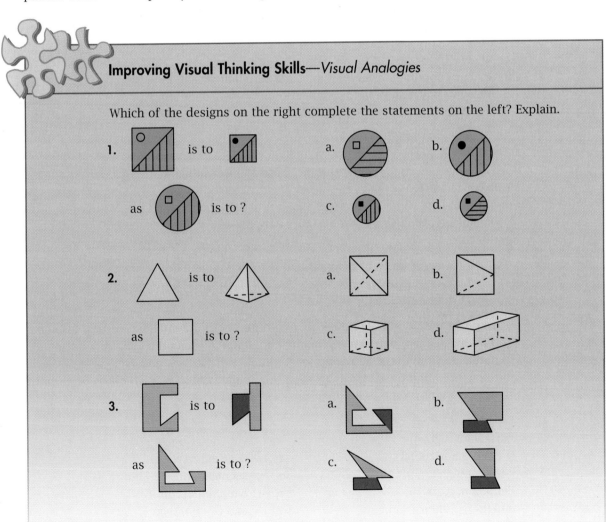

Lesson 16.6

Proof with Coordinate Geometry

Conjectures involving midpoints and parallel, perpendicular, and congruent segments can be proved nicely with coordinate geometry. Three conjectures from earlier lessons on coordinate geometry are shown below. We are going to accept these three conjectures as postulates and use them in the coordinate proofs of this lesson.

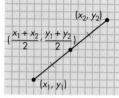

P-16 If (x_1, y_1) and (x_2, y_2) are the coordinates of the endpoints of a segment, then the coordinates of the midpoint are $(\frac{x_1 + x_2}{2}, \frac{y_1 + y_2}{2})$ (*Coordinate Midpoint Postulate*).

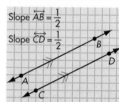

P-17 In a coordinate plane, two lines are parallel if and only if their slopes are equal (*Parallel Slope Postulate*).

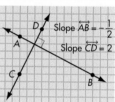

P-18 In a coordinate plane, two lines are perpendicular if and only if their slopes are negative reciprocals of each other (*Perpendicular Slope Postulate*).

Another important conjecture you made in an earlier lesson on coordinate geometry is the Distance Formula, shown below. This formula is really the Pythagorean Theorem in coordinate disguise! In fact, you will use the Pythagorean Theorem to prove the Distance Formula. Then you will use the Distance Formula as a theorem in coordinate proofs in this lesson.

If the coordinates of points A and B are (x_1, y_1) and (x_2, y_2), respectively, then $AB^2 = (x_2 - x_1)^2 + (y_2 - y_1)^2$ and $AB = \sqrt{(x_2 - x_1)^2 + (y_2 - y_1)^2}$ (*Distance Formula*).

To prove the Distance Formula, recall how to find horizontal and vertical distances. To find the distance between two points horizontal to each other, subtract one point's *x*-coordinate from the other point's *x*-coordinate and take the absolute value of the difference. To find the distance between two points vertical to each other, perform the same operations on the *y*-coordinates.

Length of a horizontal segment: If $\overline{CD}$ is a horizontal segment with endpoints (x_1, y_1) and (x_2, y_1), then $CD = |x_2 - x_1|$.

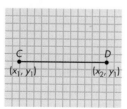

Length of a vertical segment: If $\overline{EF}$ is a vertical segment with endpoints (x_1, y_1) and (x_1, y_2), then $EF = |y_2 - y_1|$.

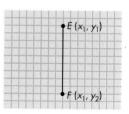

Conjecture: If the coordinates of points A and B are (x_1, y_1) and (x_2, y_2), then $AB^2 = (x_2 - x_1)^2 + (y_2 - y_1)^2$ and $AB = \sqrt{(x_2 - x_1)^2 + (y_2 - y_1)^2}$.

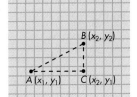

Given: Points A and B with coordinates (x_1, y_1) and (x_2, y_2)

Show: $AB = \sqrt{(x_2 - x_1)^2 + (y_2 - y_1)^2}$

Proof: Locate point C with coordinates (x_2, y_1). $\overline{BC}$ is a vertical segment because point C has the same x-coordinate as point B. $\overline{AC}$ is a horizontal segment because point C has the same y-coordinate as point A. Because $\overline{AC}$ and $\overline{BC}$ are horizontal and vertical segments, $\triangle ABC$ is a right triangle. If $\triangle ABC$ is a right triangle, then $AB^2 = AC^2 + BC^2$. $AC = |x_2 - x_1|$ and $BC = |y_2 - y_1|$. Therefore $AB^2 = AC^2 + BC^2 = |x_2 - x_1|^2 + |y_2 - y_1|^2$. Because the square of the absolute value of a number equals the square of that number, $AB^2 = (x_2 - x_1)^2 + (y_2 - y_1)^2$. Taking the positive square root of both sides of the equation, $AB = \sqrt{(x_2 - x_1)^2 + (y_2 - y_1)^2}$.

Now that the Distance Formula has been proved, you can add it to your theorem list.

Exercise Set 16.6

In Exercises 1–3, each figure demonstrates a convenient position for a polygon. Provide the missing coordinates without adding any new letters.

1. $\triangle ABC$ is isosceles.

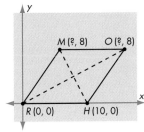

2.* Quadrilateral $ABCD$ is a parallelogram.

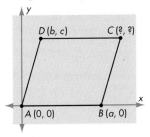

3.* Quadrilateral $ABCD$ is a rhombus.

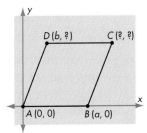

In Exercises 4–6, find the slope or length indicated.

4. Find the slope of each diagonal in rhombus $RHOM$.

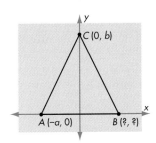

5. Find the length of each diagonal in isosceles trapezoid $ZOID$.

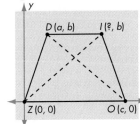

6. Find the length of the medians to the congruent sides of isosceles $\triangle TRY$.

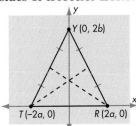

In Exercises 7–9, write a coordinate proof of each stated conjecture. Use the diagram with the given and show to help you get started. When you're finished, add each theorem to your list.

7. **Conjecture:** The midpoint of the hypotenuse of a right triangle is the same distance from the three vertices.

   Given: Right $\triangle RHT$ with M the midpoint of the hypotenuse. Because you can place the triangle anywhere you wish in the coordinate plane, locate it so that the coordinates of its vertices are simple and easy to work with. Locate right angle vertex R at the origin, vertex H at $(2a, 0)$, and vertex T at $(0, 2b)$ so that the two sides are on the axes.

   Show: $MR = MH = MT$

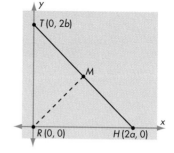

8. **Conjecture:** A midsegment of a triangle is parallel to the third side and one half the length of the third side (**Triangle Midsegment Conjecture**).

   Given: $\triangle AOK$ with M and N the midpoints of $\overline{AO}$ and $\overline{OK}$, respectively. Locate vertex O at the origin and vertex A at $(2a, 0)$ so that one side is on an axis. Locate point K at $(2b, 2c)$.

   Show: $MN = \frac{1}{2}AK$ and $\overline{MN} \parallel \overline{AK}$.

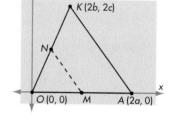

9. **Conjecture:** The midsegment of a trapezoid is parallel to the bases and is equal in length to the average of the lengths of the bases (**Trapezoid Midsegment Conjecture**).

   Given: Trapezoid $TRAP$ with $\overline{TR} \parallel \overline{AP}$ and with E and Z the midpoints of sides $\overline{PT}$ and $\overline{AR}$, respectively. Locate one vertex at the origin, with one base on the x-axis.

   Show: $EZ = \frac{1}{2}(TR + AP)$ and $\overline{EZ} \parallel \overline{TR} \parallel \overline{AP}$.

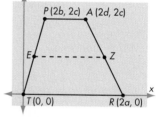

10. Prove the following conjecture: The three midsegments of a triangle divide the triangle into four congruent triangles. Use any form you'd like: a coordinate proof, a flow-chart proof, or a two-column proof. Then add the theorem to your list.

Improving Visual Thinking Skills—*Picture Patterns VI*

What is the rule for the total number of squares for the *n*th figure of this pattern?

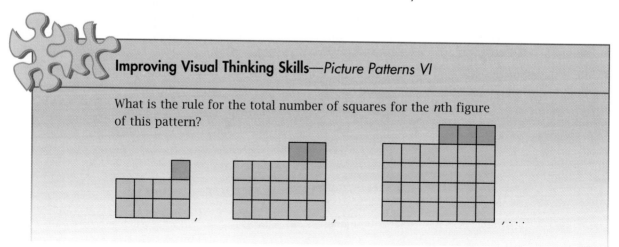

Lesson 16.7

Midsegment Conjectures

What is now proved was once only imagined.
— William Blake

You can use the Triangle Midsegment Theorem, proved in Lesson 16.6, to prove other conjectures. In this lesson you will perform investigations that lead to new conjectures. Then you will prove each conjecture, using the Triangle Midsegment Theorem. The example below shows an investigation leading to a conjecture and then to a proof of the conjecture. This example investigation is performed with patty papers, but this investigation and those in the exercise set can also be performed efficiently with a geometry computer program.

Example

Investigate: Draw a quadrilateral onto a patty paper. Fold to locate the midpoint of each side. Connect the four midpoints in order around the quadrilateral to form another quadrilateral.

Conjecture: What do you observe about the opposite sides in this new quadrilateral? State a conjecture about the quadrilateral formed by connecting the midpoints of a quadrilateral in order.

Prove: Prove your conjecture.

After completing the investigation above, you wind up with a drawing similar to that pictured at right. The opposite sides in the smaller quadrilateral appear parallel. Is this always true? You don't want to make a conjecture based on only one case, so try again, beginning with other quadrilaterals. After going through the construction several more times, you may satisfy yourself that your observations seem to hold for every quadrilateral. You are ready to make a conjecture.

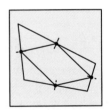

Conjecture: The quadrilateral formed by connecting the midpoints of the sides of a quadrilateral in order is a parallelogram.

How do you prove the conjecture? Start by making a diagram. You may omit the construction marks. Label your diagram. State what you are given and what you must show in terms of your diagram. Then plan and proceed with your proof.

Given: Quadrilateral *QUAD* with inscribed quadrilateral *PREL*, where *P, R, E,* and *L* are the midpoints of $\overline{UA}$, $\overline{AD}$, $\overline{DQ}$, and $\overline{QU}$, respectively.

Show: Quadrilateral *PREL* is a parallelogram.

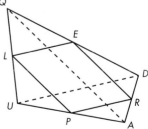

Plan: • I wish to prove *PREL* is a parallelogram. I can do this if I can prove that the opposite sides are parallel.

• In an earlier lesson, it was proved that if two lines are parallel to a third line in the plane, then they are parallel to each other.

• Therefore I construct diagonal $\overline{QA}$ because it appears to be parallel to both $\overline{ER}$ and $\overline{LP}$.

• Therefore if I can show $\overline{ER} \parallel \overline{QA}$ and $\overline{LP} \parallel \overline{QA}$, then $\overline{ER} \parallel \overline{LP}$.

• $\overline{ER}$ is a line segment connecting the midpoints of two sides of $\triangle AQD$, and $\overline{LP}$ is a line segment connecting the midpoints of two sides of $\triangle AQU$. The Triangle Midsegment Theorem says that if a line segment connects the midpoints of two sides of a triangle, then it is parallel to the third side. Therefore $\overline{ER} \parallel \overline{LP}$.

• By a similar argument, I can construct diagonal $\overline{DU}$ and show that $\overline{LE} \parallel \overline{DU}$ and $\overline{PR} \parallel \overline{DU}$. Therefore $\overline{LE} \parallel \overline{PR}$.

• Because I can show $\overline{ER} \parallel \overline{LP}$ and $\overline{LE} \parallel \overline{PR}$, then *PREL* is a parallelogram by the definition of a parallelogram.

Flow-chart proof:

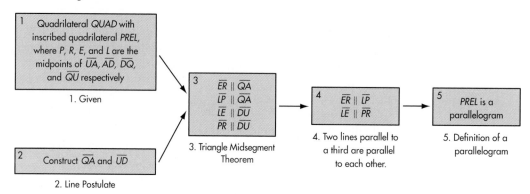

Exercise Set 16.7

Perform each investigation, using coordinate methods, patty papers, or a geometry computer program. Then make a conjecture. Finally, prove your conjecture. Once you have proved your conjectures, add them to your theorem list.

1. Investigate: Construct a rectangle. Construct the midpoint of each side. Connect the four midpoints to form another quadrilateral.

    Conjecture: What do you observe about the quadrilateral formed? From the example at the beginning of this lesson, you know that the quadrilateral is a parallelogram, but what type of parallelogram? State a conjecture about the parallelogram formed by connecting the midpoints of a rectangle.

    Prove: Prove your conjecture.

2. Investigate: Construct a rhombus. Construct the midpoint of each side. Connect the four midpoints to form another quadrilateral.

Conjecture: From the example at the beginning of this lesson, you know that the quadrilateral is a parallelogram, but what type of parallelogram? State a conjecture about the parallelogram formed by connecting the midpoints of a rhombus.

Prove: Prove your conjecture.

3. Investigate: Construct a kite (a quadrilateral with two pair of consecutive sides congruent). Construct the midpoint of each side. Connect the four midpoints to form another quadrilateral.

Conjecture: State a conjecture about the parallelogram formed by connecting the midpoints of a kite.

Prove: Prove your conjecture.

4. Investigate: Construct an isosceles trapezoid. Construct the midpoint of each side. Connect the four midpoints to form another quadrilateral.

Conjecture: State a conjecture about the parallelogram formed by connecting the midpoints of an isosceles trapezoid.

Prove: Prove your conjecture.

5. Investigate: Construct an isosceles triangle. Construct the midpoint of each side. Connect the midpoint of the base to the midpoint of each of the congruent sides to form a quadrilateral.

Conjecture: State a conjecture about the quadrilateral formed by connecting the midpoint of the base of an isosceles triangle to the midpoint of each of the congruent sides.

Prove: Prove your conjecture.

Improving Visual Thinking Skills—*Connecting Cubes*

The two objects shown at right can be placed together to form each of the shapes below except one. Which one?

a.   b. c. d.

Project
Special Proofs of Special Conjectures

You have explored and discovered many interesting geometric ideas throughout this past school year. The explanations of why these conjectures are true can be interesting, too. In this project your task is to research and present logical arguments in support of one or more of these special properties.

1. Lesson 8.4 explained why there are only three regular polygons that tile the plane. See if you can remember that proof without looking back at the text. That proof is related to another very interesting proof. In Chapter 11's Project: The Five Platonic Solids you learned that there are only five regular polyhedra. Write a convincing argument explaining why.

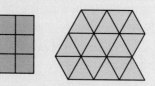

2. In Chapter 2's Project: Euler's Formula for Networks you discovered Euler's rule for determining whether a planar network can or cannot be traveled. Write a convincing argument defending Euler's rule.

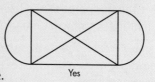

Yes

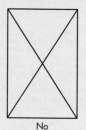

3. In Lesson 6.1, you discovered that the formula for the sum of the measures of the interior angles of an n-gon is $(n - 2)180°$. Prove this formula is correct.

No

4. In Lesson 8.2, you discovered that the composition of two reflections over intersecting lines is equivalent to one rotation. Prove this always works.

5. In Lesson 4.6, you discovered that the coordinates of the centroid of a triangle are equal to the average of the coordinates of the triangle's three vertices. Prove this is always true.

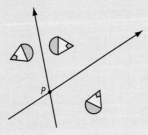

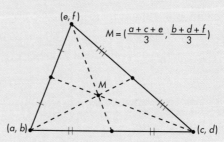

(e, f)

$M = (\dfrac{a+c+e}{3}, \dfrac{b+d+f}{3})$

M

(a, b) (c, d)

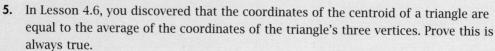

Lesson 16.8

Chapter Review

In this chapter you proved some important theorems, including the Triangle Sum Theorem, the Inscribed Angle Theorem, and the Pythagorean Theorem. You also learned special proof techniques such as indirect proofs and coordinate proofs. Finally, you combined an inductive approach to mathematics with a deductive approach by performing investigations, making discoveries, forming conjectures, and proving your conjectures. This process is at the heart of how mathematics is done.

Exercise Set 16.8

For Exercises 1–10, identify each statement as true or false. For each false statement, sketch a counterexample or explain why it is false.

1. If one pair of sides of a quadrilateral are parallel and the other pair of sides are congruent, then the quadrilateral is a parallelogram.

2. If consecutive angles of a quadrilateral are supplementary, then the quadrilateral is a parallelogram.

3. If the diagonals of a quadrilateral are congruent, then the quadrilateral is a rectangle.

4. Two exterior angles of an obtuse triangle are obtuse.

5. If two angles of one triangle are congruent to two angles of another triangle, then the remaining two angles are congruent.

6. The opposite angles of a quadrilateral inscribed within a circle are congruent.

7. The measure of an angle formed by two secants intersecting outside a circle is equal to one half the difference of the measures of the two intercepted arcs.

8. If the three sides of one triangle are proportional to the three sides of another triangle, then the two triangles are congruent.

9. The diagonals of a trapezoid bisect each other.

10. The midpoint of the hypotenuse of a right triangle is equidistant from all three vertices.

In Exercises 11–19, complete each statement.

11. A tangent is —?— to the radius drawn to the point of tangency.

12. Tangent segments from a point to a circle are —?—.

13. The perpendicular bisector of a chord passes through —?—.

14. The three midsegments of a triangle divide the triangle into —?—.

15. The parallelogram formed by connecting the midpoints of the sides of a rectangle is a —?—.

16. The parallelogram formed by connecting the midpoints of the sides of a rhombus is a —?—.

17. The parallelogram formed by connecting the four midpoints of the sides of a kite is a —?—.

18. Euclid's fifth postulate is also known as the —?— Postulate.

19. Geometry on the sphere is a model of the non-Euclidean geometry called —?—.

For Exercises 20–29, consult your list of postulates and theorems.

20. If $\triangle ABC \cong \triangle DEF$, then $\overline{AC} \cong \overline{DF}$. What reason would you give for this in a proof?

21. What is the difference between a postulate and a theorem?

22. State the following as a conditional: The segment joining the midpoints of the diagonals of a trapezoid is parallel to the bases.

23. Sometimes a proof requires the help of a construction. If you need an angle bisector in a proof, what postulate tells you that you may construct one?

24. If an altitude is needed in a proof, what postulate tells you that you may construct one?

25. Describe the procedure for an indirect proof.

26. Sometimes you can prove a conjecture as an immediate consequence of another theorem. What do you call this new theorem?

27. Upon what postulate (or postulates) does the Triangle Sum Conjecture depend?

28. Upon what postulate (or postulates) does the Pythagorean Theorem depend?

29. In a coordinate proof, what postulate do you use to prove that two lines are perpendicular?

In Exercises 30–34, devise a plan for each proof, then create a proof.

30. Refer to the figure at right.

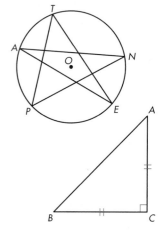

    Given: Circle O with chords $\overline{PN}$, $\overline{ET}$, $\overline{NA}$, $\overline{TP}$, $\overline{AE}$

    Show: $m\angle P + m\angle E + m\angle N + m\angle T + m\angle A = 180°$

31. Prove the conjecture.

    Conjecture: If a right triangle is isosceles, then the base angles measure 45°.

        Given: Isosceles right $\triangle ABC$ with $\overline{BC} \cong \overline{AC}$ and with right $\angle ACB$

        Show: $m\angle ABC = 45°$

32. Prove the conjecture.

    Conjecture: The sum of the measures of the five angles of a pentagon is 540°.

        Given: Pentagon $ABCDE$

        Show: $m\angle A + m\angle B + m\angle C + m\angle D + m\angle E = 540°$

33.* Prove the conjecture.

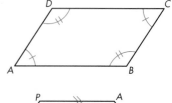

    Conjecture: If both pairs of opposite angles of a quadrilateral are congruent, then the quadrilateral is a parallelogram.

        Given: Quadrilateral $ABCD$ with $\angle DAB \cong \angle BCD$ and $\angle ADC \cong \angle ABC$

        Show: Quadrilateral $ABCD$ is a parallelogram.

34.* Prove the conjecture.

    Conjecture: The base angles of an isosceles trapezoid are congruent.

In Exercises 35–42, investigate each conjecture to determine whether or not it is true. If it is false, demonstrate a counterexample. If it is true, prove it.

35. The diagonals of a kite bisect the angles.

36. The medians to the congruent sides of an isosceles triangle are congruent.

37. If the diagonals of a parallelogram bisect the angles, then the parallelogram is a square.

38. The angle bisectors of one pair of base angles of an isosceles trapezoid are perpendicular.

39. The perpendicular bisectors to the congruent sides of an isosceles trapezoid are perpendicular.

40. The segment joining the feet of the altitudes on the two congruent sides of an isosceles triangle is parallel to the third side.

41. The bisectors of two consecutive angles of a rhombus are perpendicular.

42. The bisectors of a pair of opposite angles of a parallelogram are parallel.

In Exercises 43 and 44, draw a concept map to show how the postulates and the theorems in each group are logically related. Refer back to the proofs of the theorems in this chapter.

43.* • AIA Postulate
   • Isosceles Triangle Theorem
   • Parallel lines intercept congruent arcs on a circle.
   • Triangle Exterior Angle Theorem
   • Inscribed Angle Theorem

44.* • SSS Congruence Postulate
   • Pythagorean Theorem
   • Converse of the Pythagorean Theorem
   • AA Similarity Postulate

45. Here is another puzzling story taken from *What Is the Name of This Book?* by Raymond Smullyan: By an irresistible cannonball we shall mean a cannonball that knocks over everything in its way. By an immovable post we shall mean a post that cannot be knocked over by anything. So what happens if an irresistible cannonball hits an immovable post?

Assessing What You've Learned

If your class has been following this book in order, it may be time for a year-end assessment. That may take the form of a final exam, a project, your portfolio, a performance assessment, a presentation, or any combination of these and other assessments. The important thing is that you take the time to look back over your entire year's or semester's work to organize and review it.

If you're keeping a portfolio, you should be thinking not only of what you might add to it from this chapter, but also about organizing your selections from previous chapters. Review the entire portfolio to be sure it best represents your learning.

If you have a final exam, you can prepare by reworking selected problems from Chapter Review lessons. You should also review your notebook.

Cooperative Problem Solving

Party Time

The lunar colony's geometry class has been talking about celebrating the end of the school term with a class party. They feel it is time to use their organizational and logical-thinking skills for some fun and games! They have decided to have food and entertainment at their party. They want to serve a couple of main dishes, desserts, hot and cold beverages, and a variety of salads (pasta salad, green salad, potato salad, and fruit salad).

Your job is to help them organize their party. Where do you start? What are the tasks?

- Make a list of everything that has to be done before their party gets underway (decorations, purchasing, food preparation, drinks and ice, music, dancing, games, clean-up).
- Once you have listed all the tasks, put them in chronological order. For example, you can't make the salad until you figure out what type (or types) of salad people want, so plan a detailed menu.
- Match the tasks with people. Who is going to be responsible for what?
- You can't shop for groceries until you know how many people are attending the party, so make a guest list and approximate how much food each person will eat.
- When it comes time to prepare the food, record the approximate time it will take to complete each task. For example, if the party is a barbecue, the time you wish to start eating determines the time to light the charcoal.

Don't underestimate the complexity of the tasks involved in planning a successful party. Successful entertaining is an art that requires logical thinking and finely tuned organizational skills. Once you have organized the party, why not give it! Enjoy.

Hints for Selected Problems

You will find hints (and occasionally answers) below for problems in investigations, Take Another Look sections, and exercise sets that are marked with an asterisk (*) in the text. If you turn to a hint before you've tried to solve a problem on your own, SHAME ON YOU! You should go back and make a serious effort to solve the problem without help. But if you need additional help to solve a problem, this is the place to look.

Chapter 0

Exercise Set 0.1 (pg 5)

2. Here's the title of the sculpture.

> *Early morning calm*
> *knotweed stalks*
> *pushed into lake bottom*
> *made complete by their own reflection*

Derwent Water, Cumbria
20 February & 8–9 March, 1988

Exercise Set 0.8 (pg 31)

4.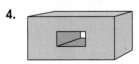

Exercise Set 0.9 (pg 34)

6a. The Aztecs created a stone calendar for keeping track of the year.

Chapter 1

Exercise Set 1.1 (pg 41)

8. I learned by trial and error that you turn wood screws clockwise to screw them into the wood and counterclockwise to remove them.

Exercise Set 1.2 (pg 44)

5. $1^2, 2^2, 3^2, 4^2, \ldots$

7. Change all fractions to the same denominator.

8. The letter pattern skips a letter.

11. $1 + 1 = 2, 1 + 2 = 3, 2 + 3 = 5, 3 + 5 = 8, \ldots$

14. Look at the differences. The differences are the numbers in Exercise 13.

15. The sequence of differences in this exercise is the same as the sequence of numbers in Exercise 5 $(4, 9, 16, 25, \ldots)$.

16. The differences are $2, 6, 18, 54, 162, \ldots$. How is this pattern generated? $3 \times 2 = 6$, $3 \times 6 = 18$, $3 \times 18 = 54$, and so on.

19. The pattern of the differences is $1, 3, 9, 27, 81, \ldots$. How is this pattern generated? $1 = 3^0$, $3 = 3^1$, $9 = 3^2$, $27 = 3^3$, $81 = 3^4$, and so on.

23. The sum of two odd numbers is always an even number.

Exercise Set 1.3 (pg 46)

6. Each new line crosses all of the previously drawn lines.

8. Two-by-three rectangle, four-by-four rectangle, six-by-five rectangle, eight-by-six rectangle, . . .

Exercise Set 1.4 (pg 54)

1.

1	2	3	4	5	6
$3 \cdot 1$	$3 \cdot 3$	$3 \cdot 5$	$3 \cdot 7$	$3 \cdot 9$	$3 \cdot 11$

9.

Points	1	2	3	4	5	6
Infinite	2	2	2	2	2	2
Finite	0	1	2	3	4	5
Total	2	3	4	5	6	7

Exercise Set 1.5 (pg 60)

11.

Rectangles	1	2	3	4	5	6
Shaded squares	2	5	9	14	20	27
Doubled	4	10	18	28	40	54
Factored	$1 \cdot 4$	$2 \cdot 5$	$3 \cdot 6$	$4 \cdot 7$	$5 \cdot 8$	$6 \cdot 9$

12.

Rectangles	1	2	3	4	5	6
Dots	3	6	10	15	21	28
Doubled	6	12	20	30	42	56
Factored	$2 \cdot 3$	$3 \cdot 4$	$4 \cdot 5$	$5 \cdot 6$	$6 \cdot 7$	$7 \cdot 8$

23. Here is the third pentagonal number shown as the sum of the third square number and the second triangular number.

nth triangular number: $\frac{n(n+1)}{2}$

$(n-1)$st triangular number: $\frac{(n-1)(n)}{2}$

nth square number: n^2

Their sum: $n^2 + \frac{n(n-1)}{2} = $ –?–

Exercise Set 1.6 (pg 66)

5. Exercises 2 and 4 are the duals of each other. That is, switch the terms *point* and *line* and the exercises are the same. Therefore their rules are the same. Exercise 1 is a pattern for the number of diagonals from each vertex, and Exercise 3 is a pattern for the total number of diagonals. Therefore you could use the rule from Exercise 1 to help find the rule in Exercise 3. There is also a connection between Exercises 2 and 3. How do they differ?

6. Let points represent people sitting at the table and let the segments connecting the points represent the conversations between them.

8. Let a point represent each team and let the segments connecting the points represent the four games played between them.

9. Use the model from Exercise 3.

10. Use the model from Exercise 1.

11. Use the model from Exercise 3. Then use "guess-and-check." Try a number (your guess) in your rule. If the answer is too small, try a larger number. Keep guessing until you narrow in on the correct guess (90 diagonals).

12. Use the model from Exercise 2. Then use "guess-and-check."

19.

T	1	2	3	4	5	6	n
Shaded squares	1	2	3	4	5	6	n
Unshaded squares	1	3	5	7	9	11	$2n-1$
Total	2	5	8	11	14	17	–?–

20.

Donut	1	2	3	4	5	6	n
Squares	8	20	36	56	–?–	–?–	–?–
Factored	$1 \cdot 8$	$2 \cdot 10$	$3 \cdot 12$	$4 \cdot 14$	$? \cdot ?$	$? \cdot ?$	$? \cdot ?$

Project: Three-Peg Puzzle (pg 69)

Rings	1	2	3	4	5	6	n
Moves	1	3	7	15	–?–	–?–	–?–

Notice that each number (1, 3, and 7) is less than a power of 2. The n will be in the exponent position in the formula for the number of necessary moves for n rings.

Exercise Set 1.7 (pg 70)

3. Try the alphabet backwards and the multiples of 2 forward.

6. The light arm moves around the star counter-clockwise, and the dot alternates dark and light.

9. Look for separate patterns for the first note of each measure, the second note of each measure, and the third note of each measure.

10. Shading alternates left/right, figure flips up/down. Two dots, no dots twice.

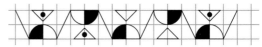

16. $1 = 1 = 1^2$
$1 + 3 = 4 = 2^2$
$1 + 3 + 5 = 9 = 3^2$
$1 + 3 + 5 + 7 = 16 = 4^2$
And so on . . . .

17.

Rectangular donut	1	2	3	4	5	6	n
Number of squares	8	14	20	26	32	-?-	-?-

19. See Exercise Set 1.6, Exercise 4.

21. See Exercise Set 1.6, Exercise 11.

Cooperative Problem Solving: Patterns at the Lunar Colony (pg 74)

On a sheet of graph paper, lay out the grid of streets as described. How many ways from Third and A to Third and B? How many ways from Third and A to Third and C? Continue in this way, finding the number of ways to go from Third and A to each of the intersections. On your graph paper, place your answers on the points of intersection. Look for a pattern in the triangle of numbers.

Chapter 2

Exercise Set 2.1 (pg 79)

3. Use only two letters to name the line. Do not use three letters. $\overleftrightarrow{ART}$ is not acceptable. You may use any two of the three named points. There are six possible ways to name this line.

20. Use only two letters to name the ray. Do not use three letters. $\overrightarrow{ABC}$ is not acceptable. The letter on the left must be the initial point A. You may use either one of the two remaining named points. There are two possible ways to name this ray.

26. Angles can be named by their vertex or by any three points, with the vertex as the middle letter.

32. Decide which vertex points define only one angle.

Exercise Set 2.2 (pg 84)

7. Subtract: $m\angle COB = m\angle COA - m\angle BOA$.

23.

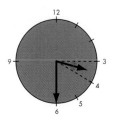

28. Place the zero-edge of your protractor against the cushion, with the center mark at the point at which the ball will bounce. Place your straightedge so that it shows the path of the ball leaving the cushion.

29.

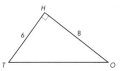

33. $m\overline{AK} = m\overline{RK}$, $m\angle A = m\angle R$

Investigation 2.3.2 (pg 90)

1. A **right angle** is an angle whose measure is -?-.

2. An **acute angle** is an angle whose measure is less than -?-.

Exercise Set 2.3 (pg 91)

10. One possible solution.

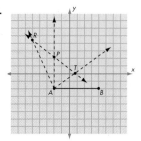

16. Measure the angle formed by the light ray and the mirror (the incoming angle). Draw an outgoing angle equal in measure to the incoming angle. Find point B' on the outgoing ray so that it is the same distance from the mirror as point B is.

Investigation 2.4 (pg 94)

5. If $\overleftrightarrow{AB}$ and $\overleftrightarrow{CD}$ intersect at point P, then $\angle APC$ and $\angle BPD$ are a pair of . . . .

6. If X, Y, and Z are consecutive collinear points and W is a point not on $\overleftrightarrow{XZ}$, then $\angle XYW$ and -?- form a **linear pair of angles**.

Exercise Set 2.4 (pg 96)

1. The expression "exactly one midpoint" means there is one midpoint and there is only one midpoint for any segment. This expression says two things. First, it claims that there *exists a midpoint*. Second, it claims that the midpoint is unique, that *there is only one midpoint*.

5. Think of a piece of spaghetti as representing a line. Place a point on the spaghetti. How many pieces of spaghetti can you place perpendicular to the spaghetti at that point?

Investigation 2.5 (pg 100)

4. Can you use *equilateral* or *equiangular* in your definition?

Exercise Set 2.5 (pg 101)

1. Triangle

10. Draw a light circle first, then place ten points on it. Connect the dots!

21.

$3x + 4y$ $3x + 4y$

28. Match each angle and side of the first triangle to a corresponding equal angle or side of the second triangle.

31. Determine which parts of the two polygons are corresponding parts. Because the two polygons are congruent, the matching sides will be of equal length. If $\overline{PR} \cong \overline{TK}$ and $TK = 34$, then $PR = 34$.

Investigation 2.6 (pg 103)

1. A **right triangle** is a triangle with one right angle.

Investigation 2.7 (pg 107)

1. A **trapezoid** is a quadrilateral with exactly one pair of parallel sides.

5. Do not use right angles in your definition.

6. Define a square three ways: as a special rhombus, as a special rectangle, or as a regular polygon.

Exercise Set 2.7 (pg 108)

29. Use your straightedge to draw the path of the ball into the cushion. The path should be at the same angle as the cue stick is to the cushion. Then use the hint for Exercise 28, Exercise Set 2.2, three times for the path of the ball off of the three cushions.

36. Because you can fold the paper in half in either direction, you can get lengths of 11″, 5.5″, and 2.75″ as well as 8.5″ and 4.25″. Now add or subtract some combination of these numbers to get 12 or a factor of 12 (3 or 4 or . . .).

Exercise Set 2.8 (pg 113)

10.

15.

26. What if they are not in the same plane?

34. To help you visualize this, hold pencils at right angles to a piece of paper.

Project: Traveling Networks (pg 116)

3. Draw a network whose points represent rooms and whose lines represent paths through doors. How many odd points are in the network? Remember, Arthur starts outside the house (odd point) and ends up outside (odd point).

4. Because Dusty must start and finish at point *B*, it follows that all the points must become even. Make all the points even by adding lines. These lines represent roads Dusty must travel twice. Choose the lines you add carefully so that you add a minimum distance to Dusty's route.

Exercise Set 2.9 (pg 123)

15. Draw a diagram. Look at what happens on the twenty-seventh, twenty-eighth, and twenty-ninth days.

17. Draw a diagram. Draw two points *A* and *B* on your paper. Locate a point that appears to be equally spaced from points *A* and *B*. The midpoint of segment *AB* is only one such point; find others. Connect the points into a line. For points in space, picture a plane between the two points.

18. Draw a diagram. Choose a point in the interior of the angle that is equally distant to the two sides of the angle. Choose a second point in the interior of the angle that is equally distant to the two sides of the angle. Connect the points.

21.

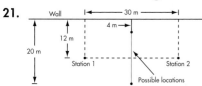

Exercise Set 2.10 (pg 127)

31. The plane intersects all six faces. Copy the figure at right and complete it by drawing the three black hidden lines.

47. Shift each *x*-coordinate three units to the right, as shown in the figure at right.

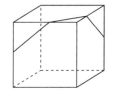

Chapter 3

Exercise Set 3.1 (pg 136)

2.

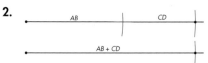

4. Begin by constructing a line *a* and a point *P*. Swing a large arc with center *P* and the same radius as the arc on the original angle. Let the arc pass through your construction line. With your compass, measure the distance between the two points where the arc crosses the first angle. With this compass setting, place the needle of your compass on the intersection point of the arc and the construction line, then swing a small arc to intersect the first arc. With your straightedge, line up this point of intersection and point *P*, then draw the second ray of your duplicate angle.

12. You duplicated a triangle in Exercise 9. Think of the quadrilateral as two triangles stuck together (draw in the diagonal). You do not need to duplicate the angles.

Exercise Set 3.2 (pg 140)

8. Bisect three times. The first perpendicular bisector divides the segment into two equal parts. The next perpendicular bisectors will break those segments into two parts for a total of four.

9. Construct one pair of intersecting arcs, then change your compass setting to construct a second pair of intersecting arcs on the same side of the line segment as the first pair.

10. The average is the sum of the two lengths divided by the number of segments (two). Construct a segment of length *AB* + *CD*. Bisect the segment to get the average length. Or take half of each, then add them.

11. Bisect $\overline{CD}$ to get the length $\frac{1}{2}CD$. Subtract this length from $2AB$.

15.

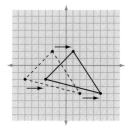

18. Imagine the mirror image of $\triangle MON$. Point *N* moves to the left side of the dashed line. Point *M* moves to the right side. What happens to point *O*?

19. Imagine that one corner of the triangle is pinned to the origin *A* and that point *Y* is rotated to the *x*-axis.

20. H, O, and X have horizontal symmetry; A, M, and N do not.

Exercise Set 3.3 (pg 143)

3. Construct the perpendicular through point B to ray TO.

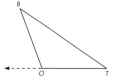

5. Construct a line with point C. Place the needle of your compass on point C and mark off two points so that C becomes the midpoint of a new line segment. Label the segment AB. Construct the perpendicular bisector of segment AB. This line should pass through point C.

9. Extend segment OT as shown in Hint 3.3.3. Then fold to construct the perpendicular to ray OT through point B.

14. What do the red octagon signs at many intersections say?

19. The figure at right shows a rectangle and a circle intersecting in zero points, one point, two points, three points, and eight points. Find the others.

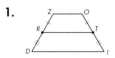

20. Point T moves from three units below the dashed line to three units above it. Point A moves from one unit below the dashed line to one unit above it. Where does point E move?

21. Try tracing $\triangle CUP$ onto tracing paper or patty paper. Rotate it $180°$ about the origin. What pattern do you see in the coordinate change?

Exercise Set 3.4 (pg 148)

5. Construct an equilateral triangle and use one of its angles to measure $60°$.

6. Construct an equilateral triangle and bisect one of its angles. This angle will measure half of $60°$, or $30°$.

12. Construct a $22\frac{1}{2}°$ (half of $45°$) angle and a $30°$ angle, then add them. You can add angles by constructing them with a common vertex and a common side.

14. Construct a segment and a line perpendicular to that segment. Bisect one of the $90°$ angles to create a $45°$ angle. $45° + 90° = 135°$

Exercise Set 3.5 (pg 150)

1.

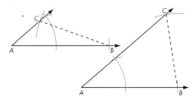

14. A, I, M, O, T, U, V, W, X, Y, and one other letter have vertical symmetry.

16. Draw the line through points (0, 0), (1, 1), (2, 2), and (3, 3). What is the equation of this line?

Exercise Set 3.6 (pg 152)

1. Construct a segment MS. Draw an arc with radius AS from point S and an arc with radius MA from point M. Connect the point of intersection of the arcs to points M and S to form your triangle.

5. Duplicate $\angle A$ and $\overline{AB}$ on one side of $\angle A$. Open the compass to length BC. If you put the compass point at point B, you'll find two possible locations to mark arcs for point C.

6. $y - x$ is the sum of the two equal sides. Find this length and bisect it to get the length of the other two legs of your triangle.

17. Start by finding points whose coordinates add up to 9, such as (3, 6) and (7, 2). Try writing an equation and graphing it.

Exercise Set 3.7 (pg 157)

1. Because the incenter is located on all three angle bisectors, it is equally distant from all three sides. Nate needs to locate the —?—.

8. The incenter is the center of the circle. To find the circle's radius, construct a line through the incenter, perpendicular to a side.

9. Because the circumcenter is located on all three perpendicular bisectors, it is equally distant from all three vertices. Therefore the circumcenter is the center of the circle that circumscribes the triangle.

13. Rose and Daisy should find the incenter by bisecting the two angles given. To find the angle bisector of the hidden angle, they will have to construct the incenter by bisecting the two angles on the eastern end of the property. They must then construct the two lines perpendicular to the legs through the incenter. The line through the incenter and any point equidistant from both perpendiculars will be the angle bisector on the inaccessible end of the property.

Exercise Set 3.8 (pg 163)

1. If $CM = 16$, then $UM = \frac{1}{2}(16) = 8$.

   If $TS = 21$, then $SM = \frac{1}{3}(21) = 7$.

11. The first step is to draw the altitude for the two visible vertices. Then construct a line perpendicular to that point through the base.

Exercise Set 3.9 (pg 168)

39. Draw a long segment and use your protractor to add y plus y plus x, then subtract (move backwards) half of z. To get one-half z, construct the perpendicular bisector of segment z.

41. Construct $\angle A$ and $\angle B$ at the opposite ends of $\overline{AB}$ ($AB = 3x$) so that their rays intersect to form a triangle.

45. $\frac{7}{2}x$ is three and one half x's.

46.

Chapter 4

Take Another Look 4.1 (pg 175)

3. $x + y = 180°$, $w + y = 180°$, . . .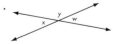

Exercise Set 4.1 (pg 176)

4. a and c form a linear pair. $a = 60°$ because it is a vertical angle.

10. B, C, D, E, H, I, O, and X have horizontal symmetry. Which of these letters have vertical symmetry?

11. Rotate the page 180° (turn it upside down). Did you notice that letters with horizontal and vertical symmetry also have 180° rotational symmetry? However, there are two letters that have rotational symmetry but that do not have horizontal or vertical symmetry. Which ones?

12. Do you recall the formula for the number of connecting segments for n points? See Lesson 1.6.

16. One such point is (9, 5).

18. If the two angles have measures of x and y, then $x = y$ and $x + y = 180°$. Then $x + x = 180°$ or $2x = 180°$, and thus $x = 90°$.

Take Another Look 4.2 (pg 181)

2. If we assume the CA Conjecture and the Vertical Angles Conjecture, then $a = b$ by the CA Conjecture but $a = c$ by the Vertical Angles Conjecture, . . .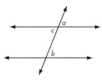

Exercise Set 4.2 (pg 181)

4.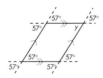

6. $z = 113°$. Why?

9. Measures a, b, c, and d are all related by parallel lines. Measures e, f, g, h, i, j, k, and s are also all related by parallel lines.

20. Start by finding how many eight-by-eight squares there are (only one). Next, find the seven-by-seven (four) and the six-by-six (nine). Continue. You should start to see a pattern.

21. There are five rotational symmetries. Therefore $360° \div 5 = -?-$.

Exercise Set 4.3 (pg 190)

8. To find the coordinates of one of the trisection points, divide each of point *B*'s coordinates by 3. How can you find the other trisection point?

25. $\dfrac{y-3}{4-2} = \dfrac{y-3}{2}$

33. According to the graph, E. Z. Ryder went uphill between 3:00 and 3:30, stopped and rested for 15 min, then continued uphill until 4:15. He then rested once again for 20 min before heading back down.

Exercise Set 4.4 (pg 198)

19.

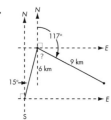

Exercise Set 4.5 (pg 203)

7. Solve for *y*. This gives you the slope and the *y*-intercept.

13. $x = 2$

16. Because the slopes are not the same, the lines intersect. You could set $-4x - 3 = 4x - 3$ and solve for *x* to find their intersection, but notice that they both have the same *y*-intercept. Therefore the *y*-intercept is their point of intersection. Although these slopes are negatives of each other, they are not negative reciprocals. Thus the lines are not perpendicular.

17. Set $\frac{4}{3}x - 5 = \frac{3}{4}x + 2$ and solve. Although these slopes are reciprocals, they are not negative reciprocals.

30. Think about the definition of the median. Point *O* should be the midpoint of $\overline{HY}$.

Exercise Set 4.6 (pg 209)

6. Centroid: $(2, \frac{2}{3})$. Because the triangle is right, the circumcenter is located on the hypotenuse and the orthocenter is located at a vertex of the triangle.

7. Because the triangle is isosceles, all four points of concurrency will fall on the line of symmetry.

8. The slope of side *JM* is $\frac{1}{2}$, therefore the slope of the perpendicular bisector of side *JM* is -2 and the altitude to $\overline{JM}$ is also -2. The slope of side *JI* is 4, therefore the slope of the perpendicular bisector of side *JI* is $-\frac{1}{4}$ and the slope of the altitude to $\overline{JM}$ is also $-\frac{1}{4}$.

Exercise Set 4.7 (pg 213)

8. $\frac{1}{5}$

12.

13. Organize your counting. For example:

ABC	ACD	ADE	AEF
ABD	ACE	ADF	
ABE	ACF		
ABF			
BCD	BDE	BEF	
BCE	BDF		
BCF			

And so on . . . .

14. If a piece of rope is more than 90 ft long, its end will hit the water.

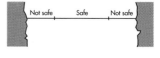

A portion of the bridge near its center is "safe" in that neither end would reach the water.

Chapter 5

Take Another Look 5.1 (pg 226)

2. If $a + b + c = 180°$ and $r + s + t = 180°$, then $a + b + c = r + s + t$. But if $a = r$ and $b = s$, then subtract equal quantities from both sides of the equation.

3. If we assume the line t is parallel to the side AC, then $x = c$ and $a = y$. But $x + b + y = 180°$. Therefore by substitution . . . .

4. If each of the three angles of an equiangular triangle measure a, then $a + a + a = 180°$. Therefore . . . .

5. Check out the sphere icon.

Exercise Set 5.1 (pg 226)

1. The sum of all three angles is $180°$. $x = 73°$

3. Ignore the $100°$ angle. You can even ignore the line that intersects the larger triangle. Find the three interior angles of the triangle ($a + b + c = 180°$). Then find z. It is the supplement of one of the interior angles of the triangle.

5. The sum of two linear pairs of angles is $360°$. So the total value for all three angles is $3 \times 360°$ minus the sum of the interior angles of the triangle.

6. The sum of a linear pair of angles is $180°$. So the total value for all three angles is $3 \times 180° - 180°$ (the sum of the interior angles of the triangle).

7. $a + 40° + 71° = 180°$. After finding a, find b. $b = 180° - 133°$. Once you have a and b, you can find the angle next to c. And so on.

9.

$a = \angle A$
$r = \angle R$
$m = \angle M$

10. Construct $\angle L$. Then bisect the supplement.

Take Another Look 5.2 (pg 234)

3. Start by finding the midpoints of the two shortest sides of the triangle. Place your compass needle at one of the midpoints. Then adjust your compass so that your pencil tip is at one of the endpoints of that side. Swing an arc through the longest side. Do the same at the other endpoint. Ah, but the big question is: Why do the two arcs intersect on the longest side?

4. With point C as center and AC as your radius, swing an arc passing through side AB.

Exercise Set 5.2 (pg 235)

1. $m\angle H + m\angle O = 158°$ (because $180° - 22° = 158°$). But because $m\angle H = m\angle O$, each angle measures half of $158°$.

7. The important thing to notice is that $d = e$. Thus $d + e + e + 66° = 180°$. Or $3e + 66° = 180°$. Next, notice that $c = 56°$ (alternate interior angles).

8. I'll bet that you couldn't find a nonisosceles triangle!

9. Look for the corresponding parts to determine congruence. $\angle G$ corresponds to $\angle N$, point A corresponds to point A.

11. The slope of $\overline{AB} = \frac{3 - 0}{1 - 6} = \frac{3}{-5}$.
The slope of $\overline{CD} = \frac{3 + 2}{4 - 1} = \frac{5}{3}$.

19. Move each point five units to the right and three units down.

20. Keep all the x-coordinates the same, but change the sign of the y-coordinates. It should look like the triangle has been flipped over the x-axis.

Take Another Look 5.3 (pg 239)

2. $a + b + c = 180°$, $b + x = 180°$

4. Is it possible for the measures of the three remote interior angles to add to less than $180°$?

5. A diagonal divides a quadrilateral into two triangles.

Exercise Set 5.3 (pg 240)

7. Find the value of the unmarked angle and order your sides according to this rule: smallest angle opposite the smallest side.

14. $a + b = 90°$, $b = 38°$

16. $135°$

23. Use the properties of parallel lines to find the corresponding parts. For example $\angle ABR \cong \angle BAE$.

25. Find the values of the slopes of the four sides. Are any of these values the same or negative reciprocals? If they are, what does this tell you about the quadrilateral?

Take Another Look 5.4 (pg 250)

2. If *PA* = *PL*, *AE* = *LE*, and *PE* = *PE*, then . . . .

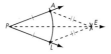

4. Because a square with sides measuring 5 cm and a (nonsquare) rhombus with sides measuring 5 cm are not congruent but have the SSSS conditions, clearly SSSS is not a shortcut for congruence between quadrilaterals. But when a diagonal is added, a rigid structure (two triangles) is created and there is a good case for finding congruence.

Exercise Set 5.4 (pg 251)

1. SAS Congruence Conjecture

4. △*ANT* ≅ △*FLE*

13. △*SUN* ≅ △*RAY* by SAS. *UN* = *YA* = 4, *RA* = *US* = 3, *m*∠*A* = *m*∠*U* = 90°

19.

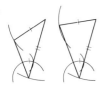

Take Another Look 5.5 (pg 256)

2. Construct the perpendicular bisector of a segment. Find two triangles that appear to be congruent (△*CAD* ≅ △*CBD*). Therefore ∠1 ≅ ∠2. But ∠5 ≅ ∠6 by the Isosceles Triangle Conjecture. Hence △*CEA* ≅ △*CEB*. Therefore . . . .

3. If two angles of one triangle are congruent to two angles of a second triangle, then the third pair of angles are congruent. Thus ASA follows.

5.

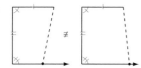

4. SAA

7. Take a closer look. *Corresponding* parts are not congruent.

9. If you separate the two triangles, it is easier to see the congruence.

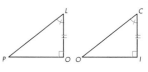

15. Because △*SIT* ≅ △*LEI* ≅ △*NTE* by SAS, $\overline{IT} \cong \overline{EI} \cong \overline{TE}$.

21. Because the angles of a triangle always sum to 180°, you can create two triangles of different sizes that have the same angles.

22. Here is one example.

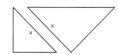

Exercise Set 5.6 (pg 260)

1. $\overline{AB} \cong \overline{BC}$ because you can show that △*ABD* ≅ △*CBD* by SAA.

3. Yes, by CPCTC because the triangles are congruent by ASA.

4. Yes, by CPCTC because the triangles are congruent by ASA.

6. Yes, by CPCTC because the triangles are congruent by SAS.

7. Yes, by CPCTC because the triangles are congruent by —?—.

8. Draw in segment $\overline{FU}$. Can you show that △*FOU* ≅ △*FRU*?

9. Draw in segment $\overline{UT}$. Can you show that △*BUT* ≅ △*ETU*?

13. You can show that $\overline{RT} \cong \overline{SA}$ if you can show that △*RET* ≅ △*AES*. What is necessary to show that triangles are congruent?

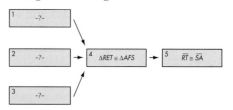

14. △*FOR* ≅ △*GET* by SAS. *FO* = *GE* = 4, *RO* = *TE* = 2, *m*∠*O* = *m*∠*E* = 90°

16. Draw congruent right triangles on the grid, with $\overline{AT}$ and $\overline{TO}$ corresponding hypotenuses.

17. $a = 72°$. Therefore $c = 144°$. Therefore $b = 36°$.

Project: Buried Treasure (pg 264)

1.

Take Another Look 5.7 (pg 267)

2. The converse: If the bisector of an angle of a triangle is either a median or an altitude, then the triangle is isosceles.

3. See Take Another Look 5.5.2.

4. If the triangle is isosceles, then the base angles are congruent. If the two legs are congruent, then the two bottom halves are congruent. The two overlapping triangles can be shown to be congruent.

Exercise Set 5.7 (pg 267)

1. If $AC = BC = 18$, then $AB = 48 - 36 = 12$. If $AB = 12$ and $AD = \frac{1}{2}AB$, then $AD = -?-$.

4. Recall that *concurrent* means that all the lines pass through the same point. Make a table. Two regions are formed by one line through a point. Four regions are formed by two lines through a point. Look for a pattern.

5. Let H be the number of hexagons and let K be the number of octagons. Then $6H + 8K = 140$. One possible solution is $6(10) + 8(10) = 140$, or 10 of each. Find others.

8. At 3:15 the hands have not yet crossed each other. The big hand (minute hand) is at the 3, but the small hand (hour hand) has moved one fourth of its way toward the 4. At 3:20 the hands have already crossed each other because the minute hand is on the 4 but the hour hand is only one third of its way from the 3 towards the 4. So the hands are together sometime between 3:15 and 3:20. Can you narrow it down? Find the time they come together to the nearest whole minute.

Exercise Set 5.8 (pg 272)

5. To reflect these points about the *y*-axis, reverse the sign of the *x*-coordinate.

32. $\triangle DMA \cong \triangle TMU$ by SAS.

35. $\angle A \cong \angle T$

38. It is given that two sides are congruent. No other information is given. The two angles at point H may look the same, but you just don't know.

40. If separate the two triangles, it is easier to see the congruence.

41. By the Triangle Sum Conjecture, $m\angle G = 67°$. But by the Isosceles Triangle Conjecture, $m\angle G = 75°$.

43. Remember that triangles are rigid.

Chapter 6

Exercise Set 6.1 (pg 281)

1. The sum of all the interior angles in a quadrilateral is $360°$. $a + 72° + 76° + 90° = 360°$. $a = 122°$

3. $d + 44° + 30° = 180°$. $c = 78° + 30°$. Can you see why?

5. $g = \frac{1}{3}[540° - (118° + 108°)]$

$360° - (108° + 130°)$
$130°$
$108°$

11. $(n - 2) \cdot 180° = 2700°$

12. Use the rule $\dfrac{(n - 2) \cdot 180°}{n}$, then substitute 10 for *n*.

Take Another Look 6.2 (pg 285)

2. If all the angles fit about a point, then their sum is $360°$.

7. Try using patty papers to show the sum. How do you have to treat the exterior angle that is in the interior of the polygon to get a sum of $360°$?

Exercise Set 6.2 (pg 286)

3. $c = \dfrac{360°}{7}$

6. $d = 39°$, $c = 180° - (39° + 39°)$, $e = \dfrac{360° - c}{2}$

11. $\dfrac{360°}{6}$

12. $24 = \dfrac{360°}{n}$

16. $\dfrac{(n - 2) \cdot 180°}{n} = 165°$

18. An obtuse angle measures greater than 90°, and the sum of exterior angles of a polygon is always 360°.

21. Look at the individual patterns of the first notes in each measure, the second notes in each measure, and so on.

24. $(x, y) \rightarrow (-x, -y)$

25. $(x, y) \rightarrow (\tfrac{1}{2}x, 2y)$

Take Another Look 6.3 (pg 292)

2. When one of the diagonals is drawn, it divides the kite into two isosceles triangles.

Exercise Set 6.3 (pg 292)

14. Construct $\angle I$ and $\angle W$ at both ends of $\overline{WI}$. Construct $\overline{IS}$. Construct a line through point S parallel to $\overline{WI}$.

17. If $ABCD$ is an isosceles trapezoid, then $\angle A \cong \angle$ -?-.

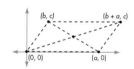

∴ $\triangle AFG \cong \triangle$ -?- by -?-.
∴ $\overline{AG} \cong$ -?- by CPCTC.

Take Another Look 6.4 (pg 299)

2.

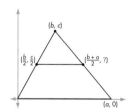

3. What is the converse of the Triangle Midsegment Conjecture? If a segment connecting points on two sides of a triangle is parallel to the third side and half the length of the third side, then the segment is a midsegment (the two points are midpoints). The use of a computer program, such as The Geometer's Sketchpad, would be helpful. Construct a triangle. Place an arbitrary point on one of the sides of the triangle. Construct a line through that point, parallel to another side of the triangle. You figure out the rest.

Exercise Set 6.4 (pg 299)

1. If $RA = 20$, then $PO = \tfrac{1}{2} RA =$ -?-. If $PR = 8$, then $PT =$ -?-. If $OA = 10$, then $TO =$ -?-. Therefore the perimeter $PO + OT + PT =$ -?-.

3. Each side of the center triangle is parallel to the opposite side of the large triangle. Mark all the congruent corresponding angles. This tells you that all four small triangles have the same three angle measures.

15. If your class feels the name should be changed to midsegment, you might try writing a convincing argument in favor of a change and send it to the National Council of Teachers of Mathematics (NCTM), suggesting the change.

Take Another Look 6.5 (pg 304)

2. Find the midpoint of each diagonal. Are they the same?

3. Converse of C-52: If the opposite angles of a quadrilateral are congruent, then the quadrilateral is a parallelogram. Converse of C-53: If the consecutive angles of a quadrilateral are supplementary, then the quadrilateral is a parallelogram. Converse of C-54: If the opposite sides of a quadrilateral are congruent, then the quadrilateral is a parallelogram.

4. Construct a diagonal in a parallelogram. Show that the two triangles are congruent. If so, then corresponding sides and angles are congruent.

Exercise Set 6.5 (pg 305)

4. $x - 3 = 17$. Solve for x. Find the value of $x + 3$. Then find the perimeter.

6. If $VF = 36$, then $VN = $ –?–. If $EI = 42$, then $NI = $ –?–. If $EF = 24$, then $VI = $ –?–. Therefore the perimeter of $\triangle NVI$ is $VN + NI + VI$.

12. Start by making a sketch of parallelogram *DROP* with the diagonals *DO* and *RP*. Recall that the diagonals of a parallelogram bisect each other. Start by constructing the longer diagonal and then find its midpoint. Construct the shorter diagonal and find its midpoint. With the midpoint of the longer diagonal as center and using the length of half the shorter diagonal as radius, construct a circle. Then finish the construction.

15.

Take Another Look 6.6 (pg 313)

2. Construct the two diagonals in a rhombus. Show that the four triangles are congruent. Then corresponding sides and angles are congruent. Thus the diagonals are angle bisectors. Then you can also show that the diagonals are perpendicular bisectors of each other.

3. Once you've discovered whether or not a rectangle has right angles on a sphere, how about investigating properties of rhombuses and parallelograms on a sphere?

Exercise Set 6.6 (pg 314)

3. $\angle P$ and $\angle PEA$ are supplementary. Therefore $48° + (y + 95°) = 180°$.

8. Consider the parallel-ogram at right.

18. Start by making a sketch of a rhombus *BAKE* with diagonals *BK* and *AE* intersecting at point *R*. Recall that the diagonals of a rhombus are perpendicular bisectors of each other and that each diagonal bisects the angles. Therefore begin by constructing $\angle B$, then bisect it. Mark off the length of diagonal *BK* on the angle bisector. Then finish the construction.

19. You are given a diagonal of a rectangle. What do you know about the diagonals of a rectangle? They are equal in length and they bisect each other.

20. The diagonals of a square are equal in length (a square is a rectangle), and the diagonals of a square are perpendicular bisectors of each other (a square is a rhombus).

23. Draw a vector diagram. For an example, refer to the hint for Exercise 15, Exercise Set 6.5.

26. The quadrilateral should have been shifted four units to the left and five units up.

27. Each *x*-coordinate becomes a *y*-coordinate, and vice versa. The new quadrilateral that appears is the reflected image of the original. The line of reflection is $y = x$.

Exercise Set 6.7 (pg 319)

27. There is more than one possible answer. $\triangle TAR \cong \triangle ARY$ by SAS. Can you find another pair of congruent triangles?

31. Sketch trapezoid *OPEN*. Construct segment *OP*. Construct $\angle O$ at point *O*. From one ray of angle *O*, measure the length of $\overline{ON}$. Construct a ray from point *N* parallel to $\overline{OP}$. Measure a segment of length *EN* on this ray. Connect the four points to form *OPEN*.

33. $1 = 1 \cdot 1$, $6 = 2 \cdot 3$, $15 = 3 \cdot 5$, $28 = 4 \cdot 7$, $45 = 5 \cdot 9$

35. One possible approach would be to start with all the different ways that four lines can intersect, in each picture add the fifth line in the different ways that it can be added, and look for some patterns.

38. Refer to the figure at the end of Investigation 6.5 for a similar example. Let 1 cm = 100 km be your scale.

Chapter 7

Investigation 7.1 (pg 329)

5. An inscribed angle is an angle whose sides are —?— and whose vertex is on the —?—.

Exercise Set 7.1 (pg 331)

21. Because $PAQB$ is a rhombus, $\overline{AB}$ and $\overline{PQ}$ are perpendicular bisectors of each other.

24. Your new circle should be one unit to the left and two units up from the original and congruent to it.

25. Because the x- and y-coordinates have been switched, your new circle should be located on the y-axis instead of on the x-axis. The new circle is a reflection of the original circle about the line $y = x$.

Take Another Look 7.2 (pg 334)

3. If a point is equally distant from the endpoints of a segment, then it lies on the perpendicular bisector.

Exercise Set 7.2 (pg 335)

3. Because the three chords are congruent, their three arcs are congruent. Therefore
$y + y + y + 72° = 360°$.

10. The perpendicular bisector of every chord must pass through the center.

Take Another Look 7.3 (pg 340)

2. Here's a proof of the Tangent Segments Conjecture.
Because $\overline{TA}$ is perpendicular to $\overline{NA}$, then
$x + a = 90°$, and because $\overline{TG}$ is perpendicular to $\overline{NG}$, then $y + b = 90°$.

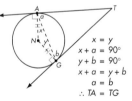

$x = y$
$x + a = 90°$
$y + b = 90°$
$x + a = y + b$
$a = b$
$\therefore\ TA = TG$

Therefore $x + a = y + b$. But because $\triangle NAG$ is isosceles, $x = y$. Thus $a = b$. Therefore $TA = TG$.

4. Try holding the eraser ends of two pencils in your hand as if they were lines intersecting in space, and rest the pencils on the surface of a ball (such as a tennis ball or a volleyball). Move the pencils around on the surface. What do you notice about the tangent segments in space?

Exercise Set 7.3 (pg 340)

1. $130° + 90° + w + 90° = 360°$

2. Because the triangle is isosceles,
$x + x + 70° = 180°$.

4. Because t is tangent, the angles formed by the tangent and the radii are each —?— and the sum of angles in any quadrilateral is —?—.

5. $CP = PA = AO = OR = 13$
$CT = TD = DS = SR = -?-$

11. From the Tangent Conjecture you know that the tangent is perpendicular to the radius at the point of tangency. Therefore, to construct a tangent through a point on a circle, perform the following steps.

Step 1   Construct a large circle. Label the center O.

Step 2   Choose a point on the circle and label it P.

Step 3   Construct $\overline{OP}$.

Step 4   Construct the perpendicular to $\overline{OP}$ through point P.

Take Another Look 7.4 (pg 345)

3.

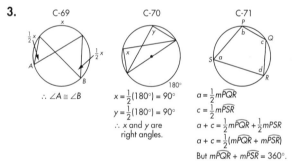

C-69

$\therefore \angle A \cong \angle B$

C-70

$x = \frac{1}{2}(180°) = 90°$
$y = \frac{1}{2}(180°) = 90°$
$\therefore\ x$ and y are right angles.

C-71

$a = \frac{1}{2}m\overset{\frown}{PQR}$
$c = \frac{1}{2}m\overset{\frown}{PSR}$
$a + c = \frac{1}{2}m\overset{\frown}{PQR} + \frac{1}{2}m\overset{\frown}{PSR}$
$a + c = \frac{1}{2}(m\overset{\frown}{PQR} + m\overset{\frown}{PSR})$
But $m\overset{\frown}{PQR} + m\overset{\frown}{PSR} = 360°$.

4. C-71: If a quadrilateral is inscribed in a circle then —?—. Converse of C-71: If the opposite angles of a quadrilateral are —?—, then the quadrilateral can be inscribed in a circle.

Exercise Set 7.4 (pg 346)

3. $c + 120° = 2(95°)$

6. Draw in the radius to the tangent to form a right triangle. The three angles of a triangle sum to 180°. Find the central angle.

12. $120° + p + 98° + p = 360°$
$98° + q + q = 180°$

14. The measure of each of the five angles is half the measure of its intercepted arc. But the five arcs add up to the complete circle (360°).

15. $a = \frac{1}{2}(70°)$

$b = \frac{1}{2}(80°)$

$y = a + b$

19. Answer will vary. One possible answer is shown at right.

20.

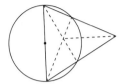

One possible location for the camera to get all the students in the photo

Students lined up for the photo.

Exercise Set 7.5 (pg 350)

4. $C = \pi D = \pi(5\pi) = 5\pi^2$ m

7. If the perimeter of the square is 24 cm, then each side measures 6 cm. If each side measures 6 cm, the diameter of the circle is 6 cm. If the diameter is 6 cm, the circumference is –?–.

12. $C = 2\pi r$

$44 = 2(3.14)r$

Exercise Set 7.6 (pg 352)

3. $\text{speed} = \dfrac{\text{distance}}{\text{time}} = \dfrac{\text{circumference}}{12 \text{ hours}}$

$\text{speed} = \dfrac{(2)(3.14)(2000 + 6400) \text{ km}}{12 \text{ hours}}$

7. Calculate the distance traveled in one revolution (the circumference) at each radius. Multiply this by the rpm to get the distance traveled in 1 min. Remember, your answer is in centimeters. You may want to divide your answer by 100,000 and multiply it by 60 to change its unit into kilometers per hour.

8. 2.5 m equals 250 cm, and 3 m equals 300 cm. A hoop with a diameter of 60 cm has a circumference of $60 \cdot \pi \approx 188.5$ cm. A hoop with a diameter of 90 cm has a circumference of $90 \cdot \pi \approx 283$ cm. This should tell you which length of tubing to choose for each size of hoop.

Exercise Set 7.7 (pg 357)

3. $m\widehat{BIG} = 360° - 150° = 210°$

$\dfrac{210}{360} = \dfrac{7}{12}$

Arc length of $\widehat{BIG}$ is $(\frac{7}{12})\pi(24)$.

6. $m\widehat{SO} = 80°$

9. $m\widehat{AR} + 70° + m\widehat{AR} + 146° = 360°$

$\dfrac{m\widehat{AR}}{360°}(2\pi r) = 40\pi$

13. Calculate the distance of one lap. It is equal to $(2 \times 100) + (2 \times 20\pi) = d$. The total distance covered is $4d$ in 6 minutes.

16. If you draw a radius from the center of the circle that contains the arc to each endpoint of the bridge, you create an equilateral triangle. Therefore the radius for the arc is 180. Therefore the arc length is $\frac{1}{6}[2\pi(180 \text{ m})]$.

17. $\frac{1}{9}(2\pi r) = 12$ meters

18. Location? Reverse the letters in each word. Time? $40\pi = (\frac{x}{360°})\pi(120°)$. The big hand moves x degrees in –?– minutes.

Exercise Set 7.8 (pg 365)

18. Draw in the radius to the tangent to form a right triangle.

21. See the hint for Exercise 15, Exercise Set 7.4.

22. $a = \frac{1}{2}(118°)$

$b = \frac{1}{2}f$

$a + b + 88° = 180°$

26. $C = \pi D$, $132 = \pi D$, $D = \dfrac{132}{\pi}$

27. Arc length of $\widehat{AB}$ is $(\frac{100°}{360°})\pi(27)$.

28. To find the length of $\widehat{DC}$, first find the degree measure of $\widehat{DL}$. $a = b + 50°$ by the Triangle Exterior Angle Conjecture. But $b = 30°$ by the Inscribed Angle Conjecture. Therefore $a = 80°$. Therefore $m\widehat{DL} = 160°$.

31. Find the circumference of the circle with
$D = 50$ m.

33. The circumference of the table is 2×100.
Calculate the diameter.

39. $\frac{2\pi(6357)}{(360)(60)} \approx 1.849 < 1.852 < 1.855 \approx \frac{2\pi(6378)}{(360)(60)}$

Chapter 8

Exercise Set 8.1 (pg 377)

7. A skater can't become the mirror image of
herself, but a second skater could mirror
her actions.

15. Think of the two dotted arcs as arcs of
concentric circles whose center is the center of
rotation. How would you find the center? The
perpendicular bisector of a chord of a circle
passes through the center of the circle.

21. (1, 2) becomes (2, 4), and (4, 4) becomes (8, 8).
Each image point is twice as far from the
origin as the original, and the points in the
image are farther away from each other than
are the corresponding points in the original.

Exercise Set 8.2 (pg 383)

4. All positive y's become negative y's, therefore
the figure is reflected over the x-axis.

13. Connect a pair of corresponding points (a point
in the original with its image point). Construct a
perpendicular somewhere on the segment.
Construct a second perpendicular (parallel to
the first) at a distance from the first equal to
half the distance between the two given figures.

Exercise Set 8.3 (pg 389)

10. In this chapter you discovered that the
composition of two reflections across
intersecting lines is equivalent to one rotation,
with the measure of the angle of rotation
equal to twice the measure of the angle
between the reflection lines. In the case of
vertical and horizontal reflection lines, the
measure of the angle between them is $90°$,
thus the two reflections are equivalent to a
–?–° rotation. Therefore if a figure has both
horizontal and vertical lines of symmetry, then
it must have —?— symmetry.

15. A regular decagon would have 10-fold
symmetry: a rotational symmetry every $36°$.

18. Find the midpoint of the segment. Connect the
midpoint to one of the endpoints with a curve.
Copy the curve and rotate it about the midpoint.

Exercise Set 8.4 (pg 399)

8.

13.

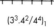

$(3^3.4^2/4^4)_1$ $(3^3.4^2/4^4)_2$

19. First visualize an approximate path. Do you
think you can bounce it off the S wall, then off
the E wall, then off the N wall, and then in? If
so, work backwards. That is, begin by
reflecting the point H over the N line (because
you expect the ball to bounce off the N wall),
then reflect that image over the E line (because
you expect the ball to bounce off the E wall),
then reflect that image over the S wall
(because you expect the ball to bounce off the
S wall). Finally, aim at this last image!

20. Work backwards. Reflect a point on the E edge
of the eight ball over the S cushion. Then
reflect this image over the N cushion. Aim at
this second image.

Exercise Set 8.6 (pg 410)

9. If you are still unsure, use patty papers to
trace the steps in the examples ("Pegasus" and
Leap Frog).

Exercise Set 8.8 (pg 423)

5. If you are still unsure, use patty papers to
trace the four steps in the Escher *Horseman*
example. If you need more practice see
Exercise 4.

6. Study the four-step example on page 423.

8. If you are still unsure, use patty papers to trace the four steps in the Garret Lum *South Pole or Bust* example.

Chapter 9

Exercise Set 9.1 (pg 436)

21. Factor 48 in two different ways. For example, $48 = 6 \times 8 = 4 \times 12$.

22. Factor 64 and draw two parallelograms in which $bh = 64$.

23. There are many solutions to the problem. For example, a rhombus with sides of length 16 and height of 4 will work.

24. $4' \times 7'$ equals $48'' \times 84''$, therefore divide each by 8 to determine the number of tiles in each dimension. (If you were planning to actually do this, it would be wise to purchase extra tiles—accidents do happen.)

26.

Take Another Look 9.2 (pg 439)

4. How about drawing in the midsegment? How about drawing a segment from the midpoint of one of the nonparallel sides to the opposite obtuse vertex?

Exercise Set 9.2 (pg 439)

2. $A = \frac{1}{2}(11)(9) = -?-$

4. Solve for b: $31.5 = \frac{1}{2}(9)b$.

8. Every side of the triangle can be a base. For every base of the triangle there is a corresponding height. Therefore the area of the triangle can be calculated in three different ways: $\frac{1}{2}(5)y = \frac{1}{2}(15)x = \frac{1}{2}(6)(9)$.

16. $84 = (\frac{1}{2})(8)(b_1 + b_2)$. Solve for $(b_1 + b_2)$: $P = 9 + 10 + (b_1 + b_2)$.

17. Find two solutions for $\frac{1}{2}bh = 54$ and draw two triangles.

18. Find two solutions for $56 = \frac{1}{2}(b_1 + b_2)h$. For example, $h = 8$, $b_1 + b_2 = 14$.

27. Refer to the diagram below.

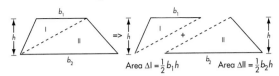
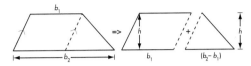

28. Refer to the diagram below.

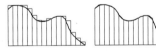

Investigation 9.3 (pg 447)

1. Divide the figure into three rectangles.

2. Use your compass to find which has the greatest perimeter. To find the area, you will have to draw an altitude in the parallelogram and the two diagonals in the kite.

3. Divide the figure into two triangles.

4. Notice how part of the figure can be cut off and reassembled to fit into another slot, creating a quadrilateral of which you know how to find the area.

5. You could select one vertex and draw in all the diagonals from that vertex, but that would result in six triangles of different sizes—(a lot of work). If you found the center of the circumscribed circle and drew in the radius to each vertex, all the triangles would be identical. Then how would you find the area of the dodecagon?

6. You can use every vertical line or every other vertical line of a sheet of lined paper to create rectangles. Then you can find the sum of the areas of all the rectangles. Or you could also divide the region into trapezoids and find the sum of the areas of all the trapezoids.

Exercise Set 9.3 (pg 448)

3. $\text{Area}_{\text{frosting}} = \text{Area}_{\text{top}} + \text{Area}_{\text{four sides}}$
$= (400 \cdot 600) + [2(180 \cdot 400) + 2(180 \cdot 600)]$
The number of liters of frosting needed is $\frac{\text{Area}_{\text{frosting}}}{1200}$.

4. Total cost is $20/yd^2. $20/yd^2 = $20/9 ft^2. Area$_{carpet}$ = $17 \cdot 27 - (6 \cdot 10 + 7 \cdot 9)$

6. First find the area of all the vertical rectangles (walls) in the first floor. Notice that the area of the front and back triangles are the same.

Take Another Look 9.4 (pg 452)

2. If the number of sides is even, try making a parallelogram. If the number of sides is odd, try making a trapezoid.

4. See Exercise 16, Exercise Set 9.5.

Exercise Set 9.4 (pg 452)

5. Remember the daisy design? Start by constructing a circle with a radius of 4 cm. Mark off six 4-cm chords around the circle.

6. In other words, you want to draw a regular pentagon circumscribed about a circle with a radius of 4 cm. Construct a circle with a 4-cm radius. Use your protractor to create five 72° angles from the center of the circle. Use your protractor to construct a tangent to the circle at each of the five equally spaced points on the circle.

12. Find the area of the large hexagon and subtract the area of the small hexagon. Because they are regular hexagons, the distance from the center to each vertex equals the length of each side.

13. Graph the two lines. The region formed by those lines, the x-axis, and the y-axis is a quadrilateral. Divide the quadrilateral into two triangles (A and B) by the line through the origin and the point of intersection of the two lines, (2, 6). Find the area of the two triangles and add them.

18. The original figure was a square ($13 \times 13 = 169$ square units). Even if you calculate the area of each component of the square, you will discover that the total area, equals A, equals $2(\frac{1}{2})(5 \times 13) + 2(\frac{1}{2})(8)(8 + 5)$, which still equals 169 square units. When the four pieces are arranged into the 8×21 rectangle, one square unit seems to be missing. Take a closer look at the diagonal of this rectangle. Do the four pieces really fit, or do they overlap a bit? The explanation is in the slopes of these four segments making up this "sort of" diagonal. Can you explain what happened with the triangle (base measuring 16 and height of 21)?

Exercise Set 9.5 (pg 455)

7.
$$A = \pi r^2$$
$$3\pi = \pi r^2$$
$$3 = r^2$$
$$r = \sqrt{3}$$

15. Calculate the area of the two circles, one with radius 3 cm and one with radius 6 cm. Compare your results.

Exercise Set 9.6 (pg 459)

3. The shaded area is three fourths of the area of the circle. $A = (0.75)\pi(16)^2$

6. The shaded area is equal to the area of the whole circle less the area of the smaller circle in the center. $A = \pi R^2 - \pi r^2 = \pi(7)^2 - \pi(4)^2$

9. $\frac{1}{3}\pi r^2 = 12\pi$

Exercise Set 9.7 (pg 464)

6. Use the formula for the lateral surface area of a cone, $A = \pi r l = \pi(3)(8)$. Then find the area of the base, $A = \pi r^2 = \pi(3)^2$. Add the two results.

7. Use the formula for finding the area of a regular hexagon to find the area of each base. To find the area of the six lateral faces, visualize unwrapping the six rectangles into one rectangle. Thus the lateral area is the height times the perimeter.

8. The total surface area equals the area of the regular pentagonal base plus the area of the five isosceles triangles. To find the area of the regular pentagon, use $A = \frac{1}{2}nas$. To find the area of one triangle, use $A = \frac{1}{2}hs$.

$$\text{Surface area} = \frac{1}{2}nas + 5\left(\frac{1}{2}\right)hs$$
$$= \left(\frac{1}{2}\right)(5)(11)(16) + (5)\left(\frac{1}{2}\right)(16)(15)$$
$$= -?-$$

9.

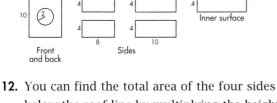

Top and bottom   Outer surface   Inner surface

10.

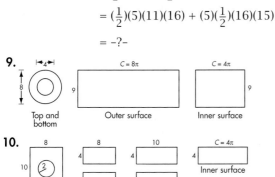

Front and back   Sides   Inner surface

12. You can find the total area of the four sides below the roof line by multiplying the height to the roof by the perimeter of the floor. You can find the total area of the end pieces under the roof by doubling the areas of the trapezoid and the triangle. $24(30 + 40 + 30 + 40) + 2[(\frac{1}{2})(12)(30 + 12)] + 2[(\frac{1}{2})(2.5)(12)]$ equals the painted area. The shingle area equals $40(15 + 6.5 + 6.5 + 15)$, or $-?-$.

To find the cost for paint, divide the painted area by 250 and round up to a whole number of gallons. Then multiply the number of gallons by $25.

To find the cost for shingles, divide the shingle area by 100 and round up to a whole number of bundles. Then multiply the number of bundles by $65.

15. Find the area of the two circles and the area of the circular city.

16. $\dfrac{\pi(4)^2}{\pi(20)^2}$

Exercise Set 9.8 (pg 467)

4. The amount of pizza per dollar is the same whether you calculate it for one pizza or for eight pizzas. The pizza size that gives the most area per dollar can quickly be found by comparing: $\dfrac{25}{10.5}$, $\dfrac{36}{13.00}$, $\dfrac{49}{16.00}$, $\dfrac{64}{20.00}$. (You don't need π to compare.)

7. Draw a circle (for the pizza) with its center. Then sketch another circle concentric to the first and whose radius is half that of the first. Isn't any point in the interior of the smaller circle closer to the center of the circle than any point in the annulus? What is the ratio of the areas (the small circle to the annulus)?

11. The shaded area equals $12(15 + 25) + (8)(20)$. The area of the vertical sides equals $(\frac{1}{2})(12)(17 + 26) + 2[(\frac{1}{2})(12)(16)] + (12)(12) + (\frac{1}{2})(12)(34 + 25)$.

To find the cost of the shingles, divide the shaded area by 100 and round up to a whole number of bundles. Then multiply the number of boxes by $35.

To find the cost of the wood stain, divide the vertical area by 150 and round up to a whole number of gallons. Then multiply the number of gallons by $15.

12.

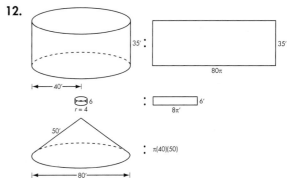

Exercise Set 9.9 (pg 470)

20. First we
constructed an
altitude from
the vertex of an

obtuse angle to the base. Then we cut off the
right triangle and moved it to the opposite
side, forming a rectangle. Because the area
hasn't changed, the area of the parallelogram
equals the area of the rectangle. Because the
area of the rectangle is given by the formula
$A = bh$, the area of the parallelogram is also
given by $A = bh$.

30.

42. $A = \pi(5.5)^2 - \pi(4)^2$

43. Use trial and error to reach your solution.

44.
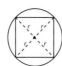

$\text{Area}_{circle} = \pi r^2$ $\text{Area}_{circle} = \pi r^2$
$\text{Area}_{square} = (2r)^2 = 4r^2$ $\text{Area}_{square} = (4)(\frac{1}{2})r^2 = 2r^2$

$\dfrac{\text{Area}_{circle}}{\text{Area}_{square}} = -?-$ $\dfrac{\text{Area}_{square}}{\text{Area}_{circle}} = -?-$

Cooperative Problem Solving: Discovering New Area Formulas (pg 474)

5.

Chapter 10

Take Another Look 10.1 (pg 478)

2. Label the length of each hypotenuse c. The area
of the entire square is $(a + b)^2 = a^2 + 2ab + b^2$.
But the square is made up of four right
triangles, whose total area is $4(\frac{1}{2})ab$, and one
square whose area is c^2. Set the two
expressions equal, $a^2 + 2ab + b^2 = 4(\frac{1}{2})ab + c^2$,
and simplify. But how do you know that the
figure with sides of length c is really a square?

Exercise Set 10.1 (pg 479)

1. $5^2 + 12^2 = 25 + 144 = 169$. Therefore
$c = \sqrt{169} = 13$.

6. $6^2 + 6^2 = 72$. Therefore $c^2 = 72$. Use your
calculator to find the square root.

8. $7^2 + b^2 = 25^2$. $b^2 = 625 - 49$. Therefore
$b = \sqrt{625 - 49} = \sqrt{576} = 24$.

11. $s^2 + s^2 = 25$. $2s^2 = 25$. $s^2 = \frac{25}{2} = 12.5$. Use your
calculator to find the square root:
$s = \sqrt{12.5} \approx 3.5$.

Take Another Look 10.2 (pg 483)

2. Picture two triangles: a triangle whose sides
are of lengths a and b and whose longest side
is of length c such that $a^2 + b^2 = c^2$, and
another triangle that is a right triangle whose
legs are of lengths a and b and whose
hypotenuse is of length x. (We want to show
that the first triangle is congruent to the
second triangle and therefore that the triangle
is a right triangle.) Because the second triangle
is a right triangle, $a^2 + b^2 = x^2$. You finish the
rest of the argument.

Exercise Set 10.2 (pg 483)

1. $8^2 + 15^2 = 64 + 225 = 289 = 17^2$. Because
8-15-17 fits into the Pythagorean formula,
the triangle is a right triangle.

6. $1.73^2 + 1.41^2 = 2.9929 + 1.9881 = 4.9810$.
$2.23^2 = 4.9729$. It may not be exactly a right
triangle, but you wouldn't be able to tell by
looking! It is very, very close to a right triangle.

8. Find the other leg: $b^2 + 5^2 = 13^2$. Then
$A = \frac{1}{2}(5)b$.

10. Because 9-12-15 fits into the Pythagorean
formula, the lengths form a right triangle.
Therefore the angle is a right angle and the
parallelogram is a rectangle.

Exercise Set 10.3 (pg 485)

2. Check the list of Pythagorean triples in Lesson
10.2 for a right triangle that has three
consecutive integers.

3. Multiply the lengths of the legs of the triangle from Exercise 2 by 2.

5. Let s represent the length of the side of the square. Then $s^2 + s^2 = 32^2$. $2s^2 = 1024$. Therefore $s^2 = 512$. But the area of the square is s^2. Therefore the area of the square is –?– square meters.

11. Let x represent the distance from the crack to the top of the pole. Then $9^2 + 12^2 = x^2$. Solve for x. The original height is $(9 + x)$ feet.

Investigation 10.4.1 (pg 488)

1. $2\sqrt{3}$	**2.** $3\sqrt{2}$	**3.** $2\sqrt{6}$
4. $4\sqrt{2}$	**5.** $2\sqrt{10}$	**6.** $4\sqrt{3}$
7. $2\sqrt{15}$	**8.** $5\sqrt{3}$	**9.** $\sqrt{83}$
10. $\sqrt{85}$	**11.** 18	**12.** 48
13. $2\sqrt{6}$	**14.** $18\sqrt{2}$	**15.** 147

Investigation 10.4.2 (pg 489)

1. $3\sqrt{2}$

Investigation 10.4.4 (pg 490)

7. $6\sqrt{3}$

Exercise Set 10.4 (pg 492)

4. Simplifying a square root means that you are finding the square root of any square factors in the number. Therefore your first task might be to find the largest square factors. Take the square root of each square. For example: $2592 = 2 \cdot 1296 = 2 \cdot 36^2$. Therefore $\sqrt{2592} = \sqrt{1296} \cdot \sqrt{2} = 36\sqrt{2}$.

Because 185 ends in 5, it is divisible by 5. $185 = 5 \cdot 37$. Because both factors are prime, there are no squares as factors. Therefore $\sqrt{185}$ cannot be simplified.

5. Be flexible! 490 ends in 0, therefore it is divisible by 10. $490 = 49 \cdot 10$. A ha, 49 is a square. Therefore $\sqrt{490} = \sqrt{49} \cdot \sqrt{10} = 7\sqrt{10}$. Because there are no squares as factors in 10, you are finished.

7. $720 = 72 \cdot 10 = (36 \cdot 2) \cdot (2 \cdot 5) = 36 \cdot 4 \cdot 5$. Therefore $\sqrt{720} = \sqrt{36} \cdot \sqrt{4} \cdot \sqrt{5} = $ –?–.

12. $(4\sqrt{3})^2 = (4\sqrt{3})(4\sqrt{3}) = 16 \cdot 3 = 48$

14. $(2\sqrt{6})(\sqrt{12}) = 2\sqrt{72} = 2\sqrt{16 \cdot 3} = 2\sqrt{16} \cdot \sqrt{3} = $ –?–

15. $(6\sqrt{8})^2 = (6\sqrt{8})(6\sqrt{8}) = 36 \cdot 8 = $ –?–

17. $b = 13$

19. Use the properties of isosceles right triangles to solve for the length of the legs, then apply the area formula for triangles.

22. $a = $ twice the shorter leg $= 2 \cdot 5 = 10$
$b = $ shorter leg times $\sqrt{3} = 5\sqrt{3}$

26. $m = \sqrt{3}\sqrt{2} = \sqrt{6}$, $k\sqrt{3} = \sqrt{3}\sqrt{2}$, $k = $ –?–

33. Find Pythagorean triples with hypotenuses of 25. 7-24-25 is one. Another would be a multiple of a 3-4-5 right triangle.

34.

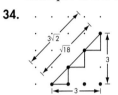

36. To construct $a\sqrt{2}$, construct an isosceles right triangle with sides of length a. The length of the hypotenuse will be $a\sqrt{2}$. To construct $a\sqrt{3}$, construct a right triangle, using $a\sqrt{2}$ and a as the length of the legs. The hypotenuse will have a length of $a\sqrt{3}$.

40. Make a list of all triples, then find all Pythagorean triples. (3, 4, 5), (3, 4, 9), (3, 4, 17) . . . (3, 8, 5), (3, 8, 9), (3, 8, 17) . . .

Exercise Set 10.5 (pg 497)

5. $A = bh$. $168 = 7b$. So $b = 24$. Therefore the triangle is the familiar 7-24-?.

8. The diameter of the semicircle is the longer leg of a 7-?-25 right triangle. If you don't recall the Pythagorean triple, you'll have to use the Pythagorean Theorem to find it. Once you've found the length of the leg, take half to find the length of the radius, then find half the area of the circle.

10. a. 9-12-15

11. 8-x-17 and 8-15-17, so $a = 30$.

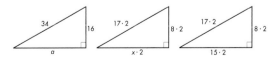

15. First find x from the triple: x-24-26 (?-12-13). Then find e from the triple 6-e-x.

19. The hypotenuse of the 30-60 right triangle with leg of length $10\sqrt{3}$ is 20. Then find w, then find y.

23. This is one way to show the relationship. Draw three 30-60-90 triangles with sides of length 1, $\sqrt{3}$, and 2. $6^2 - 3^2 = 27$, so $3\sqrt{3} = \sqrt{27}$.

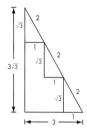

24. In the second example, the length of each side of the square is $\sqrt{2}$, thus the area of the square is 2. To find a square with an area of 5, find a length $\sqrt{5}$ in this rectangular grid. Find two squares whose sum is 5, that is, $2^2 + 1^2 = 5$. Thus you need to find a right triangle with side lengths of 1 and 2 and with a hypotenuse of length $\sqrt{5}$. To find a square with an area of 10, find a segment with a length of $\sqrt{10}$. Therefore find two squares whose sum is 10: $3^2 + 1^2 = 10$.

29. Find the hypotenuse of a right triangle formed by one leg with a length of $(50 \text{ km/hr}) \times 3 \text{ hr}$ and a second leg with a length of $(60 \text{ km/hr}) \times 4 \text{ hr}$.

30. Draw a right triangle that shows the total distance east and the total distance north.

Exercise Set 10.6 (pg 504)

3. $(\frac{1}{2})x^2 = 98$. Therefore $x^2 = 196$. Let c be the hypotenuse, then $x^2 + x^2 = c^2$. Therefore $196 + 196 = c^2$. Use a calculator from here.

4. Average speed $= \dfrac{d}{4 \text{ hours}}$

5. $x^2 + 24^2 = (36 - x)^2$
$x^2 + 24^2 = 36^2 - 72x + x^2$
$24^2 = 36^2 - 72x$
Solve for x.

6. The distance x is called the space diagonal. It is the Pythagorean Theorem in three dimensions. First find y, the length of the diagonal in the bottom of the box. Use $24^2 + 30^2 = y^2$. Then find the space diagonal. Use $18^2 + y^2 = x^2$.

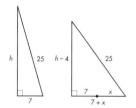

7. Find h, the height of the first triangle. The height of the second is 4 m less than the height of the first. Then find x.

8.

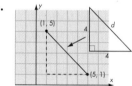

10. To lift 80 lb straight up 2 ft requires $80 \text{ lb} \cdot 2 \text{ ft} = 160 \text{ ft lb}$ of work. To do 160 ft lb of work over a distance of 8 ft requires a force of $\dfrac{160 \text{ ft lb}}{8 \text{ ft}} = -?-$.

Investigation 10.7 (pg 507)

4.

$d^2 = 4^2 + 4^2$

7. $(1 - 13)^2 + (2 - 7)^2 = d^2$. $(-12)^2 + (-5)^2 = d^2$. $144 + 25 = d^2$. Solve for d.

Exercise Set 10.7 (pg 510)

1. $(10 - 13)^2 + (20 - 16)^2 = d^2$. Solve for d. $d = 5$

6. $GH = \sqrt{(18 - 2)^2 + (6 - 2)^2}$
$GI = \sqrt{(12 - 2)^2 + (12 - 6)^2}$
$HI = \sqrt{(18 - 12)^2 + (12 - 2)^2}$

To check if the triangle is a right triangle, see if GH, GI, and HI are Pythagorean triples or compare the slopes to see if two segments are perpendicular.

13. $(x + 8)^2 = 40^2 + x^2$

14. Points A, B, G, D, E, and F divide the circle into six equal parts. If point C is selected on $\overset{\frown}{AB}$, then $\triangle ABC$ is obtuse. If point C is selected at point D or E, then $\triangle ABC$ is right. Therefore if point C is selected on $\overset{\frown}{BD}$ or $\overset{\frown}{AE}$, then $\triangle ABC$ is obtuse. If point C is selected on $\overset{\frown}{DE}$, then $\triangle ABC$ is acute.

Exercise Set 10.8 (pg 511)

1. $m\angle DOB = 180° - 105°$ because two angles of the quadrilateral are right angles.

3. When $\overline{TO}$ is constructed, it forms two 30-60 right triangles, with the tangent segments as the longer legs. Therefore the radius of the circle is the shorter leg in each 30-60 right triangle. Therefore $r = 6$.

4. The shaded area is $2[(\frac{1}{2})(8)(8\sqrt{3}) - \frac{1}{6}\pi(8)^2]$ or –?–.

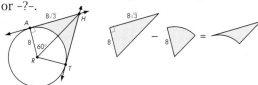

6. The shaded area is $\frac{1}{3}\pi(8)^2 - (\frac{1}{2})(8\sqrt{3})(4)$ or –?–.

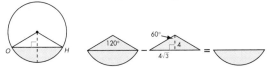

10. When $\overline{OT}$ and $\overline{OA}$ are drawn, they form a right triangle, with $OA = 15$ (length of hypotenuse) and $OT = 12$ (length of leg). Thus the length of the third side (half the chord) can be calculated.

11. Using the same diagram that you did in Exercise 10, draw in $\overline{OA}$ with length R and $\overline{OT}$ with length r. The area of the annulus is $\pi R^2 - \pi r^2$, or $\pi(R^2 - r^2)$. But $R^2 = r^2 + 18^2$, so $R^2 - r^2 = 324$.

12. The arc length of $\overset{\frown}{AC}$ is $\frac{80}{360}[2\pi(9)] = (\frac{2}{9})(18\pi) = 4\pi$. Therefore the circumference of the base of the cone is 4π. From this you can determine the radius of the base. The radius of the sector (9) becomes the slant height (the distance from the tip of the cone to the circumference of the base). The radius of the base, the slant height, and the height of the cone form a right triangle. Solve for the height.

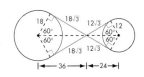

14. The belt length is $(\frac{2}{3})\pi(36) + (\frac{2}{3})\pi(24) + 2(18\sqrt{3}) + 2(12\sqrt{3})$ or –?–.

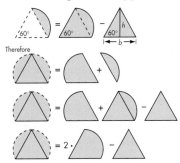

15. $C = \pi D$, so $336 = \pi D$. $D = \frac{336}{\pi}$.

But $D = x\sqrt{2}$, so $x = \frac{D}{\sqrt{2}}$. $x \approx \frac{D}{1.41}$

16. Because the central triangle is equilateral (each side measures 12 cm) the height of the equilateral triangle is $6\sqrt{3}$. Therefore the total height of the rectangle is –?–.

17. Here's one possible approach.

$2(\frac{1}{6}\pi r^2) - \frac{1}{2}bh$

Exercise Set 10.9 (pg 514)

2. Here's one approach: Construct the given angle, then bisect it. Next construct a line parallel to one of the sides of the angle at a distance OR from the side. The intersection of this line and the angle bisector is the center of the circle. Do you see why this works?

3. No, you cannot just lay your straightedge through point X and adjust it so that the edge appears to just touch the circle. That is not a construction! You need to find the point of tangency. Construct circle O and point X. Construct $\overline{XO}$. Construct midpoint M of $\overline{XO}$. With M as the center, construct a semicircle of diameter XO. Label T as the point that circle O and the semicircle intersect. Connect $\overleftrightarrow{XT}$ to create the tangent line.

4.

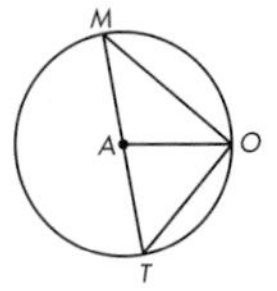

6. Start with an equilateral triangle, with the length of the legs equal to AB.

7. Construct a segment of length $2AB$ and construct an equilateral triangle with sides of length $2AB$.

8. Start with a segment of length AB. Construct a $30°$ angle at point A and a line perpendicular to $\overline{AB}$ through point B.

12. The medians (altitudes, etc.) of an equilateral triangle divide each other at the incenter/circumcenter/centroid. Therefore the longer segment on the median (two thirds as long as the median) is the radius of the circumscribed circle, and the shorter segment on the median (one third as long as the median) is the radius of the inscribed circle. If the length of a side is 6, then half the length is 3, and thus the length of the median is $3\sqrt{3}$. Therefore the radius of the circumscribed circle is $\frac{2}{3}(3\sqrt{3})$, or $2\sqrt{3}$, and the radius of the inscribed circle is $\frac{1}{3}(3\sqrt{3})$, or $\sqrt{3}$. You're on your own from here.

14. Because the altitude is $9\sqrt{3}$, the radius of the circumscribed circle is $\frac{2}{3}(9\sqrt{3})$ and the radius of the inscribed circle is $\frac{1}{3}(9\sqrt{3})$. Because the length of each altitude is $9\sqrt{3}$, half the length of a side is 9. Therefore the area of the triangle is –?–.

15. A good problem-solving technique to find ratios or percentages is to solve the problem with actual numbers (because ratios and percentages do not change for different-size triangles). For example, suppose you have an equilateral triangle, each side of which measures 6 cm (pick a number with a lot of divisors—it makes things work out nicely). If each side measures 6 cm then the altitude is $3\sqrt{3}$.

$$A_t = (\tfrac{1}{2})(6)(3\sqrt{3})$$

If the altitude is $3\sqrt{3}$, the radius of the inscribed circle is one third of $3\sqrt{3}$.

$$A_c = \pi(\sqrt{3})^2$$

Calculate the ratio of the area of the circle to the area of the triangle.

16. The area of an equilateral triangle is $\frac{1}{2}sh$. But $h = \frac{1}{2}s\sqrt{3}$.

Therefore $A = \frac{1}{4}s^2\sqrt{3}$. But $A = 25\sqrt{3}$.

Therefore $\frac{1}{4}s^2\sqrt{3} = 25\sqrt{3}$. Solve for s. The perimeter is $3s$.

18. A regular hexagon can be divided into six equilateral triangles. The radius of the inscribed circle is the height of one of the six triangles. Therefore half the length of a side of the equilateral triangle is $3\sqrt{3}$. Thus the length of each side is $6\sqrt{3}$. The area of the regular hexagon can be found with the formula $A = \frac{1}{2}ap$, where a is the apothem (9) and p is the perimeter ($6 \times 6\sqrt{3}$).

19. The radius of the inscribed circle is one third the length of the median of $\triangle PON$

26. Refer to the figure below. Distance to surface $= \sqrt{1,000^2 \text{ km}^2 + 13,000^2 \text{ km}^2} -$ radius of earth (12,750 km)

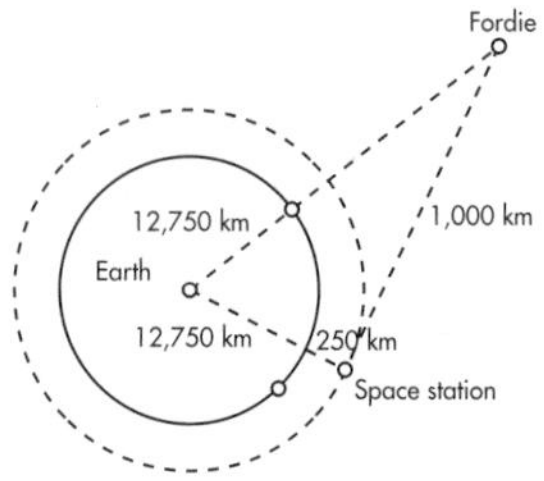

27. Break these areas into smaller parts and calculate the percentage. See the hint for Exercise 10.8.17.

28. To find the dimensions of the box on the right, see the hint for Exercise 10.8.16.

Exercise Set 10.10 (pg 517)

10. Visualize a right triangle inscribed in a semicircle. Is the midpoint of the hypotenuse the center of the semicircle? Then compare the distances from that point to all three vertices.

21. Half the shaded area is equal to the area of a quadrant (quarter) of the circle less the area of the isosceles right triangle.

28. $(45 \cdot 2)^2 + (60 \cdot 2)^2 = d^2$

29. $(12 + 12\sqrt{2})$. Why?

30. Is the length of the space diagonal greater or less than 24"? $d^2 = 12^2 + 16^2 + 14^2$

32. If the height is $12\sqrt{3}$, then the radius of the inscribed circle is $(\frac{1}{3})(12\sqrt{3})$.

35.

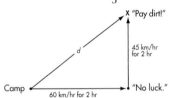

38.

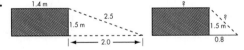

Cooperative Problem Solving: Pythagoras in Space (pg 521)

3. The smallest triangle that would enclose three congruent mutually tangent circles would be the equilateral triangle that is tangent to two circles on each side. The figure above should help.

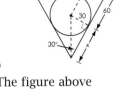

4. A drawing of two corridors intersecting at right angles shows that the limiting size is the diagonal ($\overline{AC}$) of the longest rectangular plane (*ABCD*) that can maneuver the right-angle turn. This will happen when *AE = BE*.

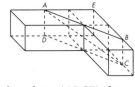

5. If you unwrap one complete loop (one of the 30), you get the diagonal of a rectangle. The diagonal is one thirtieth of the total distance. See the figure above.

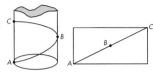

Chapter 11

Exercise Set 11.1 (pg 528)

1. Begin by writing down statements based on your observations.
A pyramid is a —?—.
The base is a —?—.
The lateral faces are —?— formed by segments that connect the base to the —?—.
Now combine these statements into a definition.

19. True. You can think of a prism as a stack of thin copies of the bases.

20. The cross sections get smaller as you get closer to the vertex of the pyramid.

27. Start by drawing a diagram. Clearly you will not be able to completely fill the box.

Exercise Set 11.2 (pg 533)

2. False. It is a sector of a circle.

14. A die is one of a pair of dice.

20.

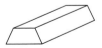

24.

Exercise Set 11.3 (pg 541)

2. $V = BH$
$= [(\frac{1}{2})(3)(4)](4)$
$= 24$

3. $V = BH$
$= [(\frac{1}{2})(3)(3 + 5)](9)$
$= -?-$

5. $V = BH$
$= [(\frac{1}{2})(\pi)(3)^2](8)$
$= -?-$

6. $\frac{90}{360}$, or $\frac{1}{4}$, of the cylinder is removed.
Therefore you need to find $\frac{3}{4}$ of the volume of the whole cylinder.

9. a. $V = (\frac{1}{2}bh)H$

15. $V = BH$
$= [\pi(\sqrt{Q})^2]T$
$= QT\pi$ (Cutie pie!
 Sorry about that.)

17.

Take Another Look 11.4 (pg 545)

3.

Exercise Set 11.4 (pg 545)

3. $V = \frac{1}{3}BH$
$= \frac{1}{3}[(\frac{1}{2})(5)(4 + 8)](15)$
$= (\frac{1}{3})(30)(15)$
$= 150$

4. $V = \frac{1}{3}BH$
$= (\frac{1}{3})(\pi 6^2)(7)$
$= (\frac{1}{3})(36\pi)(7)$
$= 84\pi$

7. $V = \frac{1}{3}BH$
$= \frac{1}{3}[(\frac{1}{2})(5)(12)](6)$
$= 60$

8. The figure is a cylinder with a cone removed.
$V = V_{\text{cylinder}} - V_{\text{cone}}$
$= \pi(6)^2(16) - (\frac{1}{3})\pi(6)^2(16)$
$= (\frac{2}{3})\pi(6)^2(16)$
$= -?-$

11. $V = \frac{1}{3}BH$
$= (\frac{1}{3})\pi b^2 \times 2b$

14. $V = \frac{1}{3}[\pi(6x)^2](8x)$
$= -?-$

18. One gallon is 0.1336816 cubic feet.

Exercise Set 11.5 (pg 550)

4. Suppose the smaller prism has dimensions 4 by 5 by 6 (volume of 120 cm³). Then the dimensions of the larger prism are 8 by 10 by 12 (volume of 960 cm³). The volume is 8 times as large. Is this always true? Try it.

7. $V = \frac{1}{3}BH$
$138\pi = (\frac{1}{3})(46\pi)H$
Solve for H. (Divide both sides by 46π, then multiply both sides by 3.)

10. $V = \frac{1}{3}BH$
$3168 = \frac{1}{3}[\frac{1}{2}(20 + 28)h](36)$
Solve for h.

12. First change 8″ to $\frac{2}{3}'$. To find the volume, multiply $(5.5)(6.5)(\frac{2}{3})$. To convert volume (cubic feet) to weight (pounds), multiply the volume by 63 pounds/cubic foot.

14. $A = \frac{1}{2}ap$

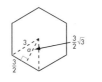

$$= \frac{1}{2}[(\frac{3}{2})(\sqrt{3})](18) = (\frac{27}{2})\sqrt{3}$$

$$V = (\frac{27}{2})(\sqrt{3})(3) \approx 68.9$$

The number of gallons is about $(68.9)(7.5)$ or –?–.

15. The swimming pool is a pentagonal prism resting on one of its lateral faces. The area of the pentagonal base can be found by dividing it into a rectangular region and a trapezoidal region.

Exercise Set 11.6 (pg 556)

4. $$V_{\text{displacement}} = \frac{7}{8}(V_{\text{block of ice}})$$

$$(35)(50)(4) = \frac{7}{8}(V_{\text{block of ice}})$$

$$(\frac{8}{7})(35)(50)(4) = V_{\text{block of ice}}$$

$$V_{\text{block of ice}} = {-?-}$$

7. The density of the clump of metal equals its mass divided by its volume.

$$0.97 = \frac{145.5}{V_{\text{displacement}}}$$

$$V_{\text{displacement}} = \frac{145.5}{0.97}$$

But $V_{\text{displacement}}$ is $(10)(10)H$.

So $\frac{145.5}{0.97} = (10)(10)H$. Solve for H.

Exercise Set 11.7 (pg 560)

1. $V = (\frac{4}{3})\pi r^3$

$$= (\frac{4}{3})\pi(3)^3$$

$$= {-?-}$$

4. $V_{\text{capsule}} = V_{\text{two hemispheres}} + V_{\text{cylinder}}$

$$= 2[(\frac{2}{3})\pi(6)^3] + [\pi(6)^2(12)]$$

$$= (\frac{4}{3})(216)\pi + (12)(36)\pi$$

$$= {-?-}$$

6. $\frac{40}{360}$, or $\frac{1}{9}$, of the hemisphere is missing. What fraction is still there?

7. $$V = (\frac{4}{3})\pi r^3$$

$$972\pi = (\frac{4}{3})\pi r^3$$

$$972(\frac{3}{4}) = r^3$$

11. $V_{\text{container}} = \pi(3)^2(10) = 90\pi$
$V_{\text{scoop}} = (\frac{4}{3})\pi(\frac{3}{2})^3 = (\frac{9}{2})\pi$

The number of scoops is $(90\pi)/[(\frac{9}{2})\pi]$ or –?–.

15. The level starts out rising quickly, then it slows down, then at the end it rises quickly again.

Project: Archimedes' Principle (pg 562)

2. First calculate the weight of the sheet less the weight of the four cut-out squares. $V_{\text{aluminum}} = (18)(32)(0.5)$ cm³ $- 4(0.5)(4)(4)$ cm³ $= $ –?– cm³. The weight is the volume of the aluminum times the density of the aluminum. Next determine the volume of the form once it is shaped into a boat. $V_{\text{boat}} = (4)(10)(24)$ cm³. The boat will float if its density, $W_{\text{aluminum}}/V_{\text{boat}}$, is less than the density of water.

3. You should get approximately 97 kg for the weight of the four balls above water. You should get approximately 8579 cm³ for the volume of the four balls. Thus the weight of the displaced water should be approximately 8836 g or 8.8 kg. One kilogram equals approximately 2.2 lb.

Take Another Look 11.8 (pg 566)

1. Cut each of the four circular regions into eight sectors. Arrange the eight pieces of each circular region into one quadrant of the sphere.

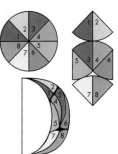

Exercise Set 11.8 (pg 566)

1. $V = \frac{4}{3}\pi r^3$

$\quad = (\frac{4}{3})\pi(9)^3$

$\quad = -?-$

$\quad S = 4\pi r^2$

$\quad = (4)\pi(9)^2$

$\quad = -?-$

4. Solve for R.

$\quad A_{\text{circle}} = \pi R^2$

$\quad 40\pi = \pi R^2$

$\quad R = \sqrt{40}$

Solve for surface area.

Exercise Set 11.9 (pg 569)

2. $V_{\text{ring}} = V_{\text{larger prism}} - V_{\text{missing prism}}$

$\quad = [(\frac{1}{2})(3\sqrt{3})(36)](2) - [(\frac{1}{2})(2\sqrt{3})(24)](2)$

$\quad = -?-$

The volume of the hole, $(48\sqrt{3})$, does not equal the volume of the ring $(60\sqrt{3})$.

3. $V_{\text{one pipe}} = V_{\text{outer cylinder}} - V_{\text{inner cylinder}}$

$\quad = \pi(3)^2(160) - \pi(2.5)^2(160)$

$\quad = \pi(160)[3^2 - (2.5)^2]$

$\quad = \pi(160)(2.75)$

Therefore the weight of 200 pipes is $(200)(V_{\text{one pipe}})(\text{density})$ or $(200)(V_{\text{one pipe}})(7.7)$ or $-?-$.

6. $V_{\text{ball}} = (\frac{4}{3})\pi(6)^3 \approx -?-$

The weight of the ball is equal to the product of the volume and the density.

8. $V_{\text{hollow ball}} = V_{\text{outer sphere}} - V_{\text{inner sphere}}$

$\quad \approx (\frac{4}{3})\pi(7)^3 - (\frac{4}{3})\pi(r)^3$

$\quad \approx -?-$

However, the volume of the hollow ball is its weight divided by its density.

$V_{\text{hollow ball}} = \frac{327.36}{0.28}$

Therefore $\frac{327.36}{0.28} = (\frac{4}{3})\pi(7^3 - r^3)$. Solve for r.

9. The volume of the object equals the area of the hexagonal base of the glass prism multiplied by the rise in the water level.

$V_{\text{object}} = [(\frac{1}{2})(\frac{5}{2})(\sqrt{3})(30)](4)$

$\quad = 150\sqrt{3}$

The density of the object is its weight divided by its volume.

$\text{Density}_{\text{object}} = \frac{5457}{150\sqrt{3}}$

$\quad \approx \frac{5457}{(150)(1.7)}$

$\quad \approx -?-$

Exercise Set 11.10 (pg 571)

20. $B = (12)(12) - (4)(6)$

21. $V = (\frac{2}{3})\pi r^3$

$\quad V = (\frac{2}{3})\pi(15)^3$

24. One fourth of the hemisphere is missing.

Therefore the volume is given by $V = \frac{3}{4}(\frac{2}{3}\pi r^3) = \frac{1}{2}\pi r^3$. If $V = 256\pi$, then $256\pi = \frac{1}{2}\pi r^3$. Solve this equation for r^3, then find the cube root.

28. $\qquad$ Initial surface area $= 4\pi(4)^2$

$\quad$ Surface area of expanded balloon $= 4\pi(8)^2$

$\quad \frac{4\pi(8)^2}{4\pi(4)^2} = -?-$

Now calculate the ratio of the volumes.

35. If it is 240 g of 22 carat gold, then $\frac{11}{12}$ of the 240 g is gold and $\frac{1}{12}$ of the 240 g is copper. The density of gold is 19.30 g/cm^3. The density of copper is 8.97 g/cm^3. If you know the mass and the density of each, you can find the volume of each.

Cooperative Problem Solving: Once Upon a Time (pg 575)

3. The octahedron is two square-based pyramids. Let each edge equal one unit. Find the volume of one pyramid and double your answer for the octahedron. The altitude of the pyramid touches the pyramid's base at the intersection of the diagonals. If you know the length of the

---

edge of the pyramid and the length of half a diagonal, you can calculate the height of the pyramid. Once you know the height of the pyramid, you can calculate the volume of the octahedron. Let each edge of the cube equal $\frac{3}{4}$. Then find the volume of the cube.

Chapter 12

Exercise Set 12.1 (pg 581)

2. $\frac{\text{shaded area}}{\text{total area}} = \frac{12}{32} = -?-$. (You might also have noticed that in every pair of columns, three of the eight squares are shaded.)

10. $y^2 = 64$, thus $y = \sqrt{64}$. (What are the two answers?)

11. First reduce $\frac{35}{56}$. Thus $\frac{10}{10 + z} = \frac{5}{8}$. Because 10 is twice as big as 5, then $10 + z = 16$.

14. $\frac{x \text{ runs}}{9 \text{ innings}} = \frac{34 \text{ runs}}{106 \text{ innings}}$

Exercise Set 12.2 (pg 590)

12. $\frac{1}{2} = \frac{3}{AL}$, $\frac{1}{2} = \frac{5}{RA}$, etc. Or $\frac{4}{3} = \frac{8}{AL}$, $\frac{4}{5} = \frac{8}{AR}$, etc.

15. All the corresponding angles are congruent. Are all these ratios equal?
$\frac{21}{14} = -?-$, $\frac{30}{20} = -?-$, $\frac{27}{18} = -?-$, $\frac{39}{26} = -?-$, $\frac{78}{52} = -?-$

18. Because the lines are parallel, the corresponding angles are congruent. Are all these ratios equal?
$\frac{2}{2 + 2^{2}/_{3}} = -?-$, $\frac{3}{3 + 4} = -?-$, $\frac{4}{9^{1}/_{3}} = -?-$

Exercise Set 12.3 (pg 596)

3. It helps to rotate $\triangle ARK$ so that you can see which sides correspond.

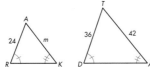

7. $\frac{96}{84} = \frac{8}{7}$, $\frac{104}{91} = \frac{8}{7}$

8.

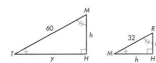

11. Because $\angle H$ and $\angle D$ are two angles inscribed in the same arc, they are congruent (Conjecture 69). $\angle T$ and $\angle G$ are congruent for the same reason. Therefore $\triangle HTU \sim \triangle DGU$ by the AA shortcut.

12. $\angle S \cong \angle T$, and $\angle A \cong \angle U$. Why? See the hint for Exercise 12.3.11.

21.

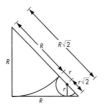

Exercise Set 12.4 (pg 603)

2. $\frac{5'}{84''} = \frac{x'}{72''}$

4.

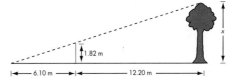

6.

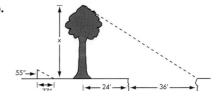

7. Because $\triangle GAR \sim \triangle DAN$, then $\frac{AD}{AG} = \frac{ND}{RG}$.

8. Because $\triangle PRE \sim \triangle POC$, then $\frac{PR}{RE} = \frac{PO}{OC}$.

Let $x = PR$. Then $\frac{x}{60} = \frac{x + 45}{90}$.

9.

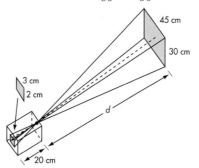

10. $\frac{9}{15} = \frac{19}{x}$

11. $\frac{y}{91} = \frac{54}{78}$

12.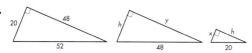

Exercise Set 12.5 (pg 610)

1. $\frac{22}{33} = \frac{12}{h}$

3. $\frac{IC}{IE} = \frac{CS}{PE}, \frac{LI}{SI} = \frac{CL}{SE}$

5. $\frac{5}{8} = \frac{x}{32}$

7. $\frac{a}{b} = \frac{p}{q}$

10. $\frac{12}{15} = \frac{x}{10-x}$

11. $\frac{9}{18} = \frac{3\sqrt{3}}{k}$

12. Because $48^2 = 24^2 + (24\sqrt{3})^2$, $\triangle ABC$ is a 30-60 right triangle. Because $m\angle A = 60$, $\triangle ABD$ is equilateral and, therefore, $AD = 24$. But because $\frac{24}{24} = \frac{x}{y}$, then $x = y = 12$.

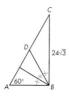

13. $\text{Area}_{\text{large triangle}} = (\frac{1}{2})(8)a$

$\text{Area}_{\text{small triangle}} = (\frac{1}{2})(8)b$

So, $\frac{\text{Area}_{\text{large triangle}}}{\text{Area}_{\text{small triangle}}} = \frac{(1/2)(8)a}{(1/2)(8)b} = \frac{a}{b}$.

But you know that the ratio of a to b equals the ratio of the corresponding sides with lengths of $8\sqrt{2}$ and 8.

So $\frac{\text{Area}_{\text{large triangle}}}{\text{Area}_{\text{small triangle}}} = \frac{a}{b} = \frac{8\sqrt{2}}{8} = \sqrt{2}$.

14. Bisect the angle between the sides of lengths $2x$ and $3x$.

15. Follow the steps and the hint for Exercise 12.5.14 above.

18. $\frac{a}{b} = \frac{c}{d}$, then $\frac{a}{b} + 1 = \frac{c}{d} + 1$, then $\frac{a}{b} + \frac{b}{b} = \frac{c}{d} + \frac{d}{d}$.

Exercise Set 12.6 (pg 616)

1. $(\frac{1}{2})^2 = \frac{\text{Area}_{\triangle MSE}}{72}$

3. Because $\frac{\text{Area}_{ZOID}}{\text{Area}_{TRAP}} = \frac{16}{25}$, then the ratio of the lengths of corresponding sides is $\frac{4}{5}$. So $\frac{4}{a} = \frac{4}{5}$ and $\frac{8}{b} = \frac{4}{5}$.

7.

Area = $6m^2$ Area = $6n^2$

12. Use the Pythagorean Theorem to find H. Once you have the value of H, set up the proportion $\frac{h}{3} = \frac{H}{12}$ and solve for h.

13. $(\frac{h}{H})^2 = \frac{9}{25}$. Find the square root of both sides.

$\frac{\text{Volume}_{\text{large prism}}}{\text{Volume}_{\text{small prism}}} = (\frac{h}{H})^3$

14. Volume of small cylinder = $\pi(9)^2(24)$

$\frac{\text{Volume}_{\text{large cylinder}}}{\text{Volume}_{\text{small cylinder}}} = (\frac{H}{24})^3$

15. $(\frac{5}{3})^3 = -?-$

16. $(\frac{h}{H})^3 = \frac{8}{125}$. Find the cube root of both sides.

17. If $(\frac{r}{R})^3 = \frac{8}{27}$, then $\frac{r}{R} = \frac{2}{3}$. Therefore $\frac{d}{D} = -?-$.

18.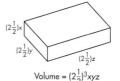

Volume = xyz Volume = $(2\frac{1}{2})^3xyz$

The volume of the larger storage facility is $(2\frac{1}{2})^3$ times as large, so the cost will be $(2\frac{1}{2})^3$ times as large. $(2\frac{1}{2})^3 = (\frac{5}{2})^3 = \frac{125}{8}$, so the cost is $(\frac{125}{8})(\$125)$ per day.

20. Because it is 4 times as large in each dimension, then it will be 4^3 times as heavy.

21. $\frac{\%\ \text{doctors in '90}}{\%\ \text{doctors in '75}} = \frac{12\%}{16\%} = 0.75$

$\frac{\text{height of graphic in '90}}{\text{height of graphic in '75}} \approx \frac{1.6}{2.1} \approx 0.76$

$\frac{\text{area of graphic in '90}}{\text{area of graphic in '75}} \approx (0.76)^2 \approx 0.58$

Project: Why Elephants Have Big Ears (pg 619)

7. The incredibly large ears give a lot more surface area with almost zero increase in volume. More surface area makes it easier to cool off.

Take Another Look 12.7 (pg 625)

1. If $\ell \parallel \overline{MN}$, then $\angle VMT \cong \angle VLU$ and $\angle VTM \cong \angle VUL$ because of corresponding angles. If the angles are congruent, then $\triangle LUV \sim \triangle MTV$ by AA similarity.

   If $\triangle LUV \sim \triangle MTV$, then $\frac{a+b}{a} = \frac{c+d}{c}$ by the definition of similar polygons.

   If $\frac{a+b}{a} = \frac{c+d}{c}$, then $\frac{a}{a} + \frac{b}{a} = \frac{c}{c} + \frac{d}{c}$.
   Then $1 + \frac{b}{a} = \frac{d}{c} + 1$, and thus $\frac{b}{a} = \frac{d}{c}$. If $\frac{b}{a} = \frac{d}{c}$, then $ad = bc$. If $ad = bc$, then $\frac{a}{b} = \frac{c}{d}$.

3. Because $\overleftrightarrow{LC} \parallel \overline{MI}$ in $\triangle MIH$, then $\frac{x}{y} = \frac{s}{t}$. Because $\overleftrightarrow{LC} \parallel \overleftrightarrow{EH}$ in $\triangle MEH$, then $\frac{x}{y} = \frac{b}{d}$.

4.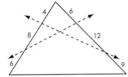

Exercise Set 12.7 (pg 626)

1. $\frac{4}{4+12} = \frac{a}{20}$

3. $\frac{60}{40} = \frac{60+c}{70}$

4. If $\frac{24}{14} = \frac{36}{d}$, then $\frac{12}{7} = \frac{36}{d}$.

6. Is $\frac{15}{36} = \frac{25}{55}$?

12. $a + b = \sqrt{(12-3)^2 + (0-9)^2}$. Once you have solved for $a + b$, then $\frac{4}{12} = \frac{a}{a+b}$.

14. Divide $\overline{IJ}$ into six parts. Then use one part as the radius of a circle. Divide the circle into six equal arcs (daisy construction).

16. If $\frac{r^2}{s^2} = \frac{49}{81}$, then $\frac{r}{s} = \frac{7}{9}$.
    If $\frac{r}{s} = \frac{7}{9}$, then $\frac{r^3}{s^3} = \frac{7^3}{9^3} = $ –?–.

17. You probably "saw" how to do it rather quickly. But why? If you visualize your segment as the hypotenuse of a right triangle, then the Extended Parallel Proportionality Conjecture explains why.

19. $\frac{x}{10} = \frac{x+12}{16}$. Then to find the volume of the truncated cone, subtract the volume of the small missing cone from the volume of the original larger cone.

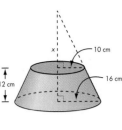

21. Connect the two 75's. If y is the length of the given segment and x is the length connecting the two 75's, then $\frac{75}{x} = \frac{100}{y}$. Solve for x in terms of y.

22. The two isosceles triangles will always be similar because their vertex angles are congruent vertical angles. Adjust the set screw so that the ratio of small part to large part is –?– to –?–.

Exercise Set 12.8 (pg 634)

18. Redraw the two triangles so that the corresponding angles are in the same position.
    $\frac{6}{5} = \frac{5}{x}$ and $\frac{9}{6} = \frac{y}{5}$

21. $\frac{2^3}{3^3} = \frac{V}{2160\pi}$

24. The fact that the two decks are similar rectangles is irrelevant. The painting contractor charges by time or square footage, and thus $\frac{\$150}{12 \times 15} = \frac{\$x}{16 \times 20}$.

28. Divide $\overline{KL}$ into five equal lengths. Then locate the point that divides $\overline{KL}$ into a 2:3 ratio.

29. $\triangle ABE \sim \triangle ADC$
    $\frac{AB}{BE} = \frac{AD}{CD}$

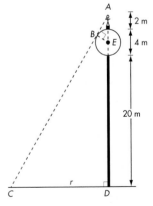

31. Because you are concerned with ratios, it doesn't make any difference what lengths you choose. Therefore let's give the square a side

of length 2. Then the area of the square is 4, the area of the circle is $\pi(1)^2$, or π, and the area of the isosceles triangle is $(\frac{1}{2})(2)(2)$, or 2. Therefore the ratio of the three areas is –?– : –?– : –?–.

The volume of the cylinder is $\pi(1)^2(2)$, or 2π,

the volume of the sphere is $(\frac{4}{3})\pi(1)^3$ or $\frac{4}{3}\pi$,

the volume of the cone is $(\frac{1}{3})\pi(1)^2(2)$ or $\frac{2}{3}\pi$.

The ratio of the three volumes is $2\pi : \frac{4}{3}\pi : \frac{2}{3}\pi$.

This can be reduced to a very nice ratio of three whole numbers.

Chapter 13

Exercise Set 13.1 (pg 647)

1. $\sin A = \frac{s}{t}$, $\cos A = \frac{r}{t}$, $\tan A = \frac{s}{r}$

6. $\sin 37° \approx 0.6018$

9. Use your calculator to find the inverse sin (or $\sin^{-1}$) of 0.5. You should get 30°.

12. $8x = 51$, $x = 6.38$

18. $\tan 30° = \frac{20}{a}$. But $\tan 30° = 0.58$.

Therefore $\frac{20}{a} = 0.58$.

25. Use $\sin 35° = \frac{b}{85}$ to find the length of the base. Then use $\cos 35° = \frac{h}{85}$ to find the height.

26. Use $\tan 55° = \frac{280}{b}$; solve for b.

Use $\tan 75° = \frac{280}{a}$; solve for a.

Exercise Set 13.2 (pg 650)

2. $\sin G = \frac{16}{16\sqrt{2}} = \frac{1}{\sqrt{2}} = \frac{\sqrt{2}}{2}$

9. $\sin 32° = \frac{a}{17}$

14. $\tan \beta = \frac{10}{17}$

15. $\frac{(1/2)d_1}{20} = \cos 56°$

21. $\tan 33° = \frac{42}{y}$

22. $\sin 44° = \frac{h}{1400}$, where h is the height of the balloon above Wendy's sextant. $\cos 44° = \frac{d}{1400}$, where d is the distance from Wendy to a position directly beneath the balloon.

23.
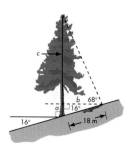

Exercise Set 13.3 (pg 657)

3. Draw in a diagonal connecting the vertices of the unmeasured angles. Find the area of the two triangles.

4. Area of regular octagon = 8 · area of isosceles triangle. Area of triangle = $\frac{1}{2}(12)(12)\sin 45°$.

5. $\frac{28}{\sin 52°} = \frac{w}{\sin 79°}$

8. $\frac{\sin 42°}{29} = \frac{\sin A}{36}$

11. $\sin 55° = \frac{h}{50}$

13. $\sin \emptyset = \frac{4.8}{55}$

16. $\sin 16° = \frac{a}{18}$

$\cos 16° = \frac{b}{18}$

$\tan 68° = \frac{c}{b}$

Take Another Look 13.4 (pg 662)

2. To prove $a^2 = b^2 + c^2 - 2bc \cos A$, begin by constructing an altitude $\overline{CD}$ in $\triangle ABC$. In $\triangle BDC$, $a^2 = x^2 + (c - y)^2$ or $a^2 = x^2 + c^2 - 2cy + y^2$ (equation 1). In $\triangle ADC$, $b^2 = x^2 + y^2$ (equation 2). Substituting equation 2 into equation 1, you get $a^2 = b^2 + c^2 - 2cy$ (equation 3). Now if you can show that $y = b \cos A$ (equation 4) and substitute equation 4 into equation 3, you're in like Flynn.

Exercise Set 13.4 (pg 663)

1. $w^2 = 36^2 + 41^2 - (2)(36)(41)(\cos 49°)$

4. $42^2 = 34^2 + 36^2 - (2)(36)(34)(\cos A)$

7. The smallest angle (A) is opposite the smallest side. $4^2 = 7^2 + 8^2 - (2)(7)(8)(\cos A)$

14. $\tan 65° = \dfrac{d}{65}$

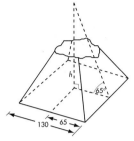

15. First find the distance (x) along his original path  from the moment Ace left it to the moment he returned to it. You could use the law of cosines: $960^2 = x^2 + 720^2 - (2)(x)(720)(\cos 20°)$. But who wants to solve a quadratic equation? Another approach is to divide the triangle into two right triangles, as shown. Find a: $\cos 20° = \dfrac{a}{720}$. Then find h: $\sin 20° = \dfrac{h}{720}$. Then use the Pythagorean Theorem to find b: $x = a + b$. When you have the distance, you can find the time it takes traveling at 720 km/hr to go that distance. Compare that time against the 2 hr 20 min for the detour. Whew!

Exercise Set 13.5 (pg 665)

1. $\tan 4.5° = \dfrac{1}{x}$

7.

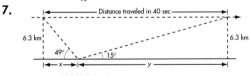

8.

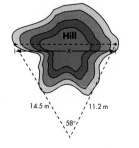

9.

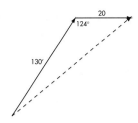

10. Find θ, then use the law of cosines.

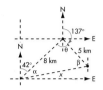

11. Refer to the graphic for the hint to Exercise 13.5.10. Find the measure of either α or β by using the law of sines.

Graphing Calculator Investigation: Trigonometric Functions (pg 671)

5. $\tan 80° \approx 5.7$, $\tan 85° \approx 11.4$, $\tan 89° \approx 57.3$, $\tan 89.9° \approx 573, \ldots$

Exercise Set 13.6 (pg 672)

3. $\sin \emptyset = s$, $\cos \emptyset = r$, $\tan \emptyset = $ -?-

28.

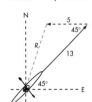

29.

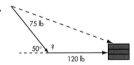

Cooperative Problem Solving: Problem Solving at Mare Imbrium (pg 675)

1.

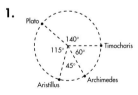

2.

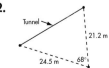

Chapter 14

Exercise Set 14.1 (pg 678)

1. B   B
    C   C
           M

Exercise Set 14.2 (pg 682)

2. If the iron tip of the walking stick was worn down, then the owner of the walking stick walks a great deal. The iron tip of the walking stick was worn down. Therefore —?—.

5. If Carolyn has study sessions, then she does well on her tests.

If Carolyn does well on her tests, then she gets good grades on her report card.

Carolyn has study sessions.

Therefore she does well on her tests. (step one)

Therefore she —?—. (step two)

7. Therefore $\triangle ABC$ is not . . . .

14. If a carp can leap over the Dragon Gate, then it must be very lucky. According to an ancient Chinese folk tale, a carp did leap over the Dragon Gate. Therefore it must have been very lucky.

Exercise Set 14.3 (pg 690)

1. Present some official identification showing the address. Other ideas?

3. If I get a job, then I will earn money.

7. IF P THEN Q

12.

17. He will be victorious.

LET P: Sherlock Holmes gathers all the clues.

LET Q: He will be victorious.

IF P THEN Q

P

THEREFORE Q

25. No valid conclusion can be drawn from the two statements.

27. Modus Tollens

Exercise Set 14.4 (pg 692)

1. $P \rightarrow \sim Q$

5. $\sim(R \rightarrow S)$

12. If international tensions are not reduced, then —?—.

14. It is not true that if world peace is not in jeopardy, then international tensions are not reduced.

17. MT

19. No valid conclusion

22. MP

23. $T \rightarrow \sim P$, T therefore $\sim P$ by MP. Then $\sim P \rightarrow \sim Q$, and $\sim P$ therefore $\sim Q$ by MP.

24. $P \rightarrow \sim Q$, Q therefore $\sim P$ by MT. Then $\sim R \rightarrow P$, and $\sim P$ therefore –?– by –?–.

Exercise Set 14.5 (pg 696)

15. Oh, no you don't! You're not giving up that quickly. Try a few statements. Assume that the statement is true—do you get a contradiction?

Investigation 14.6 (pg 699)

1. Converse: If it is a flower, then it is a rose. (false)

Inverse: If it is not a rose, then it is not a flower. (false)

Contrapositive: If it is not a flower, then it is not a rose. (true)

Exercise Set 14.6 (pg 700)

4. Valid using MP

8. No valid conclusion.

12. No valid conclusion. Although this looks like it should be a valid argument, it does not fit any of our four accepted forms of reasoning.

13. Valid using LS twice

22. Valid using LC and LS

Exercise Set 14.7 (pg 702)

1. Line 3: From lines 1 and 2, using LS

3. Line 5: From lines 3 and 4, using MP

4. Line 2: From line 1, using LC

9. LET *P:* All wealthy people are happy.
LET *Q:* Money can buy happiness.
LET *R:* Love exists.
$P \rightarrow Q$
$Q \rightarrow \sim R$
R
$\therefore \sim P$
The proof of the argument is left for you.

12. Because palm trees *A* and *C* are the only trees with an odd number of paths approaching and leaving them, then one must be the starting tree and the other must be the finishing point and thus the location of the treasure.

13. You want to create a sentence that if true, then must be false and if false, then must be true. A statement about Coldhart being sentenced to eating one of the vegetables might do the trick.

Exercise Set 14.8 (pg 706)

1. Line 7: From lines 5 and 6, using MT

10. LET *P:* The consecutive sides of a parallelogram are congruent.
LET *Q:* The parallelogram is a rhombus.
LET *R:* The diagonals are perpendicular bisectors of each other.
LET *S:* The diagonals bisect the angles.
$P \rightarrow Q$
$Q \rightarrow R$
$Q \rightarrow S$
$\sim R$
$\therefore \sim P$

Exercise Set 14.9 (pg 709)

5. Line 3: From lines 1 and 2, using MP

12. *P* only if *Q* is a style variation of IF *P* THEN *Q*.
LET *P:* Margarita and Sergio will try out for the school play.
LET *Q:* Tiffany and Tuong Cam decide to try out.
LET *R:* Dene tries out for the play.
$P \rightarrow Q$
$R \rightarrow \sim Q$
R
$\therefore \sim P$

13. LET *P:* Boris becomes a pastry chef.
LET *Q:* He gives in to his desire for chocolate mousse.
LET *R:* His waistline will suffer.
LET *S:* His dancing will suffer.

If Boris becomes a pastry chef, then if he gives in to his desire for chocolate mousse, his waistline will suffer.

IF *P* THEN (IF *Q* THEN *R*) or $P \rightarrow (Q \rightarrow R)$

Now you translate the rest.

Chapter 15

Exercise Set 15.1 (pg 722)

7. It's also called the identity property.

12. Distributive property, subtraction property of equality, —?—, —?—

14. —?—, —?—, multiplication property of equality, —?—

16. Add *c* to both sides, then divide both sides by *a,* then subtract *b* from both sides.

17. You'll need to use the square root property or the zero product property.

18. Reason for line 1: Given
Reason for line 2: Definition of a parallelogram

Exercise Set 15.2 (pg 727)

2. True. The Midpoint Postulate says that a segment has exactly one midpoint.

12. Reason for line 1: Given
Reason for line 2: AIA Postulate

13. Reason for line 1: Given
Reason for line 2: Angle Bisector Postulate
Reason for line 4: Reflexive property of congruence

Exercise Set 15.3 (pg 738)

6. Because two of the lines are perpendicular, the sum of the measures x and y equals the measure of one of the right angles. Therefore they are complementary.

Exercise Set 15.4 (pg 741)

2. See the flow-chart proof below.

7. Because y is the measure of one of a pair of congruent angles, then the other angle measures y. Likewise x is the measure of one of a pair of congruent angles; thus the other angle measures x. Because $y + y + 90° + x + x = 180°$, then $2y + 2x = 90°$, and therefore $x + y = 45°$.

Exercise Set 15.5 (pg 743)

1. See the flow-chart proof below.
4. See the flow-chart proof below.
5. See the concept map at the beginning of this lesson.

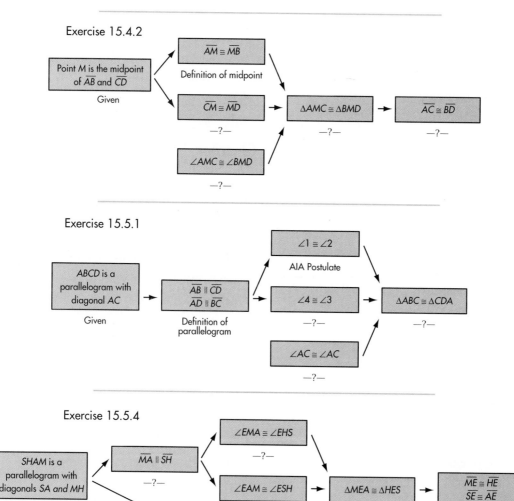

Exercise 15.4.2

Exercise 15.5.1

Exercise 15.5.4

598, 619-621
Arabic mathematicians, 643
Arc Addition Postulate, 761
archaeology, 282, 336, 512, 569, 664
arches, 295-296, 343
Archimedean screw, 666
Archimedean tilings, 398
Archimedes, 562, 637
Archimedes' principle, 562-564
architecture, 8, 9-10, 15, 17, 23, 32, 37-38,
    46, 110, 113, 192, 295-296, 343, 368,
    404-405, 437, 513, 516, 535, 582,
    630, 631, 673
Arc Length Conjecture, 355-356
arc(s)
    addition of, 761
    and angles, 333, 343-344
    and chords, 333
    defined, 328-329
    length of, 354-356
    measure of, 333, 354
    parallel lines intercepting, 345, 356
    symbol for, 328
area
    of an annulus, 457-458
    of a circle, 454-455, 457-458, 511
    defined, 434
    of a kite, 439
    maximizing, 450
    measuring, 447-450, 467-468
    of a parallelogram, 435
    and proportion, 613-614
    of a rectangle, 434-435
    of a regular polygon, 451-452
    of a sector, 457-458
    of a segment, 457-458
    of a Sierpinski gasket, 50
    surface. *See* surface area
    of a trapezoid, 438
    of a triangle, 438, 654-655
Aristotle, 28
art
    frieze patterns, 46
    geometric patterns in, 2-3, 9-10,
        26, 35
    Islamic, 2, 23-25
    knot designs, 2, 19-22
    mandalas, 2, 17-18
    murals, 593
    op art, 3, 13-16
    papercutting, 686
    perspective, 2, 27-33
    symmetry in, 3-4, 5
    See also drawing
ASA Congruence Conjecture, 254
ASA Congruence Postulate, 727
Asia, 2, 48, 386, 465, 598
    See also China; Japan
Assessing What You've Learned
    Chapter Zero, 36
    Chapter One, 72-73

Chapter Two, 130
Chapter Three, 170-171
Chapter Four, 219-220
Chapter Five, 275-276
Chapter Six, 323-324
Chapter Seven, 369-370
Chapter Eight, 431
Chapter Nine, 473
Chapter Ten, 520
Chapter Eleven, 574
Chapter Twelve, 638
Chapter Thirteen, 674
Chapter Fourteen, 714
Chapter Fifteen, 752
Chapter Sixteen, 779
associative property
    of addition, 720
    of multiplication, 720
astronomy, 356
Atlas of the Environment, 351
atoms, Platonic solids and, 536-537
auxiliary line, 743, 747, 759
average, and the centroid, 209, 210
axis, 532
Aztecs, 2, 17
Babylonians, 40, 469, 476, 643, 718
bar graphs, 120
base angles
    of an isosceles trapezoid, 291
    of an isosceles triangle, 104
    of a trapezoid, 290
base(s)
    of an isosceles triangle, 104
    of a cone, 533
    of a cylinder, 532
    of a prism, 526
    of a pyramid, 527
    of a rectangle, 435
    of solids, 461
    of a trapezoid, 290
bearing, 199, 241
Bernoulli, Jacob, 632
best-fit lines, 205-206
biconditional statements, 89
bilateral symmetry, 3, 4, 386
billiards, 85
binoculars, 294
biology, 20, 456, 619-621, 633
bisector
    of an angle, 91, 146-148, 726
    of a kite, 290
    perpendicular. *See* perpendicular
    bisectors
    of a segment, 138
    of a triangle, 155, 608-609
    of a vertex angle, 266-269
body temperature, 619
Bolyai, János, 730
Bonaparte, Napoleon, 270
bone and muscle strength, 619

Book of Kells, 19
botany, 6
Botswana, 387, 389
Brewster, Sir David, 395
Brianchon, Charles, 167
buoyancy, 562-564
business, 42, 208, 210, 219, 353, 590, 611,
    616, 617
CA Conjecture, 179, 727, 745
cameras, 287, 347, 605
Candide (Voltaire), 713
carbon molecules, 524
Cartesian coordinate geometry. *See*
    coordinate geometry
CA Theorem, 745
Celtic art, 2, 19
center, of a circle, 328
center of dilation, 588
center of gravity, 161
center of mass, 161, 165, 210
center of rotation, 375
central angle
    of a circle, 331, 333, 337, 343
    of a regular polygon, 302
centroid, 160-162, 208-209, 210
Centroid Coordinates Conjecture, 209
Chapter Review
    Zero, 34-35
    One, 70-72
    Two, 126-129
    Three, 168-170
    Four, 217-219
    Five, 272-275
    Six, 319-323
    Seven, 365-369
    Eight, 427-430
    Nine, 470-473
    Ten, 517-520
    Eleven, 571-573
    Twelve, 634-637
    Thirteen, 672-674
    Fourteen, 712-713
    Fifteen, 750-751
    Sixteen, 777-779
chemistry, 8, 68, 524, 555, 556, 557,
    570, 573
Chichén Itzá, 658
China, 2, 19, 46, 64, 76, 186, 196, 341,
    343, 349, 392, 476, 478, 494, 506,
    535, 580, 651, 686
Cholula pyramid, 548-549
chord(s), 329, 333-334, 337
Circle Area Conjecture, 454-455
circle graphs, 121
Circle Limit III (Escher), 717, 732
circle(s)
    annulus of concentric, 457-458
    arcs of. *See* arc(s)
    area of, 454-455
    central angles of, 331, 333, 337, 343
    chords of, 329, 333-334, 337

circumference of, 348-350
circumscribed, 156-157
concentric, 328, 457-458
congruent, 328
cycloid, 363
defined, 328
diameter of. *See* diameter
epicycloid, 364
equation of, 509
externally tangent, 338
inscribed, 156-157
inscribed angles of, 330, 343-345
in nature and art, 2, 17-18
nine-point, 167
proofs involving, 761-763
and Pythagorean Theorem, 509, 511
radius of, 11, 328, 334
secants of, 330
sectors of, 457-458
segments of, 457-458
tangents to. *See* tangent(s)
trigonometry using, 668-670
circumcenter, 156-157
finding, 208
interior/exterior position of, 158
properties of, 161, 162
Circumference Conjecture, 349-350
circumference/diameter ratio, 348-350
circumscribed circle, 156-157
clinometer, 215-216
clockwise rotation, 375
collinear points, 77, 157
commutative property
of addition, 720
of multiplication, 720
compass, 7, 133, 134
daisy designs with, 11-12
sector, 628
complementary angles, 95, 174
composition of isometries, 382
computers, construction of figures
with, 134
concave polygon, 98
concentric circles, 328, 457-458
concept map. *See* tree diagram;
Venn diagram
conclusion, 687
concurrent lines, 155
conditional proof, 705
conditional statement(s), 89, 687
and law of the contrapositive,
698-699
cone(s)
as geometric solid, 111-112, 533
similar, 614
surface area of, 463
volume of, 544-545
congruence
of angles, 83
ASA, 253-254, 727

defined, 83
of polygons, 99
as premise of geometric proofs, 721
SAA, 253, 254-255, 759, 760
SAS, 247, 249, 727
of segments, 83
SSS, 247, 248, 727
symbol for, 83
of triangles, 152-153, 246-256, 259,
297, 727, 765-766
congruent angles, 83
congruent circles, 328
congruent polygons, 99
congruent segments, 83
conjecture, 40-42, 139
defined, 40
proving. *See* geometric proof(s)
testing and explaining, 175, 181
See also postulates of geometry;
specific conjectures
consecutive angles
defined, 99
interior, 181
of a parallelogram, 303, 751
of a trapezoid, 291
consecutive interior angles, 181
consecutive sides, 99
consecutive vertices, 99
consequent, 705, 756
constant difference, 52-54, 57-60
construction
of angle bisectors, 146-148
of angles, 135-137, 148
with compass and straightedge, 135
defined, 134-135, 136
of a dilation design, 599-601
of an isosceles triangle, 232, 233
of a line segment, 135
of medians, 160
of a nine-point circle, 167
of parallel lines, 149-150, 180
with patty paper, 135
of a perpendicular bisector, 138-141
of perpendiculars, 142-143
of Platonic solids, 553-554
of points of concurrency, 155-159
postulates of, and proofs, 747
of a regular pentagon, 553
of a rhombus, 311-312
with right triangle or parallel lines,
277-278
See also drawing
construction and maintenance, 157, 176,
243, 246-247, 252, 269, 283, 434,
441, 448, 449, 450, 465, 472, 499,
505, 568, 569, 635, 636, 673
consumer awareness, 210, 353, 466, 467,
468, 472, 538, 561, 618, 639
contrapositive, law of the, 698-699
converse, 89, 698
Converse of the Isosceles Triangle

Conjecture, 233, 759, 760
Converse of the Parallel Lines Conjecture,
179-180
Converse of the Perpendicular Bisector
Conjecture, 140
Converse of the Pythagorean Theorem,
482-483
convex polygon, 98
Conway criterion, 419-421
Conway, John Horton, 419
cooking, 468, 538, 569, 582, 617, 628
Cooperative Problem Solving
The Centauri Challenge, 753
Construction Games from Centauri,
277-278
Designing a Lunar Colony Park,
37-38
Designing a Theater for Galileo,
371-372
Discovering New Area Formulas, 474
Games in Space, 432
Geometrivia I, 131-132
Geometrivia II, 221-222
The Geometry Scavenger Hunt,
325-326
Logic in Space, 715-716
Lunar Survival, 172
Once Upon a Time, 575-578
Party Time, 780
Patterns at the Lunar Colony, 74
Problem Solving at Mare Imbrium,
675-676
Pythagoras in Space, 521-522
Similarity in Space, 639
coordinate geometry, 186
dilation in, 589
distance in, 507-509
proofs with, 770-771
reflection in, 376
slope and, 186-190
systems of, 193-195
translation in, 375
trigonometry and, 668-670
Coordinate Midpoint Conjecture, 187
Coordinate Midpoint Postulate, 770
coplanar points, 77
corollary, 763
corresponding angles (CA), 178-180,
727, 745
Corresponding Angle Conjecture. *See*
CA Conjecture
cosine ratio, 644-645
cosines, law of, 660-662
counterclockwise rotation, 375
counterexample, 41, 89
CPCTC, 259
cube(s), as similar, 614
cubic units, 538
cyclic quadrilateral, 345
cycloid, 363
cylinder(s)
as geometric solid, 111-112, 532

numbers
    even, 55, 62
    Fibonacci, 631
    figurate, 57-64
    integers. *See* integers
    irrational, 349
    odd, 55, 62
oblique cylinder, 532, 539-540
oblique prism, 526, 539-540
oblique triangular prism, 526
obtuse angle, 90
obtuse triangle, 103, 154, 158, 166
oceanography, 598
octagon, 99
odd numbers, 55, 62, 141
oil spills, 465, 542, 547
one-point perspective, 30
op art, 3, 13-16
optics, 183, 229-230, 294
ordered-pair rule, 375
orthocenter, 156
    finding, 208, 347
    interior/exterior position of, 158
    properties of, 162, 166
orthodontia, 176
outgoing angle, 85, 229
pair share, 232
Palestinians, 389
papercutting, 686
paragraph proofs, 259, 737-738
parallel line(s), 94
    arcs on circle and, 345, 356
    construction of, 149-150
    defined, 178
    linkages of, 308-310
    postulate of, 726
    proofs involving, 745-746, 751, 756
    properties of, 178-181
    proportional segments by, 622-626
    and reflection images, 381-382
    slope of, 196-197
    symbol for, 94
Parallel Lines Conjecture, 179-180
    Converse of, 179-180
Parallelogram Area Conjecture, 435
parallelogram(s), 107-108
    area of, 435
    defined, 303
    midpoints of, 773-774
    proofs involving, 743-744, 751, 773-775
    properties of, 303-305
    special, 311-314
    See also specific parallelograms
Parallel Postulate, 726, 730-733
Parallel Proportionality Conjecture, 624-625
Parallel Slope Conjecture, 197
Parallel Slope Postulate, 770
parametric equations, 302

Parthenon, 630, 631
Pascal, Blaise, 64
Pascal's triangle, 64
Path of Life I (Escher), 579, 599-601
patterns. *See* inductive reasoning
patty papers, 133, 134
Penrose, Sir Roger, 16, 406
pentagonal numbers, 59-60
pentagon(s), 99
    in nature and art, 2, 6
    tessellation with, 403
performance assessment, 170-171
perimeter
    defined, 99
    of fractals, 49, 51
periodic curve, 363
periscope, 183
Perpendicular Bisector Conjecture, 139, 751
    Converse of, 140
perpendicular bisectors
    of a chord, 334
    constructing, 138-141, 177
    defined, 138
    of a rhombus, 312, 757
    of a triangle, 155, 266-269
perpendicular line(s), 94
    slope of, 197-198
    symbol for, 94
Perpendicular Postulate, 726
perpendiculars
    constructing, 142-143
    defined, 143
    of isosceles triangles, 266-267
    postulate of, 726
Perpendicular Slope Conjecture, 198
Perpendicular Slope Postulate, 770
Persia, 46, 64
Persian Gulf War, 542
perspective, 2, 27-33
Peru, 19, 580
photography, 287, 347, 605
physics, 20, 161, 304, 505, 506, 562-563
pi, 348-350
    symbol for, 348, 349
picture angle, 347
picture patterns, 46-51
Pima, 183
plane figures. *See* geometric figures
plane(s), defined, 76, 77
Plato, 28, 536-537, 718
Platonic solids, 536-537
    constructing, 553-554
Playfair axiom, 730
Playfair, John, 730
plumb level, 269
Poincaré disk, 732
Poincaré, Henri, 732
point(s)
    collinear, 77

    coplanar, 77
    defined, 76, 77
    locus of, 122-123
points of concurrency, 155-156, 161-167
    coordinates for, finding, 208-209
point symmetry, 4
point of tangency, 330
polar coordinates, 193-194
Polaris, 215, 216
pollution. *See* environmental science
polygon(s), 98-100
    area of, 451-452
    center of mass of, 165
    circumscribed circles and, 156-157
    classification of, 99
    concave, 98
    congruent, 99
    consecutive angles of, 99
    consecutive sides of, 99
    consecutive vertices of, 99
    convex, 98
    defined, 98
    diagonals of, 100
    drawing, 302
    duplicating, 137
    equiangular, 100, 285
    equilateral, 100
    inscribed circles and, 156-157
    regular. *See* regular polygon(s)
    sides of, 98, 99
    similar. *See* similarity
    sum of angle measures, 280-281, 284-287
    symmetries of, 387-388
    tessellations with, 396-399, 402-403, 419-421
    vertex of a, 98
    See also specific polygons
Polygon Sum Conjecture, 281
polyhedron(s)
    defined, 525
    edges of, 525
    Euler's formula for, 530-531
    faces of, 525
    with holes, 531
    models of, building, 527-528
    naming, 525
    regular, 525, 536-537
    similar, 614-615
    unfolded, 537
    vertex of a, 525
    See also specific polyhedrons
Polykleitus, 633
Polynesians, 215
Poncelet, Jean Victor, 167
pool, geometry of, 85, 380
population, 120, 123, 583
portfolio, defined, 36
postulates of geometry, 718, 725-727, 730-733, 761, 764, 770

rectangular solid(s), drawing, 111
rectifying shapes, 460
recursion, 43
recursive rules, 49
reflection
    composition of isometries and, 376, 381-382
    glide, 376, 422-423
    line of, 376
    minimal path and, 380-381
    as type of isometry, 374, 376
reflectional symmetry, 3-4, 386-388
reflexive property of equality, 720
regular dodecahedron, 525, 536-537, 553-554
regular hexagon, 11, 396
regular hexahedron, 536-537, 553
regular icosahedron, 536-537, 553
regular octahedron, 536-537, 553
regular pentagon, constructing, 553
Regular Polygon Area Conjecture, 452
regular polygon(s), 100
    area of, 451-452
    drawing, 302
    symmetries of, 387-388
    tessellation with, 396-399
regular polyhedron, 525, 536-537
regular tessellation, 396-397
regular tetrahedron, 536, 537
religion
    and Islamic art, 23
    and Roman Catholic art, 27
Renaissance, 2, 27-28
rep-tiles, 628
Reptiles (Escher), 414
resources, and consumption of, 121, 124, 617, 628
resultant vector, 304
rhombus, 108
    midpoints of, 774-775
    properties of, 311-312
Rice, Marjorie, 403
Riemann, Georg Friedrich Bernhard, 731
right angle, 90
right cone, 463, 533
right cylinder, 532, 539-540
right pentagonal prism, 526
right prism, 526, 538-540
right triangle(s), 103
    congruence shortcuts for, 765-766
    hypotenuse of, 476
    isosceles, 488-489
    legs of, 476, 644
    multiples of, 495-497
    proofs involving, 764-766, 772
    properties of. *See* Pythagorean Theorem; trigonometry
    similarity of, 764-766
    30-60 type, 490-492
rigid transformation, 374. *See* isometry
Riley, Bridget, 440

rise over run, 187-188
Romans, ancient, 186, 295, 296, 396
rotation
    and composition of isometries, 382
    tessellations by, 413-415
    as type of isometry, 374, 375
rotational symmetry, 4, 387-388
ruler, 7
Russell, Bertrand, 687
Russia, 116
SAA Congruence Conjecture, 255, 759, 760
Sahin, Nurten, 392
sailing. *See* navigation
SAS Congruence Conjecture, 249
SAS Congruence Postulate, 727
SAS Similarity Conjecture, 595
scaled drawing, 113, 589
scalene triangle, 104
Schattschneider, Doris, 403, 601
School of Athens (Raphael), 28
Scientific American, 403
scientific method, 39, 40-41
sculpture, geometry and, 26, 182, 592, 637
secant, 330
section, 114
sector of a circle, 457-458
sector compass, 628
Segment Addition Postulate, 726
segment(s)
    of a circle, 457-458
    line. *See* line segment(s)
self-similarity, 49
semicircle, 329
    inscribed angles in, 344
    measure of, 333
semiregular tessellation, 396-398
sextant, 229-231
Shah Jahan, 8
Shortest Distance Conjecture, 143
show. *See* consequent
Side-Angle Inequality Conjecture, 238
sides
    of an angle, 79
    of a parallelogram, 304
    of a polygon, 98, 99
Sierpinski gasket, 49-51
similarity, 580, 584-590
    and area, 613-614
    dilation and. *See* dilation(s)
    indirect measurement and, 602-603
    of polygons, 585-587
    and Pythagorean Theorem, 500
    ratio and proportion in, 580-581, 585-587, 595, 613-615
    segments and, 622-626
    of solids, 614-615
    spiral, 601
    symbol for, 586
    of triangles, 594-596, 602-603, 608-609, 622-624, 764-766

    and volume, 613, 614-615
similar polygons, 585
similar solids, 614
sine ratio, 644-646
sines, law of, 654-655
size. *See* measure
sketching, defined, 134
skew lines, 94
slant height, 462
slope
    best-fit lines, 205-206
    calculating, 187-189
    defined, 187
    of parallel lines, 196-197
    of perpendicular lines, 197-198
Slope-Intercept Conjecture, 201
Slope of a Line Conjecture, 188
smoking cigarettes, 204, 712
Smullyan, Raymond, 769, 779
Snell's law, 676
solid of revolution, 114, 534
solid(s), 524
    bases of, 461
    with curved surfaces, 532-533
    drawing, 111-112, 113
    faces of. *See* faces
    lateral faces of. *See* lateral faces
    Platonic. *See* Platonic solids
    of revolution, 114, 534
    section, 114
    similar, 614-615
    surface area of, 461-464
    See also polyhedron(s); *specific solids*
South America, 19
space
    defined, 78
    See also solid(s)
Spain, 15, 23
speed. *See* velocity and speed calculations
sphere(s)
    coordinates on, 194-195
    elliptic geometry and, 731-732
    geodesic lines on, 184-185
    as geometric solid, 111-112, 532
    great circles, 185, 532, 731
    as similar, 614
    surface area of, 565-566
    volume of, 559-560
Sphere Surface Area Conjecture, 566
Sphere Volume Conjecture, 560
spherical coordinates, 194-195
spiral similarity, 601
sports, 67, 85, 189, 191, 199, 210, 241, 358, 360-361, 377, 380-381, 385, 401, 429, 466, 485, 582, 619, 621, 667
square numbers, 57, 61
square pyramid, 527
square roots, 488
square(s), 108

defined, 313
in nature and art, 6
properties of, 313
symmetry of, 388
square units, 434
squaring the circle, 460
Sri Lanka, 465
SSS Congruence Conjecture, 248, 727
SSS Congruence Postulate, 727
SSS Similarity Conjecture, 594
star polygons, 317–318
statistics, 205–206, 618, 628–629
steering linkage, 308, 310
Stein, Rolf, 403
Straight Curve (Riley), 440
straightedge, 7, 133, 134
subscripts, 145
substitution property, 720
subtraction property of equality, 721
supplementary angles, 95
defined, 174
proofs involving, 742, 746
surface area
of a cone, 463
of a cylinder, 462
defined, 461
of a prism, 461–462
of a pyramid, 461, 462, 463
of a sphere, 565–566
surfaces, 461
surveying land, 40, 186, 483, 486, 666, 718
syllogism, law of, 694–695
symbols
angle, 79, 81
arc, 328
congruence, 83
glide reflection, 422
golden ratio, 630
image point label, 375
line, 77
line segment, 78
of logic, 687, 692
measure, 78, 83
parallel, 94
perpendicular, 94
pi, 348, 349
point, 77
ray, 79
same measure, 83
similarity, 586
therefore, 692
symmetric property of equality, 720
symmetry, 3
bilateral, 3, 4, 386
glide-reflectional, 387
point, 4
of polygons, 387–388
reflectional, 3–4, 386–388
rotational, 4, 387–388
translational, 387

Tagore, Rabindranath, 8
Taj Mahal, 8
tangent circles, 338
Tangent Conjecture, 339, 768
tangential velocity, 358
tangent ratio, 644–646
tangent(s), 330
external, 338
internal, 338
point of, 330
proofs involving, 768, 769
properties of, 338–340
and Pythagorean Theorem, 511
segments, 339
as term, use of, 330
Tangent Segments Conjecture, 339, 769
Tangent Theorem, 768
tangrams, 506
tatami, 404–405
Tchokwes, 19, 21
telecommunications, 65, 67, 456
temperature conversion, 200
term(s), *n*th. *See n*th term, finding
tessellation(s)
Conway criterion and, 419–421
defined, 396
demiregular, 398–399
drawing, 23–25, 408–411, 413–419, 422–425
dual of, 400
glide reflection, 422–423
Islamic art and, 23, 407
monohedral, 396
nonperiodic, 406
regular, 396–397
rotation, 413–415
semiregular, 396–398
translation, 407–409
test problems, writing, 275–276
tetrahedron, 525, 536–537
Thales of Miletus, 718, 759
Theorem of Pythagoras. *See* Pythagorean Theorem
theorems, 718
corollaries of, 763
proving. *See* geometric proofs
See also specific theorems
therefore, symbol for, 692
Third Angle Conjecture, 225, 759, 760
30-60 right triangle, 490–493
30-60 Right Triangle Conjecture, 491
Thompson, Sir D'Arcy, 633
Tibet, 598
tiling, 396. *See* tessellation(s)
Timaeus (Plato), 536
topology, 110
transformation(s), 374
dilations. *See* dilation(s)
rigid, 374. *See* isometry
transitive property of equality, 720

translation
and composition of isometries, 376, 380–381
tessellations by, 407–409
as type of isometry, 374–375
vector, 374
translational symmetry, 387
transversal line, 178
Trapezoid Area Conjecture, 438
Trapezoid Consecutive Angles Conjecture, 291
Trapezoid Midsegment Conjecture, 299, 772
trapezoid(s), 107
arch design and, 295–296
area of, 438
defined, 290
isosceles. *See* isosceles trapezoid(s)
proofs involving, 767, 769
properties of, 290–291
tree diagram, 109
Triangle Area Conjecture, 438
Triangle Exterior Angle Conjecture, 239
Triangle Inequality Conjecture, 238
Triangle Midsegment Conjecture, 297, 772
Triangle Midsegment Theorem, 772, 773–775
triangle(s), 99, 103–105
acute, 103
adjacent interior angles of, 238
altitudes of, 105, 142, 155, 608
angle bisectors of, 155
applications, modeling of, 243–245
area of, 438
congruence of, 152–153, 246–256, 259, 297, 727, 765–766
constructing, 135
determining parts of, 152
drawing, 134
equiangular, 233–234
equilateral. *See* equilateral triangle(s)
exterior angles of, 238–239
height of, 105
inequalities, 237–239
interior angles of, 238–239
isosceles. *See* isosceles triangle(s)
medians of, 104, 160–161, 608
midsegments of, 297–298, 772
obtuse, 103, 154
parallel lines and proportions of, 622–626
perpendicular bisectors of, 155
points of concurrency of, 156–157, 161–167
relationships of, 109
remote interior angles of, 238–239
right. *See* right triangle(s)
scalene, 104
in Sierpinski gasket, 49–51
similarity of, 594–596, 602–603, 608–609, 622–624, 764–766

sum of angles of, 224–225
    tessellation with, 396–399, 402
    vertex angle, 104
Triangle Sum Conjecture, 224–226, 731, 759
Triangle Sum Theorem, 731, 759
triangular numbers, 59, 64
triangular prism, 526
triangular pyramid, 527
trigonometry, 642–646
    on coordinate plane, 668–670
    cosine ratio, 644–645
    graphs of functions, 671
    law of cosines, 660–662
    law of sines, 654–655
    problem solving with, 649, 665
    ratios, 642–646
    sine ratio, 644–646
    tangent ratio, 644–646
Turkey, 392, 396
Twain, Mark, 42
two-column proof, 701, 736
two-point perspective, 30
Umbilic Torus (Ferguson), 26
undecagon, 99
unit circle, 668–670
units
    arbitrary, 78
    area and, 434
    nautical mile, 369
    and scaled drawing, 113
    volume and, 538
unsound argument, 687

valid argument, 687
valid reasoning. *See* logic
vanishing lines, 29–32
vanishing point(s), 28, 29–32
Vasarely, Victor, 3, 13, 573
vectors, 304
    resultant, 304
    translation, 374
    trigonometry with, 665
vector sum, 304
velocity
    angular, 357, 358
    tangential, 358
    vectors, 304
velocity and speed calculations, 191, 204, 304, 306, 316, 323, 339, 352, 353, 357, 358, 368, 665, 666, 667, 673, 674
Venn diagram, 109
    logic statements and, 689, 695
Vertex Angle Bisector Conjecture, 267, 734–735
vertex angle(s)
    of an isosceles triangle, 104, 266–269
    of a kite, 289
vertex (vertices)
    of a cone, 533
    defined, 79
    naming angles by, 79, 81
    of a polygon, 98
    of a polyhedron, 525
    of a pyramid, 527
    and tessellation(s), 398–399
vertical angles, 95, 174

Vertical Angles Conjecture, 174, 175, 740–741
Vertical Angles Theorem, 740–741
Vietnam Memorial, 182
visual thinking, 46, 120–123
Voltaire, François, 713
volume
    of a cone, 544–545
    of a cylinder, 539–540
    defined, 538
    displacement and density and, 555
    of a hemisphere, 559–560
    maximum, 558
    of a prism, 538–540
    problems in, 550–552, 569–570
    and proportion, 613, 614–615
    of a pyramid, 544–545
    of a sphere, 559–560
water
    and buoyancy, 562
    and volume, 524, 556
weight, 505
    buoyancy and, 562–564
What Is the Name of this Book? (Smullyan), 769, 779
work, 505, 506
World Book Encyclopedia, 35
Wright, Frank Lloyd, 9–10
writing test problems, 275–276
y-intercept, 201
yin and yang, 341
Zhoubi Suanjing, 478
zoology and animal care, 6, 15, 467, 598, 619–621

Author's Acknowledgments

What an eventful seven years since the publication of the first edition of *Discovering Geometry*! Neither Steve Rasmussen nor I had any idea of how warmly *Discovering Geometry* would be received. So first, to all the teachers who have used *Discovering Geometry,* a sincere thank you for your wonderful support and encouragement. I wish to thank my always delightful and ever-so-patient students for their insight and hard work. I also wish to thank the many students across the country who have written to me with their kind words, comments, and suggestions. And thanks to Madeleine Mulgrew and Kelvin Taylor for their hard work in bringing the first edition into so many classrooms.

This second edition, as you can see from the credits, involved a much larger team. To the staff at Key, the field testers, the advisors, the consultants, and the reviewers, I am grateful for your quality work.

There are two special people who, more than anyone else, have added their special touch to *Discovering Geometry:* Steve Rasmussen and Dan Bennett. Steve and I worked side by side when putting the first edition of *Discovering Geometry* together. It was a special time in both of our lives, and I will always treasure it. Thank you, Steve. The second edition was a different experience but an experience that I will also treasure as very special because of Dan Bennett. Dan's wonderful grasp of technology and his mathematical and writing skills helped tremendously to shape the new directions taken in this second edition. But it was his support of what was most characteristic of the original *Discovering Geometry* and his patience with me that was most rewarding. Thank you, Dan.

Photography Credits

CHAPTER 9

434: Andy Sachs/Tony Stone **437:** Courtesy Naoko Hirakura Associates **440:** ©Bridget Riley, courtesy the artist **442:** Shelburne Museum, Shelburne, VT **445:** Quilt by Mabry Benson, Kensington, CA/photo by Carlberg Jones **455:** *l* Susanne Buckler/Gamma Liaison; *r* Grant Heilman **457:** Shahn Kermani/Gamma Liaison **458:** Wernher Krutein/Gamma Liaison **461:** *far l* Don Spiro/Tony Stone; *2nd from l* Stephen Johnson/Tony Stone; *2nd from r* Alain Benainous/Gamma Liaison; *far r* Marie Ueda/Tony Stone **462:** Eslami-Rad/Gamma Liaison **465:** Bryn Campbell/Tony Stone **466:** Douglas Peebles **468:** Grant Heilman

CHAPTER 10

478: Will & Deni McIntyre/Photo Researchers **482:** Swank & Newell* **486:** Grant Heilman **488:** ©1939 Turner Entertainment, All Rights Reserved **506:** Comstock/Russ Kinne **513:** Bruno Joachim/Gamma Liaison **516:** Hillary Turner*

CHAPTER 11

524: *tl* Runk\Schoenberger/Grant Heilman; *ml* Ken Eward/Bio Grafx/Science Source/Photo Researchers; *bl* Ken Eward/Bio Grafx/Science Source/Photo Researchers; *tr* Esaias Baitel/Gamma Liaison; *br* Scott Camazine/Science Source/Photo Researchers **527:** Bonnie Kamin* **532:** Barry L. Runk/Grant Heilman **533:** ©1996, C. Herscovici, Brussels, Artists Rights Society (ARS), NY/photo courtesy Minneapolis Institute of Arts **535:** George Chan/Tony Stone **536:** Fitzwilliam Museum, Cambridge **538:** TIB/Bill Varie **540:** Hillary Turner* **543:** *l* Jeff Tinsley/NAMES Project Foundation; *r* Nancy Katz/NAMES Project Foundation **544:** Bonnie Kamin* **548:** David Hiser/Photographers Aspen **549:** Sylvain Grandadam/Photo Researchers **550:** David Sutherland/Tony Stone **552:** *t* Hillary Turner*; *b* Designed by Heather Barranco, Parsons School of Design student of instructors Constantin Boym and Laurene Leon/photo by Deborah Goletz **556:** Mark A. Leman/Tony Stone **557:** Runk\Schoenberger/Grant Heilman **559:** Bonnie Kamin* **561:** Runk\Schoenberger/Grant Heilman **562:** David Morris/Gamma Liaison **565:** Comstock/John Cooke **566:** Alan Smith/Tony Stone **567:** David Hardy/Science Photo Library/Photo Researchers **568:** Larry Lefever/Grant Heilman **570:** Paul Chesley/Tony Stone **573:** Comstock/Dr. Georg Gerster

CHAPTER 12

580: ©Bates Littlehales/National Geographic Image Collection **582:** NASA **584:** TIB/G&J Images **592:** Paul Damien/Tony Stone **596:** NASA **598:** *t* Thomas Dove/Douglas Peebles; *b* Courtesy Rossi & Rossi Ltd. **607:** Bonnie Kamin* **616:** David R. Frazier/Photo Researchers **619:** Bettmann **621:** ©1996 Demart Pro Arte, Geneva, Artists Rights Society (ARS), NY **630:** Vic Thomasson/Tony Stone **633:** *t* Kathleen Campbell/Gamma Liaison; *b* Scala/Art Resource, NY **637:** *both* Alex Webb/Magnum

CHAPTER 13

642: Grant Heilman **652:** Swank & Newell* **658:** *t* Sarah Stone/Tony Stone; *b* Murray & Associates/Tony Stone **663:** Grant Heilman **671:** Courtesy California Academy of Sciences

CHAPTER 14

686: ©Sterling/Lark Books, 50 College Street, Asheville, NC, 28801, from *The Book of Papercutting* by Chris Rich/reproduced with permission of publisher **687:** Swank & Newell* **691:** *l* M. H. Feder; *r* Kenneth L. Feder **698:** Illustration by John Tenniel **712:** Courtesy American Cancer Society **713:** Collection of Mr. & Mrs. A. Reynolds Morse, on loan to the Salvador Dali Museum, St. Petersburg, FL/©1996, Salvador Dali Museum

CHAPTER 15

725: Art Resource, NY **729:** Comstock **732:** Bonnie Kamin*

CHAPTER 16

758: ©Paula Nadelstern/photo by Bobby Hansson

* Photographed for Key Curriculum Press.

Additional Credits

Illustrations: R. Diggs, Bill Eral, Casey FitzSimons, Jerry Simpfenderfer
Technical Art: Kirk Mills
Scanning and Proofs: Digital Prepress International, San Francisco, California
Compositor: Clarinda Prepress Services, Clarinda, Iowa
Printer: R. R. Donnelley, Willard, Ohio

This book was, for the most part, set in Lucida Bright and Futura fonts. It was laid out on the Macintosh platform using the following software: QuarkXPress, Microsoft Word, Macromedia FreeHand, Adobe Illustrator, Adobe PhotoShop, and Strata StudioPro.